INTRODUCTION TO
ENGINEERING
MATHEMATICS

INTRODUCTION TO
ENGINEERING
MATHEMATICS

**Anthony Croft, Robert Davison
and Martin Hargreaves**
De Montfort University

Prentice
Hall

An imprint of **Pearson Education**

Harlow, England · London · New York · Reading, Massachusetts · San Francisco
Toronto · Don Mills, Ontario · Sydney · Tokyo · Singapore · Hong Kong · Seoul
Taipei · Cape Town · Madrid · Mexico City · Amsterdam · Munich · Paris · Milan

Pearson Education Limited
Edinburgh Gate
Harlow CM20 2JE
Tel: +44 (0)1279 623623
Fax: +44 (0)1279 431059

Website: www.pearsoned.co.uk

Cover designed by Designers & Partners of Oxford
and printed by The Riverside Printing Co. (Reading) Ltd
Typeset by P&R Typesetters Ltd, Salisbury, Wilts, UK
and printed by Biddles Ltd, *www.biddles.co.uk*

First printed 1995

ISBN 0-201-62442-7

10 9 8
06 05 04 03

A.C. To my father and mother, Jan and Thomas.
R.D. To Sharon.
M.H. To Pamela and all our cats, past and present.

Preface

Audience

This book has been written to serve the mathematical needs of students engaged in a first course on engineering. It is particularly suited to ONC and HNC/HND courses as well as the increasingly popular Foundation courses that are undertaken prior to degree study. The material included embraces the proposals contained in the report 'A Core Curriculum in Mathematics for the European Engineer' published by The European Society for Engineering Education (SEFI) in 1992. This Core Curriculum details the mathematical knowledge students should have before embarking upon an engineering degree – known as Core Level Zero. We have taken care to ensure that the entire content of this level has been adequately covered. This book is also suitable for students in Further Education studying for the Advanced GNVQ in Engineering and in particular the mandatory unit, 'Mathematics for Engineering'. Students on the first year of an engineering degree may also find the book helpful as a reference work and to use as a refresher. Very little mathematical knowledge is assumed in order to make the book useful to as wide a range of students as possible, and the style of the book makes it suitable for readers wishing to engage in self study. Therefore, the book will also be helpful to individuals undertaking assessment within employment for NVQs at level 3. Engineering examples have been drawn from across the engineering spectrum. We have made the coverage of the book comprehensive in order to suit the needs of a wide range of engineers.

Motivation

Engineers are called upon to analyse a variety of engineering systems, which can be anything from a few electronic components connected together through to a complete factory. The analysis of these systems benefits from the intelligent application of mathematics. Indeed, many cannot be analysed without the use of mathematics. Mathematics is the language of engineering. It is essential to understand how mathematics works in order to master the complex relationships present in modern engineering systems and products.

Aims

There are two main aims of the book. Firstly, we wish to provide an accessible, readable introduction to engineering mathematics. The second aim is to encourage the integration of engineering and mathematics.

Style

We have adopted a similar style to our degree level book, *Engineering mathematics – a modern foundation for electronic, electrical and control engineers,* in order that the two can be used in sequence to provide a progression of material for students wishing to take their studies further. This style is to teach through the widespread use of examples rather than abstract formal proof. This reflects our belief that engineers prefer this approach because it corresponds to the way they use mathematics in practice. Engineers have a lot to learn and therefore the need for abstract theory should be justified whenever it is included. We have included many engineering examples and have tried to make them as free standing as possible to keep the necessary engineering prerequisites to a minimum. Subsequently, the engineering examples are not central to the mathematical development and so can be missed out if desired. This provides the flexibility necessary to tailor a course to the needs of different types of engineers. However, we feel that at this level it is desirable for an engineer to be exposed to as wide a range of different engineering examples as possible. This has the obvious benefit of illustrating the reason for learning a particular piece of mathematics. A further benefit is the development of the link between mathematics and the physical world. An appreciation of this link is essential if engineers are to take full advantage of engineering mathematics. The engineering examples make the book more colourful but, more importantly, they help develop the ability to see an engineering problem and translate it into a mathematical form so that a solution can be obtained. This is one of the most difficult skills that an engineer needs to acquire. The ability to manipulate mathematical equations is of itself insufficient. It is sometimes necessary to derive the equations corresponding to an engineering problem. Interpretation of mathematical solutions in terms of the physical variables is also essential. We have tried to encourage the reader to continually refer the equations back to the physical world by asking questions such as, 'What is the effect on the pressure of doubling the flow in this equation?' and 'What should be the physical dimensions of this quantity given the dimensions of the other quantities in the equation?' These are vital questions that an engineer must be asking continually. Engineers cannot afford to get lost in mathematical symbolism.

Sometimes engineers may be required to carry out 'back of the envelope' calculations. Advanced mathematical analysis is not always appropriate or necessary. This is not to dismiss this work as unimportant but to put it in its true perspective. The ability to approximate and estimate is vital. When using a calculator it is easy to get an answer by pushing the appropriate buttons. Having confidence in the answer requires skills of estimation and approximation which are more difficult to acquire. This book will help in the acquisition and development of these skills.

Use of modern I.T. aids

One of the main developments in the teaching of engineering mathematics in recent years has been the widespread availability of sophisticated computer software and its adoption by many educational institutions. We have avoided making specific references to such software as it is continually being changed and improved. However, we would encourage its use, and so we have made general references at several points in the text. Two packages that we have found to be of particular value are the symbolic mathematics processor and graph plotter, DERIVE, and the general purpose mathematics package, MATLAB. The latter is extremely useful for teaching engineering subjects such as control, signal processing and systems. In fact, the widening availability of such software makes the learning of basic mathematical techniques more important than ever before because of the sophistication of the packages. Many features available in software packages can also be found in graphics calculators.

Format

Each chapter has its Key Points listed at the start. Important results are boxed as they occur and there are self-assessment questions and exercises after each section. These sections have been kept short in order to aid digestion. Solutions are provided at the back of the book. Where appropriate, sets of exercises for solution by graphics calculator or computer software have been included and the use of these is encouraged. Finally, Review Exercises are provided at the end of each chapter to revise the material of the chapter and also to allow the work from individual sections to be integrated.

To place engineering mathematics in a broader context we have included a number of cameos of famous engineers, important engineering projects and other related topics.

We denote engineering examples by use of the ⛩ icon and computer and calculator exercises by use of the 🖥 icon.

We hope that you, the reader, find the book useful and we wish you the best of luck!

AC, RD, MH
September 1994

Acknowledgements

We wish to thank all the staff at Addison-Wesley who have been involved in the production of this book. In particular, our special thanks go to Sarah Mallen for commissioning the book and to Jane Hogg for her enthusiasm and support throughout the project. Also thanks to Victoria Cook, Martin Tytler and Tim Pitts.

We should also like to thank Roger Lawrence and John Royle for their thorough appraisal of the manuscript and their most valuable comments.

Several of our students have helped by highlighting points that have required further explanation and by working through many of the exercises. In particular, we have appreciated the contribution from Iain McLoughlan.

Finally, the comments of many unknown referees were most useful and are gratefully acknowledged.

AC, RD, MH
September 1994

The publisher wishes to thank the following for permission to reproduce material in this book.

Figure 1, p. 39, page illustration from *An Introduction to the History of Mathematics, Sixth Edition with Cultural Connections* by Howard Eves, copyright © 1990 by Saunders College Publishing, reproduced by permission of the publisher. Figure 1, p. 123, screenshot from the DERIVE® program. DERIVE® is a registered trademark of Soft Warehouse, Inc., Honolulu, Hawaii. Figure 1, p. 167, photograph of the Thames Barrier. © National Rivers Authority, reprinted by permission of the Thames Barrier Visitors Center. Tel: 0181 854 1373. Quoted material, p. 237, from *Brunel's Britain* by Derick Beckett (David and Charles, Newton Abbot, 1980). Figure 1, p. 237, Isambard Kingdom Brunel, courtesy of the National Portrait Gallery, London. Figure 2, p. 237, Royal Albert Bridge, Saltash, courtesy of Mary Evans Picture Library. Quoted material, p. 284, by permission of CUP and taken from H.W. Dickinson, *James Watt: Craftsman and Engineer* (David and Charles, Newton Abbot, 1967). (First edition, Cambridge University Press, 1936.) Figure 1, p. 284, James Watt, courtesy of Mary Evans Picture Library. Figure 2, p. 284, Watt's model of a Surface Condenser reproduced by permission of the Science & Society Picture Library. Figure 1, p. 310, Konstantin Tsiolkovsky, courtesy of Mary Evans Picture Library. Figure 2, p. 310, Tsiolkovsky's formula for rocket motion reproduced by permission of MIR Publishers, Moscow, from the 1968 English translation of *Tsiolkovsky's selected works.* Figure 1, p. 401, model of the Channel Tunnel, courtesy of QA Photos Ltd. Figure 1, p. 413, Orbiting satellite reproduced by permission of the Science & Society Picture Library. Figures 1 and 2, p. 447, three dimensional probe head and gear measurement system, courtesy of Carl Zeiss, Oberkochem, Germany. Figure 1, p. 507, Michael Farraday, courtesy of Mary Evans Picture Library. Figure 2, p. 507, cartoon of 1881, courtesy of Punch Publications Ltd. Figure 1, p. 585, page showing Green's work, taken from the 1958 facsimile edition of G. Green's *Essay*, reproduced and printed by Wezäta-Melins Aktiebolag, Göteburg, Sweden. Figure 2, p. 585, windmill at Green's Mill and Science Centre, courtesy of Nottingham City Council. Figures 1 and 2, p. 625, mesh generated models reproduced by permission of Masson, S.A. and taken from P.L. George, *Automatic Mesh Generation Application to F.E. Methods* (J. Wiley Ltd, Paris, 1991).

Contents

1

Arithmetic

KEY POINTS

This chapter

- lists some common rules of arithmetic

- explains the term 'lowest common multiple'

- explains the term 'highest common factor'

- defines the terms 'proper fraction' and 'improper fraction'

- shows how to multiply, divide, add and subtract fractions

- explains the order in which addition, subtraction, multiplication and division are carried out

- explains the term 'ratio' and shows how to perform calculations involving ratio

- explains the term 'percentage' and shows how to perform calculations involving percentages

CONTENTS

1.1 Introduction

Arithmetic is the study of numbers and their manipulation. Its use permeates most areas of human endeavour. A clear and firm understanding of the rules of arithmetic is essential for everyday calculations in engineering. Arithmetic also serves as a springboard for tackling more abstract mathematics.

1.2 Arithmetic notation and rules

A combination of numbers and the operations of $+$, $-$, $\times$ and $\div$ is called an **arithmetic expression**. For example, $3 \div 2$ and $4 + 3 - 2$ are arithmetic expressions.

1.2.1 The plus sign, $+$, and the minus sign, $-$

A useful device for representing numbers is the **number line**. Figure 1.1 shows part of this line. Positive numbers are on the right-hand side of the line; negative numbers are on the left-hand side. Note that the minus sign is used to show a number is negative, for example -3. We sometimes write $+3$ to mean positive 3, but usually the plus sign is omitted and we write simply 3. When writing negative numbers, the negative sign is never omitted. We often refer to the **sign** of a number, meaning either positive or negative. So, for example, the sign of -4 is negative and the sign of 4 is positive.

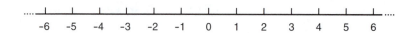

Figure 1.1   The number line.

The plus and minus signs are also used to show that two numbers or expressions must be added to or subtracted from one another. For example, $3 + 2$ means add the positive numbers 3 and 2, while $3 - 2$ means subtract the positive number 2 from the positive number 3. We say that $3 + 2$ is the **sum** of 3 and 2, and that $3 - 2$ is the **difference** of 3 and 2. Note that $3 + 2$ is the same as $2 + 3$, and so the order of the numbers is unimportant when being added. However, $2 - 3$ and $3 - 2$ have different values, and so the order of the numbers is important when subtraction takes place.

We can also add and subtract negative numbers. For example, we can calculate $3 + (-2)$ and $3 - (-2)$. In the first case $3 + (-2)$ is equivalent to $3 - 2$, that is, 1. In the second case $3 - (-2)$ is equivalent to $3 + 2$, that is, 5.

The $+/-$ button on a scientific calculator can be used to change the sign of a number. It is worth checking that you can use this facility.

Example 1.1

Evaluate (a) $5+(-3)$ (b) $-4-(+2)$ (c) $6-(-2)$ (d) $-4-(-1)$

Solution   (a) This is equivalent to $5-3$, so

$$5+(-3)=5-3=2$$

(b) This is equivalent to $-4-2$, so

$$-4-(+2)=-4-2=-6$$

(c) $6-(-2)$ is equivalent to $6+2$, so

$$6-(-2)=6+2=8$$

(d) For $-4-(-1)$ we may write $-4+1$, which, of course, equals -3.

From Example 1.1(c) and (d) we note the following important result:

> subtracting a negative number is equivalent to adding a positive number

1.2.2 The multiplication symbol, $\times$

The multiplication symbol is sometimes omitted, or replaced with a dot. For example, 3×2 may be written as $(3)(2)$, or as $3 \cdot 2$. In the last case, be careful not to confuse the $\cdot$ with a decimal point. The quantity 3×2 is known as the **product** of 3 and 2. Given a product of many numbers, we can use the dot notation. For example, $(3)(4)(5)(6)$ could be written as $3 \cdot 4 \cdot 5 \cdot 6$.

The product $(2)(1)$ is written as $2!$ and read as 'factorial 2'. The product $(3)(2)(1)$ is written as $3!$ and read as 'factorial 3'. Clearly $2! = 2$ and $3! = 6$. Similarly, $4! = (4)(3)(2)(1)$ and $5! = (5)(4)(3)(2)(1)$, so that $4! = 24$ and $5! = 120$. Most scientific calculators have a factorial button marked $n!$. Practise using this facility.

When multiplying numbers, their order is unimportant. For example, $6 \times 3 \times 4$ is identical in value to $4 \times 3 \times 6$ or $3 \times 6 \times 4$.

Consider multiplying two numbers, which may be positive or negative. The sign of the product is given by the following rule:

> (positive)(positive) = positive
> (positive)(negative) = negative
> (negative)(positive) = negative
> (negative)(negative) = positive

Example 1.2

Evaluate the following

(a) 4×5 (b) $(-4) \times (-5)$ (c) $4 \times (-3)$

(d) $(-4) \times 5 \times 6$ (e) $(-1) \times (-2) \times (-3)$

Solution

(a) When two positive numbers are multiplied, the result is positive. Hence $4 \times 5 = 20$.

(b) When two negative numbers are multiplied, the result is positive. Hence $(-4) \times (-5) = 20$.

(c) When a positive number and a negative number are multiplied, the result is negative. Hence $4 \times (-3) = -12$.

(d) We can think of $(-4) \times 5 \times 6$ as $(-4) \times (5 \times 6)$. This is $(-4) \times 30$. When a negative number and a positive number are multiplied, the result is negative. Hence $(-4) \times 30 = -120$, and so $(-4) \times 5 \times 6 = -120$.

(e) Let us consider $(-1) \times (-2)$ first of all. The product of two negative numbers is a positive number, and so $(-1) \times (-2) = 2$. We may now write $(-1) \times (-2) \times (-3)$ as $2 \times (-3)$. The product of a positive number and a negative number is a negative number, and so $2 \times (-3) = -6$. Hence $(-1) \times (-2) \times (-3) = -6$.

Example 1.3 Spring production

A machine can be used to produce 500 springs in a day. Calculate the number of springs that can be produced using 7 machines during a working week of 5 days.

Solution

The number of springs that can be produced is given by

$$500 \times 7 \times 5 = 500 \times 35 = 17\,500$$

1.2.3 The division symbol, $\div$

There are various ways of representing division. The expressions $3 \div 2$, $3/2$ and $\frac{3}{2}$ all mean 3 divided by 2. The quantity $3/2$ is sometimes referred to as the **quotient** of 3 and 2. Note that $3/2$ is different from $2/3$, and so order is important with division. If $\frac{3}{2}$ is **inverted**, we get $\frac{2}{3}$. The **reciprocal** of a number is found by inverting it, so, for example, the reciprocal of $\frac{2}{7}$ is $\frac{7}{2}$. Note that because 3 is $\frac{3}{1}$, the reciprocal of 3 is $\frac{1}{3}$. The sign of a quotient is given by the following rule:

$$\frac{\text{positive}}{\text{positive}} = \text{positive}$$

$$\frac{\text{positive}}{\text{negative}} = \text{negative}$$

$$\frac{\text{negative}}{\text{positive}} = \text{negative}$$

$$\frac{\text{negative}}{\text{negative}} = \text{positive}$$

Example 1.4

Evaluate (a) $\dfrac{-6}{2}$ (b) $8/(-4)$ (c) $(-12) \div (-4)$

Solution (a) The quotient of a negative number and a positive number is negative. Hence $\dfrac{-6}{2}$ is -3.

(b) The quotient of a positive number and a negative number is negative. Hence $\dfrac{8}{-4}$ is -2.

(c) The quotient of two negative numbers is positive. Hence $\dfrac{-12}{-4}$ is 3.

Example 1.5 Lorry loads required to clear a site

A builder needs to clear a site of rubble and top soil prior to developing the land. The total weight of material to be removed is 538 tonnes. The lorries being used can carry a maximum of 25 tonnes. How many lorry loads are needed to clear the site?

Solution The number of lorry loads required is

$$\frac{538}{25} = 21.52$$

Clearly we need 22 full lorry loads.

1.2.4 The plus or minus symbol, ±

In scientific and engineering literature we often see the symbol ± used. We write 200±10 to mean the two numbers 200−10 and 200+10, that is, the numbers 190 and 210. Suppose we say that the number of components produced by a machine is in the range 200±10. This means that the number produced can be anywhere from 190 to 210 inclusive.

Example 1.6

Evaluate

(a) 10±2   (b) 25±12   (c) −5±3

Solution

(a)   We subtract 2 from 10 to give 8 and add 2 to 10 to give 12. Note that 10 lies halfway between 8 and 12.

(b)   For 25±12, we subtract 12 from, and add 12 to, 25. This gives 13 and 37. Note that 25 lies in the middle of the range from 13 to 37.

(c)   We subtract 3 from −5 to get −8. We add 3 to −5 to get −2. So −5±3 means the numbers in question are −8 and −2. The number −5 occurs in the middle of the range from −8 to −2.

Example 1.7 Components made using a lathe

During a day, the number of components made using a lathe lies in the range 336±15. Find the maximum and minimum numbers of components that may be produced.

Solution   The maximum number produced is 336+15, that is, 351. The minimum number produced is 336−15, that is, 321.

Example 1.8 Cars produced in a factory

During a day, a factory produces between 1500 and 1560 cars. Express the range using the ± notation.

Solution   The value halfway between 1500 and 1560 is 1530. Hence we see the range is 1530±30.

1.2.5 The modulus symbol, | |

The **modulus** of a number is the size of the number, regardless of sign. It is denoted by vertical lines around the number. For example, |3| is 3 and |−3| is 3. The modulus of a number is never negative.

Example 1.9

State the modulus of each of the following numbers
(a) 17.1   (b) -36 (c) -0.04 (d) 0

Solution   (a) $|17.1|=17.1$ (b) $|-36|=36$ (c) $|-0.04|=0.04$ (d) $|0|=0$

Self-assessment questions 1.2

1.   Explain what is meant by the quotient of two numbers.

2.   Explain what is meant by the sum of two numbers.

3.   Explain what is meant by the difference of two numbers.

4.   Explain what is meant by the product of two numbers.

5.   Explain what is meant by the modulus of a number.

6.   Explain what is meant by the reciprocal of a number.

7.   The reciprocal of a negative number is a positive number. True or false?

Exercises 1.2

1.   Evaluate (a) $17+(-6)$ (b) $-6+(-2)$
(c) $-6-(+2)$ (d) $-6-2$
(e) $13-(-6)$

2.   Evaluate
(a) $12\times(-3)$ (b) $(-12)\times 3$
(c) $(-12)\times(-3)$ (d) $\dfrac{18}{-2}$ (e) $\dfrac{+8}{-2}$
(f) $\dfrac{-8}{-2}$ (g) $(-2)\times(-3)\times(-4)$
(h) $\dfrac{(-2)\times(-12)}{(-3)\times(-4)}$

3.   A machine produces 400 bolts in a day. Calculate the number of bolts produced by 3 machines during 7 days.

4.   Recall Example 1.5. How many lorry loads would be required to clear a site if the weight of material to be cleared is 746 tonnes and the lorries carry a maximum of 22 tonnes?

5.   Evaluate
(a) 10 ± 4 (b) 5 ± 1 (c) 40 ± 12
(d) -2 ± 2 (e) -10 ± 2
(f) $|3\times(-2)|$ (g) $|(-2)\cdot(-3)\cdot(-4)|$

6.   Three numbers are multiplied together. They are all negative. What is the sign of their product?

7.   Four numbers are multiplied together. Two of the numbers are positive and two are negative. What is the sign of the product?

8.   Four negative numbers are multiplied together and then divided by a negative number. What is the sign of the resulting number?

9.   A number is multiplied by itself. Can the result be negative? Explain your answer.

10.   Two numbers are multiplied together and their product is zero. What can you say about the two numbers?

11. The number of chocolate bars produced per day on a factory production line lies in the range 8000 ± 230. Find the maximum and minimum numbers of bars produced.

12. A plastic component produced by an injection moulding machine has a length of 352.7 ± 0.23 mm. Find the maximum and minimum possible lengths of this component.

13. Car production in a factory varies in the range 2450 to 2600. Write this range using the plus or minus $(\pm)$ notation.

14. A lathe operator machines cylindrical components to have a diameter within the range 35.023 mm to 35.327 mm. Write this range using the plus or minus $(\pm)$ notation.

15. Evaluate (a) $(-4) \times (-4)$
(b) $|-4| \times (-4)$ (c) $|-2| \times |-4|$
(d) -4×5 (e) $|-4 \times 6|$ (f) $5!/3!$
(g) $7 \times 6!$

1.3 Lowest common multiple and highest common factor

This section explains the terms 'lowest common multiple' and 'highest common factor'. A knowledge of prime numbers and factorization is needed, and so these are described first.

1.3.1 Primes and factorization

Most whole numbers can be written as a product of smaller whole numbers. This process is called **factorization**. For example,

$$20 = 4 \times 5, \qquad 6 = 2 \times 3, \qquad 12 = 2 \times 2 \times 3, \qquad 30 = 2 \times 3 \times 5, \qquad 16 = 2 \times 8$$

Consider $20 = 4 \times 5$. We say that 20 is a **multiple** of 4. Also, 20 is a multiple of 5. Furthermore, 4 is said to **divide exactly** into 20; 5 also divides exactly into 20. We say that 4 and 5 are **factors** of 20.

A whole number that cannot be written as the product of smaller whole numbers is called a **prime number**. For example, 5 cannot be written as a product of smaller whole numbers, and so is prime. On the other hand, $6 = 2 \times 3$, and so 6 is not a prime number. Examples of prime numbers are 2, 3, 5, 7, 11, 13 and 17. It is often useful to express a number as a product of prime numbers. This process is called **prime factorization**.

Example 1.10

Express each of the following numbers as a product of prime numbers:
(a) 36   (b) 40   (c) 23

Solution (a) We examine the number to see if 2, the smallest prime, is a factor. We then write 36 as 2×18. Then 18 is examined to see if 2 is a factor. It is, and so 18 is written as 2×9. At this stage we have $36 = 2 \times 2 \times 9$. Now 9 is examined. We note that 2 is not a factor, so we try the next prime, 3, as a factor. So 9 is written as 3×3. At this stage we have $36 = 2 \times 2 \times 3 \times 3$. All the factors are prime numbers, and so the prime factorization is complete. In summary, we have

$$36 = 2 \times 18 = 2 \times 2 \times 9 = 2 \times 2 \times 3 \times 3$$

(b)   We note that 2 is a factor of 40, and so we can write $40 = 2 \times 20$. Next, we examine 20. This may be written as 2×10, and so $40 = 2 \times 2 \times 10$. We now move on to examine 10 and write this as 2×5. In summary, we have

$$40 = 2 \times 20 = 2 \times 2 \times 10 = 2 \times 2 \times 2 \times 5$$

All the factors are prime, and so no more factorization is possible.

(c)   23 is a prime number. It cannot be expressed as the product of smaller whole numbers.

1.3.2 Lowest common multiple

Given two or more numbers, the **lowest common multiple (l.c.m.)** is the lowest number into which each of the given numbers will divide exactly. To put it another way, given a set of numbers, the l.c.m. is the smallest number that is a multiple of the original numbers. For example, 12 is the l.c.m. of 4 and 6, that is, 12 is the smallest number into which both 4 and 6 divide exactly. Similarly, 30 is the l.c.m. of 3, 10 and 15. All the given numbers, 3, 10, 15, divide into 30. It is impossible to find a smaller number such that this is true. The method of finding the l.c.m. is now illustrated.

Example 1.11

Find the l.c.m. of 10, 12 and 18.

Solution   The first step is to prime-factorize the given numbers, that is, express them as products of prime numbers:

$$10 = 2 \times 5, \quad 12 = 2 \times 2 \times 3, \quad 18 = 2 \times 3 \times 3$$

We now begin to form the l.c.m. from the factorized forms of the numbers. From the first number (10), the l.c.m. must contain the factors 2 and 5. From the second number (12), the l.c.m. must contain the factors 2, 2 and 3. From the third number (18), the l.c.m. must contain the factors 2, 3 and 3. The maximum number of 2s required by the l.c.m. is 2, found from the factors of 12. The maximum number of 3s required is 2, found from the factors of 18. The maximum number of 5s required is 1, found from the factors of 10. Hence

$$l.c.m. = 2 \times 2 \times 3 \times 3 \times 5 = 180$$

So 180 is the smallest number into which 10, 12 and 18 will divide; that is, 180 is the l.c.m. of 10, 12 and 18.

Example 1.12

Find the l.c.m. of 30, 50 and 76.

Solution   First of all, the numbers are prime-factorized:

$$30 = 2 \times 3 \times 5, \quad 50 = 2 \times 5 \times 5, \quad 76 = 2 \times 2 \times 19$$

From the factors of 76, we see that the maximum number of 2s required is 2. The maximum number of 3s required is 1, the maximum number of 5s required is 2 and the number of 19s required is 1. Hence

$$l.c.m. = 2 \times 2 \times 3 \times 5 \times 5 \times 19 = 5700$$

The l.c.m. of 30, 50 and 76 is 5700.

1.3.3 Highest common factor

Given two or more numbers, we can calculate the highest common factor. The **highest common factor (h.c.f.)** is the highest number that will divide exactly into all of the given numbers. For example, the highest common factor of 4 and 18 is 2. In other words, 2 is the highest number that divides exactly into 4 and into 18. The highest number that will divide into 12, 18 and 30 is 6. Hence 6 is the h.c.f. of 12, 18 and 30. We calculate the h.c.f. of a set of numbers by prime-factorizing the numbers.

Example 1.13

Find the h.c.f. of 18, 30 and 90.

Solution   First of all, the numbers are expressed as products of primes:

$$18 = 2 \times 3 \times 3, \quad 30 = 2 \times 3 \times 5, \quad 90 = 2 \times 3 \times 3 \times 5$$

We now see which primes are common to all numbers. Each number contains 2×3, and this is the greatest product common to all three numbers. Hence the h.c.f. is 6.

Example 1.14

Find the h.c.f. of 45, 75, 150 and 225.

Solution   Prime factorization produces

$$45 = 3 \times 3 \times 5, \quad 75 = 3 \times 5 \times 5, \quad 150 = 2 \times 3 \times 5 \times 5, \quad 225 = 3 \times 3 \times 5 \times 5$$

Common to each number is 3×5, and so the h.c.f. is 15.

Self-assessment questions 1.3

1. Explain what is meant by a prime number.

2. All prime numbers except 2 are odd. True or false?

3. If two numbers have no factors in common then how do you calculate (a) the h.c.f. and (b) the l.c.m.?

4. Explain the difference between factorization and prime factorization.

5. The l.c.m. of a set of numbers is always divisible exactly by the h.c.f. of the numbers. True or false?

6. What is the h.c.f. of a set of prime numbers?

7. How would you calculate the l.c.m. of a set of prime numbers?

Exercises 1.3

1. Calculate the l.c.m. of
    (a) 10, 18   (b) 6, 10, 15   (c) 16, 60, 90
    (d) 25, 60, 80   (e) 30, 40, 50, 75
    (f) 14, 28, 35, 70, 91

2. Calculate the h.c.f. of
    (a) 6, 9   (b) 16, 24   (c) 20, 35, 50

    (d) 24, 60, 84   (e) 42, 189, 231
    (f) 90, 153, 315, 1287

3. Determine which of the following are prime numbers.
    (a) 46   (b) 51   (c) 49   (d) 13
    (e) 143

Computer and calculator exercises 1.3

1. Computer algebra packages can be used to prime-factorize a given integer. Investigate how to do this using the package available for your use. Then prime-factorize
    (a) 1 334 025   (b) 152 125 131 763 605
    (c) 1 285 739 648 911

2. Computer algebra packages can be used to find prime numbers and, given any number, find the next prime. Find the four primes following 196.

1.4 Fractions

This section defines the terms proper fraction and improper fraction. It also explains how to express a fraction in its simplest form. Finally, it shows how to multiply, divide, add and subtract fractions.

1.4.1 Proper and improper fractions

A fraction has a **numerator** and a **denominator** of the form

$$\text{fraction} = \frac{\text{numerator}}{\text{denominator}}$$

If the denominator is greater than the numerator, the fraction is called a **proper fraction**. Otherwise, it is an **improper fraction**. For example, $\frac{2}{3}$ and $\frac{1}{7}$ are proper fractions; $\frac{3}{2}$ and $\frac{7}{7}$ are improper fractions. Sometimes whole numbers and fractions are written together, for example $3\frac{2}{5}$. Such a number is called a **mixed fraction**. Mixed fractions can always be written as improper fractions and vice versa.

Example 1.15

Determine which of the following are proper fractions and which are improper fractions:

(a) $\frac{6}{9}$ (b) $\frac{13}{1}$ (c) $\frac{6}{6}$ (d) $\frac{9}{7}$

Solution

(a)   In $\frac{6}{9}$ the denominator is greater than the numerator, and so the fraction is proper.

(b)   In $\frac{13}{1}$ the denominator is less than the numerator, and so the fraction is improper.

(c)   The denominator is equal to the numerator, and so $\frac{6}{6}$ is improper.

(d)   The denominator is less than the numerator, and so $\frac{9}{7}$ is improper.

1.4.2 Simplifying a fraction

It is usual to express a fraction in its simplest form. For example, $\frac{3}{6}$ can be written more simply as $\frac{1}{2}$. We express a fraction in its simplest form by prime-factorizing both numerator and denominator and then cancelling any common factors. This is equivalent to dividing both numerator and denominator by the common factors.

Example 1.16

Express the following fractions in their simplest forms:

(a) $\frac{63}{105}$ (b) $\frac{30}{16}$ (c) $\frac{20}{10}$

Solution (a) Prime-factorization and cancellation of common factors yields

$$\frac{63}{105} = \frac{\cancel{3} \times 3 \times \cancel{7}}{\cancel{3} \times 5 \times \cancel{7}} = \frac{3}{5}$$

Factors of 3 and 7 have been cancelled from the numerator and the denominator.

(b) Prime-factorizing and cancelling common factors gives

$$\frac{30}{16} = \frac{\cancel{2} \times 3 \times 5}{\cancel{2} \times 2 \times 2 \times 2} = \frac{15}{8}$$

A factor of 2 has been cancelled.

(c)

$$\frac{20}{10} = \frac{\cancel{2} \times 2 \times \cancel{5}}{\cancel{2} \times \cancel{5}} = \frac{2}{1} = 2$$

Factors of 2 and 5 have been cancelled from numerator and denominator. Note that the resulting denominator is 1. The original denominator has been divided by 10 to leave 1.

From (c), we see that whole numbers can be considered as fractions. In each case in the above example note that to reduce a fraction to its simplest form the numerator and the denominator are both divided by their h.c.f. Note that cancelling is equivalent to dividing both numerator and denominator by their h.c.f. Dividing both numerator and denominator of a fraction by the same quantity leaves the value of the fraction unchanged.

1.4.3 Multiplication of a fraction

A fraction may be multiplied by a whole number or another fraction. Numerators are multiplied together and denominators are multiplied together. Any factors that are common to both numerator and denominator may be cancelled.

Example 1.17

Calculate

(a) $3 \times \frac{5}{12}$ (b) $\frac{3}{4} \times \frac{5}{6}$ (c) $\frac{7}{12} \times \frac{3}{14}$

Solution (a) We can think of 3 as $\frac{3}{1}$. Hence

$$3 \times \frac{5}{12} = \frac{3}{1} \times \frac{5}{12} = \frac{3 \times 5}{1 \times 12} = \frac{3 \times 5}{12}$$

Writing 12 as 3×4, gives $\frac{3 \times 5}{3 \times 4}$. Since 3 is common to both numerator and denominator, it can be cancelled, leaving $\frac{5}{4}$. Hence

$$3 \times \frac{5}{12} = \frac{3 \times 5}{12} = \frac{\cancel{3} \times 5}{\cancel{3} \times 4} = \frac{5}{4}$$

(b)   The numerators are multiplied together, and the denominators are multiplied together:

$$\frac{3}{4} \times \frac{5}{6} = \frac{3 \times 5}{4 \times 6}$$

We write 6 as 2×3, and then can cancel a factor of 3 from the numerator and the denominator:

$$\frac{3 \times 5}{4 \times 6} = \frac{\cancel{3} \times 5}{4 \times 2 \times \cancel{3}} = \frac{5}{4 \times 2} = \frac{5}{8}$$

Thus

$$\frac{3}{4} \times \frac{5}{6} = \frac{5}{8}$$

(c)

$$\frac{7}{12} \times \frac{3}{14} = \frac{7 \times 3}{12 \times 14}$$

Factors of 3 and 7 can be cancelled from both the numerator and the denominator:

$$\frac{7 \times 3}{12 \times 14} = \frac{\cancel{7} \times \cancel{3}}{\cancel{3} \times 4 \times 2 \times \cancel{7}} = \frac{1}{4 \times 2} = \frac{1}{8}$$

Hence

$$\frac{7}{12} \times \frac{3}{14} = \frac{1}{8}$$

Example 1.18

Evaluate (a) $\frac{3}{7} \times (-\frac{7}{9})$ (b) $(-\frac{1}{2}) \times (-\frac{4}{7})$

Solution   (a)   A positive number multiplied by a negative number results in a negative number. So

$$\frac{3}{7} \times \left(-\frac{7}{9}\right) = -\left(\frac{3}{7} \times \frac{7}{9}\right) = -\left(\frac{3 \times 7}{7 \times 9}\right)$$

Note that factors of 3 and 7 can be cancelled. Therefore

$$-\left(\frac{3 \times 7}{7 \times 9}\right) = -\left(\frac{1}{3}\right) = -\frac{1}{3}$$

(b)   The product of two negative numbers is a positive number. Therefore

$$\left(-\frac{1}{2}\right) \times \left(-\frac{4}{7}\right) = \frac{1}{2} \times \frac{4}{7} = \frac{1 \times 4}{2 \times 7}$$

A factor of 2 can be cancelled and so

$$\frac{1 \times 4}{2 \times 7} = \frac{1 \times 2 \times 2}{2 \times 7} = \frac{1 \times 2}{7} = \frac{2}{7}$$

Example 1.19

The numerator and denominator of a fraction are both multiplied by the same number. Explain what happens to the value of the fraction.

Solution   We consider any fraction, say for example $\frac{5}{7}$. The numerator and the denominator are both multiplied by any number. Let us choose, say, 6. So we consider $\frac{5 \times 6}{7 \times 6}$. This is equivalent to $\frac{5}{7} \times \frac{6}{6}$. Clearly, $\frac{6}{6} = 1$, and so we have $\frac{5}{7} \times 1$, which is $\frac{5}{7}$. The fraction is unaltered. We conclude that when the numerator and the denominator of a fraction are both multiplied by the same number, the value of the fraction remains unchanged.

1.4.4 Division by a fraction

Any number, including a fraction, may be divided by a whole number or a fraction. Whole numbers can be considered as fractions with a denominator of 1.

To divide by a fraction, we invert the fraction and multiply. For example, suppose we wish to calculate $3 \div \frac{5}{6}$. We find

$$\frac{3}{\frac{5}{6}} = 3 \times \left(\frac{6}{5}\right) = \frac{3}{1} \times \frac{6}{5} = \frac{18}{5}$$

Note that the fraction $\frac{5}{6}$ has been inverted to $\frac{6}{5}$, and the division replaced by multiplication. The reason for this can be seen as follows. The value of a fraction remains unchanged if both numerator and denominator are multiplied by the same number. $\dfrac{3}{\frac{5}{6}}$ remains unchanged if both numerator and denominator are

multiplied by 6. This results in $\dfrac{3 \times 6}{\frac{5}{6} \times 6}$. But $\frac{5}{6} \times 6$ is 5, and so we get $\frac{3 \times 6}{5} = 3 \times \frac{6}{5}$

as before.

Example 1.20

Evaluate (a) $\frac{3}{5} \div 2$ (b) $2 \div \frac{3}{5}$ (c) $\frac{6}{7} \div \frac{9}{11}$

Solution (a) We treat 2 as $\frac{2}{1}$. This is inverted to $\frac{1}{2}$ and then multiplication is performed. Thus

$$\frac{3}{5} \div 2 = \frac{3}{5} \div \frac{2}{1} = \frac{3}{5} \times \frac{1}{2} = \frac{3}{10}$$

(b) The $\frac{3}{5}$ is inverted to $\frac{5}{3}$ and multiplication is then carried out. So

$$2 \div \frac{3}{5} = \frac{2}{1} \times \frac{5}{3} = \frac{10}{3}$$

(c) The $\frac{9}{11}$ is inverted and then multiplication performed. Hence

$$\frac{6}{7} \div \frac{9}{11} = \frac{6}{7} \times \frac{11}{9} = \frac{6 \times 11}{7 \times 9} = \frac{2 \times \cancel{3} \times 11}{7 \times \cancel{3} \times 3} = \frac{22}{21}$$

Example 1.21

Evaluate (a) $\dfrac{-\frac{3}{2}}{4}$ (b) $\dfrac{-\frac{1}{2}}{-\frac{1}{4}}$ (c) $\dfrac{\frac{3}{4}}{-\frac{8}{9}}$

Solution (a) The result of dividing a negative number by a positive number is negative. So

$$\frac{-\frac{3}{2}}{4} = -\left(\frac{3}{2} \div 4\right) = -\left(\frac{3}{2} \times \frac{1}{4}\right) = -\frac{3}{8}$$

(b) Dividing a negative number by a negative number results in a positive number. Therefore

$$\frac{-\frac{1}{2}}{-\frac{1}{4}} = \frac{\frac{1}{2}}{\frac{1}{4}} = \frac{1}{2} \div \frac{1}{4} = \frac{1}{2} \times \frac{4}{1} = 2$$

(c) A positive number divided by a negative number is a negative number. Thus

$$\frac{\frac{3}{4}}{-\frac{8}{9}} = -\left(\frac{\frac{3}{4}}{\frac{8}{9}}\right) = -\left(\frac{3}{4} \div \frac{8}{9}\right) = -\left(\frac{3}{4} \times \frac{9}{8}\right) = -\left(\frac{27}{32}\right) = -\frac{27}{32}$$

1.4.5 Rewriting fractions

We saw in Example 1.16 that $\frac{63}{105}$ is identical in value to $\frac{3}{5}$. Thus the same fraction may be expressed in different ways.

Example 1.22

Express $\frac{3}{7}$ so that the denominator is 35.

Solution　The original denominator is 7; the required denominator is 35. Hence the denominator needs to be multiplied by 5. To keep the value of the fraction unchanged, the numerator must also be multiplied by 5. Thus

$$\frac{3}{7} = \frac{3 \times 5}{7 \times 5} = \frac{15}{35}$$

Example 1.23

Express $\frac{15}{60}$ so that the denominator is 20.

Solution　The denominator of 60 is divided by 3 to obtain a denominator of 20. Hence

$$\frac{15}{60} = \frac{15 \div 3}{60 \div 3} = \frac{5}{20}$$

1.4.6 Adding and subtracting fractions

To add and subtract two or more fractions, we first of all rewrite the fractions so that they have the same denominator. This is known as the **common denominator**. The denominator chosen is the l.c.m. of the denominators of the fractions. For example, given $\frac{3}{4} + \frac{5}{6}$, we calculate the l.c.m. of 4 and 6. This is 12. Each fraction is then expressed with 12 as the denominator. Finally, the numerators are added/subtracted and the result divided by the common denominator.

Example 1.24

Calculate $\frac{3}{4} + \frac{5}{6}$.

Solution　The denominators are 4 and 6. The l.c.m. of 4 and 6 is 12. Each fraction is expressed with 12 as the denominator. Thus

$$\frac{3}{4} = \frac{3 \times 3}{4 \times 3} = \frac{9}{12}, \quad \frac{5}{6} = \frac{5 \times 2}{6 \times 2} = \frac{10}{12}$$

So

$$\frac{3}{4} + \frac{5}{6} = \frac{9}{12} + \frac{10}{12}$$

Next, only the numerators are added and the result is divided by the common denominator:

$$\frac{9+10}{12} = \frac{19}{12}$$

Example 1.25

Calculate $\frac{5}{7} + \frac{1}{3} - \frac{4}{9}$.

Solution   The l.c.m. of 7, 3 and 9 is 63. Each fraction is expressed with 63 as its denominator. Thus

$$\frac{5}{7} = \frac{45}{63}, \quad \frac{1}{3} = \frac{21}{63}, \quad \frac{4}{9} = \frac{28}{63}$$

Hence

$$\frac{5}{7} + \frac{1}{3} - \frac{4}{9} = \frac{45}{63} + \frac{21}{63} - \frac{28}{63}$$

$$= \frac{45+21-28}{63}$$

$$= \frac{38}{63}$$

When converting a mixed fraction to an improper fraction, we use rules of addition of fractions. Consider the following examples.

Example 1.26

Write $3\frac{2}{5}$ as an improper fraction.

Solution   We note that $3\frac{2}{5}$ is identical to $3 + \frac{2}{5}$. Then

$$3\frac{2}{5} = 3 + \frac{2}{5} = \frac{3}{1} + \frac{2}{5} = \frac{15}{5} + \frac{2}{5} = \frac{17}{5}$$

Thus writing $3\frac{2}{5}$ as an improper fraction gives $\frac{17}{5}$.

Example 1.27

Express $\frac{25}{7}$ as a mixed fraction.

Solution   We note that 7 divides 3 times into 25, with 4 remaining left over:

$$\frac{25}{7} = \frac{21+4}{7} = \frac{21}{7} + \frac{4}{7} = 3 + \frac{4}{7} = 3\tfrac{4}{7}$$

Example 1.28

Calculate $1\tfrac{5}{12} - \tfrac{2}{3}$.

Solution   We first express the mixed fraction $1\tfrac{5}{12}$ as an improper fraction, $\tfrac{17}{12}$. So

$$1\tfrac{5}{12} - \tfrac{2}{3} = \frac{17}{12} - \frac{2}{3}$$

The l.c.m. of 12 and 3 is 12. Now

$$\frac{2}{3} = \frac{2 \times 4}{3 \times 4} = \frac{8}{12}$$

So

$$\frac{17}{12} - \frac{2}{3} = \frac{17}{12} - \frac{8}{12} = \frac{17-8}{12} = \frac{9}{12}$$

The fraction $\tfrac{9}{12}$ can be simplified to $\tfrac{3}{4}$. Hence

$$1\tfrac{5}{12} - \tfrac{2}{3} = \frac{3}{4}$$

Self-assessment questions 1.4

1. Explain the role of l.c.m. when adding and subtracting fractions.
2. Explain the role of h.c.f. when simplifying a fraction.
3. All fractions have a value of less than 1. True or false?
4. All whole numbers can be regarded as fractions. True or false?

Exercises 1.4

1. Express the following in their simplest form.
    (a) $\tfrac{20}{45}$ (b) $\tfrac{16}{36}$ (c) $-\tfrac{42}{21}$ (d) $\tfrac{18}{16}$
    (e) $\tfrac{30}{30}$ (f) $\tfrac{17}{21}$ (g) $-\tfrac{49}{35}$ (h) $\tfrac{90}{30}$

2. Calculate
    (a) $\tfrac{1}{2} + \tfrac{1}{3}$ (b) $\tfrac{1}{2} - \tfrac{1}{3}$ (c) $\tfrac{2}{3} + \tfrac{3}{4}$
    (d) $\tfrac{5}{6} - \tfrac{2}{3}$ (e) $\tfrac{8}{9} + \tfrac{1}{5} + \tfrac{1}{6}$ (f) $\tfrac{4}{5} + \tfrac{3}{7} - \tfrac{9}{10}$
    (g) $2 - \tfrac{4}{5}$ (h) $\tfrac{9}{7} + \tfrac{10}{3}$ (i) $2\tfrac{3}{4} - 1\tfrac{1}{2} + \tfrac{7}{8}$

3.   Evaluate (a) $\frac{4}{5} \times \frac{3}{16}$ (b) $2 \times 3 \times \frac{1}{4}$
   (c) $\frac{3}{4} \times \frac{3}{4}$ (d) $\frac{4}{9} \times 6$ (e) $\frac{15}{16} \times \frac{4}{5}$
   (f) $\frac{2}{5} \times \frac{1}{3} \times \frac{15}{27}$ (g) $2\frac{1}{4} \times \frac{3}{4}$
   (h) $1\frac{1}{2} \times 1\frac{1}{2} \times 2\frac{1}{4}$

4.   Evaluate (a) $3 \div \frac{1}{2}$ (b) $\frac{1}{2} \div \frac{1}{4}$ (c) $\frac{6}{7} \div \frac{16}{21}$
   (d) $\dfrac{\frac{3}{4}}{4}$ (e) $5 \div \frac{10}{9}$ (f) $\frac{3}{4} \div \frac{4}{3}$ (g) $4\frac{1}{2} \div \frac{3}{7}$
   (h) $3\frac{2}{3} \div 1\frac{4}{7}$

5.   Classify the following as proper fractions or improper fractions.
   (a) $\frac{4}{5}$ (b) $\frac{9}{11}$ (c) $\frac{11}{11}$ (d) $\frac{2}{1}$ (e) $\frac{13}{9}$

6.   Express the following as mixed fractions.
   (a) $\frac{5}{2}$ (b) $\frac{7}{3}$ (c) $-\frac{11}{4}$ (d) $\frac{6}{5}$ (e) $\frac{12}{5}$
   (f) $\frac{18}{7}$ (g) $\frac{16}{3}$ (h) $\frac{83}{9}$

7.   Express the following as improper fractions.
   (a) $2\frac{1}{4}$ (b) $3\frac{1}{2}$ (c) $5\frac{2}{3}$ (d) $-3\frac{2}{5}$
   (e) $11\frac{4}{6}$ (f) $8\frac{2}{9}$ (g) $16\frac{3}{4}$ (h) $89\frac{2}{7}$

1.5 The order in which operations are performed (BODMAS)

This section explains the order in which the operations of addition, subtraction, multiplication and division are carried out.

When we are presented with, say, $3+2\times 4$, we need to decide the order in which the addition and the multiplication are carried out. Doing the addition first gives $5 \times 4 = 20$, while doing the multiplication first gives $3+8=11$, so clearly the order is important.

Given any expression, we first evaluate bracketed terms. Then division and multiplication calculations are carried out. In mathematics an alternative word for multiplication is 'of'. For example, we may wish to calculate $\frac{3}{4}$ of 70, and to do this we calculate $\frac{3}{4} \times 70$. Finally, addition and subtraction are done. The order may be remembered using the word BODMAS. This means

Brackets	First priority
Of **Division** **Multiplication**	Second priority
Addition **Subtraction**	Third priority

If an expression contains several divisions and multiplications, these are carried out from left to right. Similarly, if an expression contains several additions and subtractions, these are carried out from left to right.

Example 1.29

Evaluate the following expressions.
   (a) $3+2\times 4$ (b) $1+1-2 \div 2$ (c) $(3+1)\times 2+1$ (d) $4 \div 2+2$
   (e) $4 \div (2+2)$ (f) $6 \div 3 \times 4$

Solution   (a)   Multiplication takes place before addition, so the expression becomes $3+8$. Hence

$$3+2\times4=3+8=11$$

(b)   Division takes place before addition and subtraction, giving

$$1+1-2\div2=1+1-1$$

and so

$$1+1-2\div2=1+1-1=1$$

(c)   The bracketed expression is calculated first; therefore

$$(3+1)\times2+1=4\times2+1$$

Multiplication is performed before addition, and so

$$4\times2+1=8+1=9$$

Hence

$$(3+1)\times2+1=9$$

(d)   Division is carried out before addition; therefore

$$4\div2+2=2+2=4$$

(e)   The bracketed part is evaluated first, and so

$$4\div(2+2)=4\div4=1$$

By comparing (d) and (e), we note that the inclusion of brackets can change the value of an expression.

(f)   Multiplication and division have equal priority, and so these are performed by working from left to right. Hence in this example the division is performed first:

$$6\div3\times4=2\times4=8$$

Example 1.30

   Find the values of (a) $\frac{2}{3}$ of 96   (b) $\frac{7}{8}$ of 100

Solution   (a)   Recall that 'of' means multiply:

$$\frac{2}{3}\text{ of }96=\frac{2}{3}\times96=2\times32=64$$

(b)

$$\frac{7}{8}\text{ of }100=\frac{7}{8}\times100=\frac{7}{2}\times25=\frac{175}{2}=87.5$$

Example 1.31 Number of bricks required to build an office block

Consider Figure 1.2, which shows a plan view of an office that is to be built. The walls have a height of 10 m, including foundations. The roof is flat, and so the walls are rectangular. Derive an expression for the number of external bricks required to build the office block, assuming that one square metre of wall requires 50 bricks. A first estimation is required, so ignore any savings due to doors and windows. Evaluate this expression.

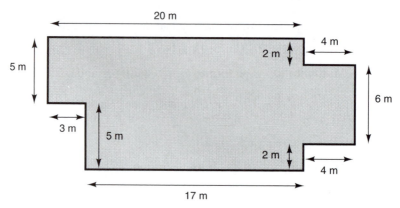

Figure 1.2   Floor plan of the office block.

Solution   First of all, we need to determine the lengths of the various walls and their number. We have

> one wall of length 20 m
>
> one wall of length 17 m
>
> one wall of length 3 m
>
> two walls of length 5 m
>
> two walls of length 2 m
>
> two walls of length 4 m
>
> one wall of length 6 m

So the total length of wall is

$$20 + 17 + 3 + 2 \times 5 + 2 \times 2 + 2 \times 4 + 6$$

Now the walls all have a height of 10 m, and so the total wall area is

$$10 \times (20 + 17 + 3 + 2 \times 5 + 2 \times 2 + 2 \times 4 + 6)$$

Now each square metre of wall requires 50 bricks, and so the total number of bricks is

$$50 \times 10 \times (20 + 17 + 3 + 2 \times 5 + 2 \times 2 + 2 \times 4 + 6)$$

$$= 50 \times 10 \times (20 + 17 + 3 + 10 + 4 + 8 + 6)$$

$$= 50 \times 10 \times 68$$

$$= 500 \times 68$$

$$= 34\,000 \text{ bricks}$$

Example 1.32

Evaluate $\dfrac{6 + 2 \times 4}{(5 - 1) \times 3}$.

Solution   We must evaluate the numerator and denominator separately:

$$6 + 2 \times 4 = 6 + 8 = 14, \qquad (5 - 1) \times 3 = 4 \times 3 = 12$$

So

$$\frac{6 + 2 \times 4}{(5 - 1) \times 3} = \frac{14}{12} = \frac{7}{6}$$

Self-assessment questions 1.5

1. List the order in which the various parts of an expression must be evaluated.

2. Division takes place before multiplication when evaluating an expression. True or false?

3. Multiplication and addition are of equal priority when evaluating an expression. True or false?

Exercises 1.5

1. Evaluate the following expressions:
    (a) $3 + 2 - 1$ (b) $4 \div 2 + 2$
    (c) $10 + 2 \div 5$ (d) $10 - 3 \times 2$
    (e) $3 - (4 \div 2) \times 3$ (f) $\dfrac{4 - 3 \times 3}{(4 - 3) \div 2}$
    (g) $(5 \times 2 + 3) \times (3 + 1 \times 2)$
    (h) $3 \times 2 \times 2 + 2 \div 3$ (i) $3 + 3 \div 2$
    (j) $(3 \div 3) \div 2$ (k) $3 \div (3 \div 2)$

2. Recall Example 1.31. Calculate the number of external bricks required to build the following buildings. As only a first estimation is required, ignore any savings due to doors and windows. In each case assume the buildings have flat roofs and 50 bricks are required to build one square metre of wall.

    (a) The office shown in Figure 1.3(a), assuming the walls have a height of 9 m.
    (b) The factory shown in Figure 1.3(b), assuming the walls have a height of 25 m.
    (c) The warehouse shown in Figure 1.3(c), assuming the walls have a height of 20 m.

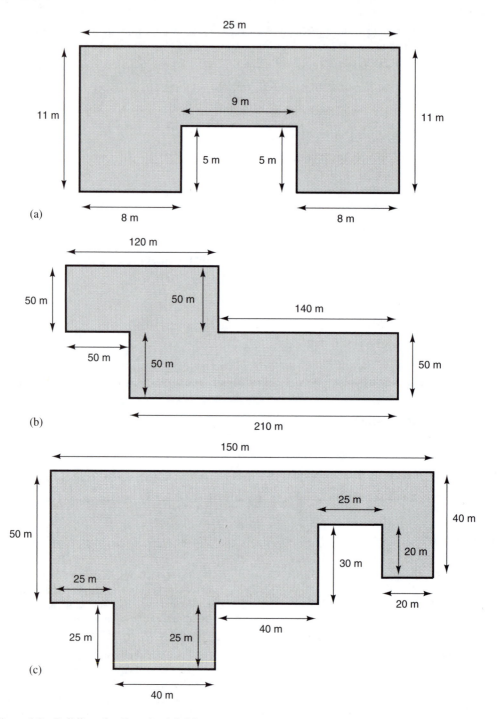

Figure 1.3 Buildings for Exercise 1.5 Q2.

3. Calculate (a) $\frac{7}{10}$ of 5   (b) $\frac{9}{14}$ of (6×4)
     (c) $\frac{2}{5}$ of 30   (d) $\frac{12}{11}$ of 90   (e) $\frac{2}{3}$ of $|-30|$

4. Evaluate
     (a) $\frac{3}{4} - \frac{2}{3} \times \frac{5}{6}$ (b) $\frac{7}{8} \times \frac{3}{14} \times \frac{2}{3} - \frac{1}{2}$

     (c) $1\frac{1}{4} \div \frac{2}{3} + \frac{1}{2}$ (d) $1\frac{1}{4} \div (\frac{2}{3} + \frac{1}{2})$

5. Evaluate
     (a) $3 - 2 + 1 - 4$ (b) $3 - |2 + 1| - 4$
     (c) $3 - |2 + 1 - 4|$ (d) $3 - 2 + |1 - 4|$

1.6 Ratio

In this section, we examine the meaning of ratio. If the number of red cars is 3 times the number of yellow cars, we say 'the red car–yellow car ratio is 3 to 1', written $3 : 1$. This is a way of stating

$$\frac{\text{number of red cars}}{\text{number of yellow cars}} = \frac{3}{1}$$

For every 4 cars, 3 are red and 1 is yellow. Similarly, if the number of type A resistors and the number of type B resistors are such that

$$\frac{\text{number of type A resistors}}{\text{number of type B resistors}} = \frac{4}{9}$$

we say 'the ratio of type A resistors to type B resistors is $4 : 9$'. For every 13 resistors, 4 are of type A and 9 are of type B.

Example 1.33

A packet contains 32 resistors. The resistors are of two types: A and B. The ratio of type A resistors to type B resistors is $1 : 7$. Calculate the numbers of each type of resistor.

Solution   For every 1 type A resistor there are 7 type B resistors. So in every 8 resistors there are 1 type A resistor and 7 type B resistors. Hence in $8 \times 4 = 32$ resistors there are $1 \times 4 = 4$ type A resistors and $7 \times 4 = 28$ type B resistors.

Example 1.34

A rod, 120 mm long, is cut so that the two pieces are in the ratio of $3 : 7$. Calculate the length of each piece.

Solution   The lengths are in the ratio of $3 : 7$. We can consider the rod as being divided into $3 + 7 = 10$ sections, 3 of which constitute the first piece while the remaining 7 sections constitute the second. Since the total length is 120 mm, each of the 10 sections must be 12 mm long. The first piece comprises 3 sections, and so is 36 mm long. The second piece comprises 7 sections, and so is 84 mm long.

A quantity can be divided into many parts and the ratio of each part to the others stated. For example, suppose a length of steel is divided into 4 sections in the ratio $2:4:1:5$. Now $2+4+1+5=12$, and so the first section is $\frac{2}{12}=\frac{1}{6}$ of the original, the second section is $\frac{4}{12}=\frac{1}{3}$ of the original, the third section is $\frac{1}{12}$ of the original and the fourth section is $\frac{5}{12}$ of the original.

Example 1.35

A mass of 40 kg is divided into 3 portions in the ratio of $3:4:8$. Calculate the mass of each portion.

Solution

We can consider the mass as being divided into a total of $3+4+8=15$ parts. The first portion has 3 of these parts, that is, $\frac{3}{15}$ of the mass.

$$\text{mass of first portion} = \frac{3}{15} \times 40 = 8 \text{ kg}$$

Similarly,

$$\text{mass of second portion} = \frac{4}{15} \times 40 = 10.67 \text{ kg}$$

$$\text{mass of third portion} = \frac{8}{15} \times 40 = 21.33 \text{ kg}$$

A ratio of, say, $1:3$, may be expressed as $2:6$, or $10:30$ for example. Usually a ratio is expressed as simply as possible, by dividing through by any common factor. Fractions can be avoided by multiplying through. For example, rather than write $\frac{1}{3}:2$, we multiply by 3 to get $1:6$. This avoids the use of fractions.

Example 1.36

Express the following ratios in their simplest forms:
(a) $6:9$ (b) $18:45$ (c) $2:4:6$ (d) $4:1\frac{1}{3}$ (e) $\frac{1}{7}:\frac{1}{5}$ (f) $0.25:8$

Solution

(a) We note that 3 is a common factor of 6 and 9. Hence $6:9$ is equivalent to $2:3$.
(b) Here 9 is a common factor of 18 and 45, so the ratio simplifies to $2:5$.
(c) We note that 2, 4 and 6 can all be divided by 2. This leads to $1:2:3$.
(d) It is usual to express ratios using whole numbers. By multiplying by 3, the ratio becomes $12:4$. This can be simplified to $3:1$.
(e) By multiplying by $7 \times 5 = 35$ we obtain $5:7$.
(f) We multiply the ratio by 4 to avoid the decimal number. Thus $0.25:8$ is equivalent to $1:32$.

Example 1.37 Metal alloys

A metal alloy is a mixture of two or more metals. Metal alloys are usually created in order to improve on the properties of their component metals. It is sometimes convenient to express the components of an alloy using ratios. Usually the ratios are expressed as 'ratios by mass'. Carry out the following calculations.

(a)   Cupronickel is an alloy of copper and nickel in the ratio $3:1$. It is used for making coins. Calculate the mass of the metallic constituents in 10 kg of cupronickel.

(b)   Brass is an alloy of copper and zinc. There are several types of brass, each containing a different ratio of copper and zinc. Cartridge brass has a ratio of copper to zinc of $7:3$. Calculate the mass of the metallic constituents in 50 kg of cartridge brass.

Solution   (a)   The cupronickel is divided into $3+1=4$ parts. The copper represents 3 of these parts, that is, $\frac{3}{4}$ of the cupronickel. Therefore

$$\text{mass of copper} = \tfrac{3}{4} \times 10 = 7.5 \text{ kg}$$

The nickel represents 1 of these parts, that is, $\frac{1}{4}$ of the cupronickel. Therefore

$$\text{mass of nickel} = \tfrac{1}{4} \times 10 = 2.5 \text{ kg}$$

(b)   The brass is divided into $7+3=10$ parts. The copper represents 7 of these parts, that is, $\frac{7}{10}$ of the brass. Therefore

$$\text{mass of copper} = \tfrac{7}{10} \times 50 = 35 \text{ kg}$$

The zinc represents $\frac{3}{10}$ of the brass. Therefore

$$\text{mass of zinc} = \tfrac{3}{10} \times 50 = 15 \text{ kg}$$

Self-assessment questions 1.6

1.   Explain the meaning of the term 'ratio'.

2.   When given a ratio as $a:b$, a and b must be whole numbers. True or false?

Exercises 1.6

1.   Express the following ratios as simply as possible using whole numbers:
(a) $\frac{1}{2}:1$ (b) $0.3:2$ (c) $\frac{1}{4}:\frac{1}{2}$
(d) $3:2:\frac{1}{2}$ (e) $0.4:0.6:2$ (f) $\frac{3}{4}:\frac{4}{3}$

2.   Resistors of types A and B are in a box in the ratio of $3:8$. There are 77 resistors in the box in total. Calculate the numbers of each type of resistor.

3.   Two densities are in the ratio of $7:5$. The lower density is 600 kg m^{-3}. Calculate the higher density.

4.   A metal rod of length 6.4 m is divided in the ratio of 1:3:4. Calculate the length of each piece.

5.   A type of brass known as Muntz metal has a ratio of copper to zinc of 6:4. Calculate the mass of the metallic constituents in 750 kg of Muntz metal.

6.   Aluminium bronze contains copper and aluminium in the ratio 19:1. Calculate the mass of copper and aluminium in an ingot weighing 380 kg.

7.   Two boxes of resistors both contain type A and type B resistors. In box 1 the ratio of type A resistors to type B resistors is 3:4; in box 2 the ratio of type A resistors to type B resistors is 5:3. The number of resistors in box 1 to the number of resistors in box 2 is in the ratio of 7:4. Calculate the ratio of the total number of type A resistors to the total number of type B resistors.

1.7 Percentages

We use percentages as a means of comparing two, or more, quantities. For example, which is the better performance: 46 out of 70 or 125 out of 200? In order to be able to answer such questions, we express quantities as percentages. This then allows easy comparisons to be made. Fractions with a denominator of 100 are called **percentages**. The symbol for percentage is %. For example,

$$\frac{1}{2} = \frac{50}{100} = 50\%$$

$$\frac{1}{4} = \frac{25}{100} = 25\%$$

$$30\% = \frac{30}{100} = \frac{3}{10}$$

$$75\% = \frac{75}{100} = \frac{3}{4}$$

$$100\% = \frac{100}{100} = 1$$

Example 1.38

Express the following as percentages:
(a) $\frac{7}{10}$ (b) $\frac{9}{20}$

Solution   We need to express each number as a fraction with a denominator of 100.

(a)   To express $\frac{7}{10}$ as a fraction with a denominator of 100, we multiply numerator and denominator by 10. So

$$\frac{7}{10} = \frac{7 \times 10}{10 \times 10} = \frac{70}{100} = 70\%$$

(b)   Given $\frac{9}{20}$, we multiply numerator and denominator by 5. So

$$\frac{9}{20} = \frac{9 \times 5}{20 \times 5} = \frac{45}{100} = 45\%$$

We note that an easy way to convert a fraction to its equivalent percentage is to multiply the fraction by 100 and label the result as a percentage.

Example 1.39

Convert (a) $\frac{27}{50}$ (b) $\frac{4}{7}$ to percentages

Solution   (a)   We multiply the fraction by 100 and label the result as a percentage. Thus

$$\frac{27}{50} \times 100 = 54$$

So $\frac{27}{50}$ is equivalent to 54%.

(b)   We multiply the fraction by 100 and label the result as a percentage:

$$\frac{4}{7} \times 100 = 57.14$$

So $\frac{4}{7}$ is 57.14%

Example 1.40

Express the following percentages as fractions in their simplest forms:
(a) 80%   (b) 35%   (c) 7%

Solution   (a)

$$80\% = \frac{80}{100} = \frac{8}{10} = \frac{4}{5}$$

(b)

$$35\% = \frac{35}{100} = \frac{7}{20}$$

(c)

$$7\% = \frac{7}{100}$$

Example 1.41

Calculate the following:

(a) 20% of 80   (b) 110% of 150   (c) 19% of 37

Solution   Here the word 'of' means multiply. The percentages are first expressed as fractions.

(a)

$$20\% \text{ of } 80 = \frac{20}{100} \times 80 = \frac{1}{5} \times 80 = 16$$

(b)

$$110\% \text{ of } 150 = \frac{110}{100} \times 150 = \frac{11}{10} \times 150 = 11 \times 15 = 165$$

(c)

$$19\% \text{ of } 37 = \frac{19}{100} \times 37 = \frac{703}{100} = 7.03$$

Example 1.42

(a)   Express 17 as a percentage of 50.
(b)   What percentage of 120 is 37?

Solution   (a)

$$\frac{17}{50} \times 100 = 34$$

Hence 17 is 34% of 50.

(b)

$$\frac{37}{120} \times 100 = 30.83$$

So 37 is 30.83% of 120.

Example 1.43

Which is the greater: 42% of 96 or 121% of 32?

Solution

$$42\% \text{ of } 96 = \frac{42}{100} \times 96 = 40.32$$

$$121\% \text{ of } 32 = \frac{121}{100} \times 32 = 38.72$$

Hence 42% of 96 is the greater.

Example 1.44

 A metal rod of length 200 mm is heated and expands to 201.7 mm. Calculate the percentage expansion.

Solution The expansion is $201.7 - 200 = 1.7$ mm. We need to express 1.7 as a percentage of 200:

$$\frac{1.7}{200} \times 100 = 0.85$$

The rod expands its length by 0.85%.

Example 1.45

 A production run produces 3500 components, of which 96% are reliable. Calculate the number of unreliable components.

Solution If 96% of the components are reliable then 4% are unreliable.

$$4\% \text{ of } 3500 = \frac{4}{100} \times 3500 = 140$$

There are 140 unreliable components.

Example 1.46 *Resistor tolerances*

 It is impossible to manufacture resistors to an exact value of resistance. In order to help designers of electronic circuits, manufacturers provide resistors that are manufactured within a tolerance band about the nominal value of the resistor. The most common tolerance bands are $\pm 20\%$, $\pm 10\%$, $\pm 5\%$, $\pm 2\%$, $\pm 1\%$ of the nominal value. A resistor with a smaller tolerance band is more expensive because the materials and the manufacturing process used are more expensive. Calculate the maximum and minimum values of the following resistors defined by their tolerance bands:
(a) $100\,\Omega \pm 10\%$ (b) $820\,\Omega \pm 1\%$ (c) $270\,\text{k}\Omega \pm 2\%$

Solution (a)

$$10\% \text{ of } 100 = \frac{10}{100} \times 100 = 10$$

Hence the maximum resistance is $100 + 10 = 110\,\Omega$. The minimum resistance is $100 - 10 = 90\,\Omega$.

(b)

$$1\% \text{ of } 820 = \frac{1}{100} \times 820 = 8.2$$

Hence the maximum resistance is $820 + 8.2 = 828.2\,\Omega$. The minimum resistance is $820 - 8.2 = 811.8\,\Omega$.

(c)

$$2\% \text{ of } 270\,000 = \frac{2}{100} \times 270\,000 = 5400$$

Hence the maximum resistance is $270\,000 + 5400 = 275.4\,\text{k}\Omega$. The minimum resistance is $270\,000 - 5400 = 264.6\,\text{k}\Omega$.

Self-assessment questions 1.7

1.   What advantages are there in expressing numbers as percentages?

2.   All percentages have a value less than 1. True or false?

3.   A percentage is a number between 1 and 100. True or false?

4.   All percentages must be written using whole numbers. True or false?

Exercises 1.7

1.   Calculate
     (a) 9% of 30   (b) 45% of 170
     (c) 37.5% of 220   (d) 125% of 300
     (e) 300% of 20

2.   The diameter of a piston is given as $9.3\,\text{cm} \pm 0.3\%$. Calculate the maximum and minimum possible values of the diameter.

3.   Express the following as percentages:
     (a) 0.2   (b) 0.16   (c) $\frac{4}{5}$ (d) $\frac{7}{8}$
     (e) 1.2   (f) $\frac{3}{11}$ (g) $1\frac{1}{4}$

4.   The volume of a container is 35 litres. It is cooled and contracts so the volume reduces to 34.5 litres. Calculate the percentage change in the volume.

5.   Calculate
     (a) 57% of 113   (b) 272% of 49.6
     (c) 83.4% of 1150   (d) 0.65% of 3000

6.   An engine speed is $2300\,\text{rev min}^{-1}$. The speed is increased by 7%. Calculate the new speed.

7.   A voltage source is described as lying in the range $300\,\text{V} \pm 2.5\%$. Calculate the maximum and minimum possible voltages of the source.

8.   A wire is heated and its length increases by 3%. If the new length is 2.7 m, find the original length of the wire.

9.   The number of components rejected in a quality check is 2% of the total production batch. If 90 components are rejected, find the number of components which are accepted.

10. The density of a metal alloy is measured to be $340\,\text{kg m}^{-3}$. The measurement is

subject to a $\pm 2\%$ error. Mass is found by multiplying the density by the volume. Calculate the maximum and minimum possible values of the mass of a block of alloy measuring 3 m by 10 m by 1 m.

11. Calculate the maximum and minimum values of the following resistors defined by their tolerance bands:
(a) $27\,\Omega \pm 5\%$ (b) $470\,\Omega \pm 10\%$
(c) $33\,k\Omega \pm 1\%$ (d) $2.7\,k\Omega \pm 20\%$
(e) $680\,k\Omega \pm 2\%$ (f) $2.2\,\Omega \pm 1\%$

Review exercises 1

1 Divide 96 in the ratio $7:3:2$.

2 Evaluate
(a) $3+2\times 2$ (b) $4-2\div 3$
(c) $6+(4\times 3)$ (d) $7\div 14+3$
(e) $(12+3)\times 2-5$

3 Express in their simplest form
(a) $\frac{12}{30}$ (b) $\frac{28}{35}$ (c) $\frac{16}{100}$ (d) $\frac{42}{36}$
(e) $\frac{14}{49}$

4 Calculate
(a) $\frac{10}{13}+\frac{1}{2}$ (b) $\frac{6}{7}-\frac{2}{3}$ (c) $3-\frac{5}{6}-\frac{5}{8}$
(d) $\frac{5}{4}-\frac{3}{2}+\frac{7}{12}$ (e) $\frac{6}{12}-2+\frac{4}{3}+\frac{5}{6}$
(f) $1\frac{3}{4}+2\frac{2}{3}$ (g) $5\frac{2}{5}-3\frac{4}{7}$

5 The voltage across a pair of plates is 300 V. It is decreased by 2.4%. Calculate the new voltage.

6 A power supply is guaranteed to provide a voltage of 12 ± 0.15 V. Calculate the range of values between which the supply voltage lies.

7 Calculate the tolerance bands of the following resistors.
(a) $1\,k\Omega \pm 1\%$ (b) $33\,\Omega \pm 5\%$
(c) $27\,\Omega \pm 1\%$ (d) $1\,M\Omega \pm 10\%$
(e) $6.8\,k\Omega \pm 5\%$ (f) $100\,\Omega \pm 2\%$

8 Calculate
(a) 32% of 95 (b) 121% of -50
(c) 52.3% of 130 (d) 120% of $\frac{3}{4}$
(e) 0.7% of 500 (f) 40% of $2\frac{1}{3}$

9 Find the h.c.f. of
(a) 33, 10, 121 (b) 24, 80, 128
(c) 60, 75, 90, 225

10 Find the l.c.m. of
(a) 2, 4, 6 (b) 2, 4, 6, 8
(c) 2, 3, 4, 6, 8 (d) 12, 15, 20

11 A bar of length 2.50 m is divided into the following ratios. Calculate the length of each piece.
(a) $7:3$ (b) $1:1$ (c) $1:2:1$ (d) $5:6$

12 Evaluate (a) $\frac{2}{3}+\frac{3}{2}$ (b) $\frac{2}{3}\div\frac{3}{2}$
(c) $\frac{2}{3}\times\frac{3}{2}$ (d) $\frac{2}{3}-\frac{3}{2}$

13 Evaluate (a) $3\times\frac{4}{9}$ (b) $3\div\frac{4}{9}$ (c) $\frac{4}{7}\times\frac{14}{9}$
(d) $\frac{3}{4}\times\frac{3}{4}\times\frac{3}{4}$ (e) $\frac{9}{16}\div 3$ (f) $\frac{16}{9}\div\frac{4}{3}$
(g) $\frac{2}{3}+\frac{4}{3}\times\frac{3}{2}$ (h) $5-4\div\frac{2}{3}$
(i) $1\times 2+1\div 2$ (j) $\frac{5}{6}\div\frac{10}{9}$
(k) $6\frac{1}{4}\div 2\frac{1}{2}$ (l) $6\frac{1}{4}\times 2\frac{1}{2}$

14 The temperature of a liquid is 135°C. The temperature is increased to 139°C. Calculate the percentage increase in the temperature.

15 Express the following as mixed fractions:
(a) $\frac{30}{7}$ (b) $\frac{25}{4}$ (c) $\frac{15}{14}$ (d) $\frac{9}{6}$

16 Evaluate
(a) $4-6\div 2$ (b) $4\div 6-2$
(c) $(4-6)\div 2$ (d) $5+3\times 2$
(e) $5\times 3+2$ (f) $|-0.75|$

2

Symbols and Indices

KEY POINTS

This chapter

- explains how symbols are used to represent physical quantities in engineering calculations

- lists the Greek alphabet

- explains the terms 'variable' and 'constant'

- introduces the 'delta' notation for changes in the value of a variable

- explains the meaning of the term 'indices' and explains how expressions involving indices can be simplified

CONTENTS

2.1 Introduction

In order to be able to apply mathematics to solve problems in engineering it is sometimes necessary to introduce symbols to represent physical quantities. **Algebra** is used to manipulate these symbols. The first part of this chapter explains how symbols are used and how they can be combined in various ways. In the second part of the chapter 'indices' are introduced; these enable many mathematical expressions to be written in a compact way.

2.2 Symbols

Very often we want to discuss the behaviour of a physical quantity such as temperature or voltage without actually specifying the particular value the quantity takes. In these circumstances we let a **symbol**, usually a letter, represent the quantity. For example, we might use the letter T to represent temperature and V to represent voltage. Similarly, we might use the letter m to represent the mass of an object, and the letter r to represent the radius of a circle.

Usually the letters we choose as symbols are chosen from the English alphabet, that is

$$a, b, c, \ldots, x, y, z \quad \text{and} \quad A, B, C, \ldots, X, Y, Z$$

In addition, frequent use is made of letters from the Greek alphabet. For reference this alphabet is listed in Figure 2.1.

A	α	alpha	I	ι	iota	P	ρ	rho
B	β	beta	K	κ	kappa	Σ	σ	sigma
Γ	γ	gamma	Λ	λ	lambda	T	τ	tau
Δ	δ	delta	M	μ	mu	Y, Y	υ	upsilon
E	ϵ, ε	epsilon	N	ν	nu	Φ	ϕ, φ	phi
Z	ζ	zeta	Ξ	ξ	xi	X	χ	chi
H	η	eta	O	o	omicron	Ψ	ψ	psi
Θ	θ, ϑ	theta	Π	π	pi	Ω	ω	omega

Figure 2.1 The Greek alphabet.

Some symbols represent one specific value, and these are called **constants**. Other symbols represent quantities that can change, and these are called **variables**. For example, the temperature T in a room may vary throughout the day, and so is a variable. On the other hand, the ratio of the circumference to the diameter of any circle is always constant, and is given the symbol π. This constant is approximately equal to 3.142. Your scientific calculator is preprogrammed with the value of π. Check that you can use this facility.

2.2.1 δ notation for change in a variable

The Greek letter delta, δ, is often used in a specific way to represent a change in the value of a quantity. If x is a variable then the quantity δx stands for a change in the value of x. We sometimes refer to δx as an **increment** of x. For example, if the value of x changes from 0.6 to 0.8, a change of 0.2, we could write $\delta x = 0.2$.

2.2.2 Subscripts and superscripts

Often it is necessary to attach additional letters or digits to symbols. A **subscript** follows, and is placed slightly below, the symbol. For example, the symbol x with a subscript of 3 would be written x_3. If we were considering several voltages in a circuit, we could refer to them as V_1, V_2, V_3 and so on. A **superscript** follows, and is placed slightly higher than the symbol. So, for example, the symbol a with a superscript of α would be written a^α. In mathematical work the precise position of all symbols is very important. Careless positioning can lead to confusion and errors. In written work particular attention should be paid to where symbols are placed in relation to others.

Example 2.1 Temperature distribution in a room

Engineers are often required to design systems to control the temperature in a room. A simple example is a central heating system in a house. In order to control the temperature of a room, it is first necessary to measure it. For simplicity, it is common to measure the temperature at one point in the room and then assume this is an indication of the temperature throughout the room.

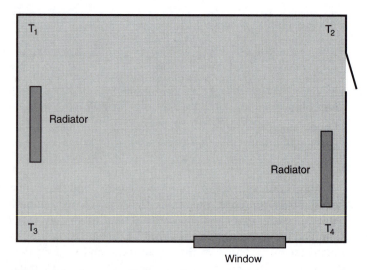

Figure 2.2   A room with two radiators.

In such a case an engineer would typically denote the room temperature by the symbol T. Consider, however, the situation depicted in Figure 2.2. There are two radiators in the room as well as a window and a door. It is clear that the temperature at various points in the room is likely to be different, even if only slightly. In such a case it is convenient to subscript the values of T to indicate the room temperature at different points. In Figure 2.2 we see that the room temperature is measured at four points; the measured values are T_1, T_2, T_3 and T_4. With this arrangement, it is possible to design a more sophisticated form of room temperature control. For example, the amount of heat from each of the two radiators could be varied by the controller to make the values of T_1, T_2, T_3 and T_4 a little more even.

2.3 Algebraic symbols and operations

All the operations of algebra are the same as those of arithmetic given in Chapter 1. A combination of variables, constants and numbers together with the operations of $+$, $-$, $\times$ and $\div$ is called an **algebraic expression**. When evaluating an algebraic expression, the normal rules of arithmetic given in Chapter 1 apply.

2.3.1 Addition

In an algebraic expression the plus sign, $+$, is used in the same way as in arithmetic. We say that $x+y$ is the **sum** of x and y. Note that when adding, the order of the symbols is not important, so that $x+y$ is the same as $y+x$.

2.3.2 Subtraction

The minus sign, $-$, is used in the same way as in arithmetic. We say that $x-y$ is the **difference** of x and y. When subtracting, the order of the symbols is very important, so that $x-y$ is not the same as $y-x$, just as $4-7$ is not the same as $7-4$.

2.3.3 Multiplication

$x \times y$ is known as the **product** of x and y. The multiplication symbol is often omitted. For example, $x \times y$ may be written as xy, and we write $3y$ for $3 \times y$. Sometimes it is convenient to use the dot notation to represent multiplication as we did in Section 1.2.2. For example, $3y{\cdot}4x$ means $3y \times 4x$. When multiplying, the order of the symbols is not important. For example, xyz is identical in value to yxz or zxy. When a combination of numbers and letters is involved, it is usual to write any numbers first. So $x \times z \times 7$ would be written $7xz$.

History of algebraic notation

The symbolism used to write mathematics as we know it today developed over centuries, with the 15th and 16th centuries seeing some notable advances.

Diophantus of Alexandria, known by subsequent historians as the Father of Higher Arithmetic, lived in Greece during the 3rd century AD. One of his many innovations was to introduce some algebraic notation into mathematics. Previously, everything had been written out in words, known as a rhetorical style. With Diophantus, there developed a syncopated style, that is, a mixture of symbols and prose.

Jordanus Nemorarius in the early 13th century was possibly the first to use letters to represent general numbers. However, his practice did not appear to be copied by contemporary mathematicians of the day.

In 1484 Chuquet produced a treatise. Although it remained unpublished until the 19th century, it was important since it introduced a notation for indices. He wrote $6x$ as .6. and $12x^3$ as .12.^{3}. His writing clearly shows that Chuquet understood the laws of indices.

The first appearance in print of $+$ and $-$ is in a book by Johann Widman in Leipzig in 1489. Their meanings were slightly different from the current ones. The symbol $+$ indicated excess and $-$ indicated deficiency. The plus sign may be a contracted form of the Latin word *et* ('and'), which was commonly used to indicate addition. Horizontal bars over words indicated that the word was an abbreviated form, and minus may have been abbreviated to $\overline{m}$ and then finally just $-$. The $+$ and $-$ signs were used with their current meaning in 1514 by the Dutch mathematician Vander Hoecke. Modern algebraic notation began to appear in the 16th century. In a book by Robert Recorde (1557) the symbol $=$ was used for the first time to represent equality. The radical sign, $\sqrt{\ }$, was introduced in 1525 by Christoff Rudolff in his book *Die Coss*. A notation for brackets was introduced by Bombelli in 1572. The French mathematician Viète (1540–1603), a great admirer of the ancient Greek mathematicians, sought to improve upon the work of Diophantus by introducing a complete symbolic algebra. He introduced the practice of using vowels to represent unknown quantities and consonants to represent known ones in a book published in 1591. However, his equations still contained a mixture of words and symbols. For example, he would write

$$3BA^2 - DA + A^3 = Z$$

as

*B*3in *A* quad $-$ *D* plano in *A* $+$ *A* cubo aequator *Z* solido

The present custom of using the later letters of the alphabet for unknowns was introduced by Descartes in 1637. The Englishman William Oughtred (1574–1660) placed great emphasis on mathematical symbols (see Figure 1), giving over 150 of them. In 1685 Wallis published his *Algebra*, which contained the first full systematic use of algebraic formulae. He is also credited with the introduction of negative and fractional indices as used today.

Leibniz (1646–1716) developed calculus and the notation dy and $\int$ he introduced still remain today. Leonard Euler (1707–1783), one of the greatest and most prolific mathematicians ever to have lived, introduced the now familiar notations: π, e, $\sum$, e^x, log x, sin x, cos x and $f(x)$.

Notæ feu fymbola quibus in fequen-

tibus utor :

Æquale= Simile *Sim.*
Majus ⊏. Proxime majus ⊏⁓.
Minus ⊐⁖ Proxime minus ⊐.
Non majus ⊏⁖. Æquale vel minus ⊏⁖
Non minus ⊐⁖. Æquale vel majus ⊐⁖.
Proportio , five ratio æqualis ∷
Major ratio ⁖. Minor ratio ⁖ .
Continuè proportionales ∴.
Commenfurabilia ⊓.
Incommenfurabilia ⊓.
Commenfurabilia potentiâ ⊓.
Incommenfurabilia potentiâ ⊓.
Rationale, ῥητὸν, R, vel ᴦ.
Irrationale, ἄλογον, ᴦ.
Medium five mediale *m*⁓
Linea fecta fecundum extremam ⎰
  & mediam rationem   ⎱ ₛ
Major ejus portio σ
Minor ejus portio τ.
Z eft A + E.   Ʒ eft a + e.
X eft A-E.   Ȣ eft a-e

          A 2   Z eft

Figure 1 *A page from* Oughtred's Clavis mathematicae *(1631), showing a number of his mathematical symbols.*

Example 2.2

Simplify $x \times 3 \times y \times 4$.

Solution $x \times 3 \times y \times 4 = 3 \times 4 \times x \times y = 12xy$. Note that the order in which the terms are written does not matter and the multiplication symbols have been omitted. It is usual to write any numbers at the beginning of the expression.

2.3.4 Division

There are various ways of representing division. The expressions $x \div y$, $\dfrac{x}{y}$ and x/y all mean x divided by y. We sometimes refer to these as the **quotient** of x and y. Here the order is important, and x/y is not the same as y/x.

2.3.5 The 'equals' symbol

The equals symbol, $=$, is used in several different ways.

First, an equals symbol is used in **equations**. The left-hand side and the right-hand side of an equation are equal only when the variable takes specific values known as **solutions** of the equation.

Example 2.3

In the equation $x + 4 = 10$ the variable is x. The left-hand side is only equal to the right-hand side when x has the value 6. If x has any value other than 6, the right-hand side is not equal to the left-hand side.

Equations like this will be discussed in Chapter 9.

Secondly, an equals symbol is used in **formulae**. Physical quantities are often related through a formula. For example, the formula for the circumference of a circle expresses the relationship between the circumference of the circle and its radius. This formula states

circumference $= 2\pi \times$ radius

or, in symbols, $C = 2\pi r$, where the symbol for the circumference is C and the symbol for the radius is r. When used in this way, the equals symbol expresses the fact that the quantity on the left is found by evaluating the expression on the right.

Example 2.4 Distance travelled by a car

Suppose a car is travelling along a road at a constant speed. If we let the symbol v stand for its speed, and the car travels for a time t, then the distance it travels, s, is given by the formula $s = vt$. The distance travelled is the product of the speed and time of travel. Note also that in this example v is a constant, and s and t are variables.

Example 2.5 Speed of a car

If a car, travelling at a constant speed v, travels a distance s in a time t then its speed is given by the formula $v = s/t$. Note that the speed is given by the quotient of s and t.

2.3.6 Identities

A third way in which an equals symbol is used is in **identities**. At first sight an identity looks rather like an equation, except an identity is true for *all* values of the variable. When we write $\dfrac{x}{y} = x/y$ we mean that the quantity on the left-hand side means exactly the same as, or is identical to, that on the right whatever the values of x and y. So, $\dfrac{x}{y} = x/y$ is called an **identity**. To distinguish this usage from other uses of the equals symbol, some books use the symbol $\equiv$. Both the left-hand side and the right-hand side of an identity are equal whatever values are substituted for the variables. Later in the book we shall meet several important identities.

2.3.7 The 'not equals' symbol, $\neq$

The symbol $\neq$ means 'is not equal to'. For example, it is correct to write expressions such as $5 \neq 6$, $7 \neq -7$.

Self-assessment questions 2.3

1. Explain what is meant by the product and quotient of the symbols r and s.

2. Explain the distinction between a variable and a constant.

3. Explain the distinction between a subscript and a superscript.

4. Describe two different uses of the equals symbol.

5. Explain the use of δ notation to represent a change in the value of a variable.

2.4 Indices

We may write the number 8 as 2·2·2 and the number 16 as 2·2·2·2. The products 2·2·2 and 2·2·2·2 can be written compactly using **indices**, or **powers**, as they are also called. We write

$$2·2·2 = 2^3 \quad \text{and} \quad 2·2·2·2 = 2^4$$

We read 2^3 as 2 raised to the power 3 and 2^4 as 2 raised to the power 4. The singular of indices is **index**.

Example 2.6

Evaluate (a) 3^3 (b) 4^5 (c) $(-1)^2$

Solution   (a) $3^3 = 3·3·3 = 27$
(b) $4^5 = 4·4·4·4·4 = 1024$
(c) Recall that the product of two negative numbers is positive, and so $(-1)^2 = (-1) \times (-1) = 1$.

Example 2.7

Evaluate

(a) 2^5 (b) $2^3·3^2$ (c) $5^2·3^5$ (d) $(-2)^3$

Solution   (a) $2^5 = 2·2·2·2·2 = 32$
(b) $2^3·3^2 = 2·2·2·3·3 = (8)(9) = 72$
(c) $5^2·3^5 = 5·5·3·3·3·3·3 = (25)(243) = 6075$
(d) $(-2)^3 = (-2)(-2)(-2) = -8$

Most scientific calculators have an x^y button. This is used to calculate expressions such as 3^7, 4^8 and so on. Use your calculator to check the solutions to Examples 2.6 and 2.7 in order to ensure that you know how to use it. Your calculator is probably also preprogrammed to calculate 10^x for any value of x. Check that you can use this facility too.

Example 2.8

Write the following products in a compact way using indices:

(a) $3·3·3·3$ (b) $(-3)(-3)(-3)(-3)$ (c) $\dfrac{1}{2·2·2·2}$ (d) $\dfrac{6·6·6}{7·7·7·7}$

Solution   (a) $3 \cdot 3 \cdot 3 \cdot 3 = 3^4$

(b) $(-3)(-3)(-3)(-3) = (-3)^4$

(c) $\dfrac{1}{2 \cdot 2 \cdot 2 \cdot 2} = \dfrac{1}{2^4}$

(d) $\dfrac{6 \cdot 6 \cdot 6}{7 \cdot 7 \cdot 7 \cdot 7} = \dfrac{6^3}{7^4}$

Example 2.9

Write the following expressions using indices

(a) xxx (b) $(-x)(-x)(-x)$ (c) $aabbbccc$ (d) $\dfrac{xxx}{yy}$ (e) $\dfrac{(-a) \cdot (-a)}{b \cdot b \cdot (-b)}$

Solution   (a) $xxx = x^3$

(b) $(-x)(-x)(-x) = (-x)^3$. Because the product of three negative numbers is negative we could also write the result as $-x^3$.

(c) $aabbbccc = a^2 b^3 c^3$

(d) $\dfrac{xxx}{yy} = \dfrac{x^3}{y^2}$

(e) $\dfrac{(-a) \cdot (-a)}{b \cdot b \cdot (-b)} = \dfrac{a^2}{-b^3}$ or $-\dfrac{a^2}{b^3}$

Example 2.10 *Volume of concrete for a foundation*

Concrete is required to provide a solid foundation for a large machine in a factory. The foundation is in the shape of a cube with side length $l = 2$ m as shown in Figure 2.3. Calculate the volume of concrete required.

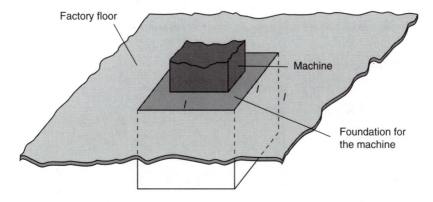

Factory floor

Machine

Foundation for the machine

Figure 2.3   The foundation for the machine is a cube of side length $l = 2$.

Solution   The volume of the cube, V, is calculated from the formula $V = l^3$, that is, by raising the side length of the cube, l, to the power 3. Therefore the volume required is $V = l^3 = 2 \times 2 \times 2 = 8$ m³.

Example 2.11 *Kinetic energy of a moving object*

 If an object of mass M is moving with a velocity v then it possesses energy due to its motion. This type of energy is known as **kinetic energy** E, and is given by the formula

$$E = \tfrac{1}{2}Mv^2$$

Note that v is raised to the power 2.

Example 2.12

Show that $(ab)^3$ is identical to a^3b^3.

Solution   We have

$$(ab)^3 = (ab)(ab)(ab)$$
$$= ababab$$
$$= aaabbb \qquad \text{(because the order does not matter)}$$
$$= a^3b^3$$

Example 2.13 *Power dissipated in a resistor*

 The power P dissipated in a resistor depends on the current I flowing through the resistor and the resistance R. The relationship is given by the formula

$$P = I^2R$$

Note that I is raised to the power 2.

Self-assessment questions 2.4

1. Explain what is meant by an 'index'. Why are indices used?

2. Indices can only be applied to positive numbers. True or false?

Exercises 2.4

1. Evaluate

    (a) 6^3 (b) $(-2)^4$ (c) 2^4 (d) $\dfrac{3}{2^3}$

    (e) $(-3)^3$ (f) $4^3 3^4$ (g) $\dfrac{6^3}{(2)(3^2)}$

    (h) 10^2 (i) 10^3 (j) 10^4

2. Write the following arithmetic expressions using index notation:

(a) $11 \cdot 11 \cdot 11$ (b) $5 \cdot 5 \cdot 6 \cdot 6 \cdot 7 \cdot 7$

(c) $(0.3)(0.3)(0.3)(0.3)$ (d) $\dfrac{5 \cdot 5 \cdot 5}{6 \cdot 6 \cdot 6}$

(e) $4 \cdot 5 \cdot 5 \cdot 6 \cdot 6 \cdot 6$ (f) $\dfrac{3 \cdot 3}{4 \cdot 4 \cdot 4 \cdot 4}$

(g) $\left(\dfrac{1}{3}\right)\left(\dfrac{1}{3}\right)\left(\dfrac{1}{3}\right)$ (h) $(-0.75)(-0.75)$

3. Write the following products in a compact way using indices:

(a) $zzzzyy$ (b) $aaabbxxxx$ (c) $\dfrac{xxx}{yyyy}$

4. Show that $(xy)^2$ is equivalent to $x^2 y^2$.

5. Show that $(-xy)^2$ is equivalent to $x^2 y^2$.

6. Show that $(4xy)^3$ is equivalent to $64x^3 y^3$.

7. Remove the brackets from each of the following:

(a) $(2a)^4$ (b) $(3x)^2$ (c) $\left(\dfrac{x}{y}\right)^2$

(d) $(-4k)^3$ (e) $(-4ab)^2$

2.5 Multiplying expressions involving indices

Consider the expression $6^2 \cdot 6^3$. Now
$$6^2 = 6 \cdot 6 \quad \text{and} \quad 6^3 = 6 \cdot 6 \cdot 6$$
and so
$$6^2 \cdot 6^3 = (6 \cdot 6) \cdot (6 \cdot 6 \cdot 6) = 6^5$$
Hence
$$6^2 \cdot 6^3 = 6^5$$
Alternatively, notice that the indices 2 and 3 could have been added to give the new index of 5, that is $6^2 \cdot 6^3 = 6^{2+3} = 6^5$. This illustrates the **first law of indices**, which is

$$a^m \cdot a^n = a^{m+n}$$

The quantity a is called the **base**. When expressions with the same base are multiplied, the indices are added.

Example 2.14

Simplify the following arithmetic expressions:

(a) $5^3 5^4$ (b) $3^6 3^2$ (c) $3^2 3^4 5^2 5^7$ (d) $4^3 5^2 4^3 5^6$ (e) $(-3)^4 (-3)^6$

Solution

To multiply two expressions having the same base and involving indices, the indices are added. So,

(a) $5^3 5^4 = 5^{3+4} = 5^7$

(b) $3^6 3^2 = 3^{6+2} = 3^8$

(c) $3^2 3^4 5^2 5^7 = 3^{2+4} 5^{2+7} = 3^6 5^9$
(d) $4^3 5^2 4^3 5^6 = 4^{3+3} 5^{2+6} = 4^6 5^8$
(e) $(-3)^4(-3)^6 = (-3)^{4+6} = (-3)^{10}$.

Example 2.15

Simplify the following arithmetic expressions:

(a) $\dfrac{4^4 4^2}{5^3 5^4}$ (b) $3 \cdot 3^3$ (c) $7^3 7^5 7^2$ (d) $3^2 3^4 3^5$ (e) $(-6)^3 5^2 (-6)^4 5^3$

Solution   (a) $\dfrac{4^4 4^2}{5^3 5^4} = \dfrac{4^{4+2}}{5^{3+4}} = \dfrac{4^6}{5^7}$

(b) $3 \cdot 3^3 = 3 \cdot (3 \cdot 3 \cdot 3) = 3^4$
(c) $7^3 7^5 7^2 = 7^{3+5+2} = 7^{10}$
(d) $3^2 3^4 3^5 = 3^{2+4+5} = 3^{11}$
(e) $(-6)^3 5^2 (-6)^4 5^3 = (-6)^{3+4} 5^{2+3} = (-6)^7 5^5$

Consider Example 2.15(b). Here we write

$$3 \cdot 3^3 = 3^4$$

This suggests that 3 is the same as 3^1. The first law of indices may now be applied:

$$3 \cdot 3^3 = 3^1 3^3 = 3^{1+3} = 3^4$$

This illustrates a general rule:

$$a = a^1$$

Raising a number to the power 1 leaves the number unchanged.

Example 2.16

Simplify

(a) $4 \cdot 4^3$ (b) $x^5 \cdot x$ (c) $y^3 \cdot y^2 \cdot y$ (d) $\dfrac{a^2 \cdot b^2 \cdot a^3}{c^2 \cdot d^3 \cdot c^3}$

Solution   (a) $4 \cdot 4^3 = 4^1 \cdot 4^3 = 4^{1+3} = 4^4$
(b) $x^5 x = x^5 x^1 = x^{5+1} = x^6$
(c) $y^3 y^2 y = y^3 y^2 y^1 = y^{3+2+1} = y^6$

(d) $\dfrac{a^2 \cdot b^2 \cdot a^3}{c^2 \cdot d^3 \cdot c^3} = \dfrac{a^{2+3} \cdot b^2}{c^{2+3} \cdot d^3} = \dfrac{a^5 b^2}{c^5 d^3}$

Example 2.17

Evaluate $(-x)^2$ and $-x^2$ when $x = 4$.

Solution   When $x = 4$,

$$(-x)^2 = (-4)^2 = (-4) \cdot (-4) = 16$$

$$-x^2 = -4^2 = -16$$

Note the crucial importance of the brackets in the previous example: $(-x)^2$ is not the same as $-x^2$.

Self-assessment questions 2.5

1. State the first law of indices.
2. Explain what is meant by a^1.

Exercises 2.5

1. Simplify the following arithmetic expressions:

   (a) $6^7 \cdot 6^2$
   (b) $3 \cdot 3^6$
   (c) $4^6 \cdot 5^2 \cdot 5^3 \cdot 4^2$
   (d) $10^2 \cdot 10^4 \cdot 100$
   (e) $5^3 \cdot 5^2$
   (f) $10^{21} \cdot 10^2$
   (g) $(-9)^{10}(-9)^6$
   (h) $6^4(-6)^3$

2. Simplify
   (a) $3^6 \cdot 2^7 \cdot 3^2 \cdot 2^5$
   (b) $4^3 \cdot 3^4 \cdot 3^2 \cdot 4$
   (c) $2 \cdot 5^2 \cdot 5 \cdot 2^3 \cdot 2$
   (d) $3 \cdot 3^6 \cdot 4^6 \cdot 4^2$

3. Simplify the following algebraic expressions:

   (a) $x^3 \cdot x^4$
   (b) $y^2 \cdot y^3 \cdot y^5$
   (c) $z^3 \cdot z^2 \cdot z$
   (d) $t^2 \cdot t^{10} \cdot t$
   (e) $a \cdot a \cdot a^2$
   (f) $t^3 t^4$
   (g) $b^6 b^3 b$
   (h) $z^7 z^7$

2.6 Dividing expressions involving indices

Consider the expression $\dfrac{4^5}{4^3}$:

$$\dfrac{4^5}{4^3} = \dfrac{4 \cdot 4 \cdot 4 \cdot 4 \cdot 4}{4 \cdot 4 \cdot 4}$$

$$= 4 \cdot 4 \qquad \text{(by cancelling 4s)}$$

$$= 4^2$$

Alternatively, note that the index 3 could have been subtracted from the index 5 to give the new index 2. That is,

$$\frac{4^5}{4^3} = 4^{5-3} = 4^2$$

This illustrates the **second law of indices**, which is

$$\frac{a^m}{a^n} = a^{m-n}$$

When expressions with the same base are divided, the indices are subtracted.

Example 2.18

Simplify the following expressions:

(a) $\dfrac{4^6}{4^4}$ (b) $\dfrac{7^9}{7^3}$ (c) $\dfrac{12^{14}}{12^{10}}$ (d) $\dfrac{(-11)^{14}}{(-11)^9}$

Solution   To divide two expressions having the same base and involving indices, the indices are subtracted. So,

(a)
$$\frac{4^6}{4^4} = 4^{6-4} = 4^2$$

(b)
$$\frac{7^9}{7^3} = 7^{9-3} = 7^6$$

(c)
$$\frac{12^{14}}{12^{10}} = 12^{14-10} = 12^4$$

(d)
$$\frac{(-11)^{14}}{(-11)^9} = (-11)^{14-9} = (-11)^5$$

Example 2.19

Simplify the following algebraic expressions:

(a) $\dfrac{x^9}{x^5}$ (b) $\dfrac{b^6}{b^5}$ (c) $\dfrac{z^6}{z^3}$ (d) $\dfrac{(-a)^6}{(-a)^4}$

Solution   (a) $\dfrac{x^9}{x^5} = x^{9-5} = x^4$

(b) $\dfrac{b^6}{b^5} = b^{6-5} = b^1 = b$

(c) $\dfrac{z^6}{z^3} = z^{6-3} = z^3$

(d) $\dfrac{(-a)^6}{(-a)^4} = (-a)^{6-4} = (-a)^2 = (-a) \times (-a) = a^2$

Consider now the expression $\dfrac{2^3}{2^3}$. We may apply the second law of indices to obtain

$$\dfrac{2^3}{2^3} = 2^{3-3} = 2^0$$

So far we have not encountered 0 as an index. We also note that any quantity divided by itself is 1, and so

$$\dfrac{2^3}{2^3} = 1$$

and so

$$2^0 = 1$$

This illustrates a general rule:

$$a^0 = 1$$

This states that any quantity to the power 0 is 1.

Example 2.20

Evaluate

(a) 12^0 (b) $(0.365)^0$ (c) x^0

Solution   (a) In this case the value of a is 12, and 12^0 is 1.
(b) $(0.365)^0 = 1$
(c) $x^0 = 1$

Self-assessment questions 2.6

1. State the second law of indices.
2. Explain what is meant by a^0.

Exercises 2.6

1. Evaluate

   (a) $\dfrac{10^6}{10^4}$ (b) $\dfrac{9^7}{9^5}$ (c) $\dfrac{(0.5)^3}{(0.5)^2}$

   (d) $\dfrac{6^3}{6^3}$ (e) $\dfrac{124^{16}}{124^{15}}$ (f) $\dfrac{(-19)^{23}}{(-19)^{21}}$

   (g) $\dfrac{(-3)^5}{(-3)^2}$

2. Simplify

   (a) $\dfrac{36^6}{36^2}$ (b) $\dfrac{17^{19}}{17^{14}}$ (c) $\dfrac{7^6 \cdot 8^4}{7^3 \cdot 8^3}$

   (d) $\dfrac{9^6 \cdot 5^{11} \cdot 7^9}{5^5 \cdot 7^6 \cdot 9^4}$ (e) $\dfrac{7^5 \cdot 12^3}{7^2 \cdot 12^2}$

   (f) $\dfrac{(-5)^4(-4)^5}{(-4)^4(-5)^2}$

3. Simplify

   (a) $\dfrac{x^6}{x^2}$ (b) $\dfrac{y^{14}}{y^{10}}$ (c) $\dfrac{t^{16}}{t^{12}}$

   (d) $\dfrac{z^{10}}{z^9}$ (e) $\dfrac{v^7}{v^0}$ (f) x^7/x^4

4. Simplify the following:

   (a) $\dfrac{10^7}{10^6}$ (b) $\dfrac{10^{19}}{10^{16}}$

   (c) $\dfrac{x^7}{x^4}$ (d) $\dfrac{x^7}{y^4}$

   (e) $\dfrac{(ab)^4}{a^2b^2}$ (f) $\dfrac{9^9 10^{10}}{10^9}$ (g) $\dfrac{x^9 y^8}{y^7 x^6}$

   (h) $\dfrac{(abc)^3}{(abc)^2}$

2.7 Negative indices

Let us now consider the expression $\dfrac{4^3}{4^5}$:

$$\frac{4^3}{4^5} = \frac{4 \cdot 4 \cdot 4}{4 \cdot 4 \cdot 4 \cdot 4 \cdot 4}$$

$$= \frac{1}{4 \cdot 4} \qquad \text{(after cancelling 4s)}$$

$$= \frac{1}{4^2}$$

Alternatively, we could apply the second law of indices to $\dfrac{4^3}{4^5}$ and subtract 5 from 3 to obtain

$$\frac{4^3}{4^5} = 4^{3-5} = 4^{-2}$$

We therefore have two expressions both equivalent to $\dfrac{4^3}{4^5}$, and hence

$$4^{-2} = \frac{1}{4^2}$$

This shows that the sign of the index is changed when the expression is inverted. We can now interpret negative indices. In general, we can say

$$a^{-m} = \frac{1}{a^m}, \quad a^m = \frac{1}{a^{-m}}$$

Example 2.21

Evaluate

(a) 3^{-2} (b) 7^{-1} (c) $(-3)^{-1}$ (d) $(-2)^{-2}$ (e) $\dfrac{1}{3^{-3}}$ (f) $\dfrac{2}{4^{-3}}$

(g) $\dfrac{3}{(-2)^{-2}}$

Solution

(a) $3^{-2} = \dfrac{1}{3^2} = \dfrac{1}{9}$

(b) $7^{-1} = \dfrac{1}{7^1} = \dfrac{1}{7}$

(c) $(-3)^{-1} = \dfrac{1}{(-3)^1} = \dfrac{1}{-3} = -\dfrac{1}{3}$

(d) $(-2)^{-2} = \dfrac{1}{(-2)^2} = \dfrac{1}{4}$

(e) $\dfrac{1}{3^{-3}} = 3^3 = 27$

(f) $\dfrac{2}{4^{-3}} = 2 \cdot 4^3 = 2(64) = 128$

(g) $\dfrac{3}{(-2)^{-2}} = 3 \times \dfrac{1}{(-2)^{-2}} = 3 \cdot (-2)^2 = 3(4) = 12$

Example 2.22

Write the following expressions using only positive indices:

(a) x^{-4} (b) $3x^{-4}$ (c) $(3x)^{-4}$ (d) $\dfrac{2}{y^{-3}}$

Solution (a) $x^{-4} = \dfrac{1}{x^4}$

(b) $3x^{-4} = \dfrac{3}{x^4}$

(c) $(3x)^{-4} = \dfrac{1}{(3x)^4}$

(d) $\dfrac{2}{y^{-3}} = 2y^3$

From (b) and (c), note that the inclusion of brackets can alter the value of an expression.

Example 2.23

Write the following expressions using only positive indices:

(a) $\dfrac{x^{-2}}{y^{-2}}$ (b) $3x^{-2}y^{-3}$

Solution (a) $\dfrac{x^{-2}}{y^{-2}} = \dfrac{1}{x^2 y^{-2}} = \dfrac{y^2}{x^2}$ (b) $3x^{-2}y^{-3} = \dfrac{3}{x^2 y^3}$

Example 2.24 Stress and strain for a wire under tension

Consider Figure 2.4, which shows a piece of wire. The length of the wire when it is unstressed is l and its cross-sectional area is A. A tension force F applied to the wire causes an extension x in the length of the wire. A quantity of interest to engineers is the **stress** in the wire. This is a measure of the amount of force per unit area being resisted by the wire. The stress σ in the wire is given by the formula

$$\sigma = \frac{F}{A}$$

The increase in length of the wire divided by the original length of the wire before a force was applied is known as the **strain** of the wire. Strain is a measure of deformation, and is the change in length per unit length of the wire. It is usually given the symbol ε, and is calculated from the formula

$$\varepsilon = \frac{x}{l}$$

A quantity often used in mechanical engineering is the **modulus of elasticity** E, which relates stress and strain. It is defined as

$$E = \frac{\sigma}{\varepsilon}$$

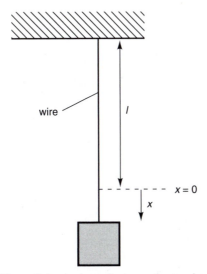

Figure 2.4 A piece of wire under tension.

Derive a formula for E in terms of F, A, x and L.

Solution Note that we can write the stress as

$$\sigma = \frac{F}{A} = \frac{F}{A^1} = FA^{-1}$$

and the strain as

$$\varepsilon = \frac{x}{l} = \frac{x}{l^1} = xl^{-1}$$

The modulus of elasticity may now be written as

$$E = \frac{\sigma}{\varepsilon} = \frac{FA^{-1}}{xl^{-1}} = \frac{Fl^1}{xA^1} = \frac{Fl}{xA}$$

An alternative way of writing E is

$$E = \frac{Fl}{x^1A^1} = Flx^{-1}A^{-1}$$

Self-assessment questions 2.7

1. The expression $-x^2$ is the same as x^{-2}. True or false?
2. Only positive numbers can be raised to a negative power. True or false?
3. Only negative numbers can be raised to a negative power. True or false?

Exercises 2.7

1. Evaluate

(a) 6^{-1} (b) 2^{-2} (c) 3^{-4}

(d) $\dfrac{4^2}{2^{-2}}$ (e) $\dfrac{2^{-2}}{4}$

(f) $(5^{-1})(5^{-2})$ (g) $(-3)^{-3}$

(h) $(-2)^3 \cdot 2^3$ (i) 10^{-1} (j) 10^{-2}

(k) 10^{-3}

2. Write the following expressions using positive indices:

(a) $x^{-2}x^{-1}$ (b) $\dfrac{3x}{x^{-4}}$ (c) $\dfrac{t^{-2}}{t^{-3}}$

(d) $(2a^2b^3)(6a^{-3}b^{-5})$ (e) $\dfrac{x^{-3}}{5^{-2}}$

(f) $\dfrac{(27)^{-1}x^{-1}}{y^{-2}}$ (g) $\dfrac{(-5)^{-3}}{-5}$

3. Evaluate

(a) $\dfrac{3}{4^{-2}}$ (b) $4 \cdot 3^{-2}$

(c) $3^{-1}9^2(27)^{-1}$ (d) $(0.25)^{-1}$

(e) $(0.2)^{-2}$ (f) $(0.1)^{-3}$

(g) $(-0.5)^{-2}$ (h) $(-2)^{-3}$

4. Simplify

(a) $t^{-6}t^3$ (b) $(-y^{-2})(-y^{-1})$

(c) $\dfrac{3y^{-2}}{6y^{-3}}$ (d) $(-2t^{-1})(-3t^{-2})(-4t^{-3})$

(e) $\dfrac{3t^{-2}}{6t^3}$ (f) $\dfrac{(2t^{-1})^3}{6t^2}$ (g) $\dfrac{(-2t)^3}{(-4t)^2}$

2.8 Multiple indices

Consider the expression $(4^3)^2$. This may be expressed as

$$(4^3)^2 = 4^3 \cdot 4^3 = 4^{3+3} = 4^6$$

Alternatively, notice that the indices 2 and 3 could have been multiplied to yield the new index of 6. That is, $(4^3)^2 = 4^{3 \times 2} = 4^6$. This illustrates the **third law of indices**:

$$(a^m)^n = a^{mn}$$

Note that m and n have been multiplied to obtain a new index, mn.

Example 2.25

Write the following expressions using a single index:

(a) $(3^2)^4$ (b) $((0.7)^6)^4$ (c) $(7^{-2})^3$ (d) $((-3)^2)^4$

Solution (a) $(3^2)^4 = 3^{2 \times 4} = 3^8$
(b) $((0.7)^6)^4 = (0.7)^{6 \times 4} = (0.7)^{24}$
(c) $(7^{-2})^3 = 7^{-2 \times 3} = 7^{-6}$
(d) $((-3)^2)^4 = (-3)^{2 \times 4} = (-3)^8$

Example 2.26

Write the following expressions using a single index:

(a) $(y^3)^2$ (b) $(y^2)^3$ (c) $(x^2)^{-3}$ (d) $((-x)^{-2})^{-3}$

Solution (a) $(y^3)^2 = y^{3 \times 2} = y^6$
(b) $(y^2)^3 = y^{2 \times 3} = y^6$
(c) $(x^2)^{-3} = x^{2 \times -3} = x^{-6}$
(d) $((-x)^{-2})^{-3} = (-x)^{-2 \times -3} = (-x)^6$. This may be further simplified by recognizing that $(-x)(-x) = x^2$; so,

$$(-x)^6 = (-x)(-x)(-x)(-x)(-x)(-x)$$

$$= x^2 x^2 x^2$$

$$= x^6$$

Example 2.27

Write the following expressions using a single index:

(a) $\left(\dfrac{t}{t^{-2}}\right)^4$ (b) $(z^{-3})^{-4}$

Solution (a) $\left(\dfrac{t}{t^{-2}}\right)^4 = (tt^2)^4 = (t^3)^4 = t^{3 \times 4} = t^{12}$
(b) $(z^{-3})^{-4} = z^{-3 \times -4} = z^{12}$

Let us now consider the expression $(2^4 5^2)^3$:

$$(2^4 5^2)^3 = (2^4 5^2)(2^4 5^2)(2^4 5^2)$$

$$= 2^4 2^4 2^4 5^2 5^2 5^2 \qquad \text{(because the order does not matter)}$$

$$= 2^{12} 5^6$$

Note that to simplify the expression $(2^4 5^2)^3$, the indices 4 and 3 can be multiplied to yield the new index on 2, and the indices 2 and 3 can be multiplied to yield the new index on 5. This example illustrates a generalization of the third law, which is:

$$(a^m b^n)^k = a^{mk} b^{nk}$$

Example 2.28

Remove the brackets from the following expressions:

(a) $(3^4 7^2)^3$ (b) $(3^2 x^3)^3$ (c) $(4^2 5^3 6^4)^3$

Solution   (a) $(3^4 7^2)^3 = 3^{4 \times 3} 7^{2 \times 3} = 3^{12} 7^6$
(b) $(3^2 x^3)^3 = 3^{2 \times 3} x^{3 \times 3} = 3^6 x^9$
(c) $(4^2 5^3 6^4)^3 = 4^{2 \times 3} 5^{3 \times 3} 6^{4 \times 3} = 4^6 5^9 6^{12}$

Example 2.29

Remove the brackets from the following expressions:

(a) $(ab^2 c^3)^2$ (b) $\left(\dfrac{b}{c^2}\right)^3$ (c) $\left(\dfrac{3a^2 b^3}{c^2}\right)^2$ (d) $(7^2 c^{-2} b^{-1})^{-2}$

Solution   (a) $(ab^2 c^3)^2 = (a^1 b^2 c^3)^2 = a^{1 \times 2} b^{2 \times 2} c^{3 \times 2} = a^2 b^4 c^6$

(b) $\left(\dfrac{b}{c^2}\right)^3 = (b^1 c^{-2})^3 = b^{1 \times 3} c^{-2 \times 3} = b^3 c^{-6} = \dfrac{b^3}{c^6}$

(c) $\left(\dfrac{3a^2 b^3}{c^2}\right)^2 = (3^1 a^2 b^3 c^{-2})^2 = 3^{1 \times 2} a^{2 \times 2} b^{3 \times 2} c^{-2 \times 2} = 3^2 a^4 b^6 c^{-4} = \dfrac{9a^4 b^6}{c^4}$

(d) $(7^2 c^{-2} b^{-1})^{-2} = 7^{2 \times -2} c^{-2 \times -2} b^{-1 \times -2} = 7^{-4} c^4 b^2 = \dfrac{c^4 b^2}{7^4}$

Self-assessment question 2.8

1.   State the third law of indices. How is the law generalized to deal with expressions of the form $(a^m b^n)^k$?

Exercises 2.8

1.   Write the following expressions using a single index:

(a) $(5^3)^5$ (b) $(3^3)^3$ (c) $(17^2)^4$

(d) $(y^3)^6$ (e) $\left(\dfrac{y^{-1}}{y^{-2}}\right)^3$

(f) $\left(\dfrac{t^{-2}}{t^4}\right)^3$ (g) $(k^{-2})^{-6}$

(h) $((-1)^4)^3$ (i) $((-1)^{-4})^{-3}$

2. Evaluate

(a) $(4^{-1})^2$ (b) $(2^2)^{-1}$ (c) $(3^2)^2$

(d) $(6^{-2})^{-1}$ (e) $\left(\dfrac{2}{5^2}\right)^{-1}$ (f) $(-2)^{-1}$

(g) $(-\tfrac{2}{3})^{-2}$

3. Remove the brackets from the following expressions:

(a) $(4^2 5^3)^3$ (b) $\left(\dfrac{3ab}{c^3}\right)^2$

(c) $\left(\dfrac{4^{-2}a^{-3}}{b^{-1}}\right)^2$ (d) $(2a^2 b)^3$

(e) $(3xy^2 z^3)^2$ (f) $\left(\dfrac{6}{ab^2}\right)^2$

(g) $\left(-\dfrac{3}{x^2}\right)^2$ (h) $\left(\dfrac{2z^2}{3t}\right)^3$

(i) $(-2x)^2$ (j) $(-2x^2)^{-2}$

(k) $\left(-\dfrac{2}{x^2}\right)^{-3}$

2.9 Fractional indices

Consider the relationship between the numbers 6 and 36. Clearly, $6^2 = 36$, so that 36 is the square of 6. Equivalently we can say that 6 is a **square root** of 36. We use the symbol $\sqrt{\ }$ to denote a square root, and write $6 = \sqrt{36}$. However, we can also write $(-6)(-6) = (-6)^2 = 36$, and so -6 is also a square root of 36. That is, $-6 = \sqrt{36}$. We can write both of these together using the 'plus or minus' symbol $\pm$. We write $\sqrt{36} = \pm 6$.

In general, a square root of a number is a number that when squared gives the original number. There are always two square roots of any nonzero number. For now, we shall simply deal with square roots of positive numbers. Square roots of negative numbers will be introduced in Chapter 28.

Similarly, the **cube root** of a number is the number that when cubed gives the original number. The symbol for a cube root is $\sqrt[3]{\ }$. Since $4^3 = 64$, we can write $4 = \sqrt[3]{64}$. All numbers, both positive and negative, possess just one cube root.

Other roots are defined in an obvious way. For example, because $5^4 = 625$, we can write $5 = \sqrt[4]{625}$.

Your scientific calculator will have the facility to calculate square roots, but only the positive value is normally given. It may be able to calculate other roots too. You should ensure that you can use such facilities correctly.

So far, we have used indices that are whole numbers. We now consider those that are fractions. Using the third law of indices, $(a^m)^n = a^{mn}$, we may write

$$(2^{\frac{1}{2}})^2 = 2^{\frac{1}{2} \times 2} = 2^1 = 2$$

So $2^{\frac{1}{2}}$ is a number that when squared gives 2. In other words, $2^{\frac{1}{2}}$ is a square root of 2:

$$2^{\frac{1}{2}} = \sqrt{2} = \pm 1.414\ldots$$

Similarly,

$$(2^{\frac{1}{3}})^3 = 2^1 = 2$$

So $2^{\frac{1}{3}}$ is a number that when cubed gives 2; that is, $2^{\frac{1}{3}}$ is the cube root of 2:

$$2^{\frac{1}{3}} = \sqrt[3]{2} = 1.2599\ldots$$

In general, $2^{\frac{1}{n}}$ is the *n*th root of 2. Even more generally, we have

$$\boxed{\qquad x^{\frac{1}{n}} \text{ is the } n\text{th root of } x \qquad}$$

Example 2.30

Evaluate, using a calculator

(a) $3^{\frac{1}{4}}$ (b) $7^{\frac{1}{3}}$ (c) $15^{\frac{1}{5}}$

Solution   (a) 1.316   (b) 1.913   (c) 1.719
Use your calculator to check these solutions.

The rules of adding, subtracting and multiplying indices that we have already met apply to fractional indices.

Example 2.31

Write the following expressions using a single index:

(a) $(3^{-2})^{\frac{1}{4}}$ (b) $3^{\frac{1}{2}}3^{\frac{1}{4}}$ (c) $x^{\frac{2}{3}}x^{\frac{5}{3}}$
(d) $yy^{\frac{2}{5}}$ (e) $a^{\frac{1}{2}}a^{\frac{1}{3}}a^2$ (f) $\sqrt{k^3}$

Solution   (a) $(3^{-2})^{\frac{1}{4}} = 3^{-2 \times \frac{1}{4}} = 3^{-\frac{1}{2}}$
(b) $3^{\frac{1}{2}}3^{\frac{1}{4}} = 3^{\frac{1}{2}+\frac{1}{4}} = 3^{\frac{3}{4}}$
(c) $x^{\frac{2}{3}}x^{\frac{5}{3}} = x^{\frac{2}{3}+\frac{5}{3}} = x^{\frac{7}{3}}$
(d) $yy^{\frac{2}{5}} = y^1 y^{\frac{2}{5}} = y^{1+\frac{2}{5}} = y^{\frac{7}{5}}$
(e) $a^{\frac{1}{2}}a^{\frac{1}{3}}a^2 = a^{\frac{1}{2}+\frac{1}{3}+2} = a^{\frac{17}{6}}$
(f) $\sqrt{k^3} = (k^3)^{\frac{1}{2}} = k^{3 \times \frac{1}{2}} = k^{\frac{3}{2}}$

Example 2.32

Simplify the following expressions:

(a) $(4\sqrt{x})^3$ (b) $\dfrac{kk^{\frac{1}{4}}}{\sqrt{k}}$ (c) $(4^{0.3}x^{0.6}\sqrt{y})^3$

Solution (a) $(4\sqrt{x})^3 = (4^1x^{0.5})^3 = 4^3x^{1.5} = 64x^{1.5}$

(b) $\dfrac{kk^{\frac{1}{4}}}{\sqrt{k}} = \dfrac{k^1k^{\frac{1}{4}}}{k^{\frac{1}{2}}} = \dfrac{k^{\frac{5}{4}}}{k^{\frac{1}{2}}} = k^{\frac{3}{4}}$

(c) $(4^{0.3}x^{0.6}\sqrt{y})^3 = (4^{0.3}x^{0.6}y^{0.5})^3 = 4^{0.3\times3}x^{0.6\times3}y^{0.5\times3} = 4^{0.9}x^{1.8}y^{1.5}$

Example 2.33 Period of a pendulum

The time taken to complete one full swing of a pendulum is known as the **period** of the pendulum. A pendulum of length l has a period T given by the formula

$$T = 2\pi\sqrt{\frac{l}{g}}$$

where g is a constant known as the acceleration due to gravity. This may be written thus:

$$T = 2\pi\sqrt{\frac{l}{g}}$$
$$= 2\pi\sqrt{lg^{-1}}$$
$$= 2\pi(lg^{-1})^{\frac{1}{2}}$$
$$= 2\pi l^{\frac{1}{2}}g^{-\frac{1}{2}}$$

Example 2.34

Evaluate

(a) $30^{\frac{2}{3}}$ (b) $5^{\frac{7}{4}}$

Solution (a) Using a calculator, we see that $30^{\frac{2}{3}} = 9.655$. Alternatively,

$$30^{\frac{2}{3}} = (30^2)^{\frac{1}{3}} = 900^{\frac{1}{3}} = 9.655$$

Hence $30^{\frac{2}{3}}$ is the cube root of 30^2.

(b) Using a calculator, we find $5^{\frac{7}{4}} = 16.719$. Alternatively,

$$5^{\frac{7}{4}} = (5^7)^{\frac{1}{4}} = (78125)^{\frac{1}{4}} = 16.719$$

So $5^{\frac{7}{4}}$ is the fourth root of 5^7.

Example 2.35 *Flow of liquid out of a large storage tank*

Consider Figure 2.5, which shows a storage tank with an opening near to the bottom of the tank. The height of liquid in the tank is h and the velocity with which the liquid flows out of the tank is v. If the cross-sectional area of the tank is large compared with the area of the opening then it can be shown that the formula for the velocity v is

$$v = \sqrt{2gh}$$

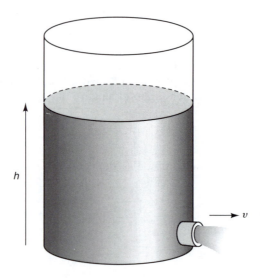

Figure 2.5 Liquid flow out of a tank.

This is known as **Torricelli's theorem**. We can rearrange this formula in many ways, for example as

$$v = (2gh)^{\frac{1}{2}} = (2g)^{\frac{1}{2}}h^{\frac{1}{2}}$$

Self-assessment questions 2.9

1. What meaning is given to the expression $7^{0.3}$?
2. All fractional indices must be positive. True or false?
3. Is 2^{-3} different from $(-2)^3$? If so, how are they different?

Exercises 2.9

1. Evaluate (a) $10^{\frac{3}{4}}$ (b) $20^{\frac{7}{8}}$
(c) $57^{0.65}$ (d) $90^{1.3}$

(e) $9^{-\frac{1}{2}}$ (f) $130^{-\frac{2}{3}}$
(g) $(1.15)^{-2.5}$

2. Remove the brackets from the following expressions:

(a) $(6^{\frac{1}{2}})^3$ (b) $(5^{\frac{1}{3}})^6$ (c) $(10^{0.6})^4$
(d) $(x^2)^{\frac{1}{3}}$ (e) $(2x^2)^{\frac{1}{3}}$ (f) $(a \cdot a^2)^{\frac{1}{2}}$
(g) $(ab^2)^{\frac{1}{2}}$

3. Remove the brackets from the following expressions:

(a) $(4^3)^{-\frac{1}{2}}$ (b) $(3^{-\frac{1}{2}})^{-\frac{1}{2}}$ (c) $(7^{\frac{2}{3}})^4$
(d) $(19^{\frac{3}{2}})^{\frac{1}{3}}$ (e) $(a^2 b^{-3})^{-\frac{3}{2}}$
(f) $\left(\dfrac{k^{-1.5}}{\sqrt{k}}\right)^{-2}$

4. Remove the brackets from the following expressions:

(a) $(5b)^{\frac{1}{6}}$ (b) $(3\sqrt{x})^3$ (c) $3(\sqrt{x})^3$
(d) $(\sqrt{3x})^3$

5. Simplify

(a) $x^{\frac{1}{2}} x^{\frac{1}{3}}$ (b) $\dfrac{x^{\frac{1}{2}}}{x^{\frac{1}{3}}}$ (c) $(x^{\frac{1}{2}})^{\frac{1}{3}}$

(d) $(8x^3)^{\frac{1}{3}}$ (e) $\sqrt{25y^2}$ (f) $\left(\dfrac{27}{t^3}\right)^{\frac{1}{3}}$

(g) $(16y^4)^{\frac{1}{4}}$ (h) $(x^{\frac{1}{4}} \cdot x^{\frac{1}{2}})^4$

(i) $\sqrt{a^2 a^6}$ (j) $\sqrt{\dfrac{a^{-4}}{a^{-1}}}$

Review exercises 2

1 A quantity p changes from 3 to 3.2. What is the value of δp?

2 A quantity x changes from 4 to 3.9. What is the value of δx?

3 A quantity x changes from x_1 to x_2. What is the value of δx?

4 Simplify

(a) $10^{11} \cdot 10^6$ (b) $10^9 \cdot 100$

(c) $6^3 \cdot 6^4 \cdot 6^5$ (d) $\dfrac{7^4}{7}$

(e) $\dfrac{49^2}{49^3}$ (f) $\dfrac{5^6}{\sqrt{5}}$

(g) $6 \cdot 6^2 \cdot \sqrt{6}$ (h) $\dfrac{3^4 \cdot 3^{\frac{2}{3}}}{3^{\frac{4}{3}}}$

5 Evaluate

(a) 16^3 (b) $16^{\frac{1}{2}}$
(c) $16^{-\frac{1}{2}}$ (d) 25^{-1}
(e) $25^{-\frac{1}{2}}$ (f) $36^{-\frac{3}{2}}$
(g) $125^{\frac{2}{3}}$ (h) $125^{-\frac{1}{3}}$
(i) $100^{\frac{3}{2}}$ (j) $(-25)^2$
(k) $(-5)^{-2}$

6 Simplify

(a) $x^3 \cdot x^7$ (b) $y^2 \cdot y^{12} \cdot y$
(c) $\dfrac{3t^6}{t^8}$ (d) $\dfrac{5y^3}{25y}$
(e) $\dfrac{(8x)^2}{16x}$ (f) $\dfrac{2xy^2}{(2xy)^2}$
(g) $\dfrac{9ab^3}{(3ab)^3}$ (h) $\sqrt{a^2 b^6 c^4}$
(i) $(k^{\frac{2}{3}})^3$ (j) $(y^{\frac{3}{4}})^8$
(k) $(64t^3)^{\frac{1}{3}}$ (l) $(64t^3)^{\frac{2}{3}}$

7 Simplify the following:

(a) $c^3 c^{-9}$ (b) $x^{16} x^{-3}$
(c) $z^{-17} z^{11}$ (d) $10^{11} 10^6$
(e) $\dfrac{10^5}{10^8}$ (f) $(y^{\frac{3}{2}})^4$
(g) $(x^{-\frac{1}{2}})^2$ (h) $\sqrt{64a^2 b^4 c^6}$

8 Explain the distinction between an identity ($\equiv$) and an equation ($=$).

9 Evaluate

(a) $(27)^{\frac{2}{3}}$ (b) $(27)^{-\frac{2}{3}}$

(c) $16^{-\frac{1}{2}}$ (d) $\dfrac{3}{16^{-\frac{1}{2}}}$

(e) $16^{\frac{3}{4}}$ (f) $\dfrac{5}{16^{\frac{3}{4}}}$

(g) $32^{\frac{2}{5}}$ (h) $32^{\frac{3}{5}}$

(i) $\sqrt{8}\sqrt{2}$ (j) $\sqrt{32}\sqrt{2}$

(k) $81^{\frac{3}{4}}$ (l) $(0.01)^{-1}$

(m) $\sqrt{0.01}$ (n) $\dfrac{10x}{x/10}$

10 Remove the brackets from

(a) $(3x)^2$ (b) $(-3x)^3$
(c) $(-3x^2)^3$ (d) $(3x)^{-2}$
(e) $(-3x)^{-3}$ (f) $(-3x^2)^{-2}$

3

Numbers and Sets

KEY POINTS

This chapter

- introduces the notation and symbols used in set theory

- shows how sets are represented using Venn diagrams

- explains the terms 'union' and 'intersection'

- explains the concept of a real number line

- defines the terms 'rational' and 'irrational'

- explains how intervals can be expressed as sets

- introduces the notation for 'greater than' and 'less than'

- explains the terms 'closed interval' and 'open interval'

- illustrates how to express a number in scientific notation

- shows how to perform calculations using scientific notation

- shows how to write a number to a given number of significant figures

- shows how to write a number to a given number of decimal places

CONTENTS

3.1 Introduction

If we can identify something that is common to several objects, it is often useful to group them together. Such a grouping is called a **set**. We can then study the whole set of objects that are linked by having a common property. Engineers may be grouped into civil engineers, chemical engineers, production engineers, mechanical engineers, electrical engineers, electronic engineers, control engineers, software engineers and others. We could then define useful sets. For example, we could consider the set C defined by

$$C = \{\text{all civil engineers aged between 30 and 50}\}$$

and

$$E = \{\text{all electronic engineers with more than 10 years' experience}\}$$

Engineers may wish to study all components of a production run that fail to meet some specified tolerance. Mathematicians may look at sets of numbers with particular properties, for example the set of all even numbers or the set of all numbers greater than zero. As we see in this chapter, numbers can be classified into various sets.

3.2 Sets

A **set** is a collection of objects, things or states. The objects may be almost anything, for example silicon chips produced by a particular machine, odd numbers, students in a class and so on. We use a capital letter to signify a set, and then either list the objects of the set or describe them. The list or description is usually enclosed in curly brackets, { }. The objects belonging to a set are called the **elements** of the set. Here are a few examples:

$A = \{\text{all odd numbers}\}$

$B = \{0, 1\}$

$C = \{\text{the resistors produced in a factory on a particular day}\}$

$D = \{\text{on, off}\}$

$E = \{0, 1, 2, 3, 4, 5, 6, 7, 8, 9\}$

A is the set of all odd numbers. Clearly, these cannot be listed, and so the elements are described. B is the set of binary digits. B has only two elements. The elements of C are the resistors produced in a factory on a particular day. These could be listed individually, but since the number is large it is not practical to do this. D lists the two possible states of a simple switch, and the elements of E are the digits used in the decimal system.

A set with a finite number of elements is called a **finite set**. B, C, D and

E are finite sets. A has an infinite number of elements, and so is not a finite set. It is known as an **infinite set**.

Two sets are **equal** if they contain exactly the same elements. For example, the sets $\{9, 10, 14\}$ and $\{10, 14, 9\}$ are equal, since they contain exactly the same elements. The order in which elements are written is unimportant. Note also that repeated elements are ignored. The set $\{2, 3, 3, 3, 5, 5\}$ is equal to the set $\{2, 3, 5\}$.

Sometimes one set is contained completely within another set. For example, if

$$X = \{2, 3, 4, 5, 6\} \quad \text{and} \quad Y = \{2, 3, 6\}$$

then all the elements of Y are also elements of X. We say that Y is a **subset** of X and write $Y \subseteq X$.

Example 3.1

Given $A = \{0, 1, 2, 3\}$, $B = \{0, 1, 2, 3, 4, 5, 6\}$ and $C = \{0, 1\}$, state which sets are subsets.

Solution

A is a subset of B; that is, $A \subseteq B$.
C is a subset of B; that is, $C \subseteq B$.
C is a subset of A; that is, $C \subseteq A$.

Example 3.2 Cars produced by a factory

A factory produces cars over a five-day period: Monday to Friday. Consider the following sets:

$A = \{\text{cars produced from Monday to Friday}\}$

$B = \{\text{cars produced from Monday to Thursday}\}$

$C = \{\text{cars produced on Friday}\}$

$D = \{\text{cars produced on Wednesday}\}$

$E = \{\text{cars produced on Wednesday and Thursday}\}$

State which sets are subsets.

Solution

B is a subset of A; that is, $B \subseteq A$.
C is a subset of A; that is, $C \subseteq A$.
D is a subset of A; that is, $D \subseteq A$.
E is a subset of A; that is, $E \subseteq A$.
D is a subset of B; that is, $D \subseteq B$.
E is a subset of B; that is, $E \subseteq B$.
D is a subset of E; that is, $D \subseteq E$.

To show that an element belongs to a particular set we use the symbol $\in$. This symbol means 'is a member of' or 'belongs to'. The symbol $\notin$ means 'is not a member of' or 'does not belong to'. For example, if $X = \{$all even numbers$\}$ then we may write $4 \in X$, $6 \in X$, $7 \notin X$ and $11 \notin X$.

Sometimes a set will contain no elements. For example, suppose we define the set K by

$$K = \{\text{all odd numbers that are divisible by 4}\}$$

Since there are no odd numbers that are divisible by 4, K has no elements. The set with no elements is called the **empty set**, and it is denoted by $\varnothing$.

It is appropriate at this stage to introduce the **universal set**. The set containing all the objects of interest is called the universal set, denoted by $\mathscr{E}$. The universal set does depend upon the context. If we are concerned only with whole numbers then $\mathscr{E}$ will be the set of whole numbers. If we are concerned only with the decimal digits then we can take $\mathscr{E} = \{0, 1, 2, 3, 4, 5, 6, 7, 8, 9\}$.

Given a set A and a universal set $\mathscr{E}$, we can define a new set, called the **complement** of A, denoted $\bar{A}$. The complement of A contains all the elements of the universal set that are not in A.

Example 3.3

Given $A = \{2, 3, 7\}$, $B = \{0, 1, 2, 3, 4\}$ and $\mathscr{E} = \{0, 1, 2, 3, 4, 5, 6, 7, 8, 9\}$ state
(a) $\bar{A}$ (b) $\bar{B}$

Solution   (a)   The elements of $\bar{A}$ are those that belong to $\mathscr{E}$ but not to A:

$$\bar{A} = \{0, 1, 4, 5, 6, 8, 9\}$$

(b)

$$\bar{B} = \{5, 6, 7, 8, 9\}$$

Sometimes a set is described in a mathematical way. Suppose that the set S contains all numbers that are divisible by 4 and 7. We can write

$$S = \{x : x \text{ is divisible by 4 and } x \text{ is divisible by 7}\}$$

The symbol ':' stands for 'such that'. We read the above as 'S is the set comprising all elements x such that x is divisible by 4 and x is divisible by 7'. As another example, consider

$$T = \{x : x \text{ is even and } x \text{ is divisible by 3}\}$$

$$V = \left\{ x : x = \frac{a}{b}, \text{ where } a \text{ is even and } b \text{ is odd} \right\}$$

Then the elements of *T* are all even numbers that are divisible by 3. The elements of *V* are all fractions, proper and improper, where the numerator is even and the denominator is odd.

3.2.1 Venn diagrams

Sets are often represented pictorially by use of **Venn diagrams**.

Example 3.4
Represent the sets $A = \{0, 1\}$ and $B = \{0, 1, 2, 3, 4\}$ using a Venn diagram.

Solution
The elements 0 and 1 are in set *A*, represented by the small circle in Figure 3.1. The large circle represents set *B*, and so contains the elements 0, 1, 2, 3 and 4. A suitable universal set in this case is the set of all integers. The universal set is shown by the rectangle.

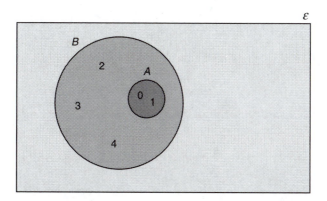

Figure 3.1   The set *A* is contained completely within B.

Note that $A \subseteq B$. This is shown in the Venn diagram by *A* being completely inside *B*.

Example 3.5
Given $A = \{0, 1\}$ and $B = \{2, 3, 4\}$, draw Venn diagrams showing
(a) *A* and *B*   (b) $\bar{A}$ (c) $\bar{B}$

Solution   (a)   Note that *A* and *B* have no elements in common. This is represented pictorially in the Venn diagram by circles that are totally separate from each other as shown in Figure 3.2.

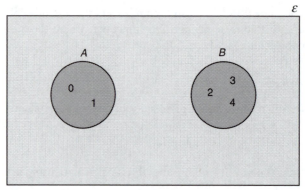

Figure 3.2   The sets *A* and *B* have no elements in common.

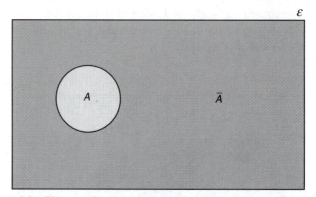

Figure 3.3   The complement of *A* contains those elements that are not in *A*.

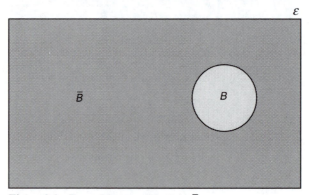

Figure 3.4   Everything outside *B* is *B̄*.

(b)   The complement of *A* is the set whose elements do not belong to *A*. The set *Ā* is shown in Figure 3.3.

(c)   The set *B̄* is shown in Figure 3.4.

Example 3.6 Bolt lengths

A company supplies bolts of a particular diameter in the following lengths: 6, 12, 20, 25, 30 and 40 mm. Given the sets $A=\{6, 20, 30\}$ and $B=\{30, 40\}$, draw Venn diagrams showing
(a) A and B (b) $\bar{A}$ (c) $\bar{B}$

Solution   (a)   The Venn diagram is shown in Figure 3.5. Note that A and B have a common element of 30. This is shown by an overlap of the Venn diagram circles.

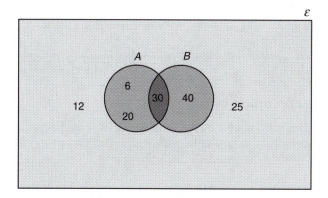

Figure 3.5   Venn diagram showing set A and set B.

(b)   The universal set is $\mathscr{E}=\{6, 12, 20, 25, 30, 40\}$, and so $\bar{A}=\{12, 25, 40\}$. The Venn diagram is shown in Figure 3.6.
(c) $\bar{B}=\{6, 12, 20, 25\}$. The Venn diagram is shown in Figure 3.7.

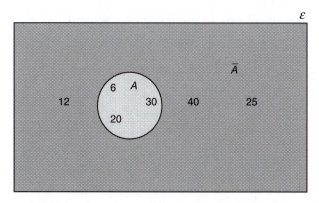

Figure 3.6   The complement of A is everything outside A.

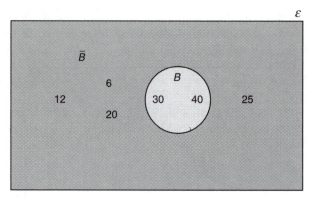

Figure 3.7 The set $\bar{B}$ comprises all elements outside B.

3.2.2 Intersection and union of sets

Given two sets, A and B, the **intersection** of A and B is the set that contains elements that are common to both A and B. We write $A \cap B$ for the intersection of A and B. Mathematically, we write this as

$$A \cap B = \{x : x \in A \text{ and } x \in B\}$$

This says that the intersection contains all the elements x such that x belongs to A and also belongs to B. Note that $A \cap B$ and $B \cap A$ are identical. The intersection of two sets can be represented by a Venn diagram as shown in Figure 3.8. The area common to both represents the intersection of the sets.

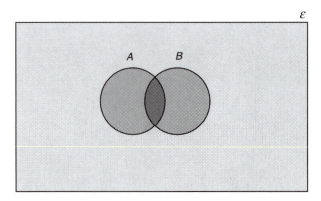

Figure 3.8 The overlapping area represents $A \cap B$.

Example 3.7

Given $A = \{3, 4, 5, 6\}$, $B = \{3, 5, 9, 10, 15\}$ and $C = \{4, 6, 10\}$, state
(a) $A \cap B$ (b) $B \cap C$

Solution   (a)   The elements common to both A and B are 3 and 5. Hence

$$A \cap B = \{3, 5\}$$

(b)   The only element common to B and C is 10. Hence

$$B \cap C = \{10\}$$

Example 3.8

Given $D = \{a, b, c\}$ and $F = \{$the entire alphabet$\}$, state $D \cap F$.

Solution   The elements common to D and F are a, b and c, and so

$$D \cap F = \{a, b, c\}$$

Note that D is a subset of F, and so $D \cap F = D$.

We can have the intersection of three or more sets. Consider the following example.

Example 3.9

Given $A = \{0, 1, 2, 3\}$, $B = \{1, 2, 3, 4, 5\}$ and $C = \{2, 3, 4, 7, 9\}$, state
(a) $A \cap B$ (b) $(A \cap B) \cap C$ (c) $B \cap C$ (d) $A \cap (B \cap C)$

Solution   (a)   The elements common to A and B are 1, 2 and 3, so

$$A \cap B = \{1, 2, 3\}$$

(b)   We need to consider the sets $(A \cap B)$ and C. $A \cap B$ is given in (a). The elements common to $(A \cap B)$ and C are 2 and 3. Hence

$$(A \cap B) \cap C = \{2, 3\}$$

(c)   The elements common to B and C are 2, 3 and 4, so

$$B \cap C = \{2, 3, 4\}$$

(d)   We look at the sets A and $(B \cap C)$. The common elements are 2 and 3. Hence

$$A \cap (B \cap C) = \{2, 3\}$$

Note from (b) and (d) that $(A \cap B) \cap C = A \cap (B \cap C)$.

The above example illustrates a general rule. For any sets A, B and C it is true that

$$(A\cap B)\cap C = A\cap(B\cap C)$$

As the position of the brackets is unimportant, they are usually omitted, and we write $A\cap B\cap C$.

Suppose that sets A and B have no elements in common. Then their intersection contains no elements. We express this as

$$A\cap B = \varnothing$$

Recall that $\varnothing$ is the empty set. In such a case we say that A and B are **disjoint** sets.

Example 3.10 Drill diameters

A manufacturer makes a range of drills with the following diameters: 1.0, 1.5, 2.0, 2.5, 3.0, 3.5, 4.0, 5.0, 6.0 mm. Consider the following sets:

$$A = \{1.0, 2.0, 3.0, 4.0, 5.0, 6.0\}$$

$$B = \{1.5, 2.5, 3.5\}$$

$$C = \{4.0, 5.0, 6.0\}$$

$$D = \{2.0, 4.0, 6.0\}$$

Calculate
(a) $A\cap B$ (b) $B\cap C$ (c) $A\cap C$ (d) $A\cap C\cap D$

Solution   (a)   There are no elements common to A and B, and so $A\cap B = \varnothing$.
(b)   There are no elements common to B and C, and so $B\cap C = \varnothing$.
(c) $A\cap C = \{4.0, 5.0, 6.0\}$
(d) $A\cap C\cap D = \{4.0, 6.0\}$

The **union** of two sets A and B is the set that contains all the elements of A together with all the elements of B. We write $A\cup B$ for the union of A and B. We can describe the set $A\cup B$ formally by

$$A\cup B = \{x : x\in A \text{ or } x\in B \text{ or both}\}$$

Thus the elements of $A\cup B$ are those quantities x such that x is a member of A or x is a member of B or x is a member of both A and B. The deeply shaded areas of Figure 3.9 represent $A\cup B$. In Figure 3.9(a) the sets intersect, whereas in Figure 3.9(b) the sets are disjoint.

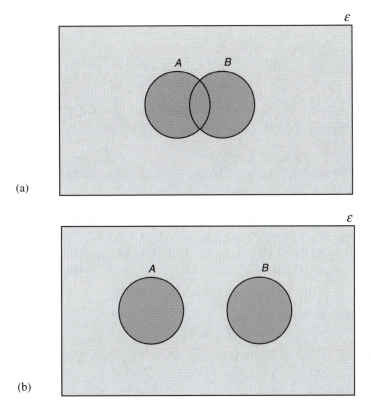

(a)

(b)

Figure 3.9 The deeply shaded areas represent the union of *A* and *B*.

Example 3.11

Given *A* = {0, 1}, *B* = {1, 2, 3} and *C* = {2, 3, 4, 5}, write down
(a) *A*∪*B*   (b) *A*∪*C*   (c) *B*∪*C*

Solution   (a)   *A*∪*B* = {0, 1, 2, 3}
            (b)   *A*∪*C* = {0, 1, 2, 3, 4, 5}
            (c)   *B*∪*C* = {1, 2, 3, 4, 5}

Recall that there is no need to repeat elements in a set. Clearly, the order of the union is unimportant, so, for example, *A*∪*B* = *B*∪*A*.

Example 3.12

Given *A* = {on, off} and *B* = {on, fast, stop}, state *A*∪*B*.

Solution

$$A \cup B = \{\text{on, off, fast, stop}\}$$

Example 3.13

Given $A = \{2, 3, 4, 5, 6\}$, $B = \{2, 4, 6, 8, 10\}$ and $C = \{3, 5, 7, 9, 11\}$, state
(a) $A \cup B$ (b) $(A \cup B) \cap C$ (c) $A \cap B$ (d) $(A \cap B) \cup C$ (e) $A \cup B \cup C$

Solution   (a) $A \cup B = \{2, 3, 4, 5, 6, 8, 10\}$
         (b)   We need to look at the sets $(A \cup B)$ and C. The elements common to both of these sets are 3 and 5. Hence $(A \cup B) \cap C = \{3, 5\}$.
         (c) $A \cap B = \{2, 4, 6\}$
         (d)   We consider the sets $(A \cap B)$ and C. We form the union of these two sets to obtain $(A \cap B) \cup C = \{2, 3, 4, 5, 6, 7, 9, 11\}$.
         (e)   The set formed by the union of all three sets will contain all the elements from all the sets:

$$A \cup B \cup C = \{2, 3, 4, 5, 6, 7, 8, 9, 10, 11\}$$

Example 3.14

Two sets A and B are not disjoint. Show by means of a Venn diagram the following sets:
(a) $A \cup B$ (b) $A \cap B$ (c) $\overline{A \cup B}$ (d) $\overline{A \cap B}$

Solution   (a)   The sets A and B overlap in the Venn diagram, showing that the sets are not disjoint. The deeply shaded area of Figure 3.10 represents $A \cup B$.

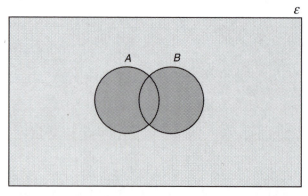

Figure 3.10   The set $A \cup B$.

         (b)   The deeply shaded area of Figure 3.11 represents $A \cap B$. It is the area common to both A and B.
         (c)   Recall that the complement of a set S is denoted by $\bar{S}$. In (a) we showed

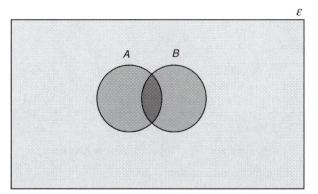

Figure 3.11   The deeply shaded area represents the set $A \cap B$.

$A \cup B$. We now require the complement of $A \cup B$. The complement, $\overline{A \cup B}$, is represented by the area not in $A \cup B$. This is the deeply shaded area of Figure 3.12.

(d)   In (b) we represented $A \cap B$. The complement, $\overline{A \cap B}$, is represented by all that area outside $A \cap B$. This is the deeply shaded area of Figure 3.13.

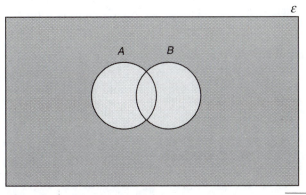

Figure 3.12   The deeply shaded area represents the set $\overline{A \cup B}$.

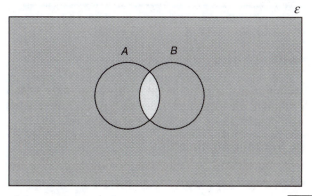

Figure 3.13   The deeply shaded area represents the set $\overline{A \cap B}$.

Self-assessment questions 3.2

1.   Given a set A, its complement $\bar{A}$ and a universal set $\mathscr{E}$, state which of the following are true and which are false:
     (a) $A\cup\bar{A}=\mathscr{E}$ (b) $A\cap\mathscr{E}=\varnothing$ (c) $A\cap\bar{A}=\varnothing$ (d) $A\cap\bar{A}=\mathscr{E}$ (e) $A\cup\varnothing=\mathscr{E}$
     (f) $A\cup\varnothing=A$ (g) $A\cup\varnothing=\varnothing$ (h) $A\cap\varnothing=A$ (i) $A\cap\varnothing=\varnothing$ (j) $A\cup\mathscr{E}=A$
     (k) $A\cup\mathscr{E}=\varnothing$ (l) $A\cup\mathscr{E}=\mathscr{E}$

2.   What is meant by (a) the intersection   (b) the union of two sets?

3.   Explain what is meant by the complement of a set.

4.   What is meant by saying A and B are disjoint sets?

Exercises 3.2

1.   Given $A=\{a, b, c, d, e, f\}$, $B=\{a, c, d, f, h\}$ and $C=\{e, f, x, y\}$, state
     (a) $A\cup B$ (b) $B\cap C$ (c) $A\cap(B\cup C)$
     (d) $C\cap(B\cup A)$ (e) $A\cap B\cap C$
     (f) $B\cup(A\cap C)$

     (e) $\overline{A\cup B}$ (f) $\overline{A\cup B}$ (g) $\overline{A\cap B}$
     (h) $\bar{A}\cap B$ (i) $\bar{B}\cup A$
     What do you notice about your answers to (c), (g) and (d), (f)?

2.   List the elements of the following sets:
     (a) $A=\{x:x$ is odd and x is greater than 0 and less than 12$\}$
     (b) $B=\{x:x$ is even and x is greater than 19 and less than 31$\}$

3.   Given $A=\{5, 6, 7, 9\}$, $B=\{0, 2, 4, 6, 8\}$ and $\mathscr{E}=\{0, 1, 2, 3, 4, 5, 6, 7, 8, 9\}$, list the elements of each of the following sets:
     (a) $\bar{A}$ (b) $\bar{B}$ (c) $\bar{A}\cup\bar{B}$ (d) $\bar{A}\cap\bar{B}$

4.   Given A and B are intersecting sets, that is, they are not disjoint, show on a Venn diagram the following sets:
     (a) $\bar{A}$ (b) $\bar{B}$ (c) $A\cup\bar{B}$ (d) $\bar{A}\cup\bar{B}$
     (e) $\bar{A}\cap\bar{B}$

5.   Given $A=\{0, 1, 2, 3, 4\}$, $B=\{3, 4, 5, 6, 7, 8\}$ and $C=\{0, 3, 6, 9, 12, 15, 18, 21, 24\}$, state
     (a) $A\cap B$ (b) $B\cup A$ (c) $C\cap A\cap B$
     (d) $C\cup(A\cap B)$ (e) $B\cap(A\cup C)$

3.3 The real number line

This section explains the concept of a real number line. The terms 'rational' and 'irrational' are explained. Intervals are expressed as sets on the real line.

3.3.1 Ordering on the real number line

Any real number can be represented by a point on a line. This line is called the **real number line**. Figure 3.14 illustrates part of the real number line. The

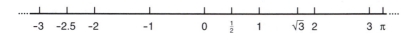

Figure 3.14   The real number line.

line extends indefinitely both to the left and to the right. Mathematically, we say that the line extends from minus infinity to plus infinity. The symbol for infinity is ∞.

The symbol $>$ means 'greater than'. So, for example, we could write $6 > 4$, $2 > -1$ and $-1 > -3$. Given a number, all numbers to the right of it on the number line are greater than the given number. The symbol $<$ means 'less than'; for example, $5 < 6$, $-1 < 0$ and $-3 < -2$ are all true statements. Closely related to 'greater than' and 'less than' are the symbols $\geqslant$ and $\leqslant$. The symbol $\geqslant$ means 'greater than or equal to'; the symbol $\leqslant$ means 'less than or equal to'. Then $5 \geqslant 4$, $4 \geqslant 4$, $3 \leqslant 6$, $3 \leqslant 3$ and $-2 \leqslant -2$ are all true statements. Statements involving the symbols $>$, $<$, $\geqslant$ and $\leqslant$ are known as **inequalities**. Inequalities can always be written in two ways: using the greater than symbols or the less than symbols. For example, $3 > 2$ could be written in the equivalent form $2 < 3$.

Sometimes two inequalities are combined into a single statement. Consider for example $3 \leqslant x < 6$. This is a compact way of writing $3 \leqslant x$ and $x < 6$. Now $3 \leqslant x$ is equivalent to $x \geqslant 3$, and so x is greater than or equal to 3 and less than 6.

3.3.2 Sets of numbers

Numbers can be divided into various sets. An obvious set is the set of whole numbers, usually denoted by $\mathbb{Z}$. Whole numbers are called **integers**.

$$\mathbb{Z} = \{\text{all whole numbers}\} = \{\ldots -4, -3, -2, -1, 0, 1, 2, 3, 4 \ldots\}$$

The dots indicate that the numbers continue indefinitely. The set of positive whole numbers is denoted by $\mathbb{N}$:

$$\mathbb{N} = \{\text{positive whole numbers}\} = \{1, 2, 3, 4, \ldots\}$$

Note that $\mathbb{N} \subseteq \mathbb{Z}$, that is, $\mathbb{N}$ is a subset of $\mathbb{Z}$.

The set comprising all numbers, representing the entire number line, is denoted by $\mathbb{R}$:

$$\mathbb{R} = \{\text{all real numbers}\}$$

Clearly, the sets $\mathbb{Z}$ and $\mathbb{N}$ are both subsets of $\mathbb{R}$.

The set of real positive numbers is denoted by $\mathbb{R}^+$:

$$\mathbb{R}^+ = \{\text{all positive numbers}\} = \{x : x > 0\}$$

Similarly, $\mathbb{R}^-$ is the set of real negative numbers:

$$\mathbb{R}^- = \{\text{all negative numbers}\} = \{x : x < 0\}$$

A **rational** number is any number of the form p/q, where p and q are whole numbers with $q \neq 0$. Hence rational numbers are simply fractions. Examples of rational numbers include $\frac{1}{2}, \frac{4}{7}, \frac{5}{8}, -\frac{2}{3}, -\frac{3}{12}$ and $\frac{17}{12}$. The set of rational numbers is denoted by $\mathbb{Q}$.

$$\mathbb{Q} = \{\text{all rational numbers}\} = \left\{ x : x = \frac{p}{q}, \ p \in \mathbb{Z}, \ q \in \mathbb{Z}, \ q \neq 0 \right\}$$

If a rational number is expressed as a decimal fraction, it either terminates or recurs infinitely. Here are some examples:

$\frac{3}{2}$ can be expressed as 1.5	these fractions terminate, that is
$\frac{3}{8}$ can be expressed as 0.375	they are of finite length
$\frac{1}{3}$ can be expressed as 0.33333...	these are infinitely
$\frac{1}{11}$ can be expressed as 0.090909...	recurring decimal fractions

A number that is not a rational number is called **irrational**. When written as a decimal fraction, an irrational number is infinite in length and non-recurring. The numbers π and $\sqrt{2}$ are both irrational. Let $\mathbb{T}$ be the set of all irrational numbers:

$$\mathbb{T} = \{\text{all irrational numbers}\}$$

Note that since any number is either rational or irrational,

$$\mathbb{Q} \cup \mathbb{T} = \mathbb{R}$$

This states that all the rational numbers together with all the irrational numbers make up the real numbers.

Since a number is either rational or irrational, but not both, no number belongs to both sets; that is, $\mathbb{Q}$ and $\mathbb{T}$ are disjoint:

$$\mathbb{Q} \cap \mathbb{T} = \varnothing$$

3.3.3 Intervals on the real line

We can represent intervals on the real line in different ways. We write $[1, 5]$ to represent the interval from 1 to 5 inclusive, that is, all the numbers from 1 to 5, including the numbers 1 and 5. An interval that includes the endpoints, such as $[1, 5]$, is called a **closed interval** and is denoted by square brackets. The closed interval $[1, 5]$ may be expressed in set notation as

$$A = \{x : x \in \mathbb{R}, \ 1 \leqslant x \leqslant 5\}$$

When drawing a closed interval on a number line we use $\bullet$ to denote the endpoints of the interval.

An interval that does not include the endpoints is called an **open interval** and is denoted by round brackets. For example, $(1, 5)$ includes all the numbers between 1 and 5, but not the numbers 1 and 5 themselves. In set notation $(1, 5)$

is given by

$$B = \{x : x \in \mathbb{R}, \ 1 < x < 5\}$$

When drawing an open interval on a number line, we use o to denote the endpoints. Note that sometimes open intervals are denoted by outward-facing square brackets, for example]1,5[.

Example 3.15

Sketch the interval on the real line given by

$$A = \{x : x \in \mathbb{R}, \ 3 \leqslant x \leqslant 5\}$$

Solution   *A* is the set of all real numbers between 3 and 5 inclusive. This is illustrated in Figure 3.15. Note the use of ● at each end of the interval to show that the interval is closed.

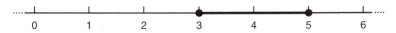

Figure 3.15   All real numbers between 3 and 5 inclusive.

Example 3.16

Sketch the intervals given by
(a) $A = \{x : x \in \mathbb{R}, \ 3 < x < 7\}$
(b) $B = \{x : x \in \mathbb{R}, \ 1 < x \leqslant 4\}$

Solution   (a)   The set *A* represents the open interval (3, 7). This interval includes all numbers from 3 to 7, but excludes the numbers 3 and 7. The interval is illustrated in Figure 3.16. Note the use of o at each end of the open interval.
            (b)   The set *B* represents an interval that is open on the left but closed on the right. We write (1, 4] to show this. This interval (1, 4] is half-open, half-closed. This interval includes the number 4 but not the number 1. It is shown in Figure 3.17. Note that a o has been used at the open end of the interval and a ● at the closed end.

Figure 3.16   The open interval (3,7).

Figure 3.17   The interval (1,4] is open on the left and closed on the right.

Example 3.17

Sketch that part of the real line defined by

$$A = \{x : x \in \mathbb{R}, \ -1 \leqslant x \leqslant 3\} \cup \{x : x \in \mathbb{R}, \ 4 \leqslant x \leqslant 5\}$$

Solution   The set A comprises all numbers from -1 to 3 inclusive, together with all numbers from 4 to 5 inclusive. This is shown in Figure 3.18.

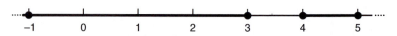

Figure 3.18   The set includes all numbers from -1 to 3 together with all numbers from 4 to 5.

Self-assessment questions 3.3

1.   Describe the following sets:
     (a) $\mathbb{N} \cap \mathbb{Q}$ (b) $\mathbb{Z} \cap \mathbb{R}^+$ (c) $\mathbb{Z} \cap \mathbb{R}$

2.   All integers are rational numbers. True or false?

3.   Explain what is meant by a rational number.

Exercises 3.3

1.   Sketch the following sets on the real number line:
     (a) $A = \{x : x \in \mathbb{R}, \ -1 \leqslant x \leqslant 7\}$
     (b) $C = \{x : x \in \mathbb{R}^+, \ x \leqslant 7\}$

2.   If t is an irrational number, is $\dfrac{1}{t}$ rational or irrational?

3.   List the elements of the following sets:
     (a) $\mathbb{Z} \cap \{x : x \in \mathbb{R}, \ 7 \leqslant x \leqslant 15\}$
     (b) $\{x : x \in \mathbb{R}, \ -5 \leqslant x \leqslant 10\} \cap \mathbb{N}$
     (c) $\mathbb{N} \cap \{0, 1\}$

4.   Sketch the following sets on the real number line:
     (a) $\mathbb{R}^+ \cap \{x : x \in \mathbb{R}, \ -2 \leqslant x \leqslant 2\}$
     (b) $\{x : x \in \mathbb{R}, \ 1 < x \leqslant 2\} \cup \{x : x \in \mathbb{R}, \ 1.5 \leqslant x \leqslant 3\}$
     (c) $\{x : x \in \mathbb{R}^-, \ -4 \leqslant x \leqslant -3\}$
     (d) $\{x : x \in \mathbb{R}, \ 0 < x \leqslant 3\} \cup \{x : x \in \mathbb{R}, \ 6 \leqslant x < 7.5\}$

5.   Classify the following intervals as open, closed or half-open–half-closed:
     (a) $(5, 8)$ (b) $(-3, -2)$ (c) $[2, 4)$
     (d) $[8, 23]$ (e) $(-\infty, \infty)$ (f) $(0, \infty)$
     (g) $[0, \infty)$

3.4 Scientific notation

This section explains the use of scientific notation. We often need to write both very small numbers, for example $0.000\,000\,3$, and very large numbers, for example

60 000 000. To do this in a compact way, we introduce **scientific notation**. A positive number written in scientific notation has the form $a \times 10^n$ where a is a number between 1 and 10. A negative number written in scientific notation has the form $-a \times 10^n$ where a is a number between 1 and 10. Recall from the work on indices that

$$10^1 = 10, \quad 10^2 = 10 \times 10 = 100, \quad 10^3 = 10 \times 10 \times 10 = 1000$$

and so on. Also recall from the laws of indices that

$$10^{-1} = \frac{1}{10} = 0.1, \quad 10^{-2} = \frac{1}{10 \times 10} = 0.01, \quad 10^{-3} = \frac{1}{10 \times 10 \times 10} = 0.001$$

and so on. Then, for example, we can write 3000 as

$$3 \times 1000 = 3 \times 10^3$$

and 0.0005 as

$$5 \times 0.0001 = 5 \times 10^{-4}$$

It is usually possible to enter numbers in scientific notation into a scientific calculator. First the number between 1 and 10 is entered, then the EXP button is pressed. The next number entered is the power of 10. Check that you can use your calculator to accept numbers in scientific notation.

Example 3.18

Write the following numbers using scientific notation:
(a) 300   (b) 100   (c) 621 000   (d) 0.3   (e) 0.000 032 1

Solution   (a) $300 = 3 \times 100 = 3 \times 10^2$

So 300 expressed in scientific notation is 3×10^2. Note that 300 has been expressed as a number between 1 and 10 multiplied by a power of 10.

(b) $100 = 1 \times 100 = 1 \times 10^2$

So 1×10^2 is scientific notation for 100. In this example, a has a value of 1, and when this is the case it is usually omitted. Hence we would simply write 10^2 as scientific notation for 100.

(c) $621\,000 = 6.21 \times 100\,000 = 6.21 \times 10^5$

Hence 6.21×10^5 is scientific notation for 621 000.

(d) $0.3 = 3 \times 0.1 = 3 \times 10^{-1}$

Hence 0.3 expressed in scientific notation is 3×10^{-1}.

(e) $0.000\,032\,1 = 3.21 \times 0.000\,01 = 3.21 \times 10^{-5}$

So 3.21×10^{-5} is scientific notation for 0.000 032 1.

Note that in each case we express the number as a number between 1 and 10 multiplied by a power of 10.

Example 3.19 Resistor values

Engineers require a large range of resistor values when designing electronic circuits. As such, they have evolved a notation for compactly specifying the resistance of a resistor. The letter R is used to denote a resistance in ohms. So, for example, 220R denotes a resistance of 220 Ω and 1R5 denotes a resistance of 1.5 Ω. Note that the R is being used in two ways. It is being used to denote a resistance in ohms and also to show the position of the decimal point. The letter K is used to denote a resistance in kilohms. So, for example, 47K denotes a resistance of 47 000 Ω or 4.7×10^4 Ω in scientific notation, and 3K3 denotes a resistance of 3300 Ω or 3.3×10^3 Ω. Similarly, the letter M is used to denote a resistance in megohms.

Example 3.20

The following is a list of some resistors from a manufacturer's catalogue. Convert each one into its value in ohms.

(a) 10R   (b) 2R2   (c) 15R   (d) 10K   (e) 33K   (f) 1M   (g) 4M7

Solution   (a) 10 Ω   (b) 2.2 Ω   (c) 15 Ω   (d) 10 000 Ω or 10^4 Ω in scientific notation
(e) 33 000 Ω or 3.3×10^4 Ω   (f) 1 000 000 Ω or 10^6 Ω
(g) 4 700 000 Ω or 4.7×10^6 Ω

We may wish to multiply and divide numbers in scientific notation. Again we use the laws of indices, in particular

$$a^m a^n = a^{m+n}, \qquad \frac{a^m}{a^n} = a^{m-n}$$

Example 3.21

Write as a single number in scientific notation
(a) $(3 \times 10^3) \times (6 \times 10^2)$ (b) $(5 \times 10^5) \times (6 \times 10^{-3})$

Solution   (a)   Recall that the order of the terms of a product is unimportant. Thus we may write $3 \times 10^3 \times 6 \times 10^2$ as $3 \times 6 \times 10^3 \times 10^2$. Now $3 \times 6 = 18$ and $10^3 \times 10^2 = 10^5$, and so

$$(3 \times 10^3) \times (6 \times 10^2) = 18 \times 10^5$$

Finally, we note that

$$18 \times 10^5 = 1.8 \times 10 \times 10^5 = 1.8 \times 10^6$$

Thus $(3 \times 10^3) \times (6 \times 10^2) = 1.8 \times 10^6$.

(b) $\qquad (5 \times 10^5) \times (6 \times 10^{-3}) = 5 \times 6 \times 10^5 \times 10^{-3}$

$$= 30 \times 10^2$$

$$= 3.0 \times 10 \times 10^2$$

$$= 3.0 \times 10^3$$

Example 3.22

Write as a single number using scientific notation:

(a) $\dfrac{9 \times 10^3}{3 \times 10}$ (b) $\dfrac{2 \times 10^3}{8 \times 10^{-2}}$

Solution (a) We calculate $\frac{9}{3}$ as 3. Next we note that $\dfrac{10^3}{10} = 10^2$, and so $\dfrac{9 \times 10^3}{3 \times 10} = 3 \times 10^2$.

The number has been expressed in scientific notation.

(b) We calculate $\frac{2}{8}$ to be 0.25. Next we note that $\dfrac{10^3}{10^{-2}} = 10^5$, and so

$\dfrac{2 \times 10^3}{8 \times 10^{-2}} = 0.25 \times 10^5$. This number is not in scientific notation, since 0.25

is not a number between 1 and 10. One further step is required:

$$0.25 \times 10^5 = 2.5 \times 10^{-1} \times 10^5 = 2.5 \times 10^4$$

and so $\dfrac{2 \times 10^3}{8 \times 10^{-2}} = 2.5 \times 10^4$.

Example 3.23 *Kinetic energy of a moving car*

Recall that the kinetic energy E of a body of mass M moving with velocity v is given by

$$E = \tfrac{1}{2} M v^2$$

Calculate the kinetic energy of the following cars:

(a) A 1000 kg car moving at a velocity of 20 m s^{-1}
(b) A 1000 kg car moving at a velocity of 40 m s^{-1}
(c) A 2000 kg car moving at a velocity of 20 m s^{-1}

Note that E has units of joules.

Solution (a)   We have $M=1000$ and $v=20$. So

$$E=\tfrac{1}{2}\times 1000\times 20^2=2\times 10^5$$

Therefore the energy of the car is 2×10^5 J.

(b)   We have $M=1000$ and $v=40$. So

$$E=\tfrac{1}{2}\times 1000\times 40^2=8\times 10^5$$

Therefore the energy of the car is 8×10^5 J.

(c)   We have $M=2000$ and $v=20$. So

$$E=\tfrac{1}{2}\times 2000\times 20^2=4\times 10^5$$

Therefore the energy of the car is 4×10^5 J.

Note that doubling the mass of a car only doubles its kinetic energy, but doubling the velocity of a car increases its kinetic energy by a factor of 4. This is the reason why fast cars are potentially so dangerous in a crash.

Self-assessment questions 3.4

1.   Only numbers greater than 10 can be expressed in scientific notation. True or false?

2.   Only positive numbers can be expressed in scientific notation. True or false?

Exercises 3.4

1.   Express the following numbers using scientific notation:
(a) 3900   (b) 40010   (c) 0.001 06
(d) −0.000 001   (e) $\frac{1}{200}$ (f) 0.76
(g) $\frac{1}{0.76}$

2.   Simplify the following expressions and express your answers using scientific notation:
(a) $2\times 10^3\times 4\times 10^2$
(b) $3.45\times 10^3\times 1.5\times 10^{-2}$
(c) $6.8\times 10^{-1}\times 9.45\times 10^{-3}$ (d) $\dfrac{6\times 10^6}{3\times 10^2}$

(e) $\dfrac{2.3\times 10^5}{9.5\times 10^4}$ (f) $\dfrac{3.56\times 10^{-3}}{2.65\times 10^{-5}}$

3.   Convert each of the following resistance values into its value in ohms:
(a) 15K   (b) 18K   (c) 22K   (d) 47K
(e) 100K   (f) 120K   (g) 150K
(h) 180K   (i) 220K   (j) 330K
(k) 470K   (l) 1M   (m) 1M2
(n) 1M5   (o) 2M2   (p) 3M3
(q) 4M7

3.5 Significant figures and decimal places

This section illustrates how to express a number to a given number of significant figures or to a given number of decimal places.

3.5.1 Decimal places

We are often asked to express a number to a given number of **decimal places**. This is usually abbreviated to d.p. The number of decimal places is the number of digits after the decimal point; for example, 3.7021 is a number given to 4 d.p.

When we are given a number and asked to write it to say 3 d.p., we need to find a number with 3 d.p. that is as close as possible to the given number. For example, expressing 3.0567 to 2 d.p. is the same as finding a number with 2 d.p. that is as close as possible to 3.0567. In this example the answer is 3.06. Thus 3.06 has 2 d.p. and is closer to 3.0567 than any other number with 2 d.p.

When asked to express a number to, say, 4 d.p., we need first to examine the first 5 d.p. The general rule is to examine to one more decimal place than requested. If the number in the 5th d.p. is a 5 or greater then the number at the 4th d.p. is increased by 1. Otherwise it is left unchanged. The following example illustrates this.

Example 3.24

(a)   Express 6.1471 to 2 d.p.
(b)   Express 6.1421 to 2 d.p.

Solution   (a)   To express a number to 2 d.p., we examine the first 3 d.p., that is, 6.147. The digit in the 3rd decimal place is a 7, and so the 4 is increased to 5. Hence to 2 d.p. we have 6.15. Thus 6.15 is a number with 2 d.p. that is as close as possible to 6.1471.

(b)   To write 6.1421 to 2 d.p., we first consider the first 3 d.p., that is, 6.142. The digit in the 3rd decimal place is a 2, and so the 4 is left unchanged. Thus to 2 d.p. 6.1421 is 6.14.

Consider again the results of (a) and (b) of the above example. When writing 6.1471 to 2 d.p., we have 6.15, which is greater than 6.1471. We say that 6.1471 has been **rounded up** to 6.15. Writing 6.1421 to 2 d.p., we have 6.14, which is smaller than 6.1421. We say that 6.1421 has been **rounded down** to 6.14.

Example 3.25

Express the following numbers to 3 d.p.:
(a) 6.942 43   (b) 0.145 782   (c) −1.679 841

Solution   To write a number to 3 d.p., we must examine the first 4 d.p.

(a)   We examine 6.9424. The fourth d.p. is a 4, and so the 3rd d.p. remains unchanged at 2. So to 3 d.p. the number is 6.942.

(b)   We ignore the 5th and 6th d.p.; we consider 0.1457. The 4th place is a 7, and so the 3rd place is rounded up from 5 to 6. To 3 d.p. the number is 0.146.

(c)   Again we ignore the 5th and 6th d.p. and consider −1.6798. The 4th place is an 8, and so the 3rd place is rounded up. However, the 3rd place is a 9, and we cannot round this up to a 10. We therefore look at the 2nd and 3rd places, that is, 79. This is rounded up to 80, giving −1.680 to 3 d.p.

3.5.2 Significant figures

Sometimes we specify a number to so many **significant figures**. We write s.f. for significant figures.

Consider the number 37.42. The most significant digit is the 3, standing for 3×10, the next most significant digit is the 7, standing for 7 units, then follows the 4, standing for $\frac{4}{10}$, and the least significant digit is the 2, standing for $\frac{2}{100}$.

When asked to write a number to so many significant figures, we first express the number in scientific notation, and work with the number between 1 and 10. For example, when asked to write 37.42 to 3 s.f., we write 37.42 as 3.742×10 and then write 3.742 to 3 s.f. We seek a 3-digit number that is as close as possible to 3.742. This is 3.74. Hence 3.742 to 3 s.f. is 3.74, and so 37.42 to 3 s.f. is $3.74 \times 10 = 37.4$.

When writing a number to a given number of significant figures, we consider one more digit than is required. So when writing to 2 s.f., we consider the first 3 most significant digits; when writing to 3 s.f., we consider the first 4 most significant digits; and so on. Note the similarity to writing to a given number of decimal places. The same rule of rounding up and rounding down applies: if the final digit is 5 or more, we round up; if the final digit is 4 or less, we round down by ignoring the final digit.

The following example illustrates the rules of significant figures.

Example 3.26

Express 47.3826 to (a) 5 s.f.   (b) 4 s.f.   (c) 3 s.f.   (d) 2 s.f.   (e) 1 s.f.

Solution   We write 47.3826 as 4.73826×10 and work with 4.73826.

(a)   When writing to 5 s.f., we look at the first 6 digits, that is 4.73826. The 6 is a number between 5 and 9, and so the 2 is rounded up to 3. Thus 4.7383 is a 5-digit number that is as close as possible to 4.73826; that is, 4.73826 to 5 s.f. is 4.7383. Then 47.3826 to 5 s.f. is 47.383.

(b)   The first 5 digits of the number are 4.7382. Since the digit 2 is less than 5, the 8 is left unchanged. Hence to 4 s.f. 4.73826 is 4.738. So 47.3826 to 4 s.f. is 47.38.

(c)   The first 4 digits are 4.738. The 8 needs to be discarded and the 3 rounded up to 4. To 3 s.f. we have 4.74. As 4.73826 to 3 s.f. is 4.74, 47.3826 to 3 s.f. is 47.4

(d)   The first three digits are 4.73. Discarding the least significant figure, that is, the 3, leaves 4.7. So 4.73826 to 2 s.f. is 4.7, and hence 47.3826 to 2 s.f. is 47.

(e)   The first 2 digits are 4.7. The 7 requires the 4 to be rounded up to 5. Thus 4.73826 to 1 s.f. is 5, and so 47.3826 to 1 s.f. is 50.

Example 3.27

Express 0.010 361 to (a) 4 s.f.   (b) 3 s.f.   (c) 2 s.f.   (d) 1 s.f.

Solution   We write 0.010 361 as 1.0361×10^{-2} and work with 1.0361.

(a)   To write to 4 s.f., we consider the first 5 digits, that is, 1.0361. The number 1.0361 to 4 s.f. is 1.036. Thus 0.010 361 to 4 s.f. is $1.036 \times 10^{-2} = 0.010\,36$.

(b)   The first 4 digits are 1.036. To 3 s.f. this becomes 1.04. So 1.0361 to 3 s.f. is 1.04, and so 0.010 361 to 3 s.f. is 0.0104.

(c)   The first 3 digits of 1.0361 are 1.03. To 2 s.f. this is 1.0. Note that the 0 must be included. So 1.0361 to 2 s.f. is 1.0, and hence 0.010 361 to 2 s.f. is $1.0 \times 10^{-2} = 0.010$. Again the final 0 must be included.

(d)   The first 2 digits of 1.0361 are 1.0. To 1 s.f. this is 1. Hence 0.010 361 to 1 s.f. is $1 \times 10^{-2} = 0.01$.

Example 3.38

Write 997 to 2 s.f.

Solution   We write 997 using scientific notation as 9.97×10^2 and work with 9.97. To write a number to 2 s.f., we consider the first 3 digits, that is, 9.97. The final digit is a 7, and so we must round up. We cannot round up a 9 to a 10, so the 9.9 is rounded up to 10. So 9.97 to 2 s.f. is 10. Hence 997 to 2 s.f. is $10 \times 10^2 = 1000$.

Self-assessment questions 3.5

1.   Writing a number to 4 d.p. is the same as writing it to 4 s.f. True or false?

2.   If a number is written to 6 s.f. then that number will always comprise 6 digits. True or false?

Exercises 3.5

1.   Write 73.093 to (a) 2 d.p.   (b) 1 d.p.   (c) 3 d.p.   (d) 2 d.p.   (e) 1 d.p.
        (c) 3 s.f.   (d) 2 s.f.   (e) 1 s.f.

3.   Write 0.0090745 to (a) 1 s.f.   (b) 2 s.f.

2.   Write 0.102546 to (a) 5 d.p.   (b) 4 d.p.   (c) 3 s.f.   (d) 4 s.f.

Review exercises 3

1 Given $A = \{17, 18, 19, 20, 21\}$, $B = \{$all odd numbers$\}$ and $C = \{$all even numbers$\}$, state
(a) $A \cap B$ (b) $A \cap C$ (c) $B \cap C$
(d) $B \cup C$ (e) $A \cup B \cup C$

2 A, B and C are three sets. If $A \cap B \neq \emptyset$, $A \cap C = \emptyset$ and $B \cap C \neq \emptyset$, draw possible Venn diagrams that illustrate A, B and C.

3 Given $A = \{-3, -1, 4, 9\}$, $B = \{-2, -1, 0, 1, 2\}$, $C = \{-1, 1, 2, 5, 7\}$ and $\mathscr{E} = \{x : x \in \mathbb{Z},\ -5 \leqslant x \leqslant 10\}$, list the elements of the following sets:
(a) $\mathscr{E}$ (b) $A \cap B$ (c) $B \cup C$ (d) $\bar{C}$
(e) $\bar{C} \cap (B \cup C)$ (f) $\overline{A \cup C}$ (g) $A \cap B \cap C$
(h) $\bar{A} \cup \bar{B}$

4 $A = \{0, 1, 2, 5, 9\}$, $B = \{1, 2, 7, 8, 9\}$ and $\mathscr{E} = \{$all decimal digits$\}$. List the elements of the following sets:
(a) $A \cup B$ (b) $A \cap B$ (c) $\bar{A}$ (d) $\bar{B}$
(e) $\overline{A \cup B}$ (f) $\bar{A} \cap \bar{B}$ (g) $\overline{A \cap B}$
(h) $\bar{A} \cup \bar{B}$
What do you conclude from (e) and (f)? What do you conclude from (g) and (h)?

5 A, B and C are three sets such that A and B are not disjoint, A and C are not disjoint, and B and C are not disjoint. No set is a subset of any other set. Figure 3.19 shows a Venn diagram of the three sets. Draw

Venn diagrams, shading the following sets:
(a) $A \cap B$ (b) $B \cup C$ (c) $B \cup (A \cap C)$
(d) $\bar{C}$ (e) $\overline{B \cap C}$ (f) $A \cap B \cap C$
(g) $\bar{B} \cap (A \cap C)$

6 The sets A, B and C are given by

$$A = \{1, 2, 3, 4, 5, 6\}, \quad B = \{4, 6\},$$

$$C = \{3, 4, 5, 6, 7, 8, 9\}$$

Illustrate A, B and C on a single Venn diagram.

7 List the elements of the following sets:
(a) $\{x : 7 \leqslant x \leqslant 10\} \cap \mathbb{N}$
(b) $\{x : x \in \mathbb{Q},\ -2 \leqslant x \leqslant 10\} \cap \mathbb{N}$
(c) $\{x : x \in \mathbb{R},\ -2 \leqslant x \leqslant 10\} \cap \mathbb{Z}$

8 Draw a Venn diagram showing sets A, B, C and D with the following properties:

$$A \cap C \neq \emptyset, \quad A \cap D \neq \emptyset,$$

$$B \cap C = \emptyset, \quad A \cap B = \emptyset,$$

$$C \cap D = \emptyset$$

Also, B is a subset of D, and there are no other subsets.

9 Write -6.5459 to (a) 2 s.f. (b) 1 s.f.
(c) 3 d.p. (d) 4 s.f. (e) 2 d.p.

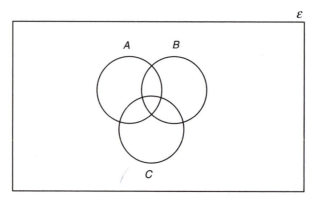

Figure 3.19 The sets A, B and C.

10 Write 9.999 to (a) 2 d.p. (b) 1 d.p.
(c) 2 s.f. (d) 1 s.f.

11 Write 0.0986 to (a) 2 d.p. (b) 1 d.p.
(c) 2 s.f. (d) 1 s.f.

12 Write the following using scientific notation:
(a) 76 (b) 76.3 (c) 763 (d) 0.76
(e) 0.763 (f) 3960 (g) 0.000 001
(h) 4 000 001

13 Evaluate the following expressions, giving your answer in scientific notation:

(a) $(4 \times 10^3) \times (9 \times 10)$

(b) $(8 \times 10^2) \times (3 \times 10^3) \times (4 \times 10^{-2})$

(c) $(9.61 \times 10^7) \times 115$ (d) $(0.036)(0.014)$

(e) $\dfrac{7.5 \times 10^3}{1.6 \times 10^4}$ (f) $-\dfrac{3.9 \times 10^{-3}}{2.71 \times 10^{-4}}$

(g) $\dfrac{3 \times 10^{-6} \times 1.7 \times 10^4}{2.96 \times 10^3 \times 9.1 \times 10^2}$

4

Working with Expressions

KEY POINTS

This chapter

- explains how to simplify algebraic expressions by adding, subtracting, multiplying and dividing

- explains how to remove brackets from expressions

- explains the term 'factor'

- shows how to factorize some common expressions, including quadratic expressions

- explains what is meant by the difference of two squares

- shows how to reduce an algebraic fraction to its simplest form

- explains how to multiply and divide algebraic fractions

- explains how to add and subtract algebraic fractions

CONTENTS

4.1 Introduction

This chapter introduces many aspects of algebra that are used throughout mathematics: simplifying expressions, adding, subtracting, multiplying and dividing algebraic fractions, and factorizing. Without grasping these algebraic building blocks, it is impossible to understand the mathematics needed for modern engineering.

4.2 Simplifying expressions

4.2.1 Simplification by addition and subtraction

It is possible to simplify certain algebraic expressions. For example, $3x + 2x$ can be simplified to $5x$, and $\frac{3}{4}y - \frac{1}{4}y = \frac{1}{2}y$. However, some expressions cannot be simplified. For example, $3x + 2t$ cannot be simplified.

Example 4.1

Simplify the following expressions:
(a) $2x + y + 5x$ (b) $3y - 2x - 2y$

Solution   (a)   The x terms are $2x$ and $5x$, and these can be added to give $7x$. Thus

$$2x + 5x + y = 7x + y$$

(b)   The y terms are $3y$ and $2y$, and subtraction yields y. Hence

$$3y - 2x - 2y = y - 2x$$

Example 4.2

Simplify
(a) $3x + t - 3t + x$ (b) $2y - y + 3y + x - 3x$

Solution   (a)   First consider the x terms: we have $3x + x = 4x$. Then consider the t terms: we have $t - 3t = -2t$. Hence

$$3x + t - 3t + x = 4x - 2t$$

(b)   In the expression $2y - y + 3y + x - 3x$ the y terms simplify to $4y$. The x terms simplify to $-2x$. Hence

$$2y - y + 3y + x - 3x = 4y - 2x$$

Example 4.3

Simplify

(a) $2x + y + xy + 2y + 3xy$ (b) $2x^2 + 3x - x^2 + x$

Solution   (a)   In the expression $2x + y + xy + 2y + 3xy$ the y terms simplify to $3y$. The xy terms simplify to $4xy$, and so

$$2x + y + xy + 2y + 3xy = 4xy + 3y + 2x$$

Note that the terms involving the product xy are quite different from terms involving just x or just y.

(b)   In the expression $2x^2 + 3x - x^2 + x$ the x^2 terms simplify to x^2. The x terms simplify to $4x$, and so

$$2x^2 + 3x - x^2 + x = x^2 + 4x$$

Example 4.4

Simplify $\dfrac{x}{2} + \dfrac{x}{4}$

Solution   Consider the expression $\dfrac{x}{2}$. Recall that division by 2 is equivalent to multiplication by $\frac{1}{2}$, and so we can write $\dfrac{x}{2} = \frac{1}{2}x$. Similarly division by 4 is equivalent to multiplication by $\frac{1}{4}$ and so $\dfrac{x}{4} = \frac{1}{4}x$. Consequently,

$$\frac{x}{2} + \frac{x}{4} = \frac{1}{2}x + \frac{1}{4}x = \frac{3}{4}x$$

Note that $\frac{3}{4}x$ can also be written as $\dfrac{3x}{4}$.

Example 4.5

Simplify (a) $\dfrac{y}{3} + 2y$ (b) $\dfrac{2t}{3} - \dfrac{t}{6}$

Solution   (a) $$\frac{y}{3} + 2y = \frac{1}{3}y + 2y = \frac{1}{3}y + \frac{6}{3}y = \frac{7}{3}y$$

Note that $\frac{7}{3}y$ can be written as $\dfrac{7y}{3}$.

(b) $$\frac{2t}{3} - \frac{t}{6} = \frac{2}{3}t - \frac{1}{6}t = \frac{4}{6}t - \frac{1}{6}t = \frac{3}{6}t = \frac{1}{2}t$$

Again, note that $\frac{1}{2}t$ is the same as $\dfrac{t}{2}$.

4.2.2 Simplification by multiplication

Recall from Section 2.3.3 that the order in which symbols are multiplied is of no consequence to the final answer, and also that the multiplication sign is usually omitted. For example, $x \times y$ is identical to $y \times x$ and would usually be written as simply xy; xyz is identical to xzy, yxz, zyx, zxy and yzx. It is worth recalling the following rules from Section 1.2.2:

> $(\text{positive}) \times (\text{positive}) = \text{positive}$
>
> $(\text{positive}) \times (\text{negative}) = \text{negative}$
>
> $(\text{negative}) \times (\text{positive}) = \text{negative}$
>
> $(\text{negative}) \times (\text{negative}) = \text{positive}$

Example 4.6

Simplify $4x^2(3x)$.

Solution　$4x^2(3x)$ means $4x^2 \times 3x$. Because we can rearrange the terms when multiplying, we can write

$$4x^2 \times 3x = 4 \cdot 3 \cdot x^2 \cdot x$$
$$= 12x^3$$

Example 4.7

Simplify
(a) $-2t(3t)t$　(b) $-4t(5t)$

Solution　(a)　$-2t(3t)t = -2 \cdot 3 \cdot t \cdot t \cdot t = -6t^3$
(b)　$-4t(5t) = -4 \cdot 5 \cdot t \cdot t = -20t^2$

Example 4.8

Simplify $4x(3xy) - 6x^2(2y)$.

Solution   First note that

$$4x(3xy)=4\cdot3\cdot x\cdot xy=12x^2y$$

and

$$6x^2(2y)=6\cdot2\cdot x^2\cdot y=12x^2y$$

So

$$4x(3xy)-6x^2(2y)=12x^2y-12x^2y=0$$

Self-assessment question 4.2

1. When adding and subtracting expressions, the order of the terms is unimportant. True or false? Discuss.

Exercises 4.2

1. Simplify the following expressions:
   (a) $3x+5x$ (b) $t+t+2t$
   (c) $4a+3b-a-b$ (d) $ab+ba$
   (e) $\dfrac{a}{2}+\dfrac{2a}{3}$ (f) $b-\dfrac{b}{3}$
   (g) $2xyz-3yzx+xy$ (h) $9x(3y)$
   (i) $2x(2x^2)-x^2(x)$ (j) $3xy+4xy-2yx$
   (k) $2ab^2-3ab(5b)$ (l) $2a+3b-a-\dfrac{b}{2}$
   (m) $(3abc)(2ab)-5a^2(b^22c)$ (n) $ab-ba$

2. Simplify the following expressions:
   (a) $\dfrac{x}{4}+\dfrac{x}{2}+\dfrac{x}{4}$ (b) $\dfrac{2y}{3}-\dfrac{y}{2}$ (c) $t-\dfrac{t}{3}$
   (d) $\dfrac{ab}{3}-\dfrac{ab}{6}$ (e) $\dfrac{3z}{7}-\dfrac{z}{7}$ (f) $4y-\dfrac{y}{2}$
   (g) $3t-\dfrac{t}{2}-\dfrac{t}{3}$

4.3 Removal of brackets

Brackets are used in many mathematical expressions. The meaning of an expression can be changed radically by simply omitting brackets or changing their position. For example, the expressions $(2x-1)+y$, $2(x-1)+y$, $2x-(1+y)$ may all look very similar but all have different meanings; the difference is due to the various positions of the brackets.

4.3.1 Removing brackets from expressions of the form
$a(b+c)$, $a(b-c)$, $(a+b)c$ **and** $(a-b)c$

Consider the expression $2(x+4)$. The 2 is multiplying all terms contained in the brackets. In other words, the 2 multiplies the x and the 4. So,

$$2(x+4)=2x+2(4)=2x+8$$

Similarly, in $2(x-4)$, the 2 multiplies the x and the 4, and so,

$$2(x-4)=2x-2(4)=2x-8$$

In the expression $(a+b)c$ both terms in the brackets multiply c to give $ac+bc$.

In general

$$a(b+c)=ab+ac, \qquad a(b-c)=ab-ac$$

$$(a+b)c=ac+bc, \qquad (a-b)c=ac-bc$$

Example 4.9

Remove the brackets from the following expressions:

(a) $3(y+2)$ (b) $4(2x-6)$ (c) $(4+x)x$ (d) $3(x+2)+5(y-6)$

Solution   Removing the brackets, we find

(a) $3(y+2)=3y+3(2)=3y+6$
(b) $4(2x-6)=4(2x)-4(6)=8x-24$
(c) $(4+x)x=4x+x^2$
(d) $3(x+2)+5(y-6)=3x+6+5y-30=3x+5y-24$

Example 4.10 *Change in the resistance of a wire*

If the resistance of a wire is R_0 at a temperature of $0\,°C$ then its resistance R_T at a temperature of $T\,°C$ is given by the formula

$$R_T=R_0(1+\alpha T)$$

where α is approximately constant for a given metal and is known as the **temperature coefficient of resistance**. Remove the brackets from the expression on the right-hand side.

Solution

$$R_T=R_0(1+\alpha T)=R_0+\alpha R_0 T$$

Example 4.11

Remove the brackets from the following:

(a) $-2(x-1)$ (b) $-(y-6)$ (c) $-2x(-x-6)$

Solution (a) The -2 multiplies both the x and the 1:

$$-2(x-1)=-2(x)-(-2)(1)=-2x-(-2)=-2x+2$$

(b) The expression $-(y-6)$ can be thought of as $-1(y-6)$. Thus

$$-(y-6)=-1(y-6)=-1(y)-(-1)(6)=-y-(-6)=-y+6$$

(c) The $-2x$ multiplies both the x and the 6:

$$-2x(-x-6)=-2x(-x)-(-2x)(6)=2x^2-(-12x)=2x^2+12x$$

Example 4.12

Remove the brackets from
(a) $(x-4)y$
(b) $(-4-z)(-b)$
(c) $2y(y-1)-3y(y+2)$

Solution (a) $(x-4)y=xy-4y$
(b) $(-4-z)(-b)=(-4)(-b)-(z)(-b)=4b-(-zb)=4b+zb$
(c)

$$2y(y-1)-3y(y+2)=2y(y)-2y(1)-3y(y)+(-3y)(2)$$
$$=2y^2-2y-3y^2-6y$$

Finally, this can be simplified to $-y^2-8y$.

4.3.2 Expressions of the form $(a+b)(c+d)$

Consider the expression $(2x+5)(3x+2)$. We wish to remove the brackets from this expression. The first set of brackets contains the terms $2x$ and 5. Both of these terms multiply the second bracket; so,

$$(2x+5)(3x+2)=2x(3x+2)+5(3x+2)$$

Expressions of the form $2x(3x+2)$ and $5(3x+2)$ have already been studied. Therefore

$$2x(3x+2)+5(3x+2)=2x(3x)+2x(2)+5(3x)+5(2)$$
$$=6x^2+4x+15x+10$$
$$=6x^2+19x+10$$

Hence

$$(2x+5)(3x+2)=6x^2+19x+10$$

In general, we have

$$(a+b)(c+d)=a(c+d)+b(c+d)=ac+ad+bc+bd$$

The process of removing brackets is also commonly known as **expanding brackets**.

Example 4.13

Remove the brackets from the following expressions:
(a) $(x+6)(3x+1)$ (b) $(4x-1)(3x+5)$ (c) $(9x-2)(-x-3)$

Solution   (a)
$$\begin{aligned}
(x+6)(3x+1)&=x(3x+1)+6(3x+1)\\
&=x(3x)+x(1)+6(3x)+6(1)\\
&=3x^2+x+18x+6\\
&=3x^2+19x+6
\end{aligned}$$

(b)
$$\begin{aligned}
(4x-1)(3x+5)&=4x(3x+5)-1(3x+5)\\
&=4x(3x)+4x(5)-1(3x)-1(5)\\
&=12x^2+20x-3x-5\\
&=12x^2+17x-5
\end{aligned}$$

(c)
$$\begin{aligned}
(9x-2)(-x-3)&=9x(-x-3)-2(-x-3)\\
&=9x(-x)+9x(-3)-2(-x)-(-2)(3)\\
&=-9x^2-27x+2x+6\\
&=-9x^2-25x+6
\end{aligned}$$

Example 4.14

Remove the brackets from (a) $(x+1)^2$ (b) $(x+1)(x-1)$.

Solution   (a)   It is important to be clear about what is meant by the expression $(x+1)^2$. This means $(x+1)$ multiplied by itself, that is, $(x+1)(x+1)$. Therefore

$$\begin{aligned}
(x+1)^2=(x+1)(x+1)&=x(x+1)+1(x+1)\\
&=x^2+x+x+1\\
&=x^2+2x+1
\end{aligned}$$

Note that $(x+1)^2$ is not the same as x^2+1^2.

(b) $(x+1)(x-1)=x(x-1)+1(x-1)=x^2-x+x-1=x^2-1$

Example 4.15 Increasing the area of a factory

A factory has a floor space of length 50 m and width 30 m. It is decided to increase the size of the factory in order to allow production to be increased. The length of the floor is increased by x m and the width by y m. The area of the floor is found by multiplying the width by the length. Derive an expression for the new floor area of the factory and remove the brackets from this expression.

Solution   The length of the floor is increased to $50+x$. The width of the floor is increased to $30+y$. Therefore the area of the floor, A, is found by multiplying the width by the length, giving

$$A=(50+x)(30+y)=1500+50y+30x+xy \text{ m}^2$$

Example 4.16 Computer language statements

Remove the brackets from the following BASIC computer language statements. Note that in the BASIC language multiplication is denoted by *.
(a) $C=((D+E)*F+G)*H$
(b)   OUTPUT$=(3*$INPUT$+($CONTROL$*$VOLUME$*1.5+2))/2$

Solution   (a) $C=((D+E)*F+G)*H$

$$=(D+E)*F*H+G*H$$

$$=D*F*H+E*F*H+G*H$$

(b)   We can still use the rules for removing brackets even though the variables now have names rather than single letters. So,

OUTPUT$=(3*$INPUT$+($CONTROL$*$VOLUME$*1.5+2))/2$

$$=3*\text{INPUT}/2+(\text{CONTROL}*\text{VOLUME}*1.5+2)/2$$

$$=3*\text{INPUT}/2+\text{CONTROL}*\text{VOLUME}*1.5/2+1$$

Self-assessment questions 4.3

1.   Why is the presence of brackets in an algebraic expression important?

2.   Explain the process of 'removing brackets' from an expression of the form $(a+b)(c+d)$.

3.   Can the position of brackets in an expression affect the meaning of the expression? If so, provide an example.

Exercises 4.3

1. Remove the brackets from the following expressions and simplify where possible:
   (a) $3(x+2)$ (b) $-3(x+2)$ (c) $x(3+x)$
   (d) $-x(3+x)$ (e) $-3(x-2)$
   (f) $-x(-x-2)$ (g) $(3+x)(2+x)$
   (h) $(3-x)(2+x)$ (i) $(-3+x)(2+x)$
   (j) $(3-2x)(2x+1)$ (k) $(3x-7)(4x+1)$
   (l) $(12x+5)(-x-3)$ (m) $(a+b)(a-b)$
   (n) $(x+1)(x+7)$ (o) $(x+3)(x+5)$
   (p) $(3+y)(y+2)$ (q) $(t-1)(t+3)$
   (r) $(z^2+2)(z+2)$ (s) $(2-v)(3+v)$
   (t) $(2x+1)(4x+3)$ (u) $(3y-1)(3y+1)$
   (v) $(1-2t)(t-7)$

2. (a) Remove the brackets from
      $(x+1) \times (x+2)$.
   (b) By considering $(x+1)(x+2)(x+3)$ as
      $[(x+1)(x+2)] \times (x+3)$ and using your
      result from (a), remove the brackets
      from $(x+1)(x+2)(x+3)$.

3. Remove the brackets from
   (a) $t(t+1)(t+2)$ (b) $2(a+2)(a-3)$

   (c) $3(t-1)(2t+1)$ (d) $(t-1)(t+1)(t-2)$
   (e) $(2x-1)(x-3)(3x+2)$ (f) $(x+3)(x+4)x$

4. Remove the brackets from the BASIC computer language statement
   $X = ((6*Y + 2*Z)*3 - 2)/4$.

5. Remove the brackets from the following expressions and simplify your answers.
   (a) $3(x+5)+2(x-1)$
   (b) $4(t-3)+3(2t+1)$
   (c) $2(3y+6)-4(y-1)+3(3y+2)$
   (d) $2(v-1)-(6-v)+3(2v+2)$
   (e) $\frac{1}{2}(x-6)+\frac{1}{4}(5+2x)$
   (f) $\frac{2}{3}(4y-9)+\frac{1}{2}(5y+8)$

6. Simplify
   (a) $(x+2)(x+3)-(x+1)(x+2)$
   (b) $(t+2)^2-t(t+3)$
   (c) $(2x+1)(x+3)-(x+2)(2x-1)$
   (d) $(3y-2)(2y+1)+(6-y)(4-2y)$
   (e) $(a+b)^2+(a-b)^2$ (f) $(a+b)^2-(a-b)^2$

4.4 Factorization

4.4.1 What is a factor?

Recall from Section 1.3.1 that a number is factorized when it is written as a product. For example, 12 may be factorized into 3×4. We see that 3 is a factor of 12; also, we see that 4 is a factor of 12. The way in which an expression or number is factorized may not be unique. Clearly, 12 could also be factorized into 2×6 or $2 \times 2 \times 3$. Algebraic expressions can also be factorized. For example, x^2+x can be written as $x(x+1)$; so x and $x+1$ are both factors of x^2+x.

4.4.2 Factorizing simple expressions

The following examples illustrate the method of factorization.

Example 4.17

Factorize
(a) $2x+4$ (b) $3y-9$ (c) $10+5t$ (d) $2y^2+y$ (e) x^3-2x
(f) $2z+4z^2$ (g) $x+\sqrt{x}$ (h) $3a^3-6a^2$

Solution   (a)   By inspection, 2 is a factor of $2x$, 2 is a factor of 4, and so 2 is a common factor; that is, $2x+4=2(x)+2(2)$. Any common factors are brought out in front of a bracket. Hence we can write $2x+4=2(x+2)$.

(b)   Clearly, 3 is a factor of $3y$, and also a factor of 9; that is, $3y-9=3(y)-3(3)$. Hence 3 is a common factor, and we can write $3y-9=3(y-3)$.

(c)   The factor common to both 10 and $5t$ is 5. So $10+5t$ may be factorized into $5(2+t)$.

(d)   The factor common to $2y^2$ and y is y. Hence $2y^2+y=y(2y+1)$.

(e)   Noting that x is a common factor allows us to write $x^3-2x=x(x^2-2)$.

(f)   Here $2z$ is a factor of both $2z$ and $4z^2$. The factorized form may now be stated: $2z+4z^2=2z(1+2z)$.

(g)   We may write x as $\sqrt{x}\times\sqrt{x}$. This helps us to see that $\sqrt{x}$ is a common factor. Hence we may write $x+\sqrt{x}=\sqrt{x}(\sqrt{x}+1)$.

(h)   Here $3a^2$ is a common factor, and so $3a^3-6a^2=3a^2(a-2)$.

It is always possible, and good practice, to check your answer by removing the brackets to see if the original expression results.

Example 4.18

Factorize

(a) $12t^2-8t^3$ (b) $4x^2+6x^4$ (c) $ax+bx$

Solution   (a)   Noting that $12t^2=4t^2(3)$ and $8t^3=4t^2(2t)$, we can see that $4t^2$ is a common factor. Hence

$$12t^2-8t^3=4t^2(3-2t)$$

The factors of $12t^2-8t^3$ are $4t^2$ and $3-2t$.

(b)   We note that $4x^2=2x^2(2)$ and $6x^4=2x^2(3x^2)$, and so

$$4x^2+6x^4=2x^2(2+3x^2)$$

The factors of $4x^2+6x^4$ are $2x^2$ and $2+3x^2$.

(c)   Clearly x is common to both ax and bx, and so $ax+bx=x(a+b)$.

4.4.3 Factorizing quadratic expressions

A **quadratic expression** is an expression having the form ax^2+bx+c where a, b and c are numbers. The number a is called the **coefficient** of x^2; the number b is called the coefficient of x; the number c is usually referred to as the **constant**

term. For example,

$$2x^2 - 13x - 7 \quad \text{and} \quad -x^2 + 2x - 1$$

are both quadratic expressions.

Consider the product $(x+1)(x+2)$. Removing the brackets yields

$$(x+1)(x+2) = x(x+2) + 1(x+2)$$
$$= x^2 + 2x + x + 2$$
$$= x^2 + 3x + 2$$

which is a quadratic expression. We see that the factors of $x^2 + 3x + 2$ are $x+1$ and $x+2$. However, if we were given $x^2 + 3x + 2$ and asked to factorize it, how would we proceed? The following examples show how to factorize quadratic expressions. It should be noted, though, that not all quadratic expressions can be factorized.

To enable us to factorize a quadratic expression, we note the following expansion:

$$(x+m)(x+n) = x(x+n) + m(x+n)$$
$$= x^2 + nx + mx + mn$$
$$= x^2 + (m+n)x + mn$$

Given a quadratic expression, we can think of the coefficient of x as $m+n$ and the constant term as mn. Once the values of m and n have been found, the factors can then be stated: $(x+m)$, $(x+n)$.

Example 4.19

Factorize $x^2 + 5x + 4$.

Solution   Suppose

$$x^2 + 5x + 4 = (x+m)(x+n) = x^2 + (m+n)x + mn$$

We seek m and n such that

$$m + n = 5, \quad \text{the coefficient of } x$$
$$mn = 4, \quad \text{the constant term}$$

From $mn = 4$ we see that the various possibilities are:

$$m = 4, \ n = 1; \quad m = 2, \ n = 2; \quad m = 1, \ n = 4$$

and also

$$m = -4, \ n = -1; \quad m = -2, \ n = -2; \quad m = -1, \ n = -4$$

We also require $m + n = 5$, and so, by careful inspection of all possibilities, we see that $m=4$, $n=1$, or $m=1$, $n=4$ are the only permitted values. Hence we see that

$$x^2 + 5x + 4 = (x+4)(x+1) \quad \text{and} \quad x^2 + 5x + 4 = (x+1)(x+4)$$

Clearly, the same factors have resulted from both solutions.

Example 4.20

Factorize

(a) $x^2 + 7x + 12$ (b) $x^2 + x - 12$ (c) $x^2 - x - 12$

Solution   (a)   Let $x^2 + 7x + 12 = (x+m)(x+n) = x^2 + (m+n)x + mn$. We seek m and n such that

$$m+n=7, \quad \text{the coefficient of } x$$

$$mn=12, \quad \text{the constant term}$$

By careful consideration of all possible values, we see that values for m and n are

$$m=3, \quad n=4$$

and

$$m=4, \quad n=3$$

Taking $m=3$, $n=4$ enables us to write

$$x^2 + 7x + 12 = (x+3)(x+4)$$

Taking $m=4$, $n=3$ enables us to write

$$x^2 + 7x + 12 = (x+4)(x+3)$$

Clearly, the same factors have resulted, and in future we shall state the factors only once. The factors of $x^2 + 7x + 12$ are $x+3$ and $x+4$.

(b)   Let $x^2 + x - 12 = (x+m)(x+n) = x^2 + (m+n)x + mn$. We seek m and n so that

$$m+n=1, \quad \text{the coefficient of } x$$

$$mn=-12, \quad \text{the constant term}$$

Consideration of all possible values shows that the correct combination is $m=4$, $n=-3$. We can now factorize $x^2 + x - 12$:

$$x^2 + x - 12 = (x+4)(x-3)$$

So the factors of $x^2 + x - 12$ are $x+4$ and $x-3$.

(c)   Suppose $x^2 - x - 12 = (x + m)(x + n) = x^2 + (m + n)x + mn$. Then m and n must satisfy

$$m + n = -1, \quad \text{the coefficient of } x$$

$$mn = -12, \quad \text{the constant term}$$

By considering possible values of m and n such that $mn = -12$, and then applying $m + n = -1$, we find that the correct combination is $m = -4$ and $n = 3$. Hence

$$x^2 - x - 12 = (x - 4)(x + 3)$$

The factors of $x^2 - x - 12$ are $x - 4$ and $x + 3$.

Example 4.21

Factorize (a) $x^2 - 9$ (b) $x^2 - 1$

Solution   (a)   Suppose $x^2 - 9 = (x + m)(x + n) = x^2 + (m + n)x + mn$. We seek m and n such that

$$m + n = 0, \quad \text{the coefficient of } x$$

$$mn = -9, \quad \text{the constant term}$$

By considering $mn = -9$ and then $m + n = 0$, we see that $m = 3$, $n = -3$. Hence

$$x^2 - 9 = (x + 3)(x - 3)$$

(b)   With the usual notation, we require

$$m + n = 0, \quad mn = -1$$

Clearly $m = 1$, $n = -1$, and so

$$x^2 - 1 = (x + 1)(x - 1)$$

The previous example highlights an important result known as the **difference of two squares**. This states that

$$\alpha^2 - \beta^2 = (\alpha + \beta)(\alpha - \beta)$$

Example 4.22

Factorize $4x^2 - 25$.

Solution Note that $4x^2 = (2x)^2$ and that $25 = 5^2$. Therefore $4x^2 - 25$ is the difference of two squares and can be factorized as

$$4x^2 - 25 = (2x)^2 - 5^2 = (2x+5)(2x-5)$$

If the coefficient of x^2 is not equal to 1 then we examine the expression to see if a numerical factor can be found. The following example illustrates this.

Example 4.23

Factorize $3x^2 - 12$.

Solution We note that 3 is a numerical factor of $3x^2 - 12$, and so

$$3x^2 - 12 = 3(x^2 - 4)$$

Now $x^2 - 4$ can be factorized in the usual way:

$$x^2 - 4 = (x+2)(x-2)$$

So, finally,

$$3x^2 - 12 = 3(x+2)(x-2)$$

Sometimes no numerical factor can be found. In such cases a slightly different approach is taken. Consider the following example.

Example 4.24

Factorize $2x^2 + 7x + 3$.

Solution To factorize a quadratic expression where the coefficient of x^2 is not 1 requires a few extra steps. First we note the coefficient of x^2; in this example it is 2. The expression is multiplied by this number:

$$2(2x^2 + 7x + 3) = 4x^2 + 14x + 6$$

Noting that $4x^2$ is $(2x)^2$, we can now write this new expression as

$$4x^2 + 14x + 6 = (2x)^2 + 7(2x) + 6$$

Let us substitute z for $2x$ to obtain

$$(2x)^2 + 7(2x) + 6 = z^2 + 7z + 6, \quad \text{where } z = 2x$$

The quadratic expression in z has a z^2 coefficient of 1, which makes factorization straightforward. After factorization, we can replace z with $2x$ to return to our original variable:

$$z^2 + 7z + 6 = (z+1)(z+6) = (2x+1)(2x+6) = 2(2x+1)(x+3)$$

So far, we have shown

$$2(2x^2 + 7x + 3) = 2(2x+1)(x+3)$$

and so

$$2x^2 + 7x + 3 = (2x+1)(x+3)$$

The factors of $2x^2 + 7x + 3$ are therefore $(2x+1)$ and $(x+3)$.

Example 4.25

Factorize
(a) $12x^2 - x - 1$ (b) $8x^2 + 18x + 9$

Solution   (a)   The coefficient of x^2 is 12, and so the given quadratic expression is multiplied by 12:

$$12(12x^2 - x - 1) = 144x^2 - 12x - 12$$

We note that $144x^2 = (12x)^2$, and so we can rewrite the quadratic as

$$144x^2 - 12x - 12 = (12x)^2 - (12x) - 12$$

Putting $z = 12x$ yields

$$(12x)^2 - (12x) - 12 = z^2 - z - 12$$

The quadratic in z can be factorized, and then $12x$ is substituted in place of z:

$$z^2 - z - 12 = (z-4)(z+3)$$
$$= (12x - 4)(12x + 3)$$
$$= 4(3x - 1)3(4x + 1)$$
$$= 12(3x - 1)(4x + 1)$$

So

$$12(12x^2 - x - 1) = 12(3x - 1)(4x + 1)$$

that is,

$$12x^2 - x - 1 = (3x - 1)(4x + 1)$$

The factors of $12x^2 - x - 1$ are $(3x - 1)$ and $(4x + 1)$.

(b)   The coefficient of x^2 is 8:

$$8(8x^2 + 18x + 9) = 64x^2 + 8(18x) + 72 = (8x)^2 + 18(8x) + 72$$

We substitute $z = 8x$ to obtain $z^2 + 18z + 72$. Now

$$z^2 + 18z + 72 = (z + 12)(z + 6)$$

$$= (8x + 12)(8x + 6)$$

$$= 4(2x + 3)2(4x + 3)$$

$$= 8(2x + 3)(4x + 3)$$

Finally,

$$8x^2 + 18x + 9 = (2x + 3)(4x + 3)$$

The factors of $8x^2 + 18x + 9$ are $(2x + 3)$ and $(4x + 3)$.

Self-assessment questions 4.4

1. Explain what is meant by the factors of an expression. Are the factors of an expression unique? Illustrate your answer with an example.

2. Explain what you understand by the phrase 'factorize an expression'.

3. Earlier in this section it was stated that not all quadratic expressions can be factorized. Give an example of such a quadratic expression.

4. State the result known as the difference of two squares.

Exercises 4.4

1. Factorize the following expressions:
   (a) $3 + 6x$ (b) $6x - 3$ (c) $8x + 8$
   (d) $6 + 3x + 9y$ (e) $2x + 4t + 6v$
   (f) $xy + xz$ (g) $ab - 2bc$ (h) $5s - 15ts$

   (i) $x^2 + 8x + 16$ (j) $x^2 - 25$
   (k) $x^2 + 2x - 15$ (l) $x^2 + 8x + 15$
   (m) $x^2 - 8x + 15$ (n) $x^2 + 3x - 15$
   (o) $x^2 - 3x - 15$

2. Factorize, if possible, the following expressions:
   (a) $x^2 + 5x + 6$ (b) $x^2 - 5x + 6$
   (c) $x^2 + x - 6$ (d) $x^2 - x - 6$
   (e) $x^2 + 9x + 14$ (f) $x^2 + 5x - 14$
   (g) $x^2 - 5x - 14$ (h) $x^2 - 9x + 14$

3. Factorize
   (a) $y^3 + y$ (b) $y^3 + y^2$ (c) $2x^4 + x^2$
   (d) $v^3 - 2v^2$ (e) $6v^3 - 2v^2$
   (f) $3a^2 - 6a + 12a^3$ (g) $abc - a^2b + 2ab^2$
   (h) $10xy^2z - 20x^2yz + 15xyz^2$

4. Factorize
(a) $2x^2 - 50$ (b) $3x^2 + 10x - 8$ (e) $4x^2 - 4x - 3$ (f) $4x^2 - 1$
(c) $8x^2 - 6x - 9$ (d) $9x^2 - 37x + 4$ (g) $2x^2 + 15x + 18$ (h) $3x^2 - x - 2$

4.5 Algebraic fractions

The rules for adding, subtracting, multiplying and dividing algebraic fractions are identical to those of numerical fractions, dealt with in Chapter 1. It is worth reviewing Section 1.4 before proceeding.

4.5.1 Expressing a fraction in its simplest form

Consider the numerical fraction $\frac{4}{10}$. To simplify this, we factorize both numerator and denominator and then cancel any common factors. Thus

$$\frac{4}{10} = \frac{2 \times 2}{2 \times 5} = \frac{2}{5}$$

The fractions $\frac{4}{10}$ and $\frac{2}{5}$ have identical values, but $\frac{2}{5}$ is in a simpler form than $\frac{4}{10}$. It is important to note that only common factors can be cancelled. We apply the same process when simplifying algebraic fractions.

Example 4.26
Simplify
(a) $\dfrac{18x^2}{6x}$ (b) $\dfrac{8x^2y^3}{4xy}$

Solution   (a)   First note that 18 can be factorized as 6×3. So there are factors of 6 and x in both numerator and denominator. These common factors are then cancelled. That is,

$$\frac{18x^2}{6x} = \frac{(6)(3)x^2}{6x} = 3x$$

       (b)   The common factors in $\dfrac{8x^2y^3}{4xy}$ are 4, x and y. These can be cancelled. That is,

$$\frac{8x^2y^3}{4xy} = 2xy^2$$

A common error is to cancel terms that are not factors from the numerator and denominator. Again we emphasize that only common factors may be cancelled.

- When simplifying algebraic fractions, only factors common to both numerator and denominator may be cancelled.
- A fraction is expressed in its simplest form by factorizing the numerator and denominator and cancelling any common factors.

Example 4.27

Simplify

(a) $\dfrac{6}{3x+9}$ (b) $\dfrac{2x}{4x^2+2x}$ (c) $\dfrac{3x^2}{15x^3+10x^2}$

Solution (a) We factorize both numerator and denominator, and cancel any factor common to both:

$$\frac{6}{3x+9} = \frac{(2)(3)}{3(x+3)} = \frac{2}{x+3}$$

So $\dfrac{2}{x+3}$ is the simplified form of $\dfrac{6}{3x+9}$.

(b)
$$\frac{2x}{4x^2+2x} = \frac{2x}{2x(2x+1)} = \frac{1}{2x+1}$$

The common factor $2x$ has been cancelled from numerator and denominator.

(c)
$$\frac{3x^2}{15x^3+10x^2} = \frac{3x^2}{5x^2(3x+2)} = \frac{3}{5(3x+2)}$$

The common factor x^2 has been cancelled from both numerator and denominator.

Example 4.28

Simplify

(a) $\dfrac{x^2-1}{x^2+5x+4}$ (b) $\dfrac{x^2+5x+6}{x^2+x-6}$

Solution   The numerator and denominator are factorized, and any common factors are cancelled.

(a)
$$\frac{x^2-1}{x^2+5x+4} = \frac{(x-1)(x+1)}{(x+1)(x+4)} = \frac{x-1}{x+4}$$

The common factor $(x+1)$ has been cancelled.

(b)
$$\frac{x^2+5x+6}{x^2+x-6} = \frac{(x+2)(x+3)}{(x+3)(x-2)} = \frac{x+2}{x-2}$$

4.5.2 Multiplication and division of algebraic fractions

To multiply two algebraic fractions together, we multiply their numerators together, and then their denominators together. That is,

$$\frac{a}{b} \times \frac{c}{d} = \frac{a \times c}{b \times d} = \frac{ac}{bd}$$

Before multiplying fractions, it is advisable to factorize all numerators and denominators, and cancel any common factors.

Division is performed by inverting the second fraction and multiplying. That is,

$$\frac{a}{b} \div \frac{c}{d} = \frac{a}{b} \times \frac{d}{c} = \frac{ad}{bc}$$

Example 4.29
   Simplify

(a)
$$\frac{3x+6}{x^2+3x+2} \times \frac{x+1}{2x+8}$$

(b)
$$\frac{4x^2+4x}{3x+6} \times \frac{x^2+2x}{2x+2}$$

Solution   (a)   Factorizing the numerators and denominators, we find

$$\frac{3x+6}{x^2+3x+2} \times \frac{x+1}{2x+8} = \frac{3(x+2)}{(x+1)(x+2)} \times \frac{x+1}{2(x+4)}$$

$$= \frac{3(x+2)(x+1)}{2(x+1)(x+2)(x+4)}$$

Common factors $(x+2)$ and $(x+1)$ can be cancelled from numerator and denominator to give

$$\frac{3}{2(x+4)}$$

Note that the cancellation can be performed before the multiplication, so that we can write

$$\frac{3x+6}{x^2+3x+2} \times \frac{x+1}{2x+8} = \frac{3(x+2)}{(x+1)(x+2)} \times \frac{x+1}{2(x+4)} = \frac{3}{2(x+4)}$$

(b)   The numerators and denominators are factorized and factors common to both are cancelled:

$$\frac{4x^2+4x}{3x+6} \times \frac{x^2+2x}{2x+2} = \frac{4x(x+1)}{3(x+2)} \times \frac{x(x+2)}{2(x+1)}$$

$$= \frac{4x}{3} \times \frac{x}{2}$$

$$= \frac{2x}{3} \times x$$

$$= \frac{2x^2}{3}$$

$$= \tfrac{2}{3}x^2$$

Example 4.30

Find $\dfrac{3}{x+2} \div \dfrac{x}{2x+4}$.

Solution   To divide, we invert the second fraction and multiply:

$$\frac{3}{x+2} \div \frac{x}{2x+4} = \frac{3}{x+2} \times \frac{2(x+2)}{x} = \frac{6}{x}$$

4.5.3 Addition and subtraction of algebraic fractions

The method of adding and subtracting algebraic fractions follows exactly that of numerical fractions. For each fraction, the numerator and denominator are factorized and any common factors are cancelled. Then the **lowest common denominator** is found. This is the simplest algebraic expression that has the given denominators as its factors. All fractions are written with this lowest common denominator as their denominator. Their sum is then found by adding the numerators and placing the result over the lowest common denominator. Consider the following examples.

Example 4.31

Express as single fractions

(a) $\dfrac{2}{x^2} + \dfrac{1}{x}$ (b) $\dfrac{3}{x} + \dfrac{2}{y}$

Solution   (a)   The term $\dfrac{1}{x}$ is rewritten as $\dfrac{x}{x^2}$, making both denominators the same but leaving the value of the fraction unaltered. The given sum becomes

$$\frac{2}{x^2} + \frac{x}{x^2}$$

Note that both original denominators are factors of the new denominator, x^2. We call this denominator the lowest common denominator. This is the simplest expression that has factors x and x^2. The sum is then found by adding the numerators, and dividing the result by the lowest common denominator:

$$\frac{2}{x^2} + \frac{1}{x} = \frac{2}{x^2} + \frac{x}{x^2} = \frac{2+x}{x^2}$$

(b)   Both fractions are written with xy as denominator. This is the simplest expression that has both x and y as its factors:

$$\frac{3}{x} + \frac{2}{y} = \frac{3y}{xy} + \frac{2x}{xy} = \frac{3y+2x}{xy}$$

Example 4.32

Express as single fractions

(a) $\dfrac{1}{x} + \dfrac{1}{x+1}$ (b) $\dfrac{3}{(x+1)^2} - \dfrac{2}{x+1}$

Solution (a) We write each fraction with $x(x+1)$ as the denominator. This is the simplest expression that has both x and $x+1$ as its factors; that is, $x(x+1)$ is the lowest common denominator:

$$\frac{1}{x} + \frac{1}{x+1} = \frac{x+1}{x(x+1)} + \frac{x}{x(x+1)}$$

$$= \frac{x+1+x}{x(x+1)}$$

$$= \frac{2x+1}{x(x+1)}$$

(b) The simplest expression that has $(x+1)^2$ and $x+1$ as its factors is $(x+1)^2$, and so $(x+1)^2$ is the common denominator:

$$\frac{3}{(x+1)^2} - \frac{2}{x+1} = \frac{3}{(x+1)^2} - \frac{2(x+1)}{(x+1)^2}$$

$$= \frac{3-2(x+1)}{(x+1)^2}$$

$$= \frac{-2x+1}{(x+1)^2}$$

Example 4.33 Capacitors in series

Consider Figure 4.1, which shows a network of three capacitors connected in series. The equivalent capacitance C of this network is given by

$$\frac{1}{C} = \frac{1}{C_1} + \frac{1}{C_2} + \frac{1}{C_3}$$

Express the right-hand side of this equation as a single fraction and hence obtain an expression for C.

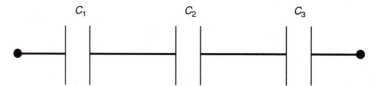

Figure 4.1 Three capacitors in series.

Solution We are given

$$\frac{1}{C} = \frac{1}{C_1} + \frac{1}{C_2} + \frac{1}{C_3}$$

The simplest expression that has C_1, C_2 and C_3 as its factors is $C_1C_2C_3$. So

$$\frac{1}{C} = \frac{C_2C_3}{C_1C_2C_3} + \frac{C_1C_3}{C_1C_2C_3} + \frac{C_1C_2}{C_1C_2C_3}$$

$$= \frac{C_2C_3 + C_1C_3 + C_1C_2}{C_1C_2C_3}$$

This is an expression for $1/C$. Finally, C is found by inverting. We obtain

$$C = \frac{C_1C_2C_3}{C_2C_3 + C_1C_3 + C_1C_2}$$

Example 4.34

Simplify to a single fraction

$$\frac{x}{x^2 + 4x + 3} - \frac{2}{x^2 + 3x + 2}$$

Solution   The denominators are factorized:

$$x^2 + 4x + 3 = (x+1)(x+3), \qquad x^2 + 3x + 2 = (x+1)(x+2)$$

The simplest expression that has both denominators as its factors is $(x+1)(x+2)(x+3)$. Thus

$$\frac{x}{x^2 + 4x + 3} - \frac{2}{x^2 + 3x + 2} = \frac{x}{(x+1)(x+3)} - \frac{2}{(x+1)(x+2)}$$

$$= \frac{x(x+2)}{(x+1)(x+2)(x+3)} - \frac{2(x+3)}{(x+1)(x+2)(x+3)}$$

$$= \frac{x(x+2) - 2(x+3)}{(x+1)(x+2)(x+3)}$$

$$= \frac{x^2 - 6}{(x+1)(x+2)(x+3)}$$

Example 4.35

Express the following as single fractions:

(a) $2 + \dfrac{1}{3x}$ (b) $1 + \dfrac{2x+1}{x+4}$

Solution   (a)   Consider 2 as $\frac{2}{1}$. Then the lowest common denominator can be found:

$$2 + \frac{1}{3x} = \frac{2}{1} + \frac{1}{3x} = \frac{6x}{3x} + \frac{1}{3x} = \frac{6x+1}{3x}$$

$$(b) \qquad 1+\frac{2x+1}{x+4}=\frac{x+4}{x+4}+\frac{2x+1}{x+4}=\frac{x+4+2x+1}{x+4}=\frac{3x+5}{x+4}$$

Self-assessment questions 4.5

1. What is meant by the term 'common denominator'? How is the lowest common denominator formed?

2. Why is it useful to factorize denominator and numerator when adding or subtracting fractions?

Exercises 4.5

1. Write the following as single fractions:

   (a) $3+\dfrac{2}{t}$ (b) $x+\dfrac{1}{x}$ (c) $\dfrac{3}{2}+\dfrac{z}{z+1}$

   (d) $a+\dfrac{2}{a}-\dfrac{1}{a^2}$

2. Write the following as single fractions:

   (a) $\dfrac{3}{t}+\dfrac{2}{t^3}$ (b) $\dfrac{x}{2}+\dfrac{2x}{3}$ (c) $\dfrac{3x}{y}-\dfrac{2}{y^2}+1$

   (d) $\dfrac{1}{x+1}+\dfrac{1}{x+2}$ (e) $\dfrac{2x}{(x+1)(x+4)}+\dfrac{3}{x+4}$

   (f) $\dfrac{6x}{(x-2)^2}-\dfrac{5}{x-2}$

   (g) $\dfrac{6}{x^2-4x+3}+\dfrac{1}{x^2+4x-21}$

   (h) $\dfrac{2x}{x^2+x-2}-\dfrac{3x}{(x-1)^2}$

   (i) $\dfrac{1}{x^2+6x+5}-\dfrac{2}{x^2+7x+10}$

3. Express in their simplest forms

   (a) $\dfrac{(-3x)^2}{6x}$ (b) $\dfrac{-24a^3b^2}{(4a)(-2b)}$

   (c) $\dfrac{(-3x)(-4x^2)(-2x^2)}{(-6x^2)(2x)}$ (d) $\dfrac{6xy}{(2x)^2}$

   (e) $\dfrac{-18z^2y}{-6yz}$ (f) $\dfrac{xyzzyx}{(xy)^2}$

   (g) $\dfrac{3x}{y}-\dfrac{6y^2x^3}{2yx^2}$

4. Simplify as much as possible

   (a) $\dfrac{x^2+4x+4}{x^2+5x+6}$ (b) $\dfrac{x^2-1}{x^2+2x+1}$

   (c) $\dfrac{2x^2+3x+1}{2x^2+5x+2}$ (d) $\dfrac{a^2-2a-8}{a^2-a-6}$

   (e) $\dfrac{6x^2+42x+60}{3x^2+18x+15}$

5. Simplify as much as possible

   (a) $(x+1)\dfrac{2x}{x^2-1}$

   (b) $\dfrac{2x-2}{x^2+3x+2}\times\dfrac{x^2+4x+4}{x^2+2x-3}$

   (c) $\dfrac{1}{x}-\dfrac{3x}{x^2+2x}$ (d) $\dfrac{x^2+xy}{x^2+2xy+y^2}$

   (e) $\left(x+\dfrac{1}{x}\right)\div\left(x-\dfrac{1}{x}\right)$

4.6 Using computer software

Sophisticated computer software is now readily available that will perform all of the algebraic manipulation met in this chapter. Packages such as DERIVE, MATHEMATICA, MAPLE and others are used widely in further and higher education. You should investigate whether you have access to such a package and, if so, how to enter, simplify and factorize algebraic expressions. Figure 4.2 shows a screen dump from DERIVE in which the package is first used to remove brackets from the expression $(a+b)^3$ and then used to factorize $x^3+6x^2+11x+6$.

1:    $(a + b)^3$

2:    $a^3 + 3\,a^2\,b + 3\,a\,b^2 + b^3$

3:    $x^3 + 6\,x^2 + 11\,x + 6$

4:    $(x + 1)\,(x + 2)\,(x+ 3)$

....

COMMAND Author   Build   Calculus   Declare   Expand   Factor   Help   Jump   soLve   Manage
                   Options   Plot   Quit   Remove Simplify   Transfer   moVe Window approX
Compute time: 0.3 seconds
Fctr (3)   Free: 100%   Derive Algebra

Figure 4.2   Screen dump from DERIVE showing removal of brackets and factorization.

Computer and calculator exercises 4.6

1. Enter the following expressions and then factorize them:
    (a) $x^3 - 10x^2 + 32x - 32$
    (b) $2x^3 + 11x^2 - 38x + 16$
    (c) $x^4 - 25x^2 + 144$ (d) $s^3 - 6s^2 - 13s + 42$

2. Factorize the denominator in the expression
    $$\frac{1}{x^3 + 9x^2 + 8x}$$

3. Express
    $$\frac{2x^2}{(x+4)(x+2)(x-13)} + \frac{x}{(x-13)(x+5)}$$
    as a single fraction.

4. Factorize $a^3 - b^3$.

Review exercises 4

1   Factorize the following expressions:
(a) $3r+15$ (b) t^6-t^4 (c) $x^2-2x-15$
(d) x^3-2x^2-15x (e) $y^5+6y^4+5y^3$
(f) $x^2-2x-35$ (g) $t^2-9t+18$
(h) $12y^2-13y-4$
(i) $\dfrac{1}{t^2}+\dfrac{1}{t}$ $\left(\text{Hint: Put } z=\dfrac{1}{t}\right)$
(j) $3y^2+9y-12$

2   Simplify
(a) $(3x)(4y^2)-(2xy)y$ (b) $\dfrac{3}{t}+\dfrac{5}{t}$

(c) $\dfrac{x-1}{x^2-1}$ (d) $\dfrac{x^2+2x+1}{x^2+3x+2}$

(e) $\dfrac{3x+21}{x^2+14x+49}$ (f) $\sqrt{x^2+6x+9}$

(g) $\dfrac{x^2-x-2}{x^2+3x}\times\dfrac{x^2+4x+3}{x^2+2x+1}$

3   Remove the brackets from the following expressions:
(a) $3(a+b)$ (b) $-6(2a-3b)$

(c) $(t+1)(t-6)$ (d) $t\left(1+\dfrac{1}{t}\right)$

(e) $\dfrac{1}{t}(3t^2+2t+6)$ (f) $(a+b)(c+d)$

(g) $(a+b)(a-b)$ (h) $(a+b)(b-a)$
(i) $(x+2)(x+3)(x-2)$

4   (a)   By substituting $x=\dfrac{1}{t}$, factorize

$$\dfrac{1}{t^2}-\dfrac{13}{t}+36$$

(b)   Write

$$\dfrac{1}{t^2}-\dfrac{13}{t}+36$$

as a single fraction, factorizing the numerator.

5   Express as single fractions in their simplest forms
(a) $\dfrac{1}{x}+\dfrac{1}{2x}$ (b) $\dfrac{1}{x}+\dfrac{2}{x}$ (c) $\dfrac{1}{x}+2x$

(d) $\dfrac{1}{x}+\dfrac{x}{2}$ (e) $\dfrac{x}{x+1}-\dfrac{3}{x+2}$

(f) $\dfrac{1}{y+1}-\dfrac{1}{y-1}$ (g) $\dfrac{3}{2x+1}+\dfrac{2}{x-4}$

(h) $4-\dfrac{3x}{y^2}+\dfrac{1}{y}$ (i) $\dfrac{x+1}{2x+1}-1$

6   Factorize
(a) $3ab+6a^2b^2$ (b) $ab+ac+xc+bx$
(c) $x^4-x^3-6x^2$ (d) $\dfrac{1}{y^2}-1$
(e) $x^{2.5}+4x^{1.5}+3\sqrt{x}$

5

Using Formulae

KEY POINTS

This chapter

- introduces some common engineering formulae

- explains how to evaluate formulae

- explains what is meant by the 'subject' of a formula

- explains how to transpose a formula

- explains and illustrates the terms 'direct proportion' and 'inverse proportion'

CONTENTS

5.1 Introduction

In engineering, physical quantities can be related to each other using a **formula**. For example, a formula exists that enables us to find the circumference of a circle if we know its radius. The plural of formula is **formulae**. In this chapter we consider how formulae are evaluated, how they can be rearranged, and why this is useful. We then introduce the related topic of proportion.

5.2 Evaluation of formulae

In engineering there are numerous formulae that relate one variable to another. The variables are represented by algebraic symbols. To evaluate a formula, we must **substitute** numbers in place of these symbols. It is important when evaluating a formula that attention is paid to the units of any physical quantities involved. Unless a consistent set of units is used a formula is not valid. The most common set of units is the SI system given in Appendix I. A knowledge of prefix conventions is also useful, and details are also given in Appendix I.

Example 5.1 Area of a helicopter landing pad

A helicopter landing pad takes the form of a circle. In general, a circle of radius r has an area A given by the formula

$$A = \pi r^2$$

where π is a constant whose value is approximately 3.1416. Most scientific calculators have a value of π preprogrammed in. Calculate the area of a landing pad with radius 5 m.

Solution   We are given that $A = \pi r^2$ and that $r = 5$ m. We replace r in the formula by 5, this process being known as **substitution**. We find

$$A = \pi r^2 = \pi(5)^2 = 25\pi = 78.54$$

The area of the landing pad is therefore 78.54 m^2. Note that in scientific notation we would write $A = 7.854 \times 10^1$ m^2.

Example 5.2 Velocity of a car moving with constant acceleration

Suppose a car is travelling with a velocity u. It then accelerates with a constant acceleration a. It can be shown that after a time t, its velocity v is given by the formula

$$v = u + at$$

Calculate the velocity of the car after 3 seconds given it is initially travelling with velocity 5 m per second and its acceleration is 2 m s^{-2}.

Solution   Here we are given $u=5$, $a=2$ and $t=3$, and so

$$v=u+at=5+2(3)=11$$

The velocity after 3 seconds is 11 m per second.

Example 5.3 Two resistors in parallel

Two resistors of resistances R_1 and R_2 are connected in parallel. The total resistance R is found from the formula

$$R=\frac{R_1R_2}{R_1+R_2}$$

Calculate the total resistance given $R_1=5\,\Omega$ and $R_2=2\,\Omega$.

Solution   The values of R_1 and R_2 are substituted into the formula for R:

$$R=\frac{R_1R_2}{R_1+R_2}$$

$$=\frac{(5)(2)}{5+2}$$

$$=\frac{10}{7}$$

$$=1.43 \quad (3 \text{ s.f.})$$

The total resistance is $1.43\,\Omega$.

Example 5.4 Moment of inertia of a hollow cylinder

The moment of inertia of an object is a measure of its resistance to rotation. It depends on both the mass of the object and the distribution of the mass about the centre of rotation. Consider Figure 5.1 which shows a hollow cylinder with internal radius r_1 and external radius r_2. The mass of the cylinder is M.

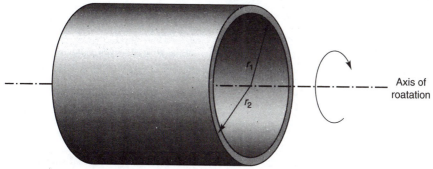

Figure 5.1   Rotating hollow cylinder.

It can be shown that the moment of inertia of the cylinder, J, is given by the formula

$$J = \tfrac{1}{2}M(r_1^2 + r_2^2)$$

Calculate the moment of inertia of the cylinder when $r_1 = 0.2$ m, $r_2 = 0.3$ m and $M = 50$ kg.

Solution   Substituting the values into the formula gives

$$J = \tfrac{1}{2}M(r_1^2 + r_2^2) = \tfrac{1}{2}50(0.2^2 + 0.3^2)$$

$$= 25(0.04 + 0.09)$$

$$= 3.25 \text{ kg m}^2$$

Self-assessment question 5.2

1.   The area of a circle is $A = \pi r^2$. A circle has a radius of 100 cm. One student calculates the area as

$$A = \pi r^2 = \pi(100)^2 = 10\,000\pi$$

Another student says that the radius is 1 m and so

$$A = \pi(1)^2 = \pi$$

Comment upon the students' solutions.

Exercises 5.2

1.   A circle with radius r has circumference C given by $C = 2\pi r$. Calculate the circumference of a circle with a radius of
     (a) 3 m   (b) 5 cm

2.   If a voltage of V V is applied to a resistor of resistance $R\,\Omega$ then a current of I A flows. The relationship between the quantities is $V = IR$. Calculate the voltage across a resistor when
     (a)   the resistance is $5\,\Omega$ and the current is 6 A
     (b)   the resistance is $3\,\Omega$ and the current is 0.7 A

3.   The area A of a rectangle is the product of the length l and the width w.
     (a)   Write a formula expressing A in terms of l and w.

(b)   Calculate the area of a rectangle whose width is 3 cm and whose length is 5 cm.
(c)   Calculate the area of a rectangle whose length is 2 m and whose width is 50 cm.

4.   The area A of a triangle is given by $A = \tfrac{1}{2}bh$, where b is the length of the base and h is the vertical height. Calculate the area of a triangle with
     (a)   base length of 6 cm and vertical height of 8 cm
     (b)   vertical height of 3 m and base length of 1.5 m

5.   A particle with initial velocity u and constant acceleration a travels a distance s

in time t. The formula for s is

$$s = ut + \tfrac{1}{2}at^2$$

Calculate the distance travelled by a particle with

(a) initial velocity of $3\,\mathrm{m\,s^{-1}}$ and acceleration of $2\,\mathrm{m\,s^{-2}}$ in $3\,\mathrm{s}$

(b) initial velocity of $4\,\mathrm{m\,s^{-1}}$ and acceleration of $1\,\mathrm{m\,s^{-2}}$ in $5\,\mathrm{s}$

6. The time T, in seconds, for a pendulum to complete a single swing is given by the formula

$$T = 2\pi\sqrt{\frac{l}{g}}$$

where l is the length of the pendulum and must be measured in metres, and g is a constant called the acceleration due to gravity and has a value $9.81\,\mathrm{m\,s^{-2}}$. Calculate the time of swing for a pendulum with

(a) length of $6\,\mathrm{m}$ (b) length of $50\,\mathrm{cm}$

7. The moment of inertia J of a solid sphere about an axis through its centre is given by

$$J = \tfrac{2}{5}MR^2$$

where M is the mass of the sphere and R is the radius of the sphere. Calculate J for the following spheres:

(a) $M = 20\,\mathrm{kg}$, $R = 0.5\,\mathrm{m}$
(b) $M = 0.3\,\mathrm{kg}$, $R = 0.025\,\mathrm{m}$
(c) $M = 0.534\,\mathrm{kg}$, $R = 0.0372\,\mathrm{m}$

8. The rotational kinetic energy E of a rotating body with moment of inertia J and angular velocity ω is given by the formula

$$E = \tfrac{1}{2}J\omega^2$$

Calculate the rotational kinetic energy of the following bodies:

(a) $J = 6\,\mathrm{kg\,m^2}$, $\omega = 50\,\mathrm{rad\,s^{-1}}$
(b) $J = 2.2\,\mathrm{kg\,m^2}$, $\omega = 23.2\,\mathrm{rad\,s^{-1}}$
(c) $J = 8.723\,\mathrm{kg\,m^2}$, $\omega = 9.283\,\mathrm{rad\,s^{-1}}$

9. The kinetic energy E of a moving object of mass M and velocity v is given by the formula

$$E = \tfrac{1}{2}Mv^2$$

Calculate the kinetic energy in the following cases:

(a) $M = 2\,\mathrm{kg}$, $v = 7\,\mathrm{m\,s^{-1}}$
(b) $M = 0.3\,\mathrm{kg}$, $v = 3.2\,\mathrm{m\,s^{-1}}$
(c) $M = 0.541\,\mathrm{kg}$, $v = 83.27\,\mathrm{m\,s^{-1}}$

Computer and calculator exercises 5.2

1. Investigate how formulae can be entered into an algebraic manipulation software package.

2. Enter the formula $E = 0.5Mv^2$ and evaluate it when $M = 8.5\,\mathrm{kg}$ and $v = 13.7\,\mathrm{m\,s^{-1}}$.

5.3 Transposition of formulae

In the formula $V = IR$ we say that V is the **subject** of the formula. A variable is the subject of a formula if it appears by itself on one side of the formula, usually the left-hand side, and nowhere else in the formula. In the example $V = IR$, if we were asked to **transpose** the formula for R, or **solve** for R, then we should have to make R the subject of the formula.

Computer algebra packages

Computer algebra packages are specially designed programs with the ability to manipulate symbols. It is only in the past few years that such sophisticated software has become available. Nowadays students have access to powerful programs that can perform mathematical tasks that a few years ago would have taken hours or even days to perform. At their simplest level, the packages can perform tasks such as removing brackets from expressions like $(a+b)(c+d)(e+f)^2$ or factorizing expressions such as x^3+3x^2-7x+2. But this is just the tip of a very big iceberg. Complicated problems in calculus can be tackled, as can the solution of large and complex systems of equations. Graphs in two and three dimensions are easily plotted. There are several such packages available for use on personal computers; some common ones are DERIVE, MATHEMATICA, REDUCE, MAPLE and MACSYMA, but there are many more. An example of a screen dump from the package DERIVE is shown in Figure 1, and illustrates the use of the package to plot a three-dimensional graph.

Computation using symbolic manipulation packages is now finding applications in a diverse range of disciplines, including engineering, biology, control and communication systems, security and coding. The principal advantage is that very long and difficult calculations are automated, to free individual research and development workers to concentrate on the underlying problems. Another advantage for the engineer is that considerable mathematical expertise has been utilized in the programming of the packages – expertise that the engineer could not be expected to develop personally in his or her professional life, but which now can be called upon through the packages.

Packages such as DERIVE can now be found in most universities and colleges. You should enquire about their availability. You will find that some time spent familiarizing yourself with a computer algebra package is time very well spent. To quote from the *DERIVE User Manual* (1990): 'Making mathematics more exciting and enjoyable is the driving force behind the development of DERIVE. The system is designed to eliminate the drudgery of performing long tedious mathematical calculations.'

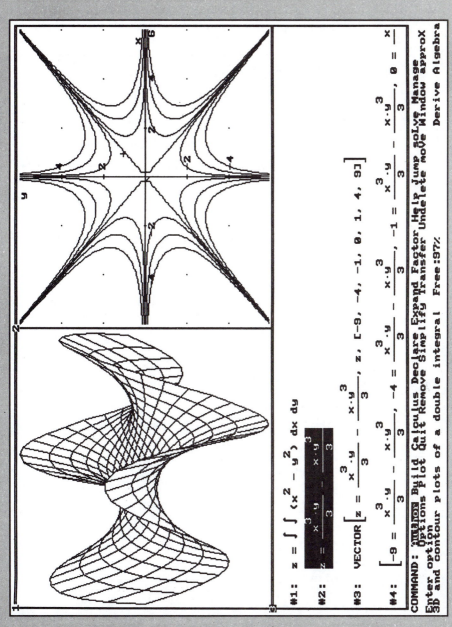

Figure 1 *A screen dump from the DERIVE program.*

5.3.1 Rules for transposing formulae

Formulae often need to be rearranged, or transposed, so as to make a different variable the subject. When transposing a formula, there are five rules that must be adhered to. When transposing a formula, you may

- add the same quantity to both sides of the formula

- subtract the same quantity from both sides of the formula

- multiply both sides of the formula by the same quantity

- divide both sides of the formula by the same quantity

- take 'functions' of both sides of the formula: for example, take the square root of both sides.

In summary, and loosely speaking, we must do precisely the same to both sides.

Example 5.5

An object has initial velocity u and constant acceleration a. After time t, the velocity v is given by $v = u + at$. Transpose the formula so that
(a) u is the subject   (b) a is the subject

Solution   (a)   We are asked to make u the subject of the formula

$$v = u + at$$

The intention is to obtain u by itself on one side of the formula. Subtracting at from both sides of the formula gives

$$v - at = u + at - at$$

$$= u$$

Hence

$$u = v - at$$

(b)   We are asked to make a the subject of the formula

$$v = u + at$$

The intention is to obtain a by itself on one side of the formula. Subtracting u from both sides gives

$$v - u = at$$

Dividing both sides by t gives

$$\frac{v-u}{t} = \frac{at}{t} = a$$

Hence

$$a = \frac{v-u}{t}$$

Example 5.6

An object with initial velocity u and constant acceleration a will cover a distance s in time t, where s is given by the formula

$$s = ut + \tfrac{1}{2}at^2$$

Rearrange the formula so that the subject is (a) u (b) a

Solution   (a) $$s = ut + \tfrac{1}{2}at^2$$

Subtracting $\tfrac{1}{2}at^2$ from both sides gives

$$s - \tfrac{1}{2}at^2 = ut$$

Dividing both sides by t gives

$$\frac{s - \tfrac{1}{2}at^2}{t} = \frac{ut}{t} = u$$

so that

$$u = \frac{s}{t} - \frac{\tfrac{1}{2}at^2}{t}$$

and finally

$$u = \frac{s}{t} - \tfrac{1}{2}at$$

(b) $$s = ut + \tfrac{1}{2}at^2$$

Subtracting ut from both sides gives

$$s - ut = \tfrac{1}{2}at^2$$

Multiplying both sides by 2 gives

$$2(s - ut) = at^2$$

Dividing both sides by t^2 gives

$$\frac{2(s - ut)}{t^2} = a$$

and finally

$$a = \frac{2(s - ut)}{t^2}$$

Example 5.7 Coulomb's law

Coulomb's law states that the force of attraction between two charged particles is given by the formula

$$F = \frac{kQ_1Q_2}{r^2}$$

where Q_1 and Q_2 are the charges of the two particles, r is the distance between the two particles, F is the force of attraction between the two particles and k is a constant.

(a)   Transpose the formula to make Q_2 the subject.
(b)   Transpose the formula to make r the subject.

Solution   (a)   We are given that

$$F = \frac{kQ_1Q_2}{r^2}$$

Multiplying both sides of the formula by r^2 gives

$$Fr^2 = kQ_1Q_2$$

Dividing both sides of the formula by kQ_1 gives

$$\frac{Fr^2}{kQ_1} = Q_2$$

and so

$$Q_2 = \frac{Fr^2}{kQ_1}$$

(b)   Multiplying the formula by r^2 gives

$$Fr^2 = kQ_1Q_2$$

Dividing both sides by F gives

$$r^2 = \frac{kQ_1Q_2}{F}$$

Taking the square root of both sides gives

$$r = \sqrt{\frac{kQ_1Q_2}{F}}$$

Example 5.8 Rate of heat transfer through a wall due to conduction

Consider Figure 5.2, which shows a section of wall with area A and thickness l. The internal temperature of the wall is T_i and the external temperature is T_e. It is useful to be able to calculate the rate of heat transfer through the wall due to conduction. For example, this sort of calculation may be needed when calculating heat losses in a house, an industrial furnace or a reaction vessel used in the chemical industry. The formula for the rate of heat transfer Q is

$$Q = \frac{kA}{l}(T_i - T_e)$$

where k is a constant called the **thermal conductivity** of the wall material. Rearrange this equation to make T_i the subject.

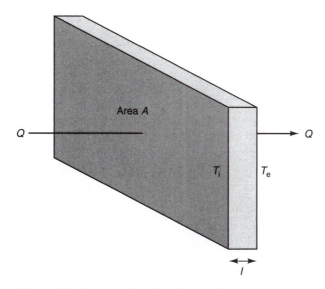

Figure 5.2 Heat transfer through a wall.

Solution

$$Q = \frac{kA}{l}(T_i - T_e)$$

Multiplying both sides by l gives

$$Ql = kA(T_i - T_e)$$

Dividing both sides by kA gives

$$\frac{Ql}{kA} = T_i - T_e$$

Adding T_e to both sides gives

$$T_i = \frac{Ql}{kA} + T_e$$

Self-assessment questions 5.3

1. Explain what is meant by the 'subject' of a formula.
2. List the rules used in transposing formulae.
3. Try to invent a formula involving x and y that it is impossible to transpose for x.

Exercises 5.3

1. Transpose the following formulae to make the given variable the subject.
   (a) $C = 2\pi r$ for r (b) $V = IR$ for R
   (c) $A = \frac{1}{2}bh$ for h (d) $V = \frac{1}{3}b^2 h$ for b
   (e) $y = \frac{\sqrt{x}}{3}$ for x (f) $T = 2\pi\sqrt{\frac{l}{g}}$ for l
   (g) $x + xy = 7$ for y (h) $x + xy = 7$ for x
   (i) $\frac{1}{R} = \frac{1}{a} + \frac{1}{b}$ for b (j) $\frac{a}{b} = \frac{c}{1+b}$ for b

2. Transpose the following formulae for t.
   (a) $c = at + b$ (b) $c = b + at^2$
   (c) $c = \frac{a}{t} + b$ (d) $c = at + bt$
   (e) $at + b = ct$ (f) $\frac{a}{t} + b = \frac{c}{t}$
   (g) $\frac{a}{t^2} + b = c$ (h) $at^2 + b = ct^2$
   (i) $at^3 + b = c$ (j) $at^n + b = c$
   (k) $\frac{a}{t} + bt = \frac{c}{t}$

3. Transpose each of the following formulae for y:
   (a) $\frac{3}{y} = x$ (b) $\frac{3}{y+1} = x$ (c) $\frac{3}{x+y} = x$
   (d) $\frac{y}{y+1} = x$ (e) $\frac{y+2}{y+1} = x$

4. The power P dissipated in a resistor is given by $P = IV$, where V is the voltage across the resistor and I is the current through the resistor. Also, for a resistor of resistance R Ohm's law states

   $$V = IR$$

   Use these two relationships to
   (a) obtain a formula for P that only involves V and R
   (b) obtain a formula for P that only involves I and R

5. Consider the steady flow of a liquid. Bernoulli's equation relates the flow velocity v, the pressure p of the liquid and the height h of the liquid above some reference level. Given two locations 1 and 2, we have

   $$\frac{p_1}{\rho g} + \frac{v_1^2}{2g} + h_1 = \frac{p_2}{\rho g} + \frac{v_2^2}{2g} + h_2$$

   where ρ is the density of the liquid. Rearrange this to obtain an expression for the velocity of the liquid at location 2. That is, obtain an expression for v_2.

6. Consider a slab of material with a difference in temperature between its two sides. The temperature on one side is T_1 and that on the other side is T_2. The rate of heat

conduction through the slab, Q, is given by

$$Q = \frac{kA(T_1 - T_2)}{d}$$

where A is the area of the slab (m²), d is the thickness of the slab (m), and k is the thermal conductivity of the slab material. Rearrange this formula to obtain an expression for T_2.

7.   A body with temperature T radiates energy R according to the Stefan–Boltzmann law. This is

$$R = kT^4$$

where k is a constant. Derive a formula for the temperature T of the body.

Computer and calculator exercises 5.3

1.   Enter the formula $P = I^2 R$ and then transpose this to find (a) I (b) R

2.   Enter the formula $s = ut + 0.5at^2$ and transpose this to find u.

5.4 Proportion

There are two types of proportionality: **direct proportion** and **inverse proportion**. We study each in turn.

5.4.1 Direct proportion

Two quantities are in direct proportion if they are always in the same ratio. This means that if one quantity is doubled then the other quantity is also doubled. If one quantity is trebled then the other quantity is also trebled, and so on. If p and q are in direct proportion, we write

$$p \propto q$$

and say p is proportional to q. The symbol $\propto$ means 'is proportional to'. If $p \propto q$ then this is equivalent to writing $p = kq$, where k is a constant, called the **constant of proportionality**. To see why these forms are equivalent, consider the following:

if $p = kq$ then $p/q = k$

If p is doubled then q must also be doubled in order to keep p/q constant. Similarly, if p is halved, then q must also be halved in order to keep p/q constant. So p and q are directly proportional if their quotient p/q is constant.

> If p is directly proportional to q then
>
> $$p = kq$$

Example 5.9

From an experiment, it is known that a variable y is proportional to x.
(a) Write down a formula relating y and x and a constant of proportionality.
(b) Suppose it is known that when $x=4$, the value of y is 12. Use this information to find the constant of proportionality.
(c) Find the value of y when $x=5$.

Solution (a) If y is proportional to x then $y \propto x$, or equivalently $y=kx$, where k is a constant of proportionality.
(b) When $x=4$, $y=12$, and so

$$12 = 4k$$

Hence the value of k must be 3, and so we can write $y=3x$.
(c) Given the formula $y=3x$, we can now find the value of y when $x=5$. That is, $y=3 \times 5=15$.

Example 5.10 Ohm's law

Consider the circuit shown in Figure 5.3. A voltage source v is connected to a resistor of resistance R. A current i flows in the circuit. Ohm's law gives the relationship between these quantities and can be written in the form

$$i = \frac{v}{R}$$

Consider the case when $R=5\,\Omega$. We then have $i=\frac{1}{5}v=0.2v$. If the voltage is 5 V then the current is 1 A. If the voltage is 50 V then the current is 10 A. The ratio of voltage to current is always the same and equal to $5:1$. We may write

$$i \propto v$$

In this example $i=0.2v$, and so the constant of proportionality is 0.2.

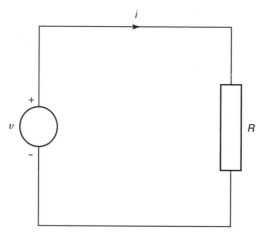

Figure 5.3 A voltage source is connected to a resistor.

Example 5.11 *Pressure due to a column of liquid*

Consider Figure 5.4. The pressure at the base of a column of liquid is directly proportional to the height of the column. If the height of the column is halved then the pressure is also halved. If p is the pressure at the base of the column and h is the height of the column then

$$p \propto h$$

or equivalently $p = kh$, where k is the constant of proportionality. It can be shown experimentally that $k = \rho g$, where ρ is the density of the liquid and g is another constant called the acceleration due to gravity. Atmospheric pressure arises due to a column of air extending upwards from the Earth. Calculating the atmospheric pressure is more difficult, however, because the density of the air varies with the height of the air column.

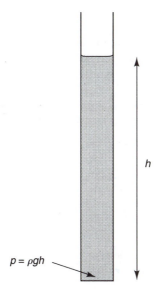

$p = \rho gh$

h

Figure 5.4 The pressure due to the liquid is proportional to the height of the column of liquid.

Example 5.12

In an experiment it is found that a variable F is directly proportional to the square of another variable w. Write down a formula relating F and w.

Solution   We are told that $F \propto w^2$, so that we can write $F = kw^2$, where k is the constant of proportionality.

5.4.2 Inverse proportion

Two quantities p and q are in inverse proportion if p is in direct proportion to $1/q$. We can write this as

$$p \propto \frac{1}{q}$$

or

$$p = k \times \frac{1}{q}$$

$$= \frac{k}{q}$$

that is

$$pq = k$$

So we see that p and q are in inverse proportion if their product pq is constant. Note that when p and q are inversely proportional, if p is doubled then q will be halved.

> If p is inversely proportional to q then
>
> $$p = \frac{k}{q}$$

Example 5.13

It is known from an experiment that a variable F is inversely proportional to the cube of the variable r. Write down a formula relating F and r.

Solution

We are told that $F \propto 1/r^3$, so that we may write

$$F = k \times \frac{1}{r^3} = \frac{k}{r^3}$$

Example 5.14 Frequency and wavelength of a wave

Consider the wave shown in Figure 5.5. Assume that this is the sort of wave arising when a stone is dropped into a pool of water. It is known as a **travelling wave** because the wave travels away from its source. Eventually a wave due to a stone will die away because the energy source is removed. However, it is possible to maintain a travelling wave in water by regularly dipping a stick

Displacement

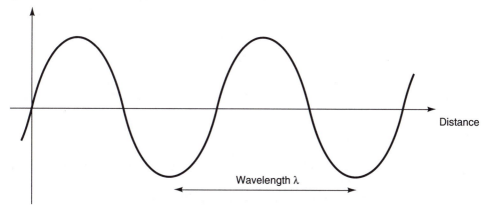

Wavelength λ

Figure 5.5   A travelling wave with wavelength λ.

into the water. The distance between two adjacent wave crests is known as the **wavelength** of the wave and is given the symbol λ. The number of wave crests passing a given point each second is known as the **frequency** and is given the symbol f. The **speed** of the wave, which is given the symbol v, is the speed with which it travels from the source. It can be shown that for a fixed wave speed the frequency of the wave is inversely proportional to the wavelength of the wave, that is,

$$f \propto \frac{1}{\lambda}$$

The constant of proportionality is the speed of the wave, v. So

$$f = v \times \left(\frac{1}{\lambda}\right) = \frac{v}{\lambda}$$

Another common type of wave is an electromagnetic wave. This has a speed of 3.00×10^8 m per second in a vacuum. This speed is conventionally given the symbol c. Carry out the following.

(a)   A radar has a wavelength of 0.03 m. The radar emits and receives electromagnetic waves. Calculate the frequency of these waves.

(b)   Determine the effect on the frequency of doubling the wavelength of the wave.

Solution   (a) $$f = \frac{c}{\lambda} = \frac{3.00 \times 10^8}{0.03} = 1 \times 10^{10} = 10^{10} \text{ Hz}$$

The frequency of the waves is 10^{10} Hz $= 10$ GHz.

(b)   Doubling the wavelength leads to a halving of the frequency, since the two quantities are inversely proportional. So the new radar wave has a frequency of

$$\frac{10^{10}}{2} = 5 \times 10^9 = 5\,\text{GHz}$$

5.4.3 Multiple proportionality

Sometimes we are told that a variable is proportional or inversely proportional to several other variables.

If F is directly proportional to both x and y then we write

$$F \propto xy, \quad \text{or equivalently} \quad F = kxy$$

If F is directly proportional to x and inversely proportional to y then we write

$$F \propto \frac{x}{y}, \quad \text{or equivalently} \quad F = k\frac{x}{y}$$

Example 5.15

A quantity P is directly proportional to $\sqrt{b}$ and inversely proportional to x.
(a)   State a formula connecting P, b and x.
(b)   When $b = 49$ and $x = 28$, the value of P is 2. Calculate the constant of proportionality.
(c)   Calculate P when $x = 9$ and $b = 13$.

Solution   (a)   Because P is directly proportional to $\sqrt{b}$ and inversely proportional to x, we can write

$$P \propto \frac{\sqrt{b}}{x}$$

or equivalently

$$P = k\frac{\sqrt{b}}{x}$$

where k is the constant of proportionality.
(b)   When $b = 49$ and $x = 28$, we are given that $P = 2$. Substituting these values into the formula for P, we find

$$2 = k\frac{\sqrt{49}}{28} = k\frac{7}{28} = \frac{k}{4}$$

Hence we see that the value of k must be 8. The formula for P is then

$$P = \frac{8\sqrt{b}}{x}$$

(c)   When $x = 9$ and $b = 13$, we find

$$P = \frac{8\sqrt{b}}{x} = \frac{8\sqrt{13}}{9} = 3.205 \quad (3 \text{ d.p.})$$

Example 5.16 Gravitation

A force of attraction exists between any pair of objects in the universe. This force f, known as the gravitational force, is directly proportional to the mass of each of the two objects and inversely proportional to the square of the distance between the objects. So

$$f \propto \frac{m_1 m_2}{r^2}$$

where m_1 and m_2 are the masses of the two objects and r is the distance between the two objects. The constant of proportionality is usually given the symbol G and is called the **universal gravitational constant**. Therefore we can write

$$f = G \frac{m_1 m_2}{r^2}$$

Self-assessment questions 5.4

1.   Explain the meaning of the symbol $\propto$.

2.   Explain the terms 'direct proportion' and 'inverse proportion'.

Exercises 5.4

1.   A quantity V is proportional to x^2 and inversely proportional to $\sqrt{y}$.
      (a)   State a formula connecting V, x and y.
      (b)   If $y = 64$ and $x = 4$ then $V = 6$. Calculate the constant of proportionality.
      (c)   Calculate V when $x = 4$ and $y = 9$.

2.   The energy received by a surface from a source of heat is inversely proportional to the square of the distance between the heat source and the surface. A surface 1 m from a heat source receives 10 W of energy.
      (a)   Calculate the energy received when the distance is changed to 5 m.
      (b)   Calculate the distance required if the surface is to receive 12 W of energy.

3.   A quantity C is proportional to a and inversely proportional to $\sqrt{b}$. If $a = 2$ and $b = 4$ then $C = 4$. Calculate C when $a = 1$ and $b = 9$.

Review exercises 5

1   Evaluate the following formulae:
(a) $C = 2\pi r$ when $r = 2.5$.
(b) $T = 2\pi\sqrt{l/g}$ when $l = 20$ and $g = 9.8$.
(c) $H = \left(\dfrac{a+b}{2}\right)^2$, when $a = 6$ and $b = 9$.
(d) $V = \sqrt{r_1^2 + r_2^2}$, when $r_1 = 2$, $r_2 = 3$.
(e) $K = \dfrac{\pi}{r+2s}$ when $r = 1$ and $s = 0.7$.

2   Consider Figure 5.6, which shows a venturi meter. The difference in heights, h, of the liquid in the two pipes can be used to calculate the flow rate of the liquid, q. The cross-sectional area of the unrestricted section is A_1, and that of the restricted section, termed the **throat**, is A_2. The formula for the flow rate q is

$$q = A_1 \sqrt{\dfrac{2gh}{(A_1/A_2)^2 - 1}}$$

Rearrange this formula to obtain an expression for h.

3   The inductance L of a solenoid is given by the formula

$$L = \dfrac{\mu N^2 A}{l}$$

where μ is the permeability of the core, N is the number of turns, A is the cross-sectional area of the solenoid and l is the length of the solenoid. Rearrange this formula to obtain an expression for N.

4   Make the specified variable subject of the formula:
(a) $H = ab + c$, a
(b) $S = 2\pi r^2 + 2\pi r h$, h
(c) $V = \sqrt{ab}$, b
(d) $V = \sqrt{a+b}$, b
(e) $m = an^2 + bc$, a
(f) $m = an^2 + bc$, n
(g) $P = \sqrt{\dfrac{a+b}{a-b}}$, a
(h) $ab + b = a$, a
(i) $\dfrac{x+y}{3} = \dfrac{x-y}{2} + 1$, x

5   A quantity L is directly proportional to a, directly proportional to b^2 and inversely proportional to c.
(a)   State L in terms of a, b, c and a constant of proportionality.
(b)   Given $L = 2$ when $a = 1$, $b = 2$ and $c = 4$, calculate L when $a = 2$, $b = 3$ and $c = 4$.

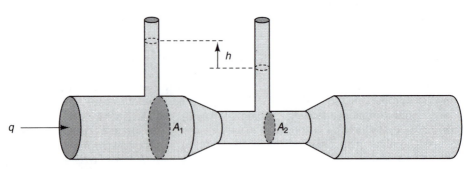

Figure 5.6   A venturi meter.

6

Functions and Their Graphs

KEY POINTS

This chapter

- explains the meaning of the term 'function'

- shows how a function can be represented by a block diagram

- introduces the symbols commonly used to denote a function

- explains what is meant by the 'argument' of a function

- explains what is meant by the 'domain' and 'range' of a function

- shows how functions can be represented graphically

- explains the meaning of the terms 'dependent variable' and 'independent variable'

- introduces graphics calculators and computer software for the plotting of graphs

CONTENTS

6.1 Introduction

Functions are among the basic building blocks of engineering mathematics. This is because they are used to describe the way in which one quantity depends upon other quantities. For example, in an alternating current circuit the voltage across an inductor depends upon time. As a further example, the drag force on an aeroplane wing depends upon many factors, including the air speed of the aeroplane. This chapter will introduce the notation used for functions and then describe how a function is represented graphically. Such a representation is frequently more useful to an engineer than an algebraic formula. You may have access to a graphics calculator or computer software that will make the sketching of graphs a relatively easy task. If this is the case, it will be beneficial to make use of it whenever a new function is introduced.

6.2 Basic concepts of functions

Consider the block shown in Figure 6.1. It may be thought of as taking an input, processing it in some way and thereby producing an output. For example, suppose that the block receives an input and multiplies it by 5. If 6 is the input then 30 will be the output. If -4 is the input then -20 will be the output. This is illustrated in Figure 6.2. If x is the input then the output will be $5x$, as shown

Figure 6.1   The block receives an input and produces an output.

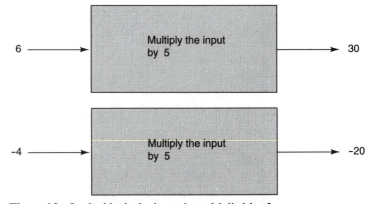

Figure 6.2   In the block the input is multiplied by 5.

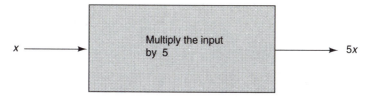

Figure 6.3 An input of x produces an output of $5x$.

in Figure 6.3. In general, provided a single output is generated for any given input, the rule that tells us how to process an input is called a **function**.

> A rule that produces a single output for any given input is called a function.

Note that if a rule yields two or more outputs for a single input then it is not a function.

It is usual to give a function a symbol. Suppose that the function that multiplies the input by 5 is given the symbol f. We then write

$$f: x \to 5x$$

to indicate that f is the function that takes an input called x and multiplies it by 5 to produce an output of $5x$.

An alternative notation that is commonly used is

$$f(x) = 5x, \quad \text{or even more simply} \quad f = 5x$$

The notation $f(x) = 5x$ indicates that the quantity in the brackets is the input to the function f, and that the output is $5x$. $f(x)$ is read 'f is a function of x', or more simply 'f of x', which means that the value of the output of f depends upon the value of the input x. A word of warning is necessary concerning the notation used here. In this context $f(x)$ does not mean the product of a variable f with a variable x.

Example 6.1

Consider the function $f(x) = \frac{1}{4}x$ depicted in Figure 6.4. The function instructs us to multiply the input by $\frac{1}{4}$. Find the output when the input is
(a) 16   (b) 18   (c) -2 (d) $\frac{1}{2}$

Solution
(a) When the input is 16, the output is $\frac{1}{4} \times 16 = 4$. We write $f(16) = 4$.
(b) When the input is 18, the output is $\frac{1}{4} \times 18 = 4.5$. We write $f(18) = 4.5$.
(c) When the input is -2, the output is $\frac{1}{4} \times -2 = -0.5$. We write $f(-2) = -0.5$.
(d) When the input is $\frac{1}{2}$, the output is $\frac{1}{4} \times \frac{1}{2} = \frac{1}{8} = 0.125$. We write $f(\frac{1}{2}) = 0.125$.

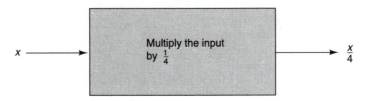

Figure 6.4   The input is multiplied by $\frac{1}{4}$.

Example 6.2

Consider the function $f(x)=3x-2$. Find the output when the input is
(a) 1   (b) 2   (c) -2 (d) 0.5

Solution   The function instructs us to multiply the input by 3 and then subtract 2.
(a)   When 1 is the input, the output is $3\times1-2=1$. We write this as

$$f:1\rightarrow1, \quad \text{or} \quad f(1)=1$$

(b)   Similarly, $f(2)=3\times2-2=4$.
(c)   When the input is -2, we have $f(-2)=3\times(-2)-2=-6-2=-8$.
(d)   Finally, $f(0.5)=3\times0.5-2=-0.5$.

Engineers usually call the pictorial representation of a function such as
that shown in Figure 6.4 a **block diagram**.

Example 6.3 *Voltage across a resistor*

 A resistor with resistance $2\,\Omega$ has a current i flowing through it, giving rise to
a voltage v across the resistor. Carry out the following.
(a)   Draw a block diagram for the resistor with current as the input and
voltage as the output.
(b)   Calculate the output when the input is 4 A.
(c)   Calculate the output when the input is 6.37 A.
(d)   Draw a block diagram for the more general case when the resistor has
a resistance of $R\,\Omega$.

Solution   (a)   Recall Ohm's law for a resistor, which is

$$v=iR$$

where R is the resistance of the resistor. For $R=2\,\Omega$ we have $v=2i$. The
function rule is therefore 'multiply the input by 2'. The block diagram
for the resistor is shown in Figure 6.5.
(b)   When $i=4$, we have $v=4\times2=8$. So the voltage across the resistor is 8 V.

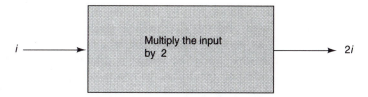

Figure 6.5 Block diagram for a 2 Ω resistor.

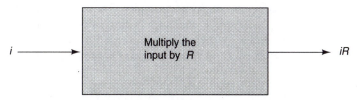

Figure 6.6 Block diagram for the resistor.

(c)   When $i=6.37$, we have $v=6.37\times2=12.74$. So the voltage across the resistor is 12.74 V.

(d)   The block diagram is shown in Figure 6.6.

Example 6.4 Hooke's law

Consider Figure 6.7, which shows a wire that is fixed at one of its ends and has a weight attached at the other. The effect of the weight is to stretch the wire and the wire is said to be 'under tension'. The weight produces a **tensile**

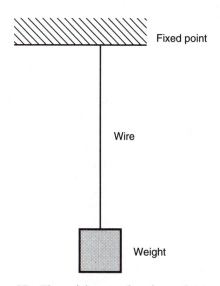

Figure 6.7 The weight puts the wire under tension.

force F in the wire. The amount by which the wire stretches is known as the **extension** of the wire, and this extension is given the symbol x. For certain materials, known as **perfectly elastic materials**, provided too much weight is not applied to the wire, when the weight is removed the wire returns to its original length. Under this condition, the relationship between F and x is given by Hooke's law. This states that

$$x = CF$$

where C is a constant of proportionality. To produce an extension x, it is necessary to apply a force F, and so it is convenient to think of F as the input to the function and x as the output. The rule for the Hooke's law function is 'multiply the input by C'. A block diagram is shown in Figure 6.8.

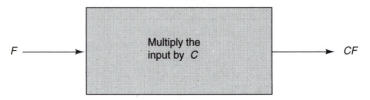

Figure 6.8   Block diagram for Hooke's law.

Example 6.5

Describe in words the rule associated with each of the following functions:

(a) $f(x) = 4x$ (b) $g(x) = 15x + 7$ (c) $h(x) = \frac{1}{2}x + 2$ (d) $y(x) = \dfrac{x+1}{2}$

(e) $z(x) = \dfrac{x}{2} + 1$

Solution   (a) f is the function that instructs us to multiply the input by 4.

(b) g is the function that instructs us to multiply the input by 15 and then add 7.

(c) h is the function that instructs us to halve the input and then add 2.

(d) y is the function that instructs us to add 1 to the input and then divide the result by 2.

(e) z is the function that instructs us to halve the input and then add 1. Note that (d) and (e) are different functions.

Self-assessment questions 6.2

1.   A function block may have two different outputs for a given input. True or false?

2.   Explain what is meant by the notation $f(x)$.

Exercises 6.2

1. Describe in words the rule associated with the function $f(x) = -12x + 12$. Find the output when the input is
    (a) 1   (b) 2   (c) -3 (d) 0   (e) 0.25

2. Write down a mathematical expression for the function given by each of the following statements:
    (a) f is the function that instructs us to subtract 5 from the square of the input;
    (b) g is the function that instructs us to divide the cube of the input by -7;
    (c) V is the function that instructs us to multiply the input by 2 and add the result to the square of the input.

3. Recall that the pressure p due to a column of liquid of height h is given by $p = \rho g h$, where ρ is the density of the liquid and is a constant, and g is the acceleration due to gravity and is a constant. Draw a block diagram for the function p, using h as the input, and describe a rule for the function.

6.3 The argument of a function

The input to a function is often called its **argument**. In the function $f(x) = 15x + 6$ the argument of f is x. However, it is not necessary to use the letter x. The functions $f(t) = 15t + 6$ and $f(z) = 15z + 6$ both instruct us to multiply the input by 15 and then add 6. Their arguments are t and z respectively.

Example 6.6

Identify the argument and describe in words the rule given by each of the following functions:

(a) $g(t) = t^2 - 7$ (b) $h(x) = x^3 - 7x^2$ (c) $y(t) = \dfrac{3}{t^2}$

Solution
(a)   The argument of the function g is t. g is the function that instructs us to square the input and then subtract 7.
(b)   The argument of the function h is x. h is the function that instructs us to cube the input and then subtract 7 times the square of the input.
(c)   The argument of the function y is t. y is the function that instructs us to divide the number 3 by the square of the input.

Example 6.7

Write down a mathematical expression for the functions given by the following statements:

(a) f is the function that instructs us to multiply the input by 4 and then add 17;

(b)   *g* is the function that squares the input and then divides the result by 5;

(c)   *h* is the function that divides the input by 5 and then squares the result.

Solution   (a) $f(x)=4x+17$. Alternatively, we could write $f: x \rightarrow 4x+17$. The argument of the function need not be x. Thus

$$f(t)=4t+17$$

and

$$f: t \rightarrow 4t+17$$

would be equally acceptable.

(b) $g(x)=\dfrac{x^2}{5}$

(c) $h(x)=\left(\dfrac{x}{5}\right)^2$

Note that g and h are quite different functions. Of course, the arguments in (b) and (c) need not be x.

Example 6.8

Describe in words the rule given by the function $f(t)=5t+1$. Then find

(a) $f(3)$ (b) $f(z)$ (c) $f(2t)$ (d) $f(t+1)$

Solution   The function instructs us to multiply the input by 5 and then add 1. So

(a)   if the input is 3, the output is $5 \times 3+1=16$; that is, $f(3)=16$;

(b)   if the input is z, the output is $5 \times z+1=5z+1$; that is, $f(z)=5z+1$;

(c)   if the input is $2t$ the output is $5 \times 2t+1=10t+1$; that is, $f(2t)=10t+1$.

(d)   here the input is $t+1$. The output is $5 \times (t+1)+1=5t+5+1=5t+6$; that is, $f(t+1)=5t+6$.

Example 6.9

Given $f(t)=t^2-7t+3$, find

(a) $f(3)$ (b) $f(0)$ (c) $f(-1)$ (d) $f(\alpha)$ (e) $f(x)$

Solution   Given $f(t)=t^2-7t+3$, then

(a) $f(3)=3^2-7(3)+3=-9$

(b) $f(0)=0^2-7(0)+3=3$

(c) $f(-1)=(-1)^2-7(-1)+3=11$

(d) $f(\alpha)=\alpha^2-7\alpha+3$

(e) $f(x)=x^2-7x+3$

Example 6.10

If $f(x)=1-7x$, find $f(x+1)$ and $f(x)+1$. Comment upon the result.

Solution   If $f(x)=1-7x$ then

$$f(x+1)=1-7(x+1)=1-7x-7=-6-7x$$

$$f(x)+1=1-7x+1=2-7x$$

It is clear that $f(x+1)$ is not the same as $f(x)+1$.

Self-assessment question 6.3

1.   What is meant by the 'argument' of a function?

Exercises 6.3

1.   Identify the argument and describe in words the rule given by each of the following functions:
   (a) $g(x)=(x-1)^3$ (b) $h(t)=t^3-1$
   (c) $y(x)=\dfrac{1}{x-1}$ (d) $f(x)=\dfrac{1}{x}-1$
   (e) $f(x)=\dfrac{x}{x-1}$

2.   If $f(x)=5x^2-2x+1$, find
   (a) $f(3)$ (b) $f(0)$ (c) $f(2x)$
   (d) $f(x^2)$ (e) $f(2x-1)$

3.   If $g(t)=t^3+7$, find
   (a) $g(-1)$ (b) $g(4)$ (c) $g(1/x)$
   (d) $g(x^2)$ (e) $g(x+1)$

4.   If $f(t)=3t+4$ find (a) $f(t+2)$ (b) $f(t-2)$
   (c) $f(t)+2$ (d) $f(t)-2$

5.   Given $u(n)=n^2+2n$, show that
   $u(n+1)-3u(n)=-2n^2-2n+3$.

6.   Given $h(t)=2t-1$ and $g(t)=t^2$, state
   (a) $h(2)$ (b) $g(3)$ (c) $h(g(3))$
   (d) $g(h(2))$

7.   If $y(x)=3x+2$, find
   (a) $y(x+1)$ (b) $y(x+2)$
   (c) $3y(x+2)-2y(x+1)$

8.   A function f is given by $f(x)=3x-2$. Find the output when x takes all integer values from -3 to 3. Present your results in a table.

6.4 Graphs of functions

Functions may also be represented pictorially using **graphs**. Suppose we wish to represent a function $y(x)$.

In order to draw a graph, two intersecting lines or **axes** are drawn, one **vertical** and one **horizontal**, as shown in Figure 6.9. The point where these axes intersect is called the **origin** and is denoted by O.

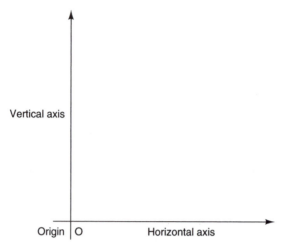

Figure 6.9 The horizontal and vertical axes intersect at the origin.

Next, it is helpful to produce a table showing the values of the function to be represented for various values of the input x.

A **scale** is drawn on the horizontal axis, and this should be sufficient to accommodate all the x values. We often call the horizontal axis the x **axis**. Similarly, a scale is drawn on the vertical axis, sufficient to accommodate the y values. We often call the vertical axis the y **axis**. It is not necessary to use the same scale on both axes. Each pair of x and y values then corresponds to a point that can be marked, or **plotted** on the diagram. We label each point as (x, y) and refer to these values as the **coordinates** of the point. Once the points have been plotted, they are joined to produce the required graph. This process is often referred to as plotting a graph of y against x. Consider the following examples.

Example 6.11

(a)   Produce a table of values of $y(x) = 2x + 4$ for values of x between -2 and 3 inclusive.

(b)   Sketch a graph of the function.

Solution (a)   We first produce a table showing the values of the function for various values of the input x. We are asked to consider x values lying between -2 and $+3$. Some of these values and the corresponding function values are shown in Table 6.1.

Table 6.1

x	-2	-1	0	1	2	3
y	0	2	4	6	8	10

(b) In order to draw a graph of the function, the two axes are drawn. A scale is drawn on the horizontal axis, and this should have sufficient range to accommodate all the *x* values. Similarly, a scale is drawn on the vertical axis, with a range sufficient to accommodate the *y* values. Each pair of *x* and *y* values in the table then corresponds to a point on the diagram. Once the points have been plotted, they are joined to produce the required graph, as shown in Figure 6.10. Note that because we wish to include the endpoints of the graph, they are shown as ● following the convention of Section 3.3.3.

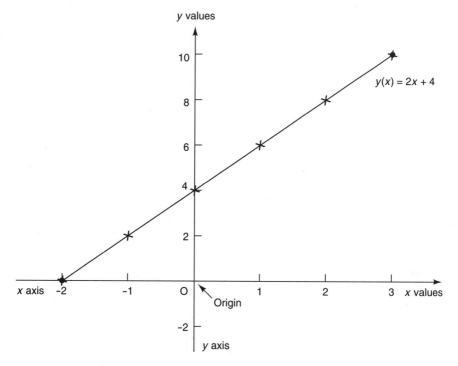

Figure 6.10   A graph showing $y(x) = 2x + 4$.

Because *x* and *y* can have a range of values, they are called **variables**; *x* is called the **independent variable** and *y* is called the **dependent variable**. This is because the value of *y* depends upon the value we have chosen for *x*. When given a function $y(x) = 2x + 4$, for example, we often use the letter *y* to refer to both the function and the output. Thus we talk about *y* **values**, meaning the output values for given *x* **values**. Note that the horizontal axis is always used for the independent variable and the vertical axis is used for the dependent variable. Naturally, in many engineering examples symbols other than x and y are used. Consider the following example.

Example 6.12

(a) Produce a table of values of the function $v(t) = 3 - 7t$ for values of t between -3 and 3.

(b) Sketch a graph of v against t.

Solution (a) We are asked to consider t values lying between -3 and $+3$. Some of these values and the corresponding function values are shown in Table 6.2.

Table 6.2

t	-3	-2	-1	0	1	2	3
v	24	17	10	3	-4	-11	-18

(b) In order to draw the graph, the axes are drawn. Each pair of values in the table corresponds to a point with coordinates (t, v). In this example the horizontal axis is called the t axis. The vertical axis is referred to as the v axis. A point on the figure has been marked to indicate each pair of values of t and v. All the points have been joined to produce the required graph, as shown in Figure 6.11.

6.4.1 The domain and range of a function

In Example 6.11, x was chosen to have values from -2 to 3 inclusive, that is, $-2 \leqslant x \leqslant 3$. The set of values that the independent variable is allowed to take is called the **domain** of the function. If a domain is not explicitly stated, it is chosen to be the largest set possible. The set of values that the dependent variable takes is called the **range** of the function. From Figure 6.10, we see that the range of $y(x)$ in Example 6.11 comprises all values between 0 and 10 inclusive. Similarly, in Example 6.12 the domain is $-3 \leqslant t \leqslant 3$ and the range is $-18 \leqslant v \leqslant 24$.

Example 6.13

(a) Plot a graph of the function $y(x) = 3x^2$ for $0 \leqslant x \leqslant 3$. State the domain and range of $y(x)$.

Solution (a) First, a table is produced showing the values that the function takes for a number of values of x in the interval $0 \leqslant x \leqslant 3$. This is shown in Table 6.3. The points with coordinates (x, y) are then plotted and joined with a smooth curve as shown in Figure 6.12.

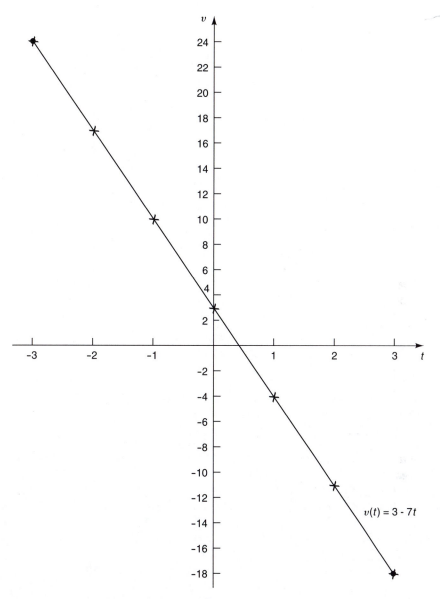

Figure 6.11   A graph showing $v(t)=3-7t$.

Table 6.3

x	0	0.5	1	1.5	2	2.5	3
y	0	0.75	3	6.75	12	18.75	27

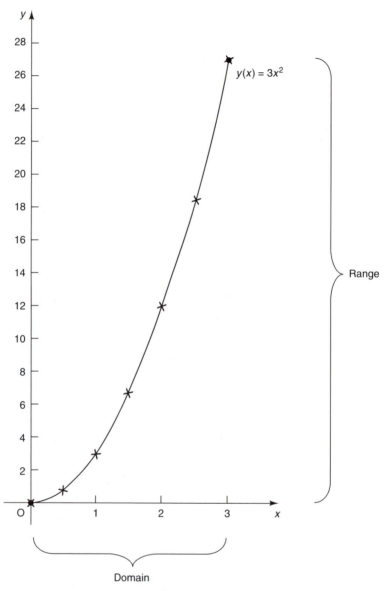

Figure 6.12   Graph of $y(x) = 3x^2$.

(b)   The domain of $y(x)$ comprises all values between 0 and 3 inclusive. The range can be seen from the graph to be all y values between 0 and 27 inclusive. We could also write the range using set notation as

$$\text{range} = \{y: y \in \mathbb{R},\ 0 \leqslant y \leqslant 27\}$$

Example 6.14

    (a)   Plot a graph of the function $g(x) = x^2 + 1$.

    (b)   State its range.

Solution   (a)   No domain is given in this example, and so in theory we should allow x to take on the largest possible set of values. Since we can evaluate g for any value of x, the domain is the whole of the real number line. That is, the domain of g is all values of x. To be practical, when plotting the graph let us choose x to lie between -4 and 4. The corresponding table of values is shown as Table 6.4. The points (x, g) have been plotted in Figure 6.13 and joined to make a smooth curve. As x increases above 4, $g(x)$ also increases. Similarly, as x decreases below -4, $g(x)$ increases.

Table 6.4

x	-4	-3	-2	-1	0	1	2	3	4
g	17	10	5	2	1	2	5	10	17

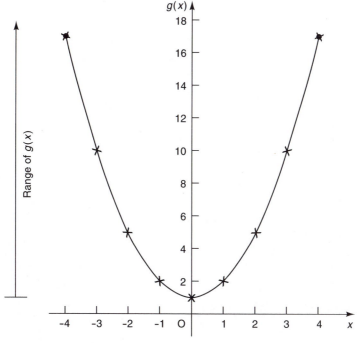

Figure 6.13   Graph of $g(x) = x^2 + 1$.

    (b)   The range of this function is the set of values that the function takes. From the graph, we see that g can take any value greater than or equal to 1.

Example 6.15 *Power dissipated in a resistor*

The power p dissipated in a resistor of resistance R when a current i flows through it is given by

$$p=i^2R$$

(a) Plot a graph of the function p when $R=4\,\Omega$ for $0\leqslant i\leqslant 4$ A.
(b) State the domain and range of the function.

Solution

(a) We have $p(i)=Ri^2=4i^2$ W. A table of values of this function is given in Table 6.5 and its graph is shown in Figure 6.14.

Table 6.5

i	0	1	2	3	4
p	0	4	16	36	64

(b) The domain of the function is $0\leqslant i\leqslant 4$. The range is $0\leqslant p\leqslant 64$.

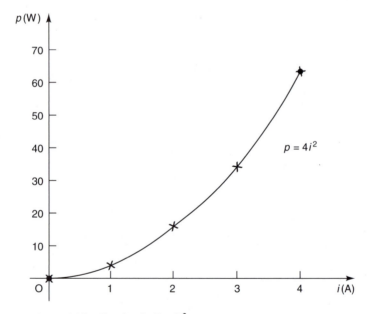

Figure 6.14 Graph of $p(i)=4i^2$.

Sometimes functions are described by different rules on different intervals in their domain. Consider the following example.

Example 6.16

The function $f(x)$ is defined by

$$f(x) = \begin{cases} 2x & 0 \leqslant x \leqslant 2 \\ 4 & 2 < x \leqslant 5 \\ 0 & \text{otherwise} \end{cases}$$

Plot a graph of $f(x)$.

Solution   In this example the function is defined by a different rule on different intervals. For example, when x lies between 0 and 2 we must use the rule $f(x) = 2x$, whereas when x lies between 2 and 5 we must use the rule $f(x) = 4$. The graph of f is shown in Figure 6.15.

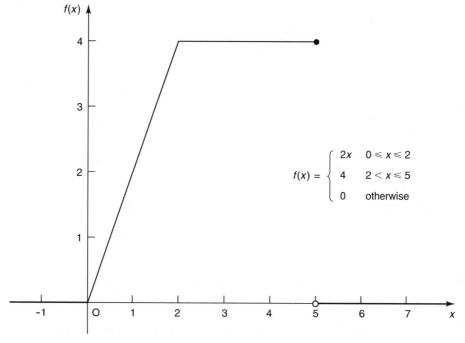

Figure 6.15   Graph of $f(x)$ for Example 6.16.

Self-assessment questions 6.4

1.   What is meant by the domain of a function?

2.   What is meant by the terms 'dependent variable' and 'independent variable'?

3.   What is meant by the range of a function?

4.   Is the dependent variable plotted on the horizontal axis or the vertical axis?

Exercises 6.4

1.   Plot a graph of the function $f(x) = -x + 2$ for values of x between -3 and $+8$ inclusive. Identify the domain and range of f.

2.   Plot a graph of the function $f(x) = x + 2$ for values of x between -3 and $+8$ inclusive. Identify the domain and range of f.

3.   On the same piece of graph paper plot graphs of the functions $g(x) = 3x + 7$ and $h(x) = 4x + 7$ for values of x between -5 and $+4$ inclusive. In each case identify the domain and range of the given function.

4.   Plot on the same piece of graph paper graphs of the functions $g(x) = x^2 + 1$ and $h(x) = x^2 - 1$ for values of x between -4 and $+4$ inclusive. In each case identify the domain and range of the given function.

5.   Plot a graph of the function $f(t) = 1/t^2$ for values of t between 0.5 and 5 inclusive. State the domain and range.

6.   Recall from Example 6.3 Ohm's law for a resistor. Plot a graph of the voltage v across a $5\,\Omega$ resistor against current i through the resistor for $0 \leqslant i \leqslant 0.5$ A. Be sure to label the axes of the graph and include the units of the quantities being plotted.

7.   When an object of mass M moves with a velocity v, it has energy due to its motion. This energy is known as kinetic energy E, and is given by $E = \frac{1}{2}Mv^2$. Plot a graph of the kinetic energy of an object of mass 5 kg against velocity for $0 \leqslant v \leqslant 4$ m per second. Ensure the axes of the graph are labelled. The SI unit of energy is the joule (J).

8.   The relationship between the frequency f and the wavelength λ of a travelling wave is given by $f = v/\lambda$, where v is the velocity of the wave and is constant. Light waves form part of the electromagnetic spectrum and have a velocity of 3.00×10^8 m per second when travelling in a vacuum. The wavelength of light extends from approximately 4×10^7 m, corresponding to violet light, through to 7×10^7 m, corresponding to red light. Plot a graph of the frequency of light against wavelength.

6.5 Use of a graphics calculator and computer software

Graphics calculators and computer software are extremely useful for plotting graphs. You should investigate whether such resources are available for your use. It is a straightforward matter to input a function, to select an appropriate domain and range, and then see the resulting graph. It is also possible to zoom in to or out from a graph to focus on particular regions of interest. This feature will enable you to solve a number of more advanced problems in your later work. Figure 6.16 shows a screen dump from the package DERIVE illustrating the function $f(x) = x^3 + 7x^2 - 3x - 4$.

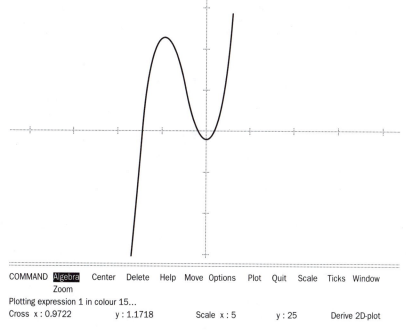

COMMAND Algebra   Center   Delete   Help   Move   Options   Plot   Quit   Scale   Ticks   Window
          Zoom
Plotting expression 1 in colour 15...
Cross x : 0.9722   y : 1.1718   Scale x : 5   y : 25   Derive 2D-plot

Figure 6.16 Screen dump from DERIVE.

Computer and calculator exercises 6.5

1. Use a graphics calculator or software to plot graphs of the following functions:

    (a) $f(x) = 15x^5$ for x between -4 and 4

    (b) $g(x) = \dfrac{2}{x^2}$ for x between 1 and 4

2. Consider $h(x) = \dfrac{1}{2+x}$. Use a graphics calculator to draw $h(x)$ for
    (a) $x=3$ to $x=6$ (b) $x=-6$ to $x=-3$
    Investigate what happens to the graph when you try to plot it over the interval $-3 \leqslant x \leqslant 3$.

Review exercises 6

1 The function $f(t)$ is defined by $f(t) = 3t^2 + 2t + 1$. Find
(a) $f(1)$ (b) $f(2)$ (c) $f(x)$
(d) $f(\alpha)$ (e) $f(1+\alpha)$ (f) $f(1/\alpha)$

2 The function $g(x)$ is defined by $g(x) = \dfrac{3}{2-x}$.

Find
(a) $g(t)$ (b) $g(\lambda)$ (c) $g(2\lambda)$

3 Plot a graph of the function $y(x) = 3 - 2x^2$ for x between -3 and 3 inclusive. Identify the domain and range of the function.

4   On the same piece of graph paper plot graphs of the functions $y = 4x - 10$ and $y = 2x - 4$. Locate the point where the two lines intersect.

5   Sketch a graph of $y(x) = \dfrac{1}{x^2}$ for $x > 0$. State the domain and range of this function.

6   The function g is defined by $g(t) = \dfrac{1}{t}$. Find

(a) $g(1)$ (b) $g(2)$ (c) $g(\alpha)$
(d) $g(2\alpha)$ (e) $g(\alpha^2)$ (f) $g(\beta - 1)$

(g) $g(\beta + 1)$ (h) $g\left(\dfrac{1}{t}\right)$

7   On the same piece of graph paper sketch graphs of $y = 5x$, $y = 5x + 1$ and $y = 5x - 1$, for $-3 \leqslant x \leqslant 3$.

8   On the same piece of graph paper sketch graphs of $y = -2x$, $y = -2x + 1$ and $y = -2x - 1$, for $-3 \leqslant x \leqslant 3$.

7

Common Engineering Functions

KEY POINTS

This chapter

- explains and illustrates what is meant by a polynomial function
- explains and illustrates what is meant by a rational function
- explains and illustrates what is meant by the modulus function
- explains and illustrates what is meant by a periodic function

CONTENTS

7.1 Polynomial functions

Consider the following functions:

$$f(x) = 4x^3 + 2x, \qquad g(x) = x^2 + 4x, \qquad h(x) = 9 - 3x^2 + x^5$$

Observe that they are all constructed using non-negative whole number powers of the independent variable x. Recall that $x^0 = 1$, and so the number 9 in $h(x)$ can be regarded as $9x^0$. All these functions are members of an important class known as **polynomial functions**.

> In general, a polynomial function $P(x)$ takes the form
>
> $$P(x) = a_0 + a_1 x + a_2 x^2 + a_3 x^3 + \dots$$
>
> where $a_0, a_1, a_2, a_3, \dots$ are all constants, called the **coefficients** of the polynomial function. The number a_0 is also known as the **constant term**.

Negative or fractional powers of the variable x are not allowed in a polynomial function. The highest power used in any polynomial function is called the **degree** of the polynomial. It is common practice to contract the term polynomial function to **polynomial**. The order in which we write down each of the terms does not matter, although it is usual to write a polynomial with its powers either increasing or decreasing. Consider the following examples.

Example 7.1

State which of the following are polynomial functions. Give the degrees of those that are.

(a) $f(x) = x^2 + x$　　(b) $f(x) = 3x^8 - 7$　　(c) $g(x) = 4x^{-1} + 2x$

(d) $h(x) = \dfrac{1}{x}$　　(e) $f(x) = -15$

Solution　　(a)　$f(x) = x^2 + x$ is a polynomial function of degree 2.
(b)　$f(x) = 3x^8 - 7$ is a polynomial function of degree 8. Note that we could write this as $f(x) = 3x^8 - 7x^0$.
(c)　$g(x)$ is not a polynomial function because negative powers are not allowed.
(d)　$h(x)$ is not a polynomial function because it could be expressed as $h(x) = x^{-1}$ and negative powers are not allowed.
(e)　$f(x) = -15$ can be written as $f(x) = -15x^0$ since $x^0 = 1$. So $f(x)$ is a polynomial function.

Polynomials with low degrees have special names.

Polynomial	Degree	Name
$ax^3 + bx^2 + cx + d$	3	cubic
$ax^2 + bx + c$	2	quadratic
$ax + b$	1	linear
a	0	constant

Constant polynomials

The graph of a constant polynomial is always a horizontal straight line. Figure 7.1 shows a graph of the polynomial $f(x) = 4$.

Linear polynomials

The graph of a linear polynomial is always a straight line. Figure 7.2 shows graphs of the polynomials $f(x) = 2x + 1$ and $g(x) = -3x + 2$.

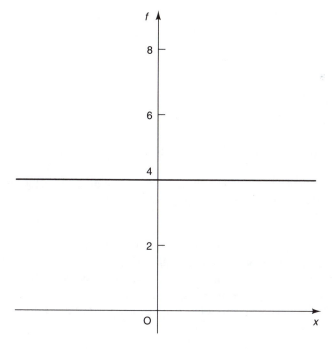

Figure 7.1 Graph of the constant polynomial $f(x) = 4$.

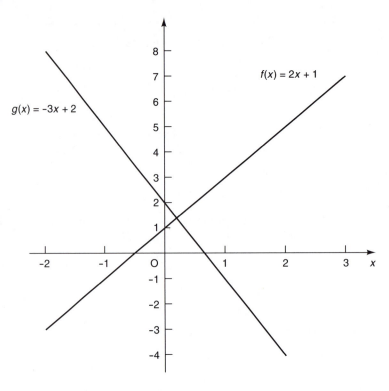

Figure 7.2 Graphs of two linear polynomials.

Example 7.2

Write down the coefficients of the polynomial $g(x) = 5x$.

Solution   The function $g(x) = 5x$ is a linear polynomial. The coefficient of x is 5. In this example the constant term is zero.

Example 7.3 *Velocity of a body travelling with constant acceleration*

The velocity v of a body travelling with a constant acceleration a is given by the function $v(t) = u + at$, where t is the time for which the body has been travelling and u is the velocity of the body at $t = 0$. The function $v(t)$ is a linear polynomial with independent variable t. The constants a and u are the coefficients of the polynomial.

Quadratic polynomials

The graph of a quadratic polynomial is a curve known as a **parabola**. Figure 7.3(a) shows a graph of the polynomial function $f(x)=x^2$. Figure 7.3(b) shows a graph of the polynomial function $g(x)=-x^2+x+1$.

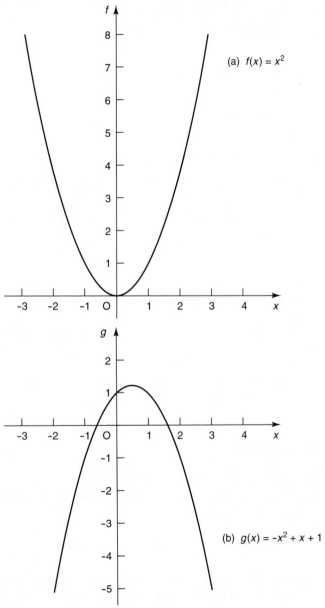

(a) $f(x) = x^2$

(b) $g(x) = -x^2 + x + 1$

Figure 7.3 Graphs of two quadratic polynomials.

Example 7.4 Energy stored in a capacitor

The energy E stored in a charged capacitor of constant capacitance C dependent on the applied voltage V, and is given by

$$E = \tfrac{1}{2}CV^2$$

This is a quadratic polynomial function. The coefficient of V^2 is $\tfrac{1}{2}C$. Th coefficient of V and the constant term are both zero. The dependent variabl is E and the independent variable is V.

Self-assessment questions 7.1

1. Explain what is meant by a polynomial function.

2. What is meant by the degree of a polynomial?

3. Polynomials of low degree have special names. Can you give the names of polynomials of degree 0, 1, 2 and 3?

4. If two polynomials each of degree 3 are multiplied together, what will be the degree of the resulting polynomial?

Exercise 7.1

1. State which of the following functions are polynomials. Give the degree of those that are.

    (a) $f(t) = t^2 - 3t$ (b) $g(x) = 3x^{-1}$
    (c) $f(x) = 1 - x$ (d) $h(x) = 17$

    (e) $f(t) = 0$ (f) $A(z) = (z-1)(z+2)$
    (g) $r(s) = (s^2 + 1)^2$ (h) $v(t) = \dfrac{3}{t+1}$
    (i) $w(b) = (b + 100)^6$
    (j) $h(v) = v^2 + 2v + \sqrt{v}$

7.2 Rational functions

A **rational function** always takes the form of a quotient of two polynomials; that is, one polynomial divided by another polynomial. The following are examples of rational functions:

$$f(x) = \frac{x^2 + 3}{2x^2 + x + 1}, \qquad g(t) = \frac{t}{t^2 + 7}, \qquad h(z) = \frac{2z^2 + z - 1}{z - 4}$$

In each case the polynomial at the top is called the **numerator** and the polynomial at the bottom is called the **denominator**. The form of the graphs of rational functions can vary widely depending upon the nature of the two polynomials

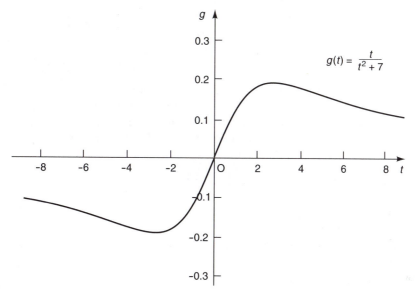

Figure 7.4 Graph of the rational function $g(t) = \dfrac{t}{t^2 + 7}$.

involved. Figure 7.4 shows the graph of the rational function

$$g(t) = \frac{t}{t^2 + 7}$$

Such graphs are often quite difficult and time-consuming to plot. However, using a graphics calculator or graph plotting computer software makes this task much easier.

Two important terms associated with rational functions are the zero and the pole. Any value that when substituted into the numerator makes the numerator zero is called a **zero** of the rational function. Any value that when substituted into the denominator makes the denominator zero is called a **pole** of the rational function. Consider the following example.

Example 7.5

State the poles and zeros of the following rational functions:

(a) $y(x) = \dfrac{x - 3}{2x - 4}$ (b) $f(t) = \dfrac{t + 1}{(t - 1)(t + 5)}$

Solution   (a) $y(x) = (x - 3)/(2x - 4)$ has a zero at $x = 3$. This is because when $x = 3$ is substituted into the numerator, the result is zero. The function has a pole at $x = 2$. This is because when $x = 2$ is substituted into the denominator the result is zero.

(b)   The function $f(t) = \dfrac{(t+1)}{(t-1)(t+5)}$ has a zero at $t=-1$. It has two poles: one at $t=1$ and one at $t=-5$.

Care must be taken when stating the domain of a rational function. As it is impossible to divide any number by zero, we must exclude any poles from the domain of the function. For example, the function $y(x)=(x-3)/(2x-4)$ has a pole at $x=2$. To state its domain precisely, we should write

$$\{x : x \in \mathbb{R}, \ x \neq 2\}$$

thereby excluding the value 2 from the domain.

Example 7.6 Capacitors in series

Consider two capacitors connected in series. The overall capacitance C of the two capacitors is given by

$$\frac{1}{C} = \frac{1}{C_1} + \frac{1}{C_2}$$

where C_1 and C_2 are the capacitances of the two capacitors. Consider the case when $C_2=1$. Then

$$\frac{1}{C} = \frac{1}{C_1} + \frac{1}{1} = \frac{1+C_1}{C_1}$$

Therefore

$$C = \frac{C_1}{1+C_1}$$

Note that as C_1 varies, the overall capacitance C varies, and so C is a function of C_1. We can write

$$C(C_1) = \frac{C_1}{1+C_1}$$

and we see that C is a rational function of C_1. This function is plotted in Figure 7.5. Note that as C_1 increases, the overall capacitance C approaches the value 1.

Self-assessment questions 7.2

1.   Explain what is meant by a rational function.

2.   Can a polynomial function be regarded as a rational function?

3.   In what circumstances is a rational function a polynomial?

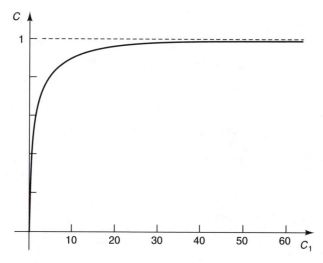

Figure 7.5 Graph of $C(C_1) = C_1/(1+C_1)$.

7.3 Some other functions

7.3.1 The modulus function

Recall that the modulus of a number is the size of that number with no regard paid to its sign. For example, the modulus of -9 is 9. The modulus of $+9$ is also 9. We write

$$|-9| = 9 \quad \text{and} \quad |9| = 9$$

The modulus function is defined by

$$f(x) = |x|$$

That is, when the input to the modulus function is x, the output is the modulus of x. A table of values of this function is given in Table 7.1 and its graph is shown in Figure 7.6.

Table 7.1 Table of values of the modulus function.

x	-4	-3	-2	-1	0	1	2	3	4
$f(x)$	4	3	2	1	0	1	2	3	4

7.3.2 Periodic functions

Any function that has a definite pattern repeated at regular intervals is said to be **periodic**. The interval over which this repetition takes place is known as the

The Thames Barrier: holding back the waves

London has always been at risk from flooding due to the river Thames bursting its banks and overflowing. As the level of the sea has risen in recent times, the probability of flooding has increased. There are records of the Thames bursting its banks dating back to 1236. In 1928 one such occurrence led to the death of 14 people. As well as the possibility of a flood causing death, there is also the potential of major damage occurring to the centre of London, in particular the financial centre, with incalculable consequences. It was these considerations that led to the decision to build the Thames Barrier (see Figure 1). It was completed in 1982, after having taken 8 years to build. The cost was £500 million, a small price to pay given the potential cost of a major flood.

The Thames Barrier is situated at Woolwich, downstream of London. It is 520 m long and has a height of 50 m, which is as high as a five-storey building. The foundations for the structure descend 24 m into the river bed at certain points. It is there-

Figure 1 *The Thames Barrier with the gates open.*

fore a massive structure, and building it was one of the major civil engineering achievements of the 20th century. The barrier consists of 10 gates connected together by piers.

Normally the gates are open so that traffic can flow up and down the river. The barrier can be closed within half an hour if there is a risk of flooding. It is then transformed into a massive dam to hold back the water and protect London. This is achieved by rotating barriers that normally lie beneath the river bed into a position where they can block the flow of water upstream. Giant hydraulic cylinders are used to rotate the barriers. It is also possible to rotate the barriers clear of the water to allow maintenance to be carried out on them. The function of the piers is to hold the barriers in place and house the machinery needed to rotate the barriers as well as maintenance cranes. Navigation lights are located on the piers to tell ships whether or not they can pass through a gate. A green arrow is used to indicate 'yes' and a red cross is used to indicate 'no'. There are service tunnels beneath the Thames to carry power and control signals to the various piers. They also allow people to reach the various piers conveniently. The control room for the whole system is located in one of the banks of the river, housing the mechanism by which the barrier is rotated both to close it and to lift it clear of the water for maintenance.

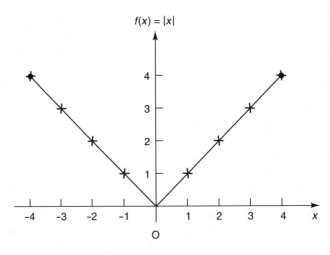

Figure 7.6   The graph of the modulus function.

period of the function. This period is given the symbol T. Note that T is a fixed quantity.

Example 7.7

Plot a graph of the function defined by

$$f(t)=\begin{cases}2t & 0<t\leqslant 1\\3-t & 1<t\leqslant 3\end{cases}\quad\text{with period } T=3$$

Solution   We are given an expression for the function over the interval between $t=0$ and $t=3$. We can use this information to plot the graph on this interval. Then, by repeating the pattern with a period of 3, the whole function can be drawn. This is shown in Figure 7.7. Note that the function value at any value of t is the same as the function value when t is increased by 3; that is, $f(t)=f(t+3)$. This forms a mathematical description of the periodicity of the function.

A function $f(t)$ is periodic with period T if for all values of t, $f(t)=f(t+T)$.

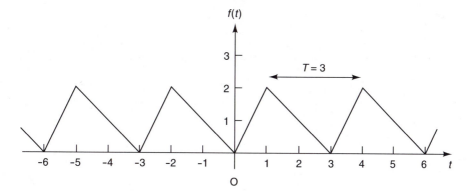

Figure 7.7 Graph of a periodic function with period $T = 3$.

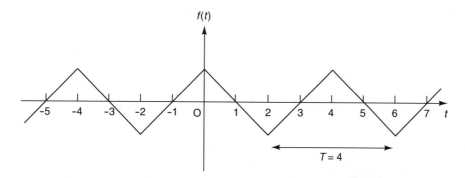

Figure 7.8 Triangular waveform.

Example 7.8 Triangular waveform

Figure 7.8 shows a triangular waveform. Engineers use the term 'waveform' as an alternative to the word 'signal', particularly when the signal is displayed on an oscilloscope. This particular signal repeats every 4 seconds, that is,

$$f(t) = f(t + 4)$$

and so the function is periodic. The period is 4 s, that is, $T = 4$.

Self-assessment questions 7.3

1. Is it true that the modulus function $f(x) = |x|$ must always give positive output values?

2. Explain what is meant by a periodic function.

Exercises 7.3

1. Sketch a graph of the periodic function with period 4 for which $f(t)=3t+2$, for $0 \leqslant t < 4$.

2. A periodic function is defined as

$$f(x)= \begin{cases} x & 0 \leqslant x \leqslant 1 \\ 2-x & 1 \leqslant x < 2 \end{cases}$$

with period $T=2$

   Sketch a graph of this function, showing three periods.

3. A function $g(x)$ is such that $g(x)=g(x+\frac{1}{2})$ for all values of x. State the period of g.

Computer and calculator exercises 7.3

1. (a) Use a graphics calculator or computer software to plot a graph of the function $y(x)=x^4+5x^3-4x^2-20x$ for values of x between -6 and 6.
   (b) Identify the values of x where the curve cuts the horizontal axis.
   (c) Count the number of peaks and troughs on the graph.

2. (a) Plot a graph of $y(x)=x^3-9x^2/2-\frac{5}{2}x+21$ for values of x between -4 and 6.
   (b) Count the number of times the curve cuts the horizontal axis and find the x coordinates at these points.

3. (a) Plot a graph of $y(x)=\dfrac{3x+2}{x^2+x-6}$ for values of x between -8 and $+8$.

   (b) Comment upon the behaviour of the graph when $x=-3$ and when $x=2$.

4. Use a graphics calculator to plot graphs of the linear functions $f(x)=7x-37$ and $g(x)=3-x$ over an appropriate domain. From your graphs, determine the point of intersection of these two lines.

5. (a) Plot a graph of the function $f(t)=-t^2+3t+2$ for values of t between 0 and 6.
   (b) Use your calculator or graph plotting computer package to find the co-ordinates of the highest point on the graph.

Review exercises 7

1 Which of the following are polynomial functions?
   (a) $f(t)=3t^{-1}$ (b) $y(x)=7$
   (c) $g(x)=-x^2-3x-1$ (d) $f(t)=t^{100}$

2 Sketch graphs of each of the following polynomial functions:

   (a) $f(x)=7$ (b) $f(x)=x+7$
   (c) $f(x)=x^2+x+7$

3 Sketch a graph of the rational function $f(x)=1/x^3$, being careful to exclude from the domain $x=0$, where the function is not defined.

8

Further Properties of Functions and Their Graphs

KEY POINTS

This chapter

- explains what is meant by the terms 'many-to-one function' and 'one-to-one function'

- explains the term 'composition' of two functions

- explains what is meant by the inverse of a function

- demonstrates how graphs of functions can be translated vertically and horizontally

- explains what is meant by the terms 'continuous function' and 'discontinuous function'

CONTENTS

8.1 Introduction

In this chapter we develop our knowledge of functions and define a number of important terms. The distinction between many-to-one functions and one-to-one functions is explained. The output from a function can be used as the input to a second function; this gives rise to the so-called composition of functions. To reverse the effect of a function, we then introduce the inverse function and show how this is calculated. We study how functions can be translated both vertically and horizontally around the (x, y) plane. Finally we explain the distinction between a continuous and a discontinuous function and introduce the concept of the limit of a function.

8.2 One-to-one functions and many-to-one functions

If each value in the range of a function corresponds to a single value in the domain, the function is said to be **one-to-one**. If any value in the range of a function corresponds to more than one value in the domain, the function is said to be **many-to-one**. Consider the following examples.

Example 8.1

Figure 8.1 shows a graph of the function $y(x)=x^2-4$. From the graph, we see, for example, that the value $y=5$ corresponds to two values in the domain, namely $x=3$ and $x=-3$. This function is therefore many-to-one. If a horizontal

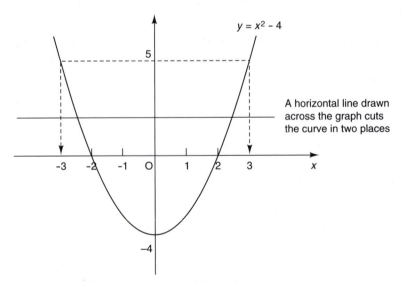

Figure 8.1   Graph of $y(x)=x^2-4$.

line drawn across the graph intersects the curve in two or more places then the function is many-to-one. This is a useful way of testing whether or not a function is many-to-one.

Example 8.2

Figure 8.2 shows a graph of the function $y(x)=x^3$. From the graph, we see that any value chosen in the range of the function corresponds to a single value in the domain. Alternatively, any horizontal line drawn on the graph intersects the curve at only one place. Thus $y(x)=x^3$ is a one-to-one function.

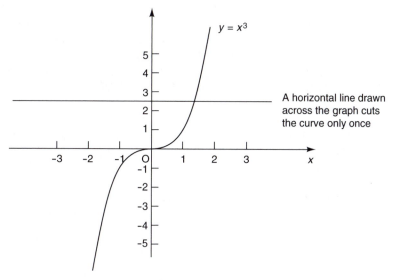

Figure 8.2   Graph of $y(x)=x^3$.

Self-assessment question 8.2

1. Give an example of a one-to-one function and an example of a many-to-one function, explaining the distinction between them.

Exercises 8.2

1. By sketching graphs of the following functions, determine which are one-to-one and which are many-to-one:

    (a) $f(x)=x^2+4$ (b) $f(x)=\dfrac{1}{x}$, with $x\neq0$

2. State which of the following functions, if any, are one-to-one:

    (c) $f(t)=3t-7$ (d) $g(x)=3x^4$

    (a) $f(t)=t+7$ (b) $f(t)=-t^2$
    (c) $f(t)=|2t|$ (d) $f(t)=4t^3$

Computer and calculator exercise 8.2

1.   By plotting the following graphs determine   (a) $y(x)=x^3-6x^2+12x-8$
     which are one-to-one and which are many-   (b) $y(x)=x^3-6x^2+10x-8$
     to-one: (c) $y(x)=x^3+6x^2+12x+8$

8.3 Composition of functions

Sometimes the output from one function is used as input to another. Consider
the two functions $f(x)=x^2$ and $g(x)=x+3$. The first function, f, is described
by the rule 'square the input', while the rule of the second function, g, is 'add
3 to the input'. Consequently, when the output from f is used as input to g,
the result will be x^2+3, as shown in Figure 8.3. We write the result as

$$g(f(x))=x^2+3$$

and call $g(f(x))$ the **composition** of g and f. On the other hand, the composition
$f(g(x))$ is depicted in Figure 8.4 We see that

$$f(g(x))=(x+3)^2$$

which is not the same as $g(f(x))$. It is therefore important when discussing
composition of functions to clearly specify the order in which the composition
is to take place.

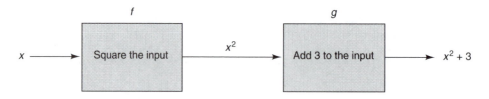

Figure 8.3   The output from f is used as the input to g.

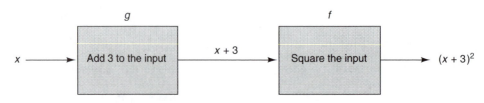

Figure 8.4   The composition $f(g(x))$.

Example 8.3

If $f(t)=3t-2$ and $g(t)=1/t$, find
(a) $f(g(t))$ (b) $g(f(t))$

Solution   (a)   To find $f(g(t))$, we must use the output from $g(t)$ as input to f. So, the input to f is $1/t$. Therefore

$$f(g(t))=f\left(\frac{1}{t}\right)=\frac{3}{t}-2$$

(b)   Similarly,

$$g(f(t))=g(3t-2)=\frac{1}{3t-2}$$

Example 8.4 Two-stage amplifier

The amplifier for a hi-fi system consists of two stages. There is a pre-amplifier defined by the function $f(v)=25v$ and there is a power amplifier defined by the function $g(v)=10v$. Carry out the following.
(a)   Draw a block diagram for each part of the system and write the function rules in each of the blocks.
(b)   Compose the functions f and g in order to obtain a function for the overall system.
(c)   Draw a block diagram for the overall system and describe the rule.

Solution   (a)   The block diagrams for the pre-amplifier and power amplifier are shown in Figure 8.5.

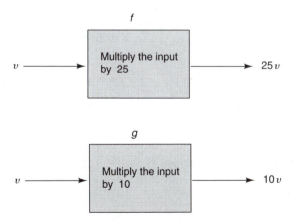

Figure 8.5   Block diagrams for the pre-amplifier and power amplifier.

(b)   The output from the pre-amplifier is $f(v)=25v$, and this is used as input to g. Hence

$$g(f(v))=10(25v)=250v$$

(c)   The overall block diagram is shown in Figure 8.6.

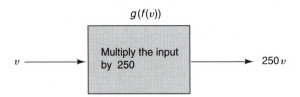

Figure 8.6   Overall block diagram.

Self-assessment questions 8.3

1.   Explain what is meant by the composition $f(g(x))$ of the two functions $f(x)$ and $g(x)$. How does this differ from $g(f(x))$?

2.   The function $f(g(x))$ is the same as $f(x)g(x)$. True or false?

3.   The function f is defined by $f(t)=1/t$. State $f(f(t))$.

Exercises 8.3

1.   Draw a diagram to illustrate the compositions of the two functions $f(x)=x^2+6$ and $g(x)=7-x$.

2.   If $f(x)=x+3$, $g(x)=x^2$ and $h(x)=6x$, find
(a) $f(g(x))$ (b) $h(f(g(x)))$
(c) $g(f(h(x)))$ (d) $f(f(x))$

3.   If $f(x)=\dfrac{x}{2x+1}$ and $g(x)=5x-7$, find $f(g(x))$ and $g(f(x))$.

4.   For which of the following pairs of functions is it true that $f(g(x))=g(f(x))$?
(a) $f(x)=x+7$, $g(x)=14+2x$

(b) $f(t)=9t$, $g(t)=\frac{1}{2}t$

5.   Recall Example 8.4. If the pre-amplifier is defined by the function $f(v)=20v$ and the power amplifier is defined by the function $g(v)=8v$ then compose these two functions in order to obtain a function for the overall system. Draw a block diagram for the system before and after composition takes place.

6.   If $f(x)=\sqrt{x}$, $g(x)=x^2$ and $h(x)=x^3$, state
(a) $f(g(x))$ (b) $f(h(x))$ (c) $f(f(x))$
(d) $g(f(x))$ (e) $g(h(x))$ (f) $g(g(x))$
(g) $h(f(x))$ (h) $h(g(x))$ (i) $h(h(x))$

8.4 Inverse of a function

We have seen that a function can be regarded as taking an input, x, and processing it in some way to produce a single output, $f(x)$, as shown in Figure 8.7. A natural question to ask is whether we can find a function that will reverse the process. In other words, can we find a function that will start with $f(x)$ and process it to produce x? This process is also shown in Figure 8.7. If we can find such a function, it is known as the **inverse function**, and it is given the symbol $f^{-1}(x)$. Do not confuse the '-1' with an index or power. Here the superscript is used purely as part of the symbol for the inverse function. We see that $f^{-1}(f(x)) = x$, as depicted in Figure 8.8.

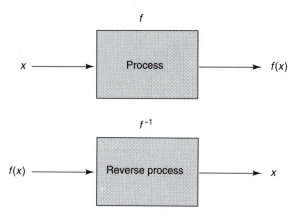

Figure 8.7   A block that takes an input x and produces an output $f(x)$, together with a block that produces an output x when the input is $f(x)$.

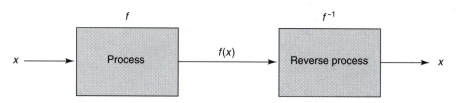

Figure 8.8 f^{-1} reverses the process in f.

Example 8.5

Find the inverse function of the function $f(x) = x + 2$.

Solution

The function $f(x) = x + 2$ is depicted in Figure 8.9. In order to reverse the process of f, the inverse function f^{-1}, when given an input $x + 2$, must produce an output of x; that is,

$$f^{-1}(x+2) = x$$

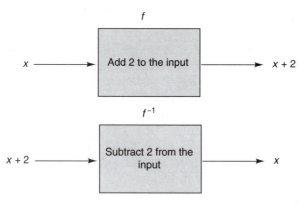

Figure 8.9   The function $f(x) = x + 2$ together with the function $f^{-1}(x) = x - 2$.

If we introduce a new variable z so that $z = x + 2$, we can write this as

$$f^{-1}(z) = x = z - 2$$

So, the rule for f^{-1} is subtract 2 from the input. Writing f^{-1} with x as its argument, we have

$$f^{-1}(x) = x - 2$$

Example 8.6

Find the inverse function of the function $f(t) = 3t - 8$.

Solution   In order to reverse the process of f, the inverse function f^{-1} when given an input $3t - 8$ must produce an output of t; that is,

$$f^{-1}(3t - 8) = t$$

If we introduce a new variable z so that $z = 3t - 8$, we can write this as

$$f^{-1}(z) = t = \frac{z + 8}{3}$$

So, the rule for f^{-1} is add 8 to input, and divide the result by 3. Writing f^{-1} with t as its argument, we have

$$f^{-1}(t) = \frac{t + 8}{3}$$

Not all functions possess an inverse function. This means that it is not always possible to find a function that will reverse the process of a given function. For example, consider the function $f(x) = x^2$. The graph of f is shown in Figure 8.10. We note that when the input to f is 2, the output is 4. When the input is -2, the output is also 4. Therefore an inverse function, if it existed, would

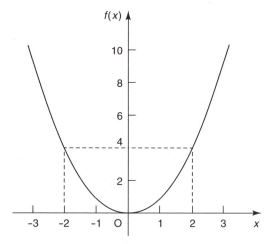

Figure 8.10 Graph of $f(x)=x^2$.

take an input of 4 and produce outputs of 2 and -2. However, this contradicts the definition of a function, which requires that there be only a single output. We say that $f(x)$ does not possess an inverse function. In fact, only a one-to-one function possesses an inverse function.

Self-assessment questions 8.4

1. Under what conditions will a function f possess an inverse function?
2. Is it possible for a function to be its own inverse? If 'yes', give an example of such a function.

Exercises 8.4

1. Find the inverse of each of the following functions:

   (a) $f(x)=7x+2$ (b) $g(x)=\dfrac{3}{x}$

   (c) $h(x)=\dfrac{1}{x-1}$

2. Which of the following functions possess an inverse function? Find the inverse functions of those that do.

   (a) $f(x)=4x-7$ (b) $f(x)=3x^2+2$

   (c) $g(x)=x^3$ (d) $f(t)=\dfrac{1}{2t+4}$

   (e) $f(x)=|x|$ (f) $f(t)=7$

3. Find $f^{-1}(x)$ when

   (a) $f(x)=3-7x$ (b) $f(x)=5+\dfrac{1}{x}$

   (c) $f(x)=x$

4. Find $y^{-1}(x)$ where

   (a) $y(x)=2-x$ (b) $y(x)=3-\frac{1}{2}x$

   (c) $y(x)=\frac{1}{2}(3-x)$ (d) $y(x)=1-\frac{2}{3}x$

   (e) $y(x)=\dfrac{3}{x+1}$ (f) $y(x)=\dfrac{3+x}{x}$

   (g) $y(x)=\dfrac{x}{3+x}$

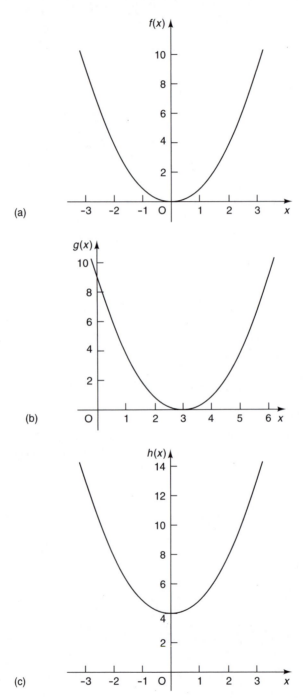

Figure 8.11   (a) Graph of $f(x)=x^2$. (b) Graph of $g(x)=(x-3)^2=f(x-3)$. (c) Graph of $h(x)=x^2+4=f(x)+4$.

8.5 Translation of functions

Figure 8.11(a) shows a graph of the function $f(x)=x^2$. Figure 8.11(b) shows a graph of $g(x)=(x-3)^2$. By inspecting the graphs, we see that $g(x)$ has the same shape as $f(x)$ but is shifted or **translated** 3 units to the right along the x axis. Furthermore, note that $g(x)=(x-3)^2=f(x-3)$.

More generally, the graph of $f(x-a)$ is the same as the graph of $f(x)$ but translated a distance a to the right. Similarly, the graph $f(x+a)$ is the same as the graph of $f(x)$ but translated a distance a to the left.

Figure 8.11(c) shows a graph of $h(x)=x^2+4$. Note that $h(x)=f(x)+4$. We see that the graph of $h(x)$ is the same as that of $f(x)$ but translated a vertical distance 4. More generally, the graph of $f(x)+a$ is the same as the graph of $f(x)$ but translated vertically by a distance a.

Example 8.7

Suppose $f(x)=5x+2$.
(a)   Sketch a graph of $f(x)$.
(b)   Find an expression for $f(x+2)$ and plot its graph.
(c)   Comment upon the two graphs.

Solution   (a)   A table of values has been drawn up, and the graph of $f(x)$ together with the table is shown in Figure 8.12.

(b)   If $f(x)=5x+2$ then $f(x+2)=5(x+2)+2=5x+12$. A table of values of this function and its graph are given in Figure 8.13.

(c)   We see that the graph of $f(x+2)$ is the same as that of $f(x)$ but is translated 2 units to the left.

Self-assessment questions 8.5

1.   Explain the distinction between the graphs of $f(x)$, $f(x+a)$ and $f(x)+a$.
2.   Explain the distinction between the graphs of $f(t+1)$ and $f(t-1)$.

Exercises 8.5

1.   On the same piece of graph paper sketch graphs of $f(x)=x^2+3$, $f(x-1)$, $f(x+1)$, $f(x)+1$ and $f(x)-1$, commenting upon your results.

2.   Sketch graphs of $f(t)$, $f(t+2)$ and $f(t-2)$ when $f(t)=1-5t$.

3.   (a)   Sketch $g(t)=\frac{1}{2}t^2+1$ for $-3\leqslant t\leqslant 3$.
       (b)   On the same graph sketch $g(t+1)$, $g(t-1)$, $g(t)+1$ and $g(t)-1$.

4.   Sketch a graph of $f(x)=x^2$ for $-4\leqslant x\leqslant 4$. On the same graph sketch $f(2x)$ and $f(\frac{1}{2}x)$.

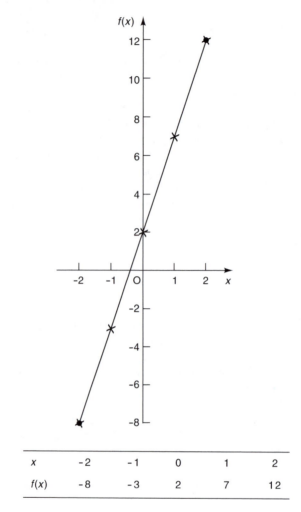

x	- 2	- 1	0	1	2
f(x)	- 8	- 3	2	7	1 2

Figure 8.12   Graph of $f(x)=5x+2$.

Computer and calculator exercises 8.5

1.   Use a graphics calculator to plot graphs of
the following functions:
(a) $f(x)$ (b) $f(x-3)$ (c) $f(x+3)$
where $f(x)=x^3+6x^2+11x+6$. In each case
identify the values of x where the curve cuts
the x axis.

2.   Plot a graph of $y(x)=x^2$. Use your calculator
to investigate the following graphs for
various values of k:
(a) $y(x)+k$ (b) $y(x+k)$ (c) $y(kx)$
(d) $y\left(\dfrac{x}{k}\right)$ (e) $ky(x)$

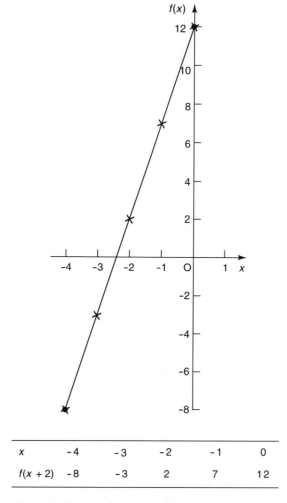

x	-4	-3	-2	-1	0
$f(x+2)$	-8	-3	2	7	12

Figure 8.13   Graph of $f(x+2)=5x+12$.

8.6 Continuous and discontinuous functions and their limits

Consider the graph of the function shown in Figure 8.14(a). Clearly, the curve can be traced out from the left to the right without moving pen from paper. Such a function is said to be **continuous**. On the other hand, if we try to trace out the curve in Figure 8.14(b), the presence of the jumps in the graph causes the pen to be lifted and moved. Such a function is said to be **discontinuous**, and the jumps are known as **discontinuities**. A discontinuity corresponds to our natural understanding of a break in the graph.

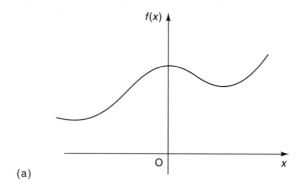

(a)

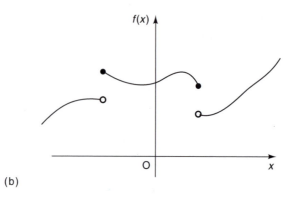

(b)

Figure 8.14   (a) A continuous function. (b) A discontinuous function.

To describe continuous and discontinuous functions mathematically requires an understanding of the **limit** of a function. We shall introduce the idea of a limit in the following example.

Example 8.8

Consider the graph of the line $y = f(x) = 6x - 5$ shown in Figure 8.15.

(a)   As the value of x gets nearer and nearer to 3, to what value does y approach?

(b)   As the value of x gets nearer and nearer to -2, to what value does y approach?

Solution   (a)   From the graph, we see that as x gets nearer and nearer to 3, the value of y gets nearer and nearer to 13. We write this concisely as

$$y \rightarrow 13 \quad \text{as} \quad x \rightarrow 3$$

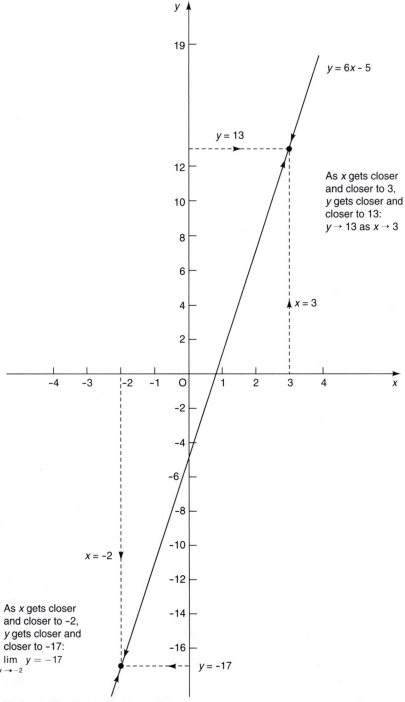

Figure 8.15   Graph for Example 8.8.

An alternative notation is

$$\lim_{x \to 3} y = 13$$

'lim' is an abbreviation of 'limit', and this last statement is read as 'the limit of y as x tends to 3 is 13'.

(b)   From the graph, we see that as x gets nearer and nearer to -2, the value of y gets nearer and nearer to -17. We write this concisely as

$$y \to -17 \quad \text{as} \quad x \to -2$$

or alternatively as

$$\lim_{x \to -2} y = -17$$

In the previous example note that it doesn't matter whether our chosen value of x is approached from the left or the right. This is not true for all functions. Consider the next example.

Example 8.9

Figure 8.16 shows a graph of the function

$$y = f(x) = \begin{cases} 2x+1 & x < 3 \\ 5 & x = 3 \\ 6 & x > 3 \end{cases}$$

This function is defined in rather a strange way. When x is less than 3, the graph takes the form of a straight line. When x is greater than 3, the graph takes the form of a horizontal straight line. At the point where $x = 3$, the value of the function is defined to be 5, as indicated by the ● on the graph.

(a)   Find the limit of the function as x tends to 0.
(b)   Find the limit of the function as x tends to 3.

Solution   (a)   From the graph, we see that as x approaches 0, the value of y approaches 1. This is true whether 0 is approached from the left or the right. We write

$$y \to 1 \quad \text{as} \quad x \to 0, \quad \text{or alternatively} \quad \lim_{x \to 0} y = 1$$

(b)   Study of the graph shows that when x is close to 3, two different sorts of behaviour are occurring. If x is approaching 3 from below, written $x \to 3^-$, that is from the left, the value of y is getting nearer and nearer to

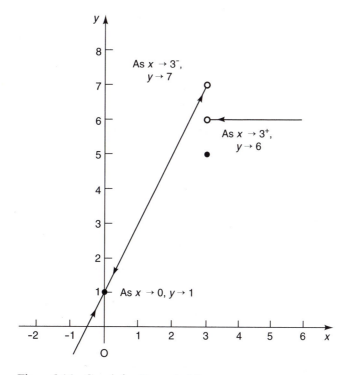

Figure 8.16 Graph for Example 8.9.

7. We write this as

$$\lim_{x \to 3^-} y = 7$$

and say that the **left-hand limit** of y is 7. If x approaches 3 from above, written $x \to 3^+$, that is from the right, the value of y is 6. We write this as

$$\lim_{x \to 3^+} y = 6$$

and say that the **right-hand limit** of y is 6.

In this example the right-hand and left-hand limits at the point where $x = 3$ are not equal. In such a case it does not make sense to talk about 'the limit' of the function. Only when the right- and left-hand limits are equal can we say 'the limit exists at that point'. So $\lim_{x \to 3} y$ does not exist, despite the existence of both left- and right-hand limits. Note that when $x = 3$, the function is defined to have a value of 5, and so the right- and left-hand limits at this particular point are quite distinct from the actual value of the function at this point.

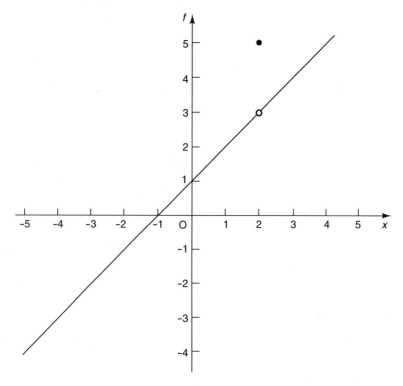

Figure 8.17   There is a discontinuity at $x=2$.

Example 8.10

Figure 8.17 shows a graph of the function

$$f(x)=\begin{cases} x+1 & x\neq 2 \\ 5 & x=2 \end{cases}$$

It is clear that there is a discontinuity in the graph when $x=2$. Calculate the following:

(a) $\lim\limits_{x\to 2^-} f(x)$ (b) $\lim\limits_{x\to 2^+} f(x)$ (c) $\lim\limits_{x\to 3^-} f(x)$ (d) $\lim\limits_{x\to 3^+} f(x)$ (e) $f(2)$

(f) $f(3)$

Comment upon these results.

Solution   (a)   As x approaches 2 from the left-hand side, we see that the value of f approaches 3; that is,

$$\lim\limits_{x\to 2^-} f(x)=3$$

(b)   As x approaches 2 from the right-hand side, we see that the value of f approaches 3; that is,

$$\lim_{x \to 2^+} f(x) = 3$$

We see that the left- and right-hand limits are now equal, and so we can say that a limit exists at the point $x = 2$; that is,

$$\lim_{x \to 2} f(x) = 3$$

(c)   As x approaches 3 from the left-hand side, we see that the value of f approaches 4; that is,

$$\lim_{x \to 3^-} f(x) = 4$$

(d)   As x approaches 3 from the right-hand side, we see that again the value of f approaches 4; that is,

$$\lim_{x \to 3^+} f(x) = 4$$

So, at the point where $x = 3$, the left- and right-hand limits are the same, and so a limit exists at this point. We can write

$$\lim_{x \to 3} f(x) = 4$$

(e)   From the graph, we see that $f(2) = 5$.
(f)   From the graph, we see that $f(3) = 4$.

Note that $\lim_{x \to 3} f(x) = f(3)$, so that at a point where the function is continuous the value of the limit of the function is the same as the value of the function. Note also that $\lim_{x \to 2} f(x) \neq f(2)$, so that at a discontinuity the function and limit values are not equal.

We can now give a mathematical description of continuous and discontinuous functions.

> A function is continuous at the point $x = a$ if
>
> $$\lim_{x \to a} f(x) = f(a)$$
>
> that is, the limit value matches the function value at a point of continuity.

Informally, we can say that the graph of a continuous function is in one piece.

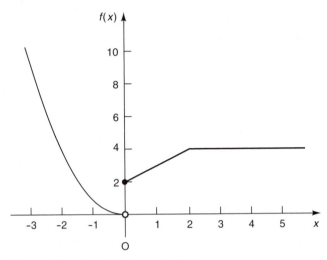

Figure 8.18 Graph for Example 8.11.

Example 8.11

Figure 8.18 shows a graph of the function

$$f(x) = \begin{cases} x^2 & x<0 \\ x+2 & 0 \leqslant x \leqslant 2 \\ 4 & x>2 \end{cases}$$

Find

(a) $\lim\limits_{x \to 0^-} f(x)$ (b) $\lim\limits_{x \to 0^+} f(x)$ (c) $\lim\limits_{x \to 0} f(x)$ (d) $\lim\limits_{x \to 2^-} f(x)$

(e) $\lim\limits_{x \to 2^+} f(x)$ (f) $\lim\limits_{x \to 2} f(x)$ (g) $f(0)$ (h) $f(2)$

Solution

(a)   From the graph, we see that $\lim\limits_{x \to 0^-} f(x)=0$.

(b)   Similarly, $\lim\limits_{x \to 0^+} f(x)=2$.

(c)   Clearly, from (a) and (b), the left- and right-hand limits are not equal, and so a limit does not exist at the point $x=0$.

(d)   From the graph we see that $\lim\limits_{x \to 2^-} f(x)=4$.

(e)   Similarly, $\lim\limits_{x \to 2^+} f(x)=4$.

(f)   In this case the left- and right-hand limits as x approaches 2 are equal, and so

$$\lim\limits_{x \to 2} f(x)=4$$

(g) $f(0)=2.$
(h) $f(2)=4.$

We note that because no limit exists at $x=0$, the function cannot be continuous there. At $x=2$ we note that $\lim_{x\to 2} f(x)=4=f(2)$, and so the function value and limit are the same. Therefore the function is continuous at $x=2$.

Example 8.12 Square waveform

Figure 8.19 shows a square waveform. The waveform is periodic with a period $T=4$; that is, $f(t)=f(t+4)$. Note that the waveform is discontinuous with discontinuities at $t=0, \pm 2, \pm 4 \dots$.

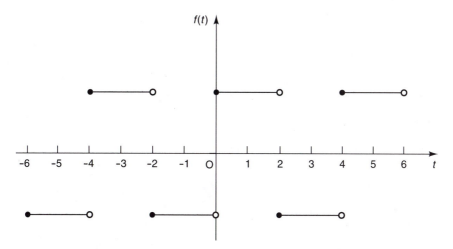

Figure 8.19 Square waveform.

Self-assessment questions 8.6

1.   Is it true that a periodic function is always continuous? Give reasons for your choice of answer.

2.   Is it true that the function $f(x)=|x|$ is discontinuous at $x=0$?

Exercises 8.6

1.   Sketch a graph of the function

$$f(x)=\begin{cases} -2 & x\leqslant 1 \\ 3 & x>1 \end{cases}$$

Locate the position of any discontinuities.

2.   Sketch a graph of the function $g(x)=\dfrac{1}{x-3}$ for $0\leqslant x\leqslant 6$. Locate the position of any discontinuities.

3. Locate the position of any discontinuities of the function

$$f(t) = \begin{cases} 1+t & t<0 \\ 2 & 0 \leqslant t \leqslant 2 \\ 3t-4 & t>2 \end{cases}$$

4. The function $g(t)$ is defined by

$$g(t) = \begin{cases} t^2+3 & t<1 \\ -t+7 & t \geqslant 1 \end{cases}$$

Find, if possible,

(a) $\lim_{t \to 1^-} g(t)$ (b) $\lim_{t \to 1^+} g(t)$

(c) $\lim_{t \to 1} g(t)$ (d) $\lim_{t \to 0^-} g(t)$

(e) $\lim_{t \to 0^+} g(t)$ (f) $\lim_{t \to 0} g(t)$

5. The function $f(t)$ is defined by

$$f(t) = \begin{cases} 2t+k & 0 \leqslant t \leqslant 3 \\ t^2 & t>3 \end{cases}$$

where k is a constant. If f is a continuous function, find the value of k.

6. The function $h(v)$ is defined by

$$h(v) = \begin{cases} 0 & v<0 \\ v & 0 \leqslant v<1 \\ \frac{1}{2}v^2 & v \geqslant 1 \end{cases}$$

(a) Sketch a graph of $h(v)$.

(b) Find (i) $\lim_{v \to 0.5^+} h(v)$ (ii) $\lim_{v \to 0.5^-} h(v)$

(iii) $\lim_{v \to 1^-} h(v)$ (iv) $\lim_{v \to 1^+} h(v)$

Review exercises 8

1 State which of the following functions are many-to-one and which are one-to-one:

(a) $y(x)=x^4$ (b) $y(t)=\dfrac{1}{t^4}$

(c) $v(t)=t^3+6$ (d) $h(t)=(t+6)^3$
(e) $A(b)=(b+1)^2$

2 Given $x(t)=\dfrac{2}{t}$, $y(t)=\dfrac{t+1}{3}$ and $z(t)=1-t$,

find

(a) $x^{-1}(t)$ (b) $y^{-1}(t)$ (c) $z^{-1}(t)$
(d) $x(y(t))$ (e) $y(z(t))$ (f) $x(y(z(t)))$

3 The function $f(t)$ is defined by

$$f(t) = \begin{cases} 4 & t<1 \\ 2t+7 & 1 \leqslant t \leqslant 3 \\ 5 & t>3 \end{cases}$$

Sketch a graph of $f(t)$ and state the position of any discontinuities.

4 (a) Sketch $g(x)=x^3$ for $-2 \leqslant x \leqslant 2$.

(b) On the same axes sketch $g(x)+1$, $g(x+1)$, $g(x-1)$ and $g(x)-1$.

5 The functions $g(x)$ and $h(x)$ are given by $g(x)=\frac{1}{2}x+1$ and $h(x)=4x-2$. The function $q(x)$ is given by $g(h(x))$. State
(a) $g^{-1}(x)$ (b) $h^{-1}(x)$ (c) $q(x)$
(d) $q^{-1}(x)$ (e) $h^{-1}(g^{-1}(x))$

6 The function $g(x)$ is defined by

$$g(x) = \begin{cases} 2x+1 & 0 \leqslant x<2 \\ 7-x & 2 \leqslant x \leqslant 4 \\ x & x>4 \end{cases}$$

State
(a) $g(1)$ (b) $g(2)$ (c) $g(4)$
(d) $\lim_{x \to 1^-} g(x)$ (e) $\lim_{x \to 1^+} g(x)$
(f) $\lim_{x \to 2^-} g(x)$ (g) $\lim_{x \to 2^+} g(x)$
(h) $\lim_{x \to 4^-} g(x)$ (i) $\lim_{x \to 4^+} g(x)$
State any points of discontinuity.

9

Solving Equations and Inequalities

KEY POINTS

This chapter

- explains what is meant by an 'equation' and a 'solution' of an equation

- gives the rules that must be obeyed when trying to solve an equation

- explains what is meant by a 'linear equation' and a 'quadratic equation' and shows how they can be solved

- explains what is meant by a pair of 'simultaneous equations' and shows how to solve them

- explains how expressions involving inequalities, $>$, $\geqslant$, $<$ and $\leqslant$, can be solved algebraically and graphically

CONTENTS

9.1 Introduction

Equations are used to express relationships between physical quantities. Frequently an engineer needs to be able to determine the value of a particular quantity from knowledge of several others. This is where the ability to solve an equation is required. In this chapter we shall describe several sorts of equations and how they can be solved. Sometimes expressions are related using one of the inequalities $<$, $>$, $\leqslant$ or $\geqslant$. The inequality is solved by finding all possible values of the unknown quantity for which the inequality is true.

9.2 Equations

An **equation** is a mathematical statement that two quantities are equal. An equation always contains an **unknown quantity** that we wish to find. For example, in the equation

$$x - 1 = 7$$

the unknown quantity is represented by the symbol x. To **solve** this equation means to find all the values of x that can be substituted into the equation so that the left-hand side equals the right-hand side. Any such value is known as a **solution**, or **root**, of the equation. It is clear that the value $x = 8$ is a solution of this equation.

Note that some equations you will meet have no solutions at all.

In order to solve an equation, we must try to make the unknown quantity the **subject** of the equation. This means that we must attempt to obtain the unknown quantity on its own on the left-hand side. To do this, we may apply the same five rules used for transposing formulae that were given in Section 5.3.1. These are

---

- add the same quantity to both sides
- subtract the same quantity from both sides
- multiply both sides by the same quantity
- divide both sides by the same quantity
- take functions of both sides; for example square both sides

---

A useful summary of these rules is 'whatever we do to one side of an equation, we must also do to the other'. We now consider the solution of several different types of equation.

9.3 Solving linear equations

The simplest sort of equations to solve are those in which the unknown quantity, x, appears only to the first power, that is as x, and not as x^2, x^3, $x^{\frac{1}{2}}$ etc. Such equations are called **linear equations**. The standard form of a linear equation is

$$ax + b = 0$$

where a and b are given numbers and x is the unknown whose value we wish to find. For example,

$$3x + 7 = 0, \quad 4x - 7 = 0, \quad -\tfrac{1}{2}x + 13 = 0$$

are all linear equations. Linear equations often appear in a nonstandard form. For example,

$$3x - 7 = 2, \quad -5x + 7 = 21x - 3, \quad 9 + 2x = 1, \quad 3x + 7 = \tfrac{2}{3}$$

are all linear equations. Where necessary, they can all be rewritten in the form $ax + b = 0$.

On the other hand, the equation $x^2 = 17$ is not a linear equation because the unknown quantity occurs to the power 2. Such an equation is said to be **nonlinear**.

Consider the following examples.

Example 9.1

Solve the equation $x + 7 = 13$.

Solution

Starting with $x + 7 = 13$, we can subtract 7 from both sides to give

$$x + 7 - 7 = 13 - 7$$

so that

$$x = 6$$

and the solution is therefore $x = 6$.

Note that it is easy to check the answer by substituting $x = 6$ into the original equation to see if both sides are the same. That is, substituting $x = 6$, we find

on the left-hand side: $x + 7 = 6 + 7 = 13$

on the right-hand side: 13

and so when $x = 6$ both sides are the same.

Example 9.2

Solve the equation $3(x-7)=2(x-4)$.

Solution   First of all, the brackets are removed:

$$3x-21=2x-8$$

Subtracting $2x$ from both sides gives

$$3x-2x-21=2x-2x-8$$

$$x-21=-8$$

Finally, adding 21 to both sides gives

$$x=13$$

Again, the answer can be checked by substitution. That is,

on the left-hand side: $3(x-7)=3(13-7)=3(6)=18$

on the right-hand side: $2(x-4)=2(13-4)=2(9)=18$

and so when $x=13$ both sides are the same.

Example 9.3

Solve the equation $3x-2=7x+15$.

Solution   We must attempt to make x the subject of this equation. We can subtract $7x$ from both sides to give

$$3x-7x-2=7x-7x+15$$

$$-4x-2=15$$

Then, adding 2 to both sides gives

$$-4x-2+2=15+2$$

$$-4x=17$$

Finally, dividing both sides by -4 gives

$$x=\frac{17}{-4}=-\frac{17}{4}$$

Example 9.4

Solve the equation

$$\frac{3}{x-2}=\frac{1}{2x-3}$$

Solution   When algebraic fractions are involved, it is often useful to try to remove such fractions. For example, multiplying both sides by $2x - 3$ gives

$$\frac{3}{x-2} \times (2x-3) = \frac{1}{2x-3} \times (2x-3)$$

Cancelling the factors on the right produces

$$\frac{3(2x-3)}{x-2} = 1$$

Then, multiplying both sides by $x - 2$ gives

$$\frac{3(2x-3)}{x-2} \times (x-2) = 1 \times (x-2)$$

$$3(2x-3) = x-2$$

$$6x - 9 = x - 2$$

$$5x = 7$$

so that

$$x = \frac{7}{5}$$

Example 9.5 Linear expansion of a material

Most solid materials expand when their temperature is increased. If an object of length l_0 undergoes a change in temperature δT then the new length of the object, l, is given by the formula

$$l = l_0(1 + \alpha \, \delta T)$$

where α is a constant called the **coefficient of linear expansion**.

Suppose a steel girder has a length of 50.000 m when it is measured at a temperature of 20 °C. The temperature rises and the girder expands to a length of 50.005 m. Calculate the new temperature of the girder, given that the coefficient of linear expansion of steel is $1.5 \times 10^{-5} \, °\text{C}^{-1}$.

Solution   The original length of the girder is 50.000 m, and so $l_0 = 50$ m. After the temperature has risen, the length becomes 50.005 m, and so $l = 50.005$ m. Taking $\alpha = 1.5 \times 10^{-5}$, we have

$$50.005 = 50(1 + 1.5 \times 10^{-5} \, \delta T)$$

$$= 50 + 50 \times 1.5 \times 10^{-5} \, \delta T$$

Subtracting 50 from both sides gives

$$0.005 = 50 \times 1.5 \times 10^{-5} \, \delta T$$

Dividing both sides by $50 \times 1.5 \times 10^{-5}$ gives

$$\delta T = \frac{0.005}{50 \times 1.5 \times 10^{-5}} = 6.67\,°C$$

We conclude that the temperature rise is $6.67\,°C$, and so the new temperature equals $20 + 6.67 = 26.67\,°C$.

Note from the previous examples that a linear equation possesses only one root.

Self-assessment questions 9.3

1. List the rules that can be applied to solve an equation.

2. Explain what is meant by the 'root' of an equation.

Exercises 9.3

1. Solve the equations
   (a) $2x - 1 = 0$ (b) $x + 2 = 0$ (c) $5x = 0$
   (d) $1 - 10x = 0$ (e) $1 - \dfrac{ax}{b} = 0$, where a
   and b are constants   (f) $5x - 7 = 0$
   (g) $2t + 17 = 0$

2. Solve the equations
   (a) $x + 3 = 7$ (b) $x - 4 = 3$ (c) $s - 2 = 5$
   (d) $t - 8 = 21$ (e) $3x - 8 = 0$ (f) $8 = 3x$
   (g) $-3x = 8x + 2$

3. Solve the equations
   (a) $3x + \frac{3}{4} = 2x - 1$
   (b) $2t - 7 = \dfrac{t}{2} + 3$
   (c) $\dfrac{3}{t} = 17$
   (d) $\dfrac{4}{x} = \dfrac{3}{x} + 2$
   (e) $\dfrac{x}{3} = \dfrac{2x - 5}{2}$
   (f) $2(3x - 5) = 18$
   (g) $3(5 - x) = 2(x + 3)$
   (h) $3(p - 2) + 4(9 - p) = 26$

4. Refer to Example 9.5. Calculate the temperature change required to increase the length of a 25.000 m steel girder to 25.008 m. The coefficient of linear expansion of steel is $1.5 \times 10^{-5}\,°C^{-1}$.

5. Refer to Example 9.5. Calculate the temperature change required to decrease the length of a pure iron bar from 5.000 m to 4.992 m. The coefficient of linear expansion for pure iron is $1.2 \times 10^{-5}\,°C^{-1}$.

6. Find the roots of the following equations:
   (a) $5t + 3 = t - 2$ (b) $3z + 17 = 2z - 8$
   (c) $9x - 5 = 19x$

7. Solve
   (a) $\dfrac{3}{x} = \dfrac{2}{x + 1}$ (b) $\dfrac{4}{x + 2} = \dfrac{7}{2x + 1}$.
   (c) $\dfrac{9}{3x - 1} - \dfrac{2}{x + 5} = 0$
   (d) $\dfrac{5}{3x - 1} + \dfrac{6}{x + 1} = 0$
   (e) $(x + 1)(x + 4) = (x - 2)(x + 3)$
   (f) $(x - 1)(2x + 3) = (2x + 1)(x + 2)$
   (g) $\dfrac{x - 1}{x + 4} = \dfrac{x + 3}{x - 2}$

9.4 Solving quadratic equations

All **quadratic equations** can be written in the standard form

$$ax^2 + bx + c = 0$$

where a, b and c are numbers, and x is the unknown whose value(s) we wish to find. For example,

$$3x^2 + 7x - 2 = 0, \quad x^2 - x - 3 = 0, \quad 0.5x^2 + 8x - 8 = 0$$

are all quadratic equations. To ensure the presence of the x^2 term, the number a cannot be zero. However, b and c may be zero, so that

$$3x^2 - 7 = 0, \quad 8x^2 - 7x = 0, \quad 4x^2 = 0$$

are all quadratic equations.

Frequently quadratic equations occur in a nonstandard form. For example,

$$3x^2 + 2x = 7, \quad -x^2 - 3x = 14, \quad x^2 = 8 - 8x$$

are all quadratic equations. Where necessary, they can be rewritten in standard form as

$$3x^2 + 2x - 7 = 0, \quad -x^2 - 3x - 14 = 0, \quad x^2 + 8x - 8 = 0$$

9.4.1 Solution by factorization

It may be possible to solve a quadratic equation by factorization using the method described for factorizing quadratic expressions in Section 4.4.3, although you should be aware that not all quadratic equations can be factorized. Consider the following example.

Example 9.6

Solve the equation $x^2 + 3x + 2 = 0$ by factorization.

Solution   The left-hand side is first factorized to produce

$$x^2 + 3x + 2 = (x + 1)(x + 2) = 0$$

Now, when the product of two quantities equals zero, at least one of the two must equal zero. In this case either $(x + 1)$ is zero or $(x + 2)$ is zero. It follows that

$$x + 1 = 0, \quad \text{giving} \quad x = -1$$

or

$$x + 2 = 0, \quad \text{giving} \quad x = -2$$

The quadratic equation has two solutions: $x = -1$ and $x = -2$. You should always check that your answers are correct by substituting these values back into the original equation.

Example 9.7

Solve the quadratic equation $6x^2 + x - 2 = 0$.

Solution   The left-hand side can be factorized to produce

$$6x^2 + x - 2 = (2x - 1)(3x + 2) = 0$$

It follows that either

$$2x - 1 = 0, \quad \text{giving} \quad x = \tfrac{1}{2}$$

or

$$3x + 2 = 0, \quad \text{giving} \quad x = -\tfrac{2}{3}$$

Example 9.8

Solve the quadratic equation $x^2 + 8x = 0$.

Solution   Factorizing the left-hand side, we find

$$x^2 + 8x = x(x + 8) = 0$$

It follows that either $x = 0$ or $x = -8$.

Example 9.9

Solving the quadratic equation $x^2 - 9 = 0$.

Solution   Factorizing the left-hand side, we find

$$x^2 - 9 = (x + 3)(x - 3) = 0$$

It follows that either $x = -3$ or $x = 3$.

Example 9.10

Solve the quadratic equation $x^2 + 4x + 4 = 0$.

Solution   Factorization yields

$$x^2 + 4x + 4 = (x + 2)(x + 2) = 0$$

In this example the factors are identical. We have

$$x+2=0$$

$$x=-2$$

Since each factor produces a root $x=-2$, we say that the equation has a **repeated root**.

Example 9.11 *Motion of a projectile*

Consider Figure 9.1, which shows a projectile fired with a speed V_0 from a gun at the origin, which is pointed at an angle of $45°$ to the horizontal. The relationship between the vertical height of the projectile, y, and the horizontal distance travelled by the projectile, x, is given by

$$y=x-\frac{gx^2}{V_0^2}$$

where g is a constant known as the acceleration due to gravity. Find an expression for the horizontal distance travelled by the projectile when it reaches the ground.

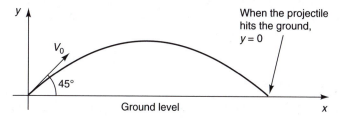

Figure 9.1 A projectile fired at an angle of $45°$.

Solution

We are given

$$y=x-\frac{gx^2}{V_0^2}$$

Note that when the projectile hits the ground, $y=0$. So we can write

$$0=x-\frac{gx^2}{V_0^2}$$

We see that this is a quadratic equation. It may be factorized and solved as follows:

$$0=x-\frac{gx^2}{V_0^2}=x\left(1-\frac{gx}{V_0^2}\right)$$

so that either $x=0$ or

$$1 - \frac{gx}{V_0^2} = 0$$

That is,

$$\frac{gx}{V_0^2} = 1$$

and so

$$x = \frac{V_0^2}{g}$$

There are two solutions. The solution $x=0$ corresponds to the position at which the projectile is fired. The solution $x=V_0^2/g$ corresponds to the position at which the projectile hits the ground after travelling through the air. We conclude that the horizontal distance travelled by the projectile is V_0^2/g.

When you find that a quadratic equation is difficult or impossible to factorize, you can always make use of a formula that determines the solutions. This is the topic of the next section.

9.4.2 Solution by formula

Recall that the standard form of a quadratic equation is $ax^2+bx+c=0$, where a, b and c are numbers. Usually quadratic equations have two solutions, and it can be shown that these solutions are given by the following formula.

> If $ax^2+bx+c=0$ then
>
> $$x = \frac{-b \pm \sqrt{b^2 - 4ac}}{2a}$$

The first solution is obtained by taking the positive square root, and the second by taking the negative square root.

Note that if b^2-4ac is a positive number, we can take its square root and this formula will produce two solutions of the quadratic equation. We say the equation has **distinct real roots**.

If $b^2-4ac=0$, there will be a single root, known as a **repeated root**. The value of the repeated root is $-b/2a$. An example was given in Example 9.10.

Finally, if $b^2 - 4ac$ is a negative number, we say that the equation possesses **complex roots**. These require special treatment and are described in Chapter 28.

The equation $ax^2 + bx + c = 0$ has

- distinct real roots if $b^2 > 4ac$

- a repeated root if $b^2 = 4ac$

- complex roots if $b^2 < 4ac$

Example 9.12

Solve the quadratic equation $4x^2 - 8x - 21 = 0$ using the formula.

Solution   In this example $a = 4$, $b = -8$ and $c = -21$. Using the formula, we find

$$x = \frac{-(-8) \pm \sqrt{(-8)^2 - 4(4)(-21)}}{(2)(4)}$$

$$= \frac{8 \pm \sqrt{64 + 336}}{8}$$

$$= \frac{8 \pm \sqrt{400}}{8}$$

$$= \frac{8 \pm 20}{8}$$

$$= \frac{28}{8} \text{ and } -\frac{12}{8}$$

The two solutions are therefore $x = \frac{7}{2}$ and $x = -\frac{3}{2}$.

Example 9.13

Solve the equation $2x^2 - 4x - 7 = 0$ using the formula.

Solution   Using the formula with $a = 2$, $b = -4$ and $c = -7$, we find

$$x = \frac{-(-4) \pm \sqrt{(-4)^2 - 4(2)(-7)}}{(2)(2)}$$

$$= \frac{4 \pm \sqrt{16 + 56}}{4}$$

$$= \frac{4 \pm \sqrt{72}}{4}$$

$$= \frac{4 \pm 8.485}{4}$$

$$= \frac{12.485}{4} \text{ and } \frac{-4.485}{4}$$

$$= 3.121 \text{ and } -1.121$$

The two solutions are therefore $x = 3.121$ and $x = -1.121$.

9.4.3 Completing the square

A third technique for solving quadratic equations is known as **completing the square**. First consider the expression $(x+a)^2$. We can write this as

$$(x+a)^2 = (x+a)(x+a) = x^2 + 2ax + a^2$$

and so, by subtracting a^2 from both sides we find

$$(x+a)^2 - a^2 = x^2 + 2ax$$

For example, by comparing $x^2 + 6x$ with $x^2 + 2ax$, we see that by choosing $a = 3$ we can write

$$x^2 + 6x = (x+3)^2 - 9$$

Similarly, comparing $x^2 - 4x$ with $x^2 + 2ax$ and choosing $a = -2$, we can write

$$x^2 - 4x = (x-2)^2 - 4$$

Now consider a quadratic expression, say, for example, $x^2 + 6x + 10$. This may be written as

$$x^2 + 6x + 10 = (x+3)^2 - 9 + 10$$

$$= (x+3)^2 + 1$$

Similarly, $x^2 - 4x - 5$ may be written as

$$x^2 - 4x - 5 = (x-2)^2 - 4 - 5$$

$$= (x-2)^2 - 9$$

Writing quadratic expressions in this form is known as 'completing the square'. Its use in solving quadratic equations is illustrated in the following examples.

Example 9.14

Solve the equation $x^2 + 8x + 12 = 0$ by completing the square.

Solution   First, consider only the first two terms, $x^2 + 8x$. Note that we can write these as $(x+4)^2 - 16$. This enables us to rewrite the equation as

$$x^2 + 8x + 12 = (x+4)^2 - 16 + 12 = 0$$

from which

$$(x+4)^2 = 4$$

Taking the square root of both sides of the equation gives

$$x + 4 = \pm 2$$

and so $x = -2$ and $x = -6$.

Example 9.15

Solve the equation $x^2 + 6x + 2 = 0$ by completing the square.

Solution   First note that the terms $x^2 + 6x$ can be written $(x+3)^2 - 9$, so that we can rewrite the equation as

$$x^2 + 6x + 2 = (x+3)^2 - 9 + 2 = 0$$

that is

$$(x+3)^2 = 7$$

so that

$$x + 3 = \pm \sqrt{7}$$

from which $x = -3 + \sqrt{7}$ and $x = -3 - \sqrt{7}$. Note that it is usually quite acceptable to leave your solutions in this form.

Self-assessment questions 9.4

1. What is the distinction between a quadratic equation and a linear equation?
2. Describe all the techniques you have met to solve quadratic equations.
3. Under what conditions would you expect to find a single solution of a quadratic equation?

Exercises 9.4

1. Solve the following equations by factorization:
   (a) $x^2 + 16x = 0$
   (b) $4x^2 + 16x = 0$
   (c) $t^2 - 2t = 0$
   (d) $x^2 - 2x - 3 = 0$
   (e) $x^2 - x - 6 = 0$
   (f) $t^2 + t - 72 = 0$
   (g) $x^2 + 9x = -20$
   (h) $x^2 - x - 42 = 0$
   (i) $x^2 - 4x - 21 = 0$
   (j) $x^2 + 8x + 16 = 0$
   (k) $x^2 - 9x + 18 = 0$
   (l) $x^2 + 11x + 10 = 0$
   (m) $x^2 - 12x + 32 = 0$
   (n) $x^2 + 3x - 18 = 0$
   (o) $9x^2 + 3x - 2 = 0$
   (p) $4x^2 - 11x - 3 = 0$
   (q) $x^2 - 121 = 0$
   (r) $3x^2 + 11x - 4 = 0$
   (s) $8x^2 - 6x = 9$
   (t) $12x^2 = -3 - 20x$
   (u) $10x = 8 - 12x^2$
   (v) $8x^2 + 22x = -15$

2. Solve the following equations by using the formula:
   (a) $3x^2 - 7x - 2 = 0$
   (b) $t^2 + 3t - 8 = 0$
   (c) $2x^2 - 7x - 3 = 0$
   (d) $x^2 + 5x - 66 = 0$
   (e) $s^2 - 3s - 7 = 0$
   (f) $x^2 + 5x + 1 = 0$
   (g) $x^2 + 5x - 1 = 0$
   (h) $x^2 - 5x + 1 = 0$
   (i) $x^2 - 5x - 1 = 0$
   (j) $2x^2 - 7x + 2 = 0$
   (k) $2x^2 - 7x - 2 = 0$
   (l) $2x^2 + 7x + 2 = 0$
   (m) $2x^2 + 7x - 2 = 0$
   (n) $\frac{1}{2}x^2 + 4x + 1 = 0$
   (o) $\frac{1}{2}x^2 + 4x - 1 = 0$
   (p) $\frac{1}{2}x^2 - 4x + 1 = 0$
   (q) $\frac{1}{2}x^2 - 4x - 1 = 0$

3. A rectangular room has a length that is 3 m longer than its width. The area of the floor is the product of the length and the width, and is equal to 70 m². Calculate the dimensions of the floor.

4. Consider Figure 9.2, which shows a projectile launched with a velocity V_0 at an angle 26.6° to the horizontal. The relationship between the vertical height of the projectile, y, and the horizontal distance travelled by the projectile, x, is given by

$$y = 0.5x - 0.625\frac{gx^2}{V_0^2}$$

where g is a constant called the acceleration due to gravity, whose value is 9.81 m s^{-2}. Calculate the horizontal distance travelled by the projectile when it hits the ground, given that $V_0 = 10$ m per second.

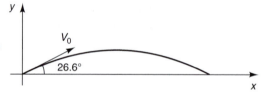

Figure 9.2 A projectile fired at an angle of 26.6°.

5. Solve the following equations by completing the square:
   (a) $x^2 - 6x + 9 = 0$
   (b) $x^2 - x - 1 = 0$

6. Solve the following quadratic equations using any appropriate method:
   (a) $t^2 - 3t - 7 = 0$
   (b) $x^2 - 7x + 10 = 0$
   (c) $x^2 + 8x - 128 = 0$

9.5 Simultaneous linear equations

Sometimes equations contain more than one unknown quantity. When this is the case, there will usually be more than one equation involved. For example, in the two equations

$$2x + y = 5, \quad 3x - 2y = 4$$

there are two unknowns: x and y. In order to solve these equations, we must find values of x and y that satisfy both equations at the same time. The two equations are known as **simultaneous equations**.

9.5.1 Solving simultaneous equations

One way of solving simultaneous equations is by **elimination**. This involves using one equation to find an expression for one of the unknowns. This expression can be used to eliminate that unknown from the second equation. Consider the following examples.

Example 9.16

Solve the simultaneous equations

$$4x + y = 11 \tag{9.1}$$

$$5x + 4y = 22 \tag{9.2}$$

Solution From Equation (9.1) we can obtain an expression for y, that is $y = 11 - 4x$. This can be substituted into Equation (9.2) in order to remove y from this equation. We find

$$5x + 4(11 - 4x) = 22$$

Solving this equation to obtain x, we find

$$5x + 44 - 16x = 22$$

so that

$$-11x = -22$$

that is

$$x = 2$$

Knowing x, we can then obtain y, because $y = 11 - 4x$. We find

$$y = 11 - 4(2) = 3$$

The solution of the simultaneous equations is therefore $x = 2$ and $y = 3$. It is easy to check that this solution is correct by substituting these values into the original equations.

Example 9.17

Solve the simultaneous equations

$$3x + 4y = 18 \tag{9.3}$$

$$4x + 5y = 23 \tag{9.4}$$

Solution   In this example we shall demonstrate an alternative method of solution to that in Example 9.16. We shall multiply both sides of Equation (9.3) by 4, and both sides of Equation (9.4) by 3, in order to make the coefficients of x in both equations the same. Recall that it is quite permissible to multiply both sides of an equation by the same number. We find

$$12x + 16y = 72 \tag{9.5}$$

$$12x + 15y = 69 \tag{9.6}$$

If we now subtract Equation (9.6) from Equation (9.5), we find

$$0x + y = 3$$

so that $y = 3$. Substituting this value for y into Equation (9.3), we find

$$3x + 4(3) = 18$$

so that $3x = 6$, that is, $x = 2$.

Example 9.18 Kirchhoff's circuit laws

We have already examined Ohm's law and used it to analyse a simple electrical circuit. In order to analyse more complicated electrical circuits, it is necessary to use the laws first established by Kirchhoff. The first of these laws is known as **Kirchhoff's current law**, and is

> The sum of the currents entering a junction in a circuit is equal to the sum of the currents leaving the junction

A junction in a circuit is any point in the circuit at which the currents can come together or split. This law is a consequence of the conservation of charge.

The second of the laws is known as **Kirchhoff's voltage law**, and is

> The sum of the potential differences (voltages) around any closed loop in a circuit must be zero

Positive potential differences occur as a result of voltage supplies and negative potential differences occur due to voltage drops across components such as resistors. We shall now analyse some electrical circuits to demonstrate the use of these laws. Consider the circuit shown in Figure 9.3. Calculate the currents flowing in the various parts of the circuit. For convenience, these currents have already been labelled in Figure 9.3.

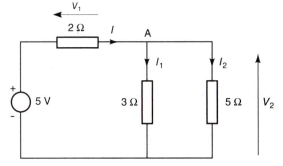

Figure 9.3 Circuit for Example 9.18.

Solution First, applying Kirchhoff's current law at the point A, we note that

$$I = I_1 + I_2$$

Using Ohm's law, the voltage drop across the 2 Ω resistor, V_1, is therefore given by

$$V_1 = 2I = 2(I_1 + I_2)$$

The voltage drop across each of the parallel resistors is the same and equal to V_2. Again, from Ohm's law, we find

$$V_2 = 3I_1 \quad \text{and also} \quad V_2 = 5I_2$$

So we can write $3I_1 = 5I_2$, or equivalently

$$3I_1 - 5I_2 = 0 \qquad (9.7)$$

Finally, using Kirchhoff's voltage law, we have

$$5 - V_1 - V_2 = 0$$

which we can write as

$$V_1 + V_2 = 5$$

Writing V_1 as $2(I_1 + I_2)$ and V_2 as $3I_1$, we have

$$2(I_1 + I_2) + 3I_1 = 5$$

that is,

$$5I_1 + 2I_2 = 5 \qquad (9.8)$$

Equations (9.7) and (9.8) are the two simultaneous equations that must be solved to find I_1 and I_2. Multiplying Equation (9.7) by 5 and Equation (9.8) by 3 makes both coefficients of I_1 the same. That is,

$$15I_1 - 25I_2 = 0 \qquad (9.9)$$

$$15I_1 + 6I_2 = 15 \qquad (9.10)$$

Subtracting Equation (9.9) from Equation (9.10), we find

$$31I_2 = 15$$

so that

$$I_2 = \frac{15}{31} = 0.484 \text{ A}$$

Substitution into Equation (9.7) yields

$$3I_1 - 5(0.484) = 0$$

$$3I_1 = 2.420$$

$$I_1 = 0.807 \text{ A}$$

Finally, we note that the current through the $2\,\Omega$ resistor is

$$I_1 + I_2 = 0.807 + 0.484 = 1.291 \ A$$

We have succeeded in finding the currents in the various parts of the circuit.

Self-assessment question 9.5

1. Explain what is meant by a pair of simultaneous equations.

Exercises 9.5

1. Solve the following simultaneous equations by elimination:
    (a) $5x - y = 4$, $2x + y = 7$
    (b) $6x - 2y = 0$, $3x + 4y = 8$
    (c) $7x + y = 2$, $5x - 3y = 12$
    (d) $y - 3x = 12$, $x = 3$
    (e) $2x - y = 5$, $x + 2y = 10$
    (f) $3x + 2y = 11$, $x + 3y = 6$
    (g) $x - y = -5$, $2x + 3y = 5$
    (h) $-2x + 7y = 2$, $\frac{1}{2}x + y = -\frac{1}{2}$
    (i) $2x + y = 3.5$, $5x - 3y = 6$
    (j) $2y + x = 1$, $\frac{1}{2}x + 4y = 5$
    (k) $4x + 5y = 5$, $x - y = -1.45$
    (l) $3x + y = 4$, $\frac{1}{2}x + \frac{1}{2}y = 0.75$
    (m) $x + 20y = 229$, $7x - 5y = 8$
    (n) $\frac{1}{3}x + \frac{1}{2}y = 1.7$, $\frac{1}{4}x - \frac{1}{9}y = 0.4$

2. Solve the following simultaneous equations:
    (a) $3A + B = 1$, $2A - B = 2$
    (b) $4A + 2B = 1$, $A - 5B = -1$

3. Calculate the currents flowing in the circuit shown in Figure 9.4.

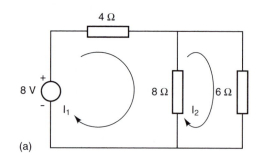

(a)

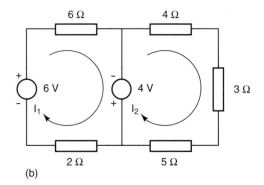

(b)

Figure 9.4   Circuit for Exercise 9.5 Q3.

9.6 Solving linear and quadratic equations using graphs

Equations can be solved quite simply by drawing a graph. The following two examples show how linear and quadratic equations can be solved graphically, but the method can be applied to more general equations as well.

Example 9.19

    (a)   Sketch a graph of the function $y = 2x - 4$ for values of x between $x = -2$ and $x = 4$.

    (b)   Use the graph to solve the equations
    (i) $2x - 4 = 0$ (ii) $2x - 4 = 3$

Solution   (a)   As usual, a table of values is drawn up in order to sketch the graph. Table 9.1 gives x values and the corresponding y values. You should check the table for yourself. The corresponding graph is shown in Figure 9.5. Note that y is a linear function of x, and so its graph is a straight line.

Table 9.1

x	-2	-1	0	1	2	3	4
y	-8	-6	-4	-2	0	2	4

    (b)   (i)   Because the straight line shown has equation $y = 2x - 4$, we can solve the equation $2x - 4 = 0$ by looking for the value of x where $y = 0$. From the graph, we see that $y = 0$ when the straight line cuts the horizontal axis at $x = 2$.

        (ii)   Because $y = 2x - 4$, we can solve the equation $2x - 4 = 3$ by looking on the graph for the point where $y = 3$. We see that at this point $x = 3.5$.

Example 9.20

    (a)   Sketch a graph of the equation $y = x^2 - 3x$ for values of x between -1 and 4.

    (b)   From the graph, solve the equations
    (i) $x^2 - 3x = 0$ (ii) $x^2 - 3x = 2.5$ (iii) $x^2 - 3x = -3$

Solution   Table 9.2 gives x values and corresponding y values. The graph is shown in Figure 9.6.

Table 9.2

x	-1	0	1	2	3	4
y	4	0	-2	-2	0	4

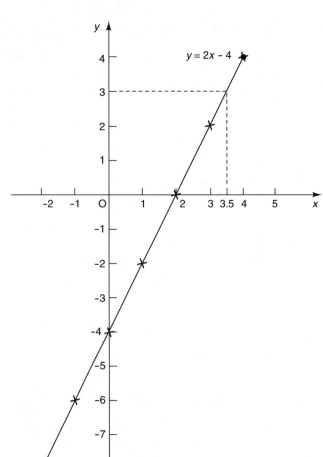

Figure 9.5   Graph of $y=2x-4$.

(b)   (i)   Because the graph shown has equation $y=x^2-3x$, we can solve
              the equation $x^2-3x=0$ by looking for values of x where $y=0$.
              From the graph, we see that $y=0$ when the curve cuts the horizontal
              axis at $x=0$ and $x=3$.

      (ii)   We can solve the equation $x^2-3x=2.5$ by looking for the values
              of x where $y=2.5$. From the graph, we see that $y=2.5$ when x is
              approximately equal to 3.7 and when x is approximately equal to
              $x=-0.7$. These are estimates of the solutions. Clearly, the accuracy
              to which we can find a solution depends upon how well we can
              draw and read the graph. If necessary, once an approximate solution
              has been found, the graph can be redrawn on a larger scale in
              order to obtain a better estimate.

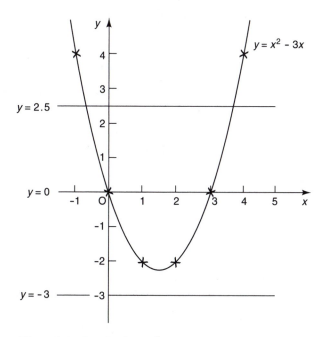

Figure 9.6 Graph of $y=x^2-3x$.

(iii) We can try to solve the equation $x^2-3x=-3$ by looking for the values of x where $y=-3$. From the graph, we see that there are no such values. Consequently, this equation has no real roots.

Example 9.21 Motion of a projectile

Recall Example 9.11. The horizontal distance travelled, x, and the vertical height, y, of a projectile fired at an angle of 45° are related by

$$y=x-\frac{gx^2}{V_0^2}$$

Consider the case when a projectile is fired with an initial velocity V_0 of 100 m per second. Assume g has a value of 10 m s^{-2} for ease of calculation.
(a) Sketch the trajectory of the projectile.
(b) Calculate the values of x at which the projectile has a height of 225 m.

Solution (a) First, we construct a table of values in order to sketch the graph. Check these values for yourself. In order to construct this table, we need to substitute the values $g=10$ and $V_0=100$ in the equation. So,

$$y=x-\frac{10x^2}{100^2}=x-\frac{x^2}{1000}$$

A table of values is given in Table 9.3. The trajectory of the projectile is shown in Figure 9.7.

Table 9.3

x	0	100	200	300	400	500	600	700	800	900	1000
y	0	90	160	210	240	250	240	210	160	90	0

(b)   We see that the projectile has a vertical height of 225 m when x is approximately 340 m and 660 m.

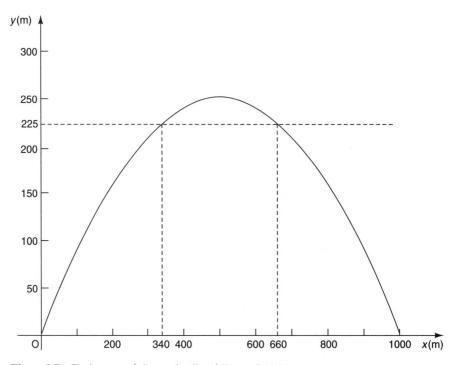

Figure 9.7   Trajectory of the projectile of Example 9.21.

Exercises 9.6

1.   Sketch a graph of the function $y = 5x + 2$. Use your graph to solve the following equations:
(a) $5x + 2 = 0$ (b) $5x = 1$

2.   Sketch a graph of the function $y = 3x^2 + 2x - 2$ for values of x between -4 and 4. Use your graph to solve the following equations:
(a) $3x^2 + 2x - 2 = 14$ (b) $3x^2 + 2x = 8$
(c) $3x^2 + 2x = 0$

3. Recall Example 9.21. A projectile is fired from the origin at an angle of $45°$ and has an initial velocity of $300 \, m$ per second. Sketch the trajectory of the projectile. Calculate the values of x at which the projectile has a vertical height of $300 \, m$.

4. (a) Sketch a graph of $y = 3 + 2x - x^2$ for $x = -2$ to $x = 4$.

(b) Hence solve
   (i) $3 + 2x - x^2 = 0$
   (ii) $5 + 2x - x^2 = 0$
   (iii) $x^2 - 2x - \frac{1}{2} = 0$

Computer and calculator exercise 9.6

1. Use a graphics calculator to plot a graph of $y = x^3 + 2x^2 - x - 2$. Hence find all roots of the equation $x^3 + 2x^2 - x - 2 = 0$.

9.7 Solving simultaneous equations using graphs

Consider the following examples.

Example 9.22

Solve the simultaneous equations:
$$2x + y = 4, \qquad -2x + y = 2$$

Solution In order to solve these using graphs, we first rewrite them in the form
$$y = -2x + 4, \qquad y = 2x + 2$$

Next we plot each equation on the same graph. The table of values is shown in Table 9.4 and the graphs are shown in Figure 9.8. The solutions of the simultaneous equations are the coordinates of the point where the two lines cross. The lines cross at the point P where $x = \frac{1}{2}$ and $y = 3$. The solution of the simultaneous equations is therefore $x = \frac{1}{2}$ and $y = 3$.

Table 9.4

x	-3	0	3
$-2x + 4$	10	4	-2
$2x + 2$	-4	2	8

Example 9.23

(a) Plot the graphs of $y = 2x^2$ and $y = 5 - \frac{5}{2}x$ for values of x between -3 and 3.

(b) Use the graphs to find approximate solutions of the equation
$$2x^2 + \frac{5}{2}x - 5 = 0.$$

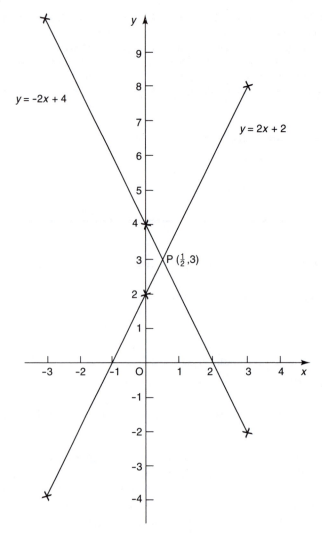

Figure 9.8 Graphs of $y=-2x+4$ and $y=2x+2$.

Solution   (a)   A table of values of $y=2x^2$ and $y=5-\frac{5}{2}x$ is given in Table 9.5 and the corresponding graphs have been plotted in Figure 9.9.

Table 9.5

x	-3	-2	-1	0	1	2	3
$2x^2$	18	8	2	0	2	8	18
$5-\frac{5}{2}x$	12.5	10	7.5	5	2.5	0	-2.5

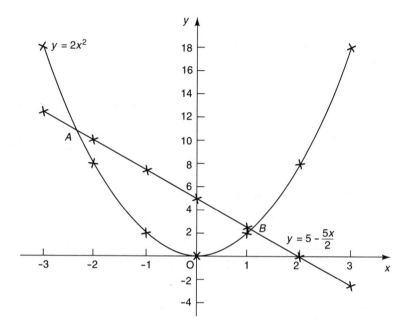

Figure 9.9 Graphs of $y=2x^2$ and $y=5-\frac{5}{2}x$.

(b)   Now consider the points where the line and curve intersect. These points have been labelled A and B.

At point A the y coordinate of the line and the y coordinate of the curve must be the same, and so the x coordinate of A must satisfy

$$2x^2 = 5 - \tfrac{5}{2}x$$

This is equivalent to

$$2x^2 + \tfrac{5}{2}x - 5 = 0$$

Hence the x coordinate of A is a solution of the given equation. From the graph, we see that x is approximately -2.3.

Similarly, at point B the y coordinate of the line and the y coordinate of the curve must be the same, and so the x coordinate of B must also satisfy

$$2x^2 = 5 - \tfrac{5}{2}x$$

Hence the x coordinate of B is also a solution of the given equation. From the graph, we see that here x is approximately 1.

In summary, we have found two approximate solutions: $x = -2.3$ and $x = 1$.

Exercises 9.7

1. Plot graphs and then solve the following pairs of simultaneous equations:
   (a) $3x+2y=5$, $x-y=0$
   (b) $2x-y=16$, $x+2y=3$
   (c) $5x-y=-18$, $4x+y=-18$

2. Plot graphs of the lines $y=4x+4$ and $y=4x-3$. Comment upon the possible solutions of the simultaneous equations
$$y-4x=4, \quad y-4x=-3$$

3. Plot graphs of $y=x^3+1$ and $y=x^2$. Hence solve the equation $x^3+1=x^2$.

4. Plot graphs of $y=2x^2$ and $y=-2x+3$. Hence solve the equation $2x^2+2x-3=0$.

Computer and calculator exercises 9.7

1. Use a graphics calculator or software to plot appropriate graphs in order to solve the following pairs of simultaneous equations:
   (a) $17x+9y=2$, $5x+8y=11$
   (b) $5I_1+4I_2=9$, $6I_1+7I_2=11$

2. Use a graphics calculator or software to find the values of x where the curves $y=7x+1$ and $y=3x^2+8x-9$ intersect. Deduce any solutions of the equation $3x^2+x-10=0$.

9.8 Solution of inequalities

Recall from Section 3.3.1 that the symbol $>$ means 'is greater than' and the symbol $\geqslant$ means 'is greater than or equal to'. Similarly, the symbol $<$ means 'is less than' and the symbol $\leqslant$ means 'is less than or equal to'. Thus we may state

$$7<11, \quad -3>-7, \quad 5^2\leqslant100, \quad 81\geqslant81, \quad \text{etc.}$$

The symbols $>$, $\geqslant$, $<$ and $\leqslant$, are known as **inequalities**. They obey simple rules when used in conjunction with arithmetic operations:

Rule 1 Adding or subtracting the same quantity from both sides of an inequality leaves the inequality sign unchanged.

For example, given that

$$5>2$$

we could add 4 to both sides to obtain

$$5+4>2+4$$

and hence

$$9>6$$

Similarly, given that

$$9 < 18$$

we can subtract 5 from both sides to give

$$4 < 13$$

Rule 2 Multiplying or dividing both sides by a **positive** number leaves the inequality sign unchanged.

For example, given

$$5 > 2$$

we could multiply both sides by 3 to obtain

$$5 \times 3 > 2 \times 3$$

that is,

$$15 > 6$$

Similarly, since

$$9 < 18$$

we can divide both sides by 3 to give

$$3 < 6$$

Rule 3 Multiplying or dividing both sides by a **negative** number reverses the inequality.

For example, given that

$$5 > 2$$

we could multiply both sides by -2, and reverse the inequality to get

$$5 \times (-2) < 2 \times (-2)$$

and so

$$-10 < -4$$

A common mistake is to forget to reverse the inequality when multiplying inequalities by negative numbers.

When we are asked to solve an inequality, the inequality will contain an unknown variable, x say. To solve means to find all values of x for which the inequality is true. Consider the following examples.

Example 9.24

Solve the inequality $7x - 2 > 0$.

Solution

$$7x - 2 > 0$$

$$7x > 2$$

$$x > \tfrac{2}{7}$$

Hence all values of x greater than $\tfrac{2}{7}$ satisfy $7x - 2 > 0$.

Example 9.25

Solve the inequality $8x - 7 \geqslant 2x$.

Solution

$$8x - 7 \geqslant 2x$$

$$8x - 2x \geqslant 7$$

$$6x \geqslant 7$$

$$x \geqslant \tfrac{7}{6}$$

Hence all values of x greater than or equal to $\tfrac{7}{6}$ satisfy $8x - 7 \geqslant 2x$.

Example 9.26

Find the range of values of x satisfying

$$x - 3 < 2x + 5$$

Solution

$$x - 3 < 2x + 5$$

$$x < 2x + 8$$

$$-x < 8$$

Then, multiplying both sides by -1, and remembering to reverse the inequality, we find

$$x > -8$$

Hence all values of x greater than -8 satisfy $x - 3 < 2x + 5$.

The modulus sign is sometimes used in conjunction with inequalities. For example, $|x| < 1$ means all numbers whose actual size, irrespective of sign, is less than 1. This means any value between -1 and 1. Thus

$$|x| < 1 \quad \text{means} \quad -1 < x < 1$$

Similarly, $|y| > 2$ means all numbers whose actual size, irrespective of sign, is greater than 2. This means any value greater than 2 and any value less than -2. Thus

$$|y| > 2 \quad \text{means} \quad y > 2 \text{ and } y < -2$$

Example 9.27

Solve the inequality $|2x + 1| < 3$.

Solution   If $|2x + 1| < 3$ then this is equivalent to

$$-3 < 2x + 1 < 3$$

We treat both parts of the inequality separately. First, consider

$$-3 < 2x + 1$$

Subtracting 1 from both sides gives

$$-4 < 2x$$

Dividing both sides by 2 gives

$$-2 < x$$

Now consider the second part: $2x + 1 < 3$. Subtracting 1 from both sides gives

$$2x < 2$$

Dividing both sides by 2 gives

$$x < 1$$

Putting the results of examining both parts together, we see that

$$-2 < x < 1$$

is the required solution. This means that any value of x greater than -2 but less than 1 satisfies the inequality $|2x + 1| < 3$.

Self-assessment questions 9.8

1. State the rules that can be applied when trying to solve inequalities.

2. Express each of the intervals $|t| < 5$ and $|t| \leqslant 3$ without using the modulus sign.

Exercises 9.8

1.   Find the range of values of x satisfying
     $2x-1<x-4$.

2.   For what values of x are the following
     inequalities satisfied:
     (a) $2x+1>5$ (b) $-2x<6$
     (c) $-3x>5x+2$ (d) $7x-12\geqslant 2x-6$

3.   Solve the following inequalities:
     (a) $3s+17\geqslant 2$ (b) $-5t-11\leqslant 7$

4.   Solve the following inequalities:
     (a) $1-x\leqslant 3$ (b) $\frac{1}{2}x+\frac{1}{3}\geqslant 0$
     (c) $\frac{1}{4}(2x-1)<-2$ (d) $4y+7\leqslant -1$
     (e) $4-\frac{1}{2}x\leqslant 3-x$ (f) $t+1\geqslant 1-t$

5.   Solve the following inequalities:
     (a) $|x+1|<5$ (b) $|y+2|\leqslant 4$
     (c) $|y-1|<9$ (d) $|3x+7|<2$

9.9 Graphical solution of inequalities

Just as graphs were used to solve equations, they can be used to solve inequalities
as well.

Example 9.28

Find the range of values of x for which $2x-3<0$.

Solution   Consider the function $y=2x-3$. The values of x we require will be those for
which $y<0$. The graph of $y=2x-3$ is shown in Figure 9.10. From the graph,
we see that y is negative whenever $x<1.5$.

Example 9.29

Find the range of values of x for which

$$(x-2)(2x+1)>0$$

Solution   Consider $y=(x-2)(2x+1)=2x^2-3x-2$. The values of x we require will be
those values for which $y>0$. To assist us, we sketch a graph of $y=2x^2-3x-2$.
A table of values is shown in Table 9.6, and the corresponding graph is shown
in Figure 9.11. From this, we can look for the appropriate x values. From the
graph, we see that the required values are

$$x<-\tfrac{1}{2}\quad\text{and}\quad x>2$$

Table 9.6

x	-2	-1	0	1	2	3	4
y	12	3	-2	-3	0	7	18

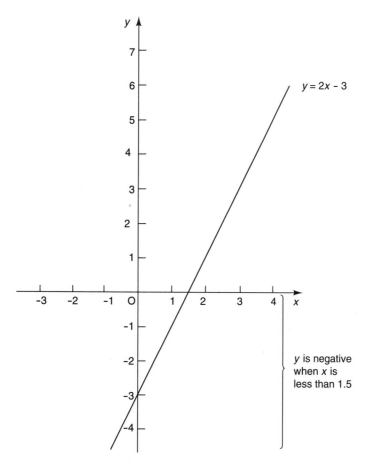

Figure 9.10   Graph of $y = 2x - 3$.

Example 9.30

Find the range of values of x for which

$$2x - 1 < x^2 - 4 < 12$$

Solution   In this example there are two inequalities to be satisfied simultaneously:

$$2x - 1 < x^2 - 4, \quad \text{that is,} \quad x^2 - 2x - 3 > 0$$

and

$$x^2 - 4 < 12, \quad \text{that is,} \quad x^2 - 16 < 0$$

Consider the first inequality. A table of values of the function $y = x^2 - 2x - 3$ is given in Table 9.7 and the corresponding graph is shown in Figure 9.12. From

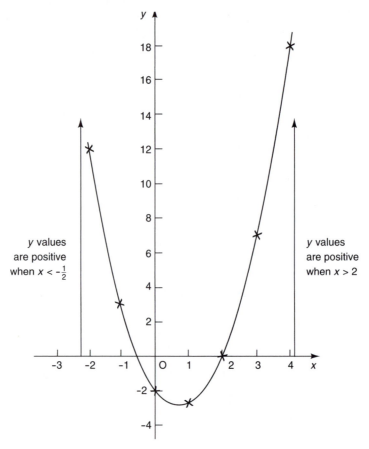

y values are positive when $x < -\frac{1}{2}$

y values are positive when $x > 2$

Figure 9.11 Graph of $y = 2x^2 - 3x - 2$.

the graph, to ensure that $x^2 - 2x - 3 > 0$, the required ranges of values are $x < -1$ and $x > 3$.

Table 9.7

x	−6	−5	−4	−3	−2	−1	0	1	2	3	4	5
$y = x^2 - 2x - 3$	45	32	21	12	5	0	−3	−4	−3	0	5	12
$y = x^2 - 16$	20	9	0	−7	−12	−15	−16	−15	−12	−7	0	9

For the second inequality, we need to sketch $y = x^2 - 16$. Again, the table of values is given in Table 9.7. The graph is shown in Figure 9.12. From this graph, we note that $x^2 - 16 < 0$ when $-4 < x < 4$.

In order to satisfy both inequalities simultaneously, consider Figure 9.13, which summarizes both earlier graphs. We see that both inequalities are true only when

$$-4 < x < -1, \quad \text{or} \quad 3 < x < 4$$

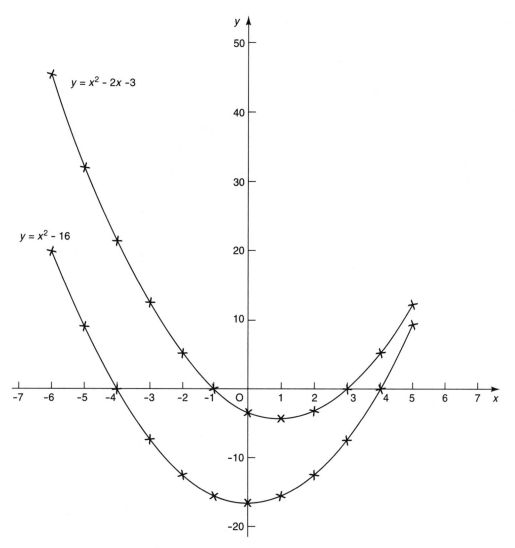

Figure 9.12 Graphs of $y = x^2 - 2x - 3$ and $y = x^2 - 16$.

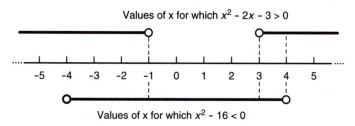

Figure 9.13 Values of x for which $x^2 - 16 < 0$ and values of x for which $x^2 - 2x - 3 > 0$.

Exercises 9.9

1. Use graphs to find the range of values of x that satisfy the following inequalities:
   (a) $5x+5<0$ (b) $2x-3<7$
   (c) $t+7<0$ (d) $3t-7>2$

2. Find the range of values of x for which
   (a) $(2x+1)(x-2)<0$
   (b) $2x^2-3x>2$
   (c) $(x+8)(x-3)<3x$
   (d) $x^2+x+1>0$

3. (a) Find the set of values of x for which $x^2-3x-4>0$.
   (b) Find the set of values of x for which $x+5>0$.
   (c) Hence find the set of values of x for which

   $$\frac{x^2-3x-4}{x+5}<0$$

   (Hint: Consider this as $f(x)/g(x)$. The fraction will be negative when f and g have opposite signs.)

4. Find the range of values of x for which
   (a) $(2x-1)(x+1)<0$
   (b) $(x+1)(x+2)>0$
   (c) $(x-3)(x+3)<0$

   (d) $(x+3)(x-3)<0$
   (e) $\dfrac{x+3}{x-3}>0$
   (f) $\dfrac{x+3}{x-3}<0$
   (g) $-6<2x+1<-2$

5. For what values of x is

   $$\frac{(2x-1)(x+2)}{(2x+1)(x-2)}>0$$

6. For what values of x are the following negative:
   (a) $(x-2)(x-1)$
   (b) $(2x+1)(x+2)$
   (c) $(3x-2)(x+1)$
   (d) $\dfrac{x+1}{2x+1}$
   (e) $\dfrac{3x+4}{2x-4}$
   (f) $\dfrac{2x+5}{x+6}$
   (g) $\dfrac{(x-1)(x-3)}{(x-2)(x-4)}$

Computer and calculator exercise 9.9

1. Use a graphics calculator to plot

   $$y=\frac{(x-7)(x+8)}{x-4}$$

   Deduce the values of x for which

   $$\frac{x^2+x-56}{x-4}$$

   is negative.

Review exercises 9

1 Solve the following equations:
(a) $t+11=17t$ (b) $v+3=12v-2$
(c) $\dfrac{1}{3-s}=\dfrac{4}{s+1}$ (d) $\dfrac{3}{2t-1}=\dfrac{2}{t+6}$

2 Solve, if possible, the following quadratic equations:
(a) $2x^2-3x-35=0$
(b) $4x^2-28x+49=0$
(c) $x^2+7x+13=0$
(d) $14x^2+47x-7=0$
(e) $6x^2+6x-1=0$
(f) $7-2x-3x^2=0$

3 Find the values of x that satisfy the following inequalities:
(a) $19x-8<0$ (b) $4x+3\leqslant7$
(c) $-3<x+2<7$ (d) $\dfrac{2}{x+1}<1$

4 Factorize the expression x^4-13x^2+36. Hence solve the equation $x^4-13x^2+36=0$.

5 Solve the following pairs of simultaneous equations:
(a) $4x+9y=80,\ -3x+y=2$

(b) $7s+11t=4,\ 8s-4t=-12$
(c) $b-4a=1,\ 2a+3b=-11$

6 Plot graphs of $y=1/x$ and $y=x+1$ for $0<x\leqslant5$. Hence find a solution of the equation

$$\frac{1}{x}=x+1$$

in the interval $(0, 5]$.

7 Plot graphs of $y=x+1/x^2$ and $y=-\frac{2}{5}x+4$ for $0<x\leqslant5$. Hence find two solutions of the equation

$$x+\frac{1}{x^2}=-\frac{2}{5}x+4$$

in the interval $(0, 5]$.

8 (a)   Sketch $y=5-x^2$ for $-3\leqslant x\leqslant3$.
(b)   On the same axes draw $y=-\frac{2}{3}x$ and $5y=4x+2$.
(c)   Use the graph to solve
(i) $5-x^2=0$ (ii) $x^2-\frac{2}{3}x-5=0$
(iii) $5x^2+4x-23=0$

10

Angles and Their Trigonometric Ratios

KEY POINTS

This chapter

- defines the units 'degree' and 'radian'

- shows how to convert from radians to degrees and vice versa

- defines the trigonometric ratios: sine, cosine, tangent

- explains the use of a scientific calculator in calculating ratios and angles

- explains the term 'quadrant'

- extends the definition of the trigonometric ratios to angles greater than 90°

- shows how to solve certain trigonometric equations

CONTENTS

10.1 Introduction

Trigonometry is concerned with angles and various ratios, known as sine, cosine and tangent. This chapter explains the units 'degree' and 'radian' and then goes on to define the trigonometric ratios. Initially these are defined with reference to angles of a right-angled triangle, but the definition is extended to include angles of any size. Finally, equations involving the trigonometric ratios are solved.

10.2 Angles and their measurement in degrees and radians

Fundamental to work in trigonometry is the measurement of angle. Just as length can be measured in many units, for example metres, inches and millimetres, so too can angle be measured in different units. The two main units used to measure angle are the degree and the radian.

10.2.1 Measuring angles in degrees

If we turn through a complete circle, so that we end up in the position where we started, then we have turned through 360 **degrees**, denoted 360°. Hence turning through 1° is the same as turning through $\frac{1}{360}$ of a circle. Figure 10.1 shows some typical angles measured in degrees.

An angle of 90° is usually referred to as a **right-angle** and is denoted by a small rectangular box as shown in Figure 10.1(b). Two lines that intersect at right-angles are called **perpendicular lines**. A triangle containing a right-angle is known as a **right-angled triangle**. An angle between 0° and 90° is called an

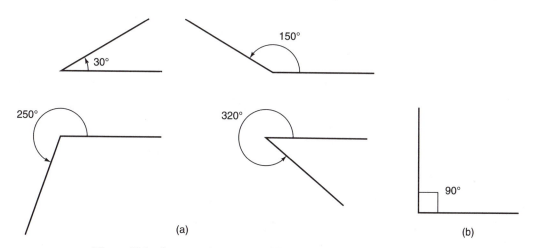

Figure 10.1 Some angles measured in degrees.

acute angle. An angle between 90° and 180° is called an **obtuse** angle. Angles greater than 180° are termed **reflex** angles.

Example 10.1

Draw an angle of 180°.

Solution   Figure 10.2 shows an angle of 180°.

An angle of 180° is sometimes referred to as a **straight-line angle**.

180°

Figure 10.2   An angle of 180° is a straight line angle.

Example 10.2 Bridge lattices

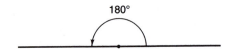   Figure 10.3 shows several bridge lattices made up of steel girders that form part of a larger steel bridge. For each one calculate the unknown angle x.

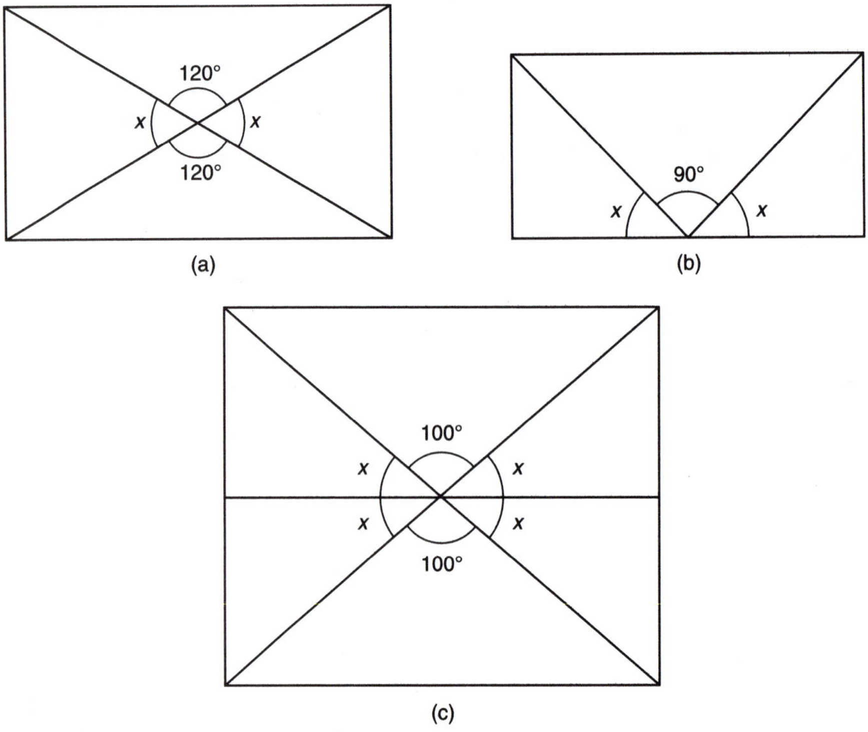

120°

x   x

120°

(a)

90°

x   x

(b)

100°

x   x

x   x

100°

(c)

Figure 10.3   Bridge lattices for Example 10.2.

Solution   (a)   A complete circle represents 360°. We are given that two of the angles take up $120 + 120 = 240°$. Therefore the two unknown angles must take up $360 - 240 = 120°$. So we can write

$$x + x = 120$$
$$2x = 120$$
$$x = 60°$$

(b)   Angles forming a straight line take up 180°. We are given that one of the angles takes up 90°. Therefore the two unknown angles take up $180 - 90 = 90°$. So we can write

$$x + x = 90$$
$$2x = 90$$
$$x = 45°$$

(c)   A complete circle represents 360°. We are given that two of the angles take up $100 + 100 = 200°$. Therefore the four unknown angles must take up $360 - 200 = 160°$. So we can write

$$x + x + x + x = 160$$
$$4x = 160$$
$$x = 40°$$

10.2.2 Measuring angles in radians

The second common unit used to measure angle is the radian. Like the degree, it is defined by reference to a circle. Consider a circle, centre O, radius r, and let an arc AB of length r be marked on the circumference as shown in Figure 10.4. Note that the arc AB is part of the circle and is curved. The angle AOB is 1 radian. We say angle AOB is **subtended** by the arc AB. So 1 **radian** is the angle subtended by an arc whose length is equal to the radius. Radian (often abbreviated to 'rad') is short for 'radius angle'.

An arc length of 1 radius subtends an angle of 1 radian.
An arc length of 2 radii subtends an angle of 2 radians.
An arc length of n radii subtends an angle of n radians.

Consider the case where $n = 2\pi$. An arc length of $2\pi r$ subtends an angle of 2π radians. We note that the circumference of a circle is $2\pi r$, and conclude that a circle subtends 2π radians. A circle also subtends 360°, and so

$$360° = 2\pi \text{ radians}$$
$$180° = \pi \text{ radians}$$

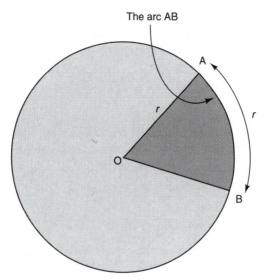

Figure 10.4   The arc AB and the radius OA have equal length.

Using the above equation, it is possible to convert from degrees to radians and vice versa. Recall that π represents the constant that is approximately equal to 3.142 and is available on most scientific calculators.

Example 10.3
　　Express the following angles in radians:
　　(a) 90°　(b) 270°　(c) 45°　(d) 121°

Solution　We know that $180° = \pi$ radians.
　　(a)　　　　$180° = \pi$ radians
　　　　Dividing by 2 we get

$$90° = \frac{\pi}{2} \text{ radians} = 1.571 \text{ radians}$$

　　　　Note that angles measured in radians are often expressed as multiples of π.
　　(b)　　　　$180° = \pi$ radians
　　　　Multiplying by $\frac{3}{2}$ gives:

$$270° = \frac{3\pi}{2} \text{ radians} = 4.712 \text{ radians}$$

　　(c)　　　　$180° = \pi$ radians
　　　　Dividing by 4, we obtain

$$45° = \frac{\pi}{4} \text{ radians} = 0.785 \text{ radians}$$

(d)
$$180° = \pi \text{ radians}$$

$$1° = \frac{\pi}{180} \text{ radians}$$

$$121° = 121 \times \frac{\pi}{180} \text{ radians} = 2.112 \text{ radians}$$

Some common angles measured in both degrees and radians are shown in Figure 10.5.

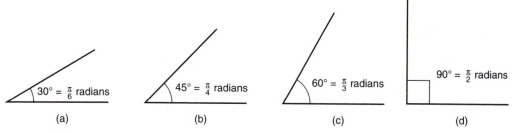

$30° = \frac{\pi}{6}$ radians (a)

$45° = \frac{\pi}{4}$ radians (b)

$60° = \frac{\pi}{3}$ radians (c)

$90° = \frac{\pi}{2}$ radians (d)

Figure 10.5 Some common angles measured in both degrees and radians.

Example 10.4

Express the following angles in degrees:
(a) 3 radians   (b) 1.23 radians   (c) 1 radian

Solution   (a)
$$\pi \text{ radians} = 180°$$

$$1 \text{ radian} = \frac{180°}{\pi}$$

$$3 \text{ radians} = 3 \times \frac{180°}{\pi} = 171.9°$$

(b)
$$\pi \text{ radians} = 180°$$

$$1 \text{ radian} = \frac{180°}{\pi}$$

$$1.23 \text{ radians} = 1.23 \times \frac{180°}{\pi} = 70.5°$$

(c) π radians $= 180°$

$$1 \text{ radian} = \frac{180°}{\pi} = 57.3°$$

We see that 1 radian is about 57°.

Self-assessment questions 10.2

1.   Radian is a shortened form of 'radius angle'. Why is this an apt description of a radian?

2.   Derive an equation connecting degrees and radians.

Exercises 10.2

1.   Convert the following angles to degrees:
     (a) $\frac{1}{3}\pi$ radians   (b) 0.4π radians
     (c) 0.4 radians   (d) 7 radians
     (e) 0.12 radians

2.   Convert the following angles to radians:
     (a) 135°   (b) 300°   (c) 240°   (d) 67°
     (e) 214°

3.   A circle has a radius of 3 cm.
     (a)   An arc has length 4.5 cm. What angle does the arc subtend at the centre of the circle? Give your answer in radians.
     (b)   An angle of 2.4 radians is subtended at the centre by an arc. Calculate the length of the arc.

4.   Figure 10.6 shows several bridge lattices made up of steel girders, which form part of a larger steel bridge in each case. For each one, calculate the unknown angle x.

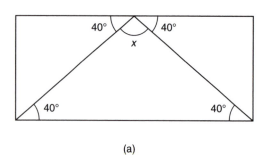

(a)

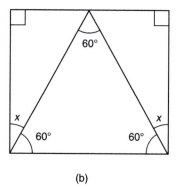

(b)

Figure 10.6   Bridge lattices for Exercise 10.2 Q4.

10.3 Definition of sine, cosine and tangent

There are three important trigonometric ratios: sine, cosine and tangent. They can be defined in terms of the lengths of the sides of a right-angled triangle. Figure 10.7 shows a right-angled triangle ABC, often abbreviated to △ABC. The side opposite the right-angle is known as the **hypotenuse.** In Figure 10.7 this is AC. The side opposite a particular angle is known as the **opposite side**. In Figure 10.7 the opposite side to angle BAC is BC. The angle BAC is written as ∠BAC or, more simply, as *A*. The side next to a particular angle is known as the **adjacent side.** In Figure 10.7 the adjacent side to angle *A* is AB. Similarly, the opposite side to angle *C* is AB and the adjacent side to angle *C* is BC.

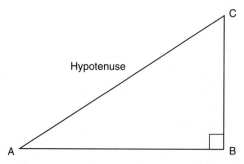

Figure 10.7 The triangle ABC has a right angle at B.

10.3.1 The sine ratio

We define the sine of an angle as the ratio of the side opposite the angle to the hypotenuse. Sine is usually shortened to sin.

$$\text{sine of an angle} = \frac{\text{opposite side}}{\text{hypotenuse}}$$

Referring to Figure 10.7, we see

$$\sin A = \frac{BC}{AC}, \quad \sin C = \frac{AB}{AC}$$

Since the hypotenuse is the longest side in a right-angled triangle, the sine of an angle is never greater than 1.

Example 10.5

In △ABC of Figure 10.7 suppose we are told that AB = 3 m, BC = 4 cm and AC = 5 cm. Calculate (a) sin *A*   (b) sin *C*

Isambard Kingdom Brunel: an engineering colossus

Isambard Kingdom Brunel (see Figure 1) was born in 1806 at Portsea. He came from a family of engineers and benefited greatly from the tuition his father gave him. He had the same ability in mathematics and drawing that his father had. Brunel went to a boarding school in Hove, and then in 1820 his father sent him to France to continue his education, which included the study of mathematics at Collège-Quatre in Paris which was famous for its mathematicians. Later in life he commented:

'I must strongly caution you against studying practical mechanics among French authors – take them for abstract science and study their statics, dynamics, geometry etc. etc., to your heart's content, but never read any of their works on mechanics any more than you would search their modern authors for religious principles. A few hours spent in a blacksmith's and wheelwright's shop will teach you more practical mechanics – read English books for practice – there is little enough to be learnt in them but you will not have to unlearn that little.'

Figure 1 *Isambard Kingdom Brunel: an engineering colossus.*

This comment characterizes Brunel's approach to engineering. He was prepared to use existing knowledge, if available, but he did not allow it to confine his imagination and was prepared to be inventive and daring when necessary. Brunel was involved in so many areas of engineering that it is difficult to comprehend this range from the perspective of the late 20th century, with its emphasis on specialization in one narrow area of knowledge. Such an approach to engineering would have been anathema to Brunel. He designed bridges, docks, stations, harbours, hospitals, buildings, tunnels, railways and ships, but just as importantly he was involved in the whole of the design, for example right down to the detail of the signalling systems for the railways.

Brunel designed many bridges. One example was the Royal Albert Bridge at Saltash (see Figure 2), which was opened in 1859. This has a length of 2200 feet. The two main spans have a length of 455 feet and are supported by a large central pier. One of the major problems involved in the building of this bridge was the construction of the central pier. The central cast iron columns are supported on a circular column of masonry, which extends to a depth of 96 feet in order to make contact with the underlying bedrock below the mud of the river bed. The pier took three and a half years to construct owing to the complexities of the work, which required innovative new techniques such as using compressed air to pump out water from the hollow masonry column once it was in place.

After Brunel's death in 1859 Daniel Gooch wrote in his diary:

'By his death, the greatest of England's engineers was lost, the man of the greatest originality of thought and power of execution, bold in his plans, but right. The commercial world thought him extravagant; but although he was so, great things are not done by those who sit down and count the cost of every thought and act.'

Derick Beckett in his book, *Brunel's Britain* (David and Charles, Newton Abbot, 1980) wrote:

'For thirty years he devoted all his energy to the art and practice of professional engineering, making full use of the technology currently available and in particular, experimental work. He had little time for state intervention in the form of design rules and was contemptuous of state honours. In contrast to his father, he did not file any patent specifications and spoke at length on the disadvantages of patents with regard to the progress of technology'

Figure 2 *Brunel's Royal Albert Bridge at Saltash.*

Solution   (a) $\sin A = \dfrac{BC}{AC} = \dfrac{4}{5} = 0.8$

(b) $\sin C = \dfrac{AB}{AC} = \dfrac{3}{5} = 0.6$

Note that the sine of an angle is the ratio of two lengths, and so has no units. Given an angle, we can use a scientific calculator to find the sine of it. The SIN or SINE button is used to do this. When entering an angle in your calculator, you need to check whether the angle is given in degrees or radians. Most scientific calculators have a MODE button that switches between degrees and radians.

Example 10.6

Use a calculator to find (a) $\sin 25°$ (b) $\sin 50°$ (c) $\sin 0.3$ (d) $\sin 1.127$

Solution   For (a) and (b) the calculator MODE must be set to degrees:
(a) $\sin 25° = 0.4226$ (b) $\sin 50° = 0.7660$
For (c) and (d) the MODE must be set to radians:
(c) $\sin 0.3 = 0.2955$ (d) $\sin 1.127 = 0.9031$

Given the sine of an angle, we can use a scientific calculator to find the angle itself. The INV SINE button or SIN^{-1} is used, meaning inverse sine. On some calculators you need to press SECOND FUNCTION followed by SINE. If $\sin A = 0.1234$, we write $A = \sin^{-1}(0.1234)$ and say A is the inverse sine of 0.1234. This means A is the angle whose sine is 0.1234. The -1 is not a power but simply a notation to denote the inverse.

> If
>
> $\sin A = x$
>
> then A is the angle whose sine is x. We write
>
> $A = \sin^{-1} x$

Note that $\sin^{-1} x$ is often written as arcsin x.

Example 10.7

Find the angle A when (a) $\sin A = 0.3200$ (b) $\sin A = 1.0000$
(c) $\sin A = 0.7516$

Solution   (a)   We are given $\sin A = 0.3200$. Using the inverse sine buttons, we find
$A = \sin^{-1}(0.3200) = 18.7°$, or equivalently $A = 0.3257$ radians.

(b)   We are given $\sin A = 1.0000$. Hence $A = \sin^{-1}(1.0000) = 90°$ or equivalently
1.5708 radians.

(c)   From $\sin A = 0.7516$, we know that $A = \sin^{-1}(0.7516) = 48.7°$ or
equivalently 0.8505 radians.

Example 10.8 Roof angles

Figure 10.8 shows two roof sections. Calculate the slope of the roof, θ, in each
case.

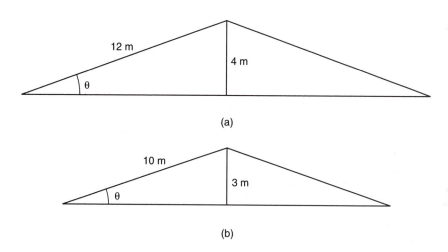

(a)

(b)

Figure 10.8   Roof sections for Example 10.8.

Solution   (a)   We have

$$\sin \theta = \frac{4}{12} = \frac{1}{3}$$

$$\theta = \sin^{-1}\left(\tfrac{1}{3}\right)$$

Hence

$$\theta = 19.5°$$

(b)   We have

$$\sin\theta = \frac{3}{10} = 0.3$$

$$\theta = \sin^{-1}(0.3)$$

Hence

$$\theta = 17.5°$$

A related ratio is the cosecant of an angle. Cosecant is usually shortened to cosec. It is defined as

$$\mathrm{cosec}\,\theta = \frac{1}{\sin\theta}$$

Example 10.9
        Evaluate (a) cosec 0.3   (b) cosec 40°

Solution   (a)   Here the angle is in radians:

$$\mathrm{cosec}\,0.3 = \frac{1}{\sin 0.3} = \frac{1}{0.2955} = 3.384$$

(b)
$$\mathrm{cosec}\,40° = \frac{1}{\sin 40°} = \frac{1}{0.6428} = 1.556$$

10.3.2 The cosine ratio

The cosine of an angle is the ratio of the side adjacent to the angle to the hypotenuse:

$$\text{cosine of an angle} = \frac{\text{adjacent side}}{\text{hypotenuse}}$$

Cosine is usually shortened to cos. Referring to $\triangle ABC$ in Figure 10.7, we see

$$\cos A = \frac{AB}{AC}, \quad \cos C = \frac{BC}{AC}$$

Like the sine of an angle, the cosine of an angle is never greater than 1.

Example 10.10

In $\triangle ABC$ in Figure 10.7 we are told that $AB = 3$ cm, $BC = 4$ cm and $AC = 5$ cm. Calculate (a) $\cos A$ (b) $\cos C$

Solution   (a) $$\cos A = \frac{AB}{AC} = \frac{3}{5} = 0.6$$

(b) $$\cos C = \frac{BC}{AC} = \frac{4}{5} = 0.8$$

The cosine of an angle can be found using the COS button of a scientific calculator.

Example 10.11

Use a calculator to find (a) $\cos 30°$ (b) $\cos 56°$ (c) $\cos 0.7$ (d) $\cos 1.231$

Solution   For (a) and (b) the calculator must be set to degree mode:
(a) $\cos 30° = 0.8660$ (b) $\cos 56° = 0.5592$
For (c) and (d) the calculator must be set to radian mode:
(c) $\cos 0.7 = 0.7648$ (d) $\cos 1.231 = 0.3333$

Given the cosine of an angle, we can use a scientific calculator to find the angle itself. The INV COS buttons or COS^{-1} button on a calculator can be used to do this. If $\cos A = 0.6$ we say $A = \cos^{-1}(0.6)$; that is, A is the inverse cosine of 0.6.

If

$$\cos A = x$$

then A is the angle whose cosine is x. We write

$$A = \cos^{-1} x$$

Note that $\cos^{-1} x$ is often written as arccos x.

Example 10.12

Find A given (a) $\cos A = 0.6213$ (b) $\cos A = 0.2396$

Solution   (a) $\cos A = 0.6213$

$A = \cos^{-1}(0.6213) = 51.6°$ or 0.9004 radians

(b) $\cos A = 0.2396$

$A = \cos^{-1}(0.2396) = 76.1°$ or 1.3288 radians

A related ratio is the secant of an angle. Secant is usually shortened to sec. The secant is defined as

$$\sec \theta = \frac{1}{\cos \theta}$$

Example 10.13

Evaluate (a) sec 1.3   (b) sec 60°

Solution   (a) $\sec 1.3 = \dfrac{1}{\cos 1.3} = \dfrac{1}{0.2675} = 3.7383$

(b) $\sec 60° = \dfrac{1}{\cos 60°} = \dfrac{1}{0.5} = 2$

10.3.3 The tangent ratio

The tangent of an angle is the ratio of the side opposite the angle to the side adjacent to the angle:

$$\text{tangent of an angle} = \frac{\text{opposite side}}{\text{adjacent side}}$$

Tangent is often shortened to tan. Referring to Figure 10.7, we see

$$\tan A = \frac{BC}{AB} \qquad \tan C = \frac{AB}{BC}$$

Example 10.14

In $\triangle ABC$ of Figure 10.7 we are told that $AB = 3$ cm, $BC = 4$ cm and $AC = 5$ cm. Calculate (a) $\tan A$ (b) $\tan C$

Solution (a)
$$\tan A = \frac{BC}{AB} = \frac{4}{3} = 1.3333$$

(b)
$$\tan C = \frac{AB}{BC} = \frac{3}{4} = 0.7500$$

Note that the tangent of an angle may exceed 1, unlike the sine and cosine. The TAN button may be used to find the tangent of an angle.

Example 10.15

Use a scientific calculator to find (a) $\tan 30°$ (b) $\tan 45°$ (c) $\tan 1.3$.

Solution The calculator mode is set to degrees for (a) and (b), and to radians for (c):
(a) $\tan 30° = 0.5774$ (b) $\tan 45° = 1$ (c) $\tan 1.3 = 3.6021$

Given the tangent of an angle, we can use a scientific calculator to determine the angle itself. The INV TAN buttons or TAN^{-1} button can be used to do this. If $\tan A = 0.3456$ then we write $A = \tan^{-1}(0.3456)$. This means that A is the angle whose tangent is 0.3456.

> If
>
> $$\tan A = x$$
>
> then A is the angle whose tangent is x. We write
>
> $$A = \tan^{-1} x$$

Note that $\tan^{-1} x$ is often written as arctan x.

Example 10.16

Use a calculator to find A given (a) $\tan A = 0.3142$ (b) $\tan A = 2.1461$

Solution   (a) $\tan A = 0.3142$

$$A = \tan^{-1}(0.3142) = 17.4°, \quad \text{or equivalently} \quad 0.3044 \text{ radians}$$

(b) $\tan A = 2.1461$

$$A = \tan^{-1}(2.1461) = 65.0°, \quad \text{or equivalently} \quad 1.1347 \text{ radians}$$

Example 10.17 Bridge lattices

Figure 10.9 shows several bridge lattices made up of steel girders that form part of a larger steel bridge. For each one, calculate the unknown angle x between the girders marked in the figures.

Solution   (a)   In order to calculate the lengths of the sides of the right-angled triangle containing x, we note the side opposite x is half the length of the smaller side of the outer rectangle, that is, it has a length of 3 m. The side adjacent to x is half the length of the other side of the outer rectangle, that is, 5 m. Hence we have

$$\tan x = \frac{3}{5} = 0.6$$

$$x = \tan^{-1}(0.6)$$

Therefore

$$x = 31.0°$$

(b)   By examining Figure 10.9(b), we see

$$\cos x = \frac{3}{4} = 0.75$$

$$x = \cos^{-1}(0.75)$$

Hence

$$x = 41.4°$$

(c)   By examining Figure 10.9(c), we see

$$\sin x = \frac{12}{14} = \frac{6}{7}$$

$$x = \sin^{-1}\left(\tfrac{6}{7}\right)$$

Hence

$$x = 59.0°$$

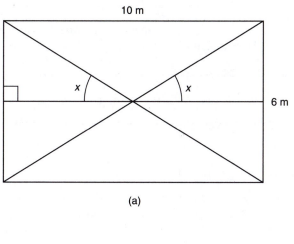

(a)

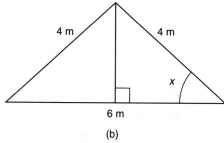

(b)

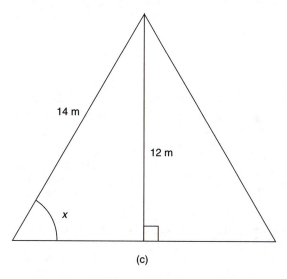

(c)

Figure 10.9 Bridge lattices for Example 10.17.

A related ratio is the cotangent of an angle. Cot is a shortened form of cotangent. The cotangent is defined as

$$\cot \theta = \frac{1}{\tan \theta}$$

Example 10.18

Evaluate (a) cot 1.2 (b) cot 30°

Solution   (a) $\cot 1.2 = \dfrac{1}{\tan 1.2} = \dfrac{1}{2.5722} = 0.3888$

(b) $\cot 30° = \dfrac{1}{\tan 30°} = \dfrac{1}{0.5774} = 1.7321$

Self-assessment questions 10.3

1. Define the trigonometric ratios sine, cosine and tangent.

2. Explain what is meant by the inverse trigonometric ratios.

3. Explain the ratios cosecant, secant and cotangent.

4. Explain what is meant by the hypotenuse of a right-angled triangle.

Exercises 10.3

1. Use a scientific calculator to evaluate
    (a) sin 37°   (b) cos 1.2   (c) tan 50°
    (d) sin 0.75   (e) cos 75°   (f) tan 0.03
    (g) cosec 0.95   (h) sec 84°   (i) cot 1

2. Evaluate (a) $\sin^{-1}(0.4261)$
    (b) $\cos^{-1}(0.3210)$ (c) $\tan^{-1}(2.5000)$
    (d) $\sin^{-1}(0.6513)$ (e) $\cos^{-1}(0.4693)$
    (f) $\tan^{-1}(0.5060)$

3. By using the definitions of sine, cosine and tangent, show that

    $$\frac{\sin A}{\cos A} = \tan A$$

4. Figure 10.10 shows several roof sections from a variety of buildings. Calculate the slope of the roof, θ, in each case.

5. Figure 10.11 shows several bridge lattices made up from steel girders that form part of a larger steel bridge. For each one calculate the unknown angle x.

6. Show that

    $$\cot \theta = \frac{\cos \theta}{\sin \theta}$$

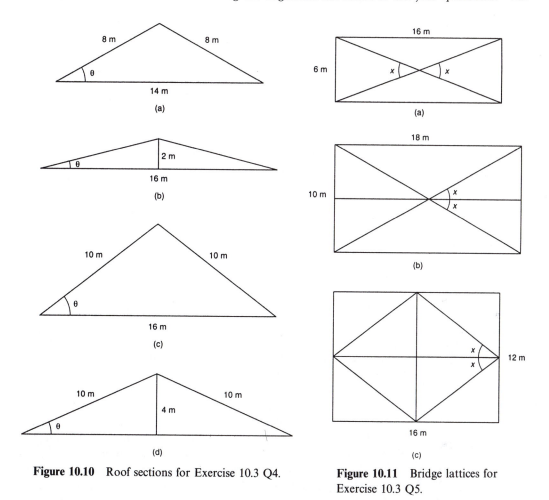

Figure 10.10 Roof sections for Exercise 10.3 Q4.

Figure 10.11 Bridge lattices for Exercise 10.3 Q5.

10.4 Extending the trigonometric ratios to the four quadrants

In Section 10.3 the trigonometric ratios were defined in terms of the lengths of the sides of a right-angled triangle. Thus only the trigonometric ratios of acute angles can be defined in this way. However, the definition of the ratios can be extended to include angles larger than 90°. This involves looking at the projections of a revolving arm onto the x and y axes.

10.4.1 The four quadrants

Figure 10.12 shows the x and y axes at right-angles, intersecting at the origin O. Recall that each axis has a positive and a negative part. The positive x axis

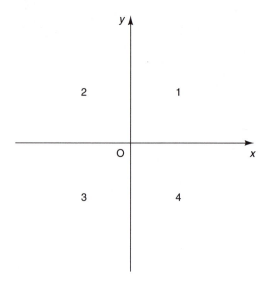

Figure 10.12   The *x* and *y* axes divide the plane into four quadrants.

is to the right of O; the positive *y* axis is above O. The *x* and *y* axes divide the plane into four **quadrants**. The four quadrants, numbered 1, 2, 3 and 4 are shown in Figure 10.12.

   We now consider a rotating arm OC. The arm is fixed at O, and can rotate anticlockwise. We consider the angle between the arm and the positive *x* axis. Figure 10.13 shows the arm in each of the four quadrants. The angle θ is the angle between the positive *x* axis and the arm OC, measured anticlockwise. When the arm is in the first quadrant, $0° \leqslant \theta < 90°$. In the second quadrant $90° \leqslant \theta < 180°$, in the third quadrant $180° \leqslant \theta < 270°$ and in the fourth quadrant $270° \leqslant \theta < 360°$. If θ is measured in radians then we have $0 \leqslant \theta < \frac{1}{2}\pi$, $\frac{1}{2}\pi \leqslant \theta < \pi$, $\pi \leqslant \theta < \frac{3}{2}\pi$ and $\frac{3}{2}\pi \leqslant \theta < 2\pi$ for the four quadrants.

   On occasion, angles are measured in a clockwise direction from the positive *x* axis. In such cases these angles are conventionally taken to be negative. Figure 10.14 shows angles of $-60°$ and $-120°$. Note that for $-60°$ the arm is in the same position as for $300°$. Similarly, for $-120°$ the arm is in the same position as for $240°$.

10.4.2 Extended definition of the trigonometric ratios

   In Section 10.3 we defined the trigonometric ratios by reference to the sides of a right-angled triangle. With such a definition, it is impossible to determine the ratios of angles larger than $90°$. In order to remedy this shortcoming, the definition of the trigonometric ratios needs to be extended to include angles greater than $90°$. This is done using the four quadrants and the rotating arm OC.

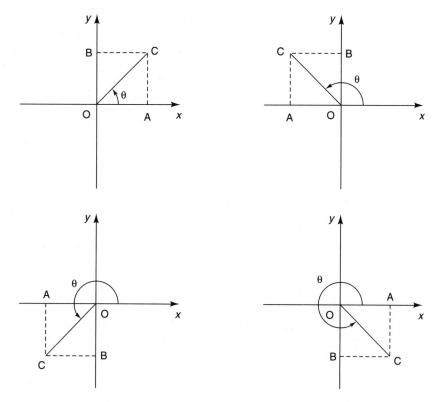

Figure 10.13   The arm OC rotates into each of the four quadrants.

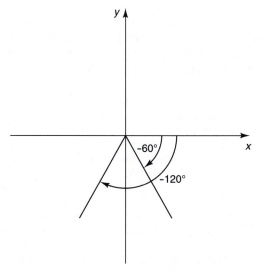

Figure 10.14   Angles of $-60°$ and $-120°$.

The length of the arm OC is always considered to be positive. The arm can be **projected** onto the x and y axes. The projection onto the x axis is OA; the projection onto the y axis is OB. This is illustrated in Figure 10.13. Note that the x projection OA and the y projection OB may be positive or negative, depending upon which quadrant the arm OC happens to be in. Table 10.1 lists the signs of the x and y projections for the four possible quadrants.

Table 10.1

Quadrant in which OC lies	Sign of the x projection	Sign of the y projection
First	positive	positive
Second	negative	positive
Third	negative	negative
Fourth	positive	negative

We are now ready to define the trigonometrical ratios for angles of any size. The extended definitions are as follows:

$$\sin\theta = \frac{\text{projection of OC onto } y \text{ axis}}{\text{arm OC}} = \frac{\text{OB}}{\text{OC}}$$

$$\cos\theta = \frac{\text{projection of OC onto } x \text{ axis}}{\text{arm OC}} = \frac{\text{OA}}{\text{OC}}$$

$$\tan\theta = \frac{\text{projection of OC onto } y \text{ axis}}{\text{projection of OC onto } x \text{ axis}} = \frac{\text{OB}}{\text{OA}}$$

Note that the ratios may be negative. For example, when $90° \leqslant \theta < 180°$, that is, the second quadrant, the x projection is negative, and so $\cos\theta$ will be negative. We can use Table 10.1 to draw up a table (Table 10.2) giving the sign of the trigonometrical ratios in each of the four quadrants.

Table 10.2

Quadrant in which θ lies	$\sin\theta$	$\cos\theta$	$\tan\theta$
First	positive	positive	positive
Second	positive	negative	negative
Third	negative	negative	positive
Fourth	negative	positive	negative

The information in Table 10.2 can be represented as in Figure 10.15. The ratios that are positive are listed for each quadrant. All the ratios are positive

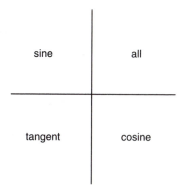

Figure 10.15 Quadrants in which the trigonometric ratios are positive.

in the first quadrant, only sine is positive in the second quadrant, only tangent is positive in the third quadrant and only cosine is positive in the fourth quadrant.

Example 10.19

State the signs of (a) sin 300°   (b) cos 170°   (c) tan 200°   (d) sin 500°
(e) sin 2.3   (f) cos 3   (g) tan 4.1

Solution
(a)   An angle of 300° lies in the fourth quadrant. Hence sin 300° is negative.
(b)   An angle of 170° is in the second quadrant, and so cos 170° is negative.
(c)   An angle of 200° is in the third quadrant, and so tan 200° is positive.
(d)   When the arm OC rotates one full revolution, it sweeps out an angle of 360°. An angle of 500° can be considered as one full revolution, plus an additional 140°. This means that the arm will be in the second quadrant, and so sin 500° is positive.
(e)   2.3 lies between $\frac{1}{2}\pi$ and π; that is, the angle is in the second quadrant. Hence sin 2.3 is positive.
(f)   3 lies between $\frac{1}{2}\pi$ and π, that is, in the second quadrant, and so cos 3 is negative.
(g)   4.1 lies between π and $\frac{3}{2}\pi$, that is, in the third quadrant, and so tan 4.1 is positive.

Example 10.20

The angle A is such that sin $A > 0$ and cos $A < 0$. In which quadrant does A lie?

Solution   The sine of an angle is positive in the first and second quadrants. The cosine of an angle is negative in the second and third quadrants. For both conditions to be satisfied, we must have A in the second quadrant.

Note that adding or subtracting 360° or 2π radians to an angle will not alter the position of the arm OC, and so the trigonometric ratios of the angle will remain unaltered. For example, $\sin 20° = \sin 380°$, $\cos 100° = \cos 460°$, $\tan 210° = \tan(-150°)$, $\sin 1.2 = \sin(2\pi + 1.2) = \sin 7.4832$.

Self-assessment questions 10.4

1. What is meant by the projection of an arm OC onto the x and y axes?

2. Explain how the trigonometric ratios of angles greater than 90° can be found using the extended definitions. Is it possible to find the trigonometric ratios of angles greater than 360° using the extended definitions?

Exercises 10.4

1. Verify, using a calculator that
   (a) $\sin 135° = \sin 45°$
   (b) $\sin 220° = -\sin 40°$
   (c) $\cos 210° = -\cos 30°$
   (d) $\tan 230° = \tan 50°$
   (e) $\cos 130° = -\cos 50°$
   (f) $\tan 170° = -\tan 10°$
   (g) $\sin 310° = -\sin 50°$
   (h) $\cos 320° = \cos 40°$
   (i) $\tan 315° = -\tan 45°$

2. Verify that

$$\sin 30° = \sin(360° + 30°)$$
$$= \sin(720° + 30°)$$
$$= \sin(1080° + 30°)$$

3. An angle θ is such that $\cos\theta > 0$ and $\tan\theta < 0$. In which quadrant does θ lie?

4. An angle x is such that $\tan x > 0$ and $\sin x < 0$. In which quadrant does x lie?

10.5 Trigonometric equations

We are now ready to solve some trigonometrical equations. The following examples illustrate the methods used.

Example 10.21

Find all values of θ from 0° to 360° with (a) $\sin\theta = 0.3612$ (b) $\cos\theta = -0.7522$
(c) $\tan\theta = -1.2118$

Solution   (a)   We note that $\sin\theta$ is positive, and so θ must be in the first or second quadrant. Using a calculator, we have

$$\theta = \sin^{-1}(0.3612) = 21.174°$$

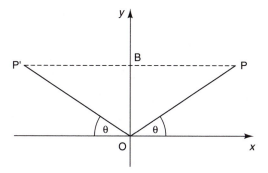

Figure 10.16   Arms OP and OP′ both have *y* projection of OB.

Clearly, this is the solution in the first quadrant. We now seek the solution in the second quadrant. Figure 10.16 shows the arms OP and OP′, both with the same *y* projection, OB.

The arm OP makes an angle of 21.174° with the *x* axis. From symmetry, OP′ makes an angle of 21.174° with the negative *x* axis. Hence OP′ makes an angle of $180° - 21.174° = 158.826°$ with the positive *x* axis. Since OP and OP′ have the same *y* projection, the sine of their angles is also the same. Hence

$$\sin 158.826° = \sin 21.174° = 0.3612$$

The values of θ are thus 21.17° and 158.83°.

(b)   Since $\cos \theta$ is negative the solutions lie in the second and third quadrants. Using a calculator, we see

$$\theta = \cos^{-1}(-0.7522) = 138.781°$$

This is the solution in the second quadrant. We now seek the solution in the third quadrant. Figure 10.17 shows OP making an angle of 138.781° with the positive *x* axis. Clearly,

$$\angle\, \text{POA} = 180° - 138.781° = 41.219°$$

Arm OP′ is drawn so that OP and OP′ have the same *x* projection. Then, by symmetry, $\angle\, \text{AOP}' = 41.219°$, and so the arm OP′ makes an angle of $180° + 41.219° = 221.219°$.

Since OP and OP′ have the same *x* projection, their angles have the same cosine. Hence

$$\cos 138.781° = \cos 221.219° = -0.7522$$

The solutions are 138.78° and 221.22°.

(c)   Since $\tan \theta$ is negative, there are solutions in the second and fourth quadrants. Using a calculator we find

$$\theta = \tan^{-1}(-1.2118) = -50.470°$$

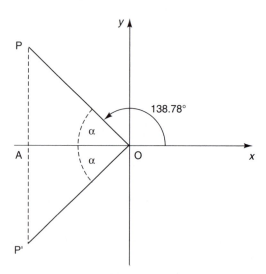

Figure 10.17   Arms OP and OP′ have the same *x* projection.

We require angles between 0° and 360°. We know that adding 360° to an angle leaves the position of the arm unchanged. So adding 360° to $-50.470°$ gives 309.530°, and hence tan 309.530° $= -1.2118$. Hence the angle in the fourth quadrant is 309.530°.

We now seek the solution in the second quadrant. Figure 10.18 shows OP making an angle of 309.530° with the *x* axis. The *x* projection is OA, which is positive, and the *y* projection is OB, which is negative. We draw OP′ in the second quadrant. The *x* projection is OA′, which is negative, and the *y* projection is OB′, which is positive. Note that OA and OA′ have the same length but opposite signs, as do OB and OB′. Then, clearly,

$$\frac{OB}{OA} = \frac{OB'}{OA'}$$

Recall that the tangent of an angle is defined as

$$\frac{y \text{ projection}}{x \text{ projection}}$$

Thus the tangents of the angles of OP and OP′ are the same. But OP′ makes an angle of 309.530° $- 180° = 129.530°$ with the *x* axis. Hence

$$\tan 129.530° = \tan 309.530° = -1.2118$$

Hence the required solutions are 129.53° and 309.53°. In summary, to find the angle in the second quadrant, we simply subtract 180° from the angle in the fourth quadrant.

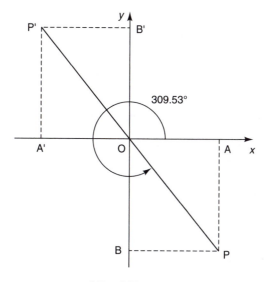

309.53°

Figure 10.18 $\dfrac{\text{OB}}{\text{OA}} = \dfrac{\text{OB}'}{\text{OA}'}.$

Example 10.22

Find all values of t between 0 and 4π such that $\sin t = 0.3612$.

Solution We use radian measure. Using a scientific calculator, we have

$$t = \sin^{-1}(0.3612) = 0.3696 \text{ radians}$$

This is the solution between 0 and $\frac{1}{2}\pi$. Now, using the same method as in Example 10.21, we find the angle in the second quadrant to be

$$\pi - 0.3696 = 2.7720 \text{ radians}$$

Hence we have $t = 2.7720$ radians. This is the solution between $\frac{1}{2}\pi$ and π. Thus, between 0 and 2π we have solutions $t = 0.3696$ radians and $t = 2.7720$ radians.

Adding 2π radians to an angle does not change the position of the arm. Hence the solutions between 2π and 4π are obtained by adding 2π to the existing solutions; that is,

$$t = 2\pi + 0.3696 = 6.6528 \text{ radians}, \qquad t = 2\pi + 2.7720 = 9.0552 \text{ radians}$$

The values of t are 0.3696, 2.7720, 6.6528 and 9.0552 radians. Note that there are other values of t such that $\sin t = 0.3612$, but these lie outside the range 0 to 4π.

Example 10.23

Find all values of t between 0 and 2π such that $\sin 2t = 0.3612$.

Solution   Let $\theta = 2t$, so that $\sin \theta = 0.3612$. As t varies from 0 to 2π radians, θ varies from 0 to 4π radians. So, initially, we seek values of θ between 0 and 4π radians with $\sin \theta = 0.3612$. From Example 10.22, this leads to

$$\theta = 0.3696, \ 2.7720, \ 6.6528, \ 9.0552 \text{ radians}$$

and so the required values of t are

$$t = \tfrac{1}{2}\theta = 0.1848, \ 1.3860, \ 3.3264, \ 4.5276 \text{ radians}$$

Self-assessment questions 10.5

1.   If $\sin \theta = 0.7899$, in which quadrants do the solutions lie?

2.   If $\sin \theta = -0.4367$, in which quadrants do the solutions lie?

3.   If $\cos \theta = -0.8966$, in which quadrants do the solutions lie?

4.   If $\cos \theta = 0.3456$, in which quadrants do the solutions lie?

5.   If $\tan \theta = 2.4568$, in which quadrants do the solutions lie?

6.   If $\tan \theta = -1.4683$, in which quadrants do the solutions lie?

Exercises 10.5

1.   Solve the following trigonometric equations, giving all solutions between $0°$ and $360°$
(a) $\sin t = 0.7834$ (b) $\sin t = -0.2391$
(c) $\cos z = 0.5921$ (d) $\cos z = -0.8941$
(e) $\tan x = 1.5691$ (f) $\tan x = -0.3461$

2.   Solve the following equations, giving all solutions between 0 and 2π:
(a) $\sin x = 0.6701$ (b) $\sin z = -0.2222$
(c) $\cos t = 0.3421$ (d) $\cos t = -0.8933$
(e) $\tan y = 2.5$ (f) $\tan y = -0.3560$

3.   Solve the following trigonometric equations:
(a) $\cos 3t = 0.1249$ for $0 \leqslant t \leqslant 2\pi$
(b) $\tan 2t = 1.2056$ for $0 \leqslant t \leqslant 2\pi$
(c) $\sin 4t = 0.5000$ for $0 \leqslant t \leqslant \pi$

(d) $\sin (3t - 2) = -0.6500$ for $0 \leqslant t \leqslant \pi$
(e) $\cos (2t + 3) = -0.7516$ for $0 \leqslant t \leqslant \pi$
(f) $\tan (\tfrac{1}{2}t - 1) = -1.2300$ for $0 \leqslant t \leqslant 2\pi$
(g) $2 \sin (3t + 4) = 1.5680$ for $0 \leqslant t \leqslant \pi$

4.   Solve the following equations, giving all solutions between $0°$ and $360°$:
(a) $\sin (2\theta) = 0.6780$
(b) $2 \sin \theta = -0.6780$
(c) $\cos (3\theta) = -0.7894$
(d) $2 \cos (2\theta) = 1.5682$
(e) $\cos (\theta + 20°) = -0.5551$
(f) $\cos (3\theta - 50°) = 0.7845$
(g) $\tan (4\theta) = 1.4666$
(h) $\tan (\tfrac{\theta}{3}) = -0.5882$
(i) $\tan (3\theta + 40°) = 1.4433$

Review exercises 10

1 Convert the following angles in degrees to radians:
(a) $10°$ (b) $215°$ (c) $1000°$
(d) $-130°$ (e) $500°$

2 Convert the following angles in radians to degrees:

(a) 3.6 (b) $-\pi$ (c) $-\dfrac{\pi}{4}$ (d) 2.16

(e) 5

3 A circle has a radius of 12 cm. An arc of the circle has a length of 20 cm. Calculate the angle subtended by the arc at the centre of the circle. Give your answer in radians.

4 An arc of length 12 cm subtends an angle of 1.2 radians at the centre of a circle. Calculate the radius of the circle.

5 Figure 10.7 shows a right-angled triangle $\triangle ABC$. If $AC = 9$ cm and $BC = 6$ cm, calculate
(a) $\angle A$ (b) $\angle C$

6 Solve the following equations:
(a) $\sin 2t = 0.6$ for $0 \leqslant t \leqslant \pi$
(b) $\cos 2t = 0.6$ for $0 \leqslant t \leqslant \pi$
(c) $\tan 2t = 0.6$ for $0 \leqslant t \leqslant \pi$
(d) $\sin (\tfrac{1}{2}t) = 0.5$ for $0 \leqslant t \leqslant \tfrac{3}{2}\pi$
(e) $\cos (\tfrac{1}{2}t) = 0.5$ for $-\tfrac{1}{2}\pi \leqslant t \leqslant \tfrac{1}{2}\pi$
(f) $\tan (\tfrac{1}{2}t) = 0.5$ for $-\pi \leqslant t \leqslant \pi$
(g) $3 \cos (t + 0.7) = 2$ for $0 \leqslant t \leqslant 2\pi$

7 Solve the following equations, giving all solutions between $0°$ and $360°$:
(a) $\sin \theta = 0.7$ (b) $\sin \theta = -0.7$
(c) $\cos \theta = 0.7$ (d) $\cos \theta = -0.7$
(e) $\tan \theta = 0.7$ (f) $\tan \theta = -0.7$
(g) $\sin (\theta + 20°) = 0.7$
(h) $\cos (\theta + 30°) = -0.7$ (i) $\tan 2\theta = 0.7$

11

Partial Fractions

KEY POINTS

This chapter

- explains the terms 'proper' and 'improper' as applied to algebraic fractions

- shows how to calculate partial fractions of a proper fraction when the denominator contains only linear factors

- explains the term 'repeated linear factor'

- shows how to calculate the partial fractions of a proper fraction containing repeated linear factors

- explains how to find the partial fractions of a proper fraction containing a quadratic factor in the denominator

- illustrates how to calculate the partial fractions of improper fractions

CONTENTS

11.1 Introduction

A complicated algebraic fraction may be broken down into the sum of much simpler fractions, called **partial fractions**. For example, it can be shown that

$$\frac{3x+5}{x^2+3x+2} = \frac{2}{x+1} + \frac{1}{x+2}$$

We say that the partial fractions of $\dfrac{3x+5}{x^2+3x+2}$ are

$$\frac{2}{x+1} \quad \text{and} \quad \frac{1}{x+2}$$

It is often easier to deal with a few relatively easy fractions than with one complicated fraction. This chapter explains the technique of expressing a fraction as a sum of its partial fractions.

11.2 Proper and improper fractions

We can write an algebraic fraction in the form:

$$\text{algebraic fraction} = \frac{\text{numerator}}{\text{denominator}} = \frac{\text{polynomial expression}}{\text{polynomial expression}}$$

The **degree** of a polynomial is the highest power occurring in the polynomial. For example, $3x^4 + 7x^2 - 8$ is a polynomial of degree 4. When presented with a fraction, we can note the degree of the numerator, say n, and the degree of the denominator, say d.

A fraction is **proper** if $d > n$, that is, the degree of the denominator is greater than the degree of the numerator. If $d \leqslant n$ then it is **improper**.

Example 11.1

Classify the following fractions as either proper or improper. In each case state the degree of both numerator and denominator.

(a) $\dfrac{x^2+3x+2}{x^3+1}$

(b) $\dfrac{3t^3+4t+1}{t^5+6}$

(c) $\dfrac{(y+1)(y+2)}{y^2+y+2}$

(d) $\dfrac{3z^4+z^3-2z+3}{z^3+3z+7}$

(e) $\dfrac{(x+1)^3}{x^2+2x+3}$

Solution   (a)   The degree of the numerator, n, is 2. The degree of the denominator, d, is 3. Since $d>n$, the fraction is proper.

(b)   Here $n=3$ and $d=5$. The fraction is proper, since $d>n$.

(c)   Expanding $(y+1)(y+2)$ produces y^2+3y+2, that is, a polynomial of degree 2, and so $n=2$. The degree of the denominator, d, is 2. Therefore $d=n$, and the fraction is improper.

(d) $n=4$, $d=3$ and so $d<n$. The fraction is improper.

(e)   Expanding the numerator produces a polynomial of degree 3. The degree of the denominator is 2, and so $d<n$ and the fraction is improper.

The denominator of an algebraic fraction can be factorized into a product of linear and quadratic factors. **Linear factors** are those of the form $ax+b$, for example, $3x+2$, $2x-7$ and $4-x$. **Quadratic factors** are those of the form ax^2+bx+c, for example x^2+x+1. Quadratic factors cannot be factorized.

Self-assessment questions 11.2

1.   Is it possible to have an algebraic fraction that is neither proper nor improper? Give a reason for your answer.

2.   (a)   Suppose a proper fraction is inverted. What kind of fraction results: proper or improper?

(b)   A fraction is improper. It is inverted. What kind of fraction results?

Exercise 11.2

1.   For each fraction, state the degree of both numerator and denominator. Hence decide whether the fraction is proper or improper.

(a) $\dfrac{3x+2}{x^2+9}$

(b) $\dfrac{2x^3-1}{4x^3+x^2+9}$

(c) $\dfrac{1}{t+1}$

(d) $\dfrac{3y^2+9y-1}{4y+7}$

(e) $\dfrac{10z^2+1}{3z^3+2z-1}$

(f) $\dfrac{t}{2}$

(g) $\dfrac{z-1}{z+1}$

11.3 Linear factors

In this section we show how to calculate the partial fractions of proper fractions whose denominators may be expressed as a product of linear factors. The steps needed to calculate partial fractions are as follows.

(a)   First, factorize the denominator.

(b)   Each factor of the denominator produces a partial fraction. A factor of the form $ax+b$ produces a partial fraction of the form $\dfrac{A}{ax+b}$, where A is an unknown constant.

(c)   Evaluate the unknown constants of the partial fractions by evaluation using a specific value of x or by equating coefficients.

A linear factor $ax+b$ produces a partial fraction of the form $\dfrac{A}{ax+b}$

Example 11.2

Express as partial fractions $\dfrac{5x-4}{x^2-x-2}$.

Solution   The first step is to factorize the denominator.

$$x^2-x-2=(x+1)(x-2)$$

The factor $(x+1)$ is a linear factor, and so produces a partial fraction of the form $\dfrac{A}{x+1}$. The factor $x-2$ is also a linear factor, and so produces a partial fraction of the form $\dfrac{B}{x-2}$. Hence

$$\frac{5x-4}{x^2-x-2}=\frac{5x-4}{(x+1)(x-2)}=\frac{A}{x+1}+\frac{B}{x-2}$$

where A and B are constants still to be found. Writing the right-hand side using a common denominator, we have

$$\frac{5x-4}{(x+1)(x-2)}=\frac{A(x-2)+B(x+1)}{(x+1)(x-2)}$$

Multiplying both sides by $(x+1)(x-2)$, we obtain

$$5x-4=A(x-2)+B(x+1) \tag{11.1}$$

We shall first demonstrate how to find A and B by **evaluating using a specific value of** x. By appropriate choice of the value of x, the right-hand side simplifies considerably.

Evaluating Equation (11.1) when $x=2$ gives

$$6 = A(0) + B(3)$$

$$3B = 6$$

$$B = 2$$

Evaluating Equation (11.1) when $x = -1$ yields

$$-9 = A(-3) + B(0)$$

$$-3A = -9$$

$$A = 3$$

Substituting in the values for A and B yields

$$\frac{5x-4}{x^2-x-2} = \frac{3}{x+1} + \frac{2}{x-2}$$

Therefore the partial fractions of $\dfrac{5x-4}{x^2-x-2}$ are

$$\frac{3}{x+1} \quad \text{and} \quad \frac{2}{x-2}$$

The constants A and B can also be found by **equating coefficients**. Consider again Equation (11.1). We have

$$5x - 4 = A(x-2) + B(x+1)$$

$$= Ax - 2A + Bx + B$$

$$= (A+B)x + B - 2A$$

Comparing the coefficients of x on the left- and right-hand sides gives

$$5 = A + B$$

Comparing the constant terms on each side gives

$$-4 = B - 2A$$

These simultaneous equations in A and B can be solved to yield $A=3$, $B=2$. Often a combination of comparing coefficients and evaluation is used.

Example 11.3

Express as partial fractions $\dfrac{x-1}{6x^2+5x+1}$.

Solution   The denominator is first factorized:

$$6x^2 + 5x + 1 = (2x+1)(3x+1)$$

Each factor produces a partial fraction:

$$\frac{x-1}{6x^2+5x+1} = \frac{x-1}{(2x+1)(3x+1)} = \frac{A}{2x+1} + \frac{B}{3x+1}$$

We multiply by $(2x+1)$ and $(3x+1)$ to obtain

$$x - 1 = A(3x+1) + B(2x+1) \tag{11.2}$$

By putting $x = -\frac{1}{3}$ in Equation (11.2), we obtain

$$-\tfrac{4}{3} = A(0) + B(\tfrac{1}{3})$$

$$B = -4$$

By putting $x = -\frac{1}{2}$ in Equation (11.2), we obtain

$$-\tfrac{3}{2} = A(-\tfrac{1}{2}) + B(0)$$

$$A = 3$$

Hence the given fraction can be expressed as its partial fractions:

$$\frac{x-1}{6x^2+5x+1} = \frac{3}{2x+1} - \frac{4}{3x+1}$$

Example 11.4

Express as partial fractions $\dfrac{s+4}{s^2+s}$.

Solution   The denominator is first factorized:

$$s^2 + s = s(s+1)$$

Note that both s and $s+1$ are linear factors. Each factor leads to a partial fraction:

$$\frac{s+4}{s^2+s} = \frac{s+4}{s(s+1)} = \frac{A}{s} + \frac{B}{s+1}$$

By multiplying through by s and $s+1$, we obtain

$$s + 4 = A(s+1) + Bs \tag{11.3}$$

Evaluating Equation (11.3) when $s=0$ gives

$$4 = A(0+1) + B(0)$$

$$A = 4$$

Evaluating Equation (11.3) when $s = -1$ gives

$$3 = A(0) + B(-1)$$

$$B = -3$$

Hence

$$\frac{s+4}{s^2+s} = \frac{4}{s} - \frac{3}{s+1}$$

Self-assessment question 11.3

1. A fraction has been expressed as its partial fractions. If all the partial fractions are added together, the original fraction will be formed. True or false?

Exercise 11.3

1. Express as partial fractions

(a) $\dfrac{x+8}{x^2+4x}$ (b) $\dfrac{y-10}{y^2-4}$ (c) $\dfrac{9x+16}{x^2+3x+2}$ (h) $\dfrac{3s+1}{s^2-2s-3}$ (i) $\dfrac{3+t}{t^2-1}$

(d) $\dfrac{x+14}{x^2+3x-4}$ (e) $\dfrac{11x-23}{x^2-x-6}$ (j) $\dfrac{4s+1}{2s^2+3s+1}$ (k) $\dfrac{x-7}{x^2+5x+6}$

(f) $\dfrac{2x+6}{8x^2+18x+9}$ (g) $\dfrac{x+12}{6x^2+7x-3}$ (l) $\dfrac{7y-6}{3y^2-4y-4}$ (m) $\dfrac{6x^2+13x+6}{x^3+3x^2+2x}$

11.4 Repeated linear factors

When factorizing a denominator, sometimes the same factor occurs more than once; for example, in

$$\frac{1}{x^2+2x+1} = \frac{1}{(x+1)(x+1)} = \frac{1}{(x+1)^2}$$

the factor $(x+1)$ occurs twice. As a further example, in

$$\frac{1}{x^3+3x^2+3x+1} = \frac{1}{(x+1)(x+1)(x+1)} = \frac{1}{(x+1)^3}$$

the factor $(x+1)$ occurs three times. A linear factor that occurs more than once is called a **repeated linear factor**.

If a factor occurs twice, this will generate two partial fractions; if it occurs three times, it will generate three partial fractions, and so on. The factors $(ax+b)^2$ in a denominator generate partial fractions of the form

$$\frac{A}{ax+b} + \frac{B}{(ax+b)^2}$$

Repeated linear factors $(ax+b)^2$ generate partial fractions of the form

$$\frac{A}{ax+b} + \frac{B}{(ax+b)^2}$$

Example 11.5

Express as partial fractions $\dfrac{3x+7}{x^2+4x+4}$

Solution The denominator is factorized to $(x+2)^2$. Hence $(x+2)$ is a repeated linear factor. So

$$\frac{3x+7}{x^2+4x+4} = \frac{3x+7}{(x+2)^2} = \frac{A}{x+2} + \frac{B}{(x+2)^2}$$

By multiplying through by $(x+2)^2$, we get

$$3x+7 = A(x+2)+B \qquad\qquad \textbf{(11.4)}$$

Equating the x coefficients of Equation (11.4) gives

$$A = 3$$

Evaluating Equation (11.4) with $x = -2$ gives

$$1 = B$$

So we may write

$$\frac{3x+7}{x^2+4x+4} = \frac{3}{x+2} + \frac{1}{(x+2)^2}$$

Example 11.6

Express as partial fractions $\dfrac{6x-10}{(x^2-6x+9)(x+1)}$

Solution   The denominator factorizes to $(x-3)^2(x+1)$. Hence

$$\frac{6x-10}{(x^2-6x+9)(x+1)} = \frac{6x-10}{(x-3)^2(x+1)} = \frac{A}{x-3} + \frac{B}{(x-3)^2} + \frac{C}{x+1}$$

Multiplying through by $(x-3)^2(x+1)$ gives

$$6x-10 = A(x-3)(x+1) + B(x+1) + C(x-3)^2 \qquad \textbf{(11.5)}$$

Evaluating Equation (11.5) with $x=3$ gives

$$8 = B(4)$$

$$B = 2$$

Evaluating Equation (11.5) with $x=-1$ gives

$$-16 = C(-4)^2$$

$$C = -1$$

Putting $B=2$ and $C=-1$ into Equation (11.5) gives

$$6x-10 = A(x-3)(x+1) + 2(x+1) - (x-3)^2$$

$$= A(x^2-2x-3) + 2(x+1) - (x^2-6x+9)$$

$$= (A-1)x^2 + (8-2A)x - 7 - 3A$$

Equating the x^2 coefficient of both sides, we see

$$0 = A - 1$$

$$A = 1$$

and so

$$\frac{6x-10}{(x^2-6x+9)(x+1)} = \frac{1}{x-3} + \frac{2}{(x-3)^2} - \frac{1}{x+1}$$

Exercise 11.4

1.   Express as partial fractions

(a) $\dfrac{2t+11}{t^2+6t+9}$

(b) $\dfrac{2x^2+3}{(x+1)^2(x+2)}$

(c) $\dfrac{3x+1}{x^2+4x+4}$

(d) $\dfrac{2-x}{x^2+10x+25}$

(e) $\dfrac{x^2+x+1}{(x-1)(x^2-1)}$

(f) $\dfrac{s-4}{s^2-4s+4}$

(g) $\dfrac{10-3y}{9-6y+y^2}$

(h) $\dfrac{s-3}{s^2-8s+16}$

(i) $\dfrac{6z+39}{z^2+14z+49}$

(j) $\dfrac{x^2+1}{(x+1)^3}$

(k) $\dfrac{2s+1}{(s-1)^3}$

11.5 Quadratic factors

Sometimes a denominator is factorized, producing quadratic factors that cannot be factorized any further. For example,

$$x^3 + x^2 - 5x - 2 = (x - 2)(x^2 + 3x + 1)$$

The quadratic expression $x^2 + 3x + 1$ cannot be factorized further. In general, a quadratic factor of the form $ax^2 + bx + c$ that cannot be factorized produces a partial fraction of the form

$$\frac{Ax + B}{ax^2 + bx + c}$$

A quadratic factor of the form $ax^2 + bx + c$ produces a partial fraction of the form

$$\frac{Ax + B}{ax^2 + bx + c}$$

Example 11.7

Express as partial fractions $\dfrac{5x^2 + 4x + 11}{(x^2 + x + 4)(x + 1)}$.

Solution   The denominator has already been factorized. The quadratic factor $x^2 + x + 4$ cannot be factorized further. Each factor produces a partial fraction. Thus

$$\frac{5x^2 + 4x + 11}{(x^2 + x + 4)(x + 1)} = \frac{Ax + B}{x^2 + x + 4} + \frac{C}{x + 1}$$

Multiplying through by $(x^2 + x + 4)(x + 1)$ produces

$$5x^2 + 4x + 11 = (Ax + B)(x + 1) + C(x^2 + x + 4) \qquad \text{(11.6)}$$

Evaluating Equation (11.6) with $x = -1$ gives

$$12 = 4C$$

$$C = 3$$

Putting $C = 3$ in Equation (11.6) gives

$$5x^2 + 4x + 11 = (Ax + B)(x + 1) + 3(x^2 + x + 4)$$

Removing brackets results in

$$5x^2 + 4x + 11 = Ax^2 + Ax + Bx + B + 3x^2 + 3x + 12$$

which simplifies to

$$2x^2 + x - 1 = Ax^2 + Ax + Bx + B \qquad \textbf{(11.7)}$$

Equating coefficients of x^2 in Equation (11.7) yields

$$A = 2$$

Equating constant terms of Equation (11.7) gives

$$B = -1$$

Hence we can write

$$\frac{5x^2 + 4x + 11}{(x^2 + x + 4)(x + 1)} = \frac{2x - 1}{x^2 + x + 4} + \frac{3}{x + 1}$$

Exercise 11.5

1. Express as partial fractions

    (a) $\dfrac{10x^2 + x + 7}{x^3 + x}$ (b) $\dfrac{-x^2 - 8}{2x^3 - x^2 + 4x}$

    (c) $\dfrac{8x^2 + 5}{(x^2 + 1)(2x^2 + 1)}$

    (d) $\dfrac{3x^3 + 7x - 3}{(x^2 + x + 1)(x^2 + 3)}$

    (e) $\dfrac{11x^3 - 16x^2 + 14x - 67}{(2x^2 + x + 9)(3x^2 + 4)}$

    (f) $\dfrac{s^3 + s^2 + s + 3}{(s^2 + 1)(s^2 + 3)}$ (g) $\dfrac{5y^2 + 3y + 2}{(2y^2 + 1)(y + 3)}$

    (h) $\dfrac{3x^2 - 2x - 2}{(x^2 - x - 1)(x - 1)}$ (i) $\dfrac{2t^2 + 3t}{(t^2 + 1)(t + 2)}$

    (j) $\dfrac{x^3 + 2x^2 + 5x + 5}{(x^2 + 1)(x^2 + 4x + 4)}$

11.6 Partial fractions of improper fractions

The techniques of calculating partial fractions in Sections 11.3–11.5 have all been applied to proper fractions. We now turn to the calculation of partial fractions of improper fractions. When calculating the partial fractions of improper fractions, an extra term needs to be included. The extra term is a polynomial of degree $n - d$, where n is the degree of the numerator and d is the degree of the denominator. Recall that a polynomial of degree 0 is a constant, a polynomial of degree 1 has the form $Ax + B$, a polynomial of degree 2 has the form $Ax^2 + Bx + C$, and so on. For example, if the numerator has degree 3 and the denominator has degree 2, the partial fractions will include a polynomial of degree 1, that is, a term of the form $Ax + B$. If the numerator and denominator are of the same degree, the fraction is improper. The partial fractions will include a polynomial of degree 0, that is, a constant term.

Example 11.8

Express as partial fractions $\dfrac{3x^2 + 2x}{x + 1}$

Solution   The degree of the numerator, n, is 2; the degree of the denominator, d, is 1. Thus the fraction is improper. The partial fractions will include a polynomial of degree $n - d$, that is, a polynomial of degree 1. A polynomial of degree 1 has the form $Ax + B$. The denominator comprises only one factor: $x + 1$. Hence

$$\frac{3x^2 + 2x}{x + 1} = Ax + B + \frac{C}{x + 1}$$

Multiplying both sides by $x + 1$ yields

$$3x^2 + 2x = (Ax + B)(x + 1) + C$$

Evaluation when $x = -1$ gives

$$3 - 2 = C$$

$$C = 1$$

Hence

$$3x^2 + 2x = (Ax + B)(x + 1) + 1 = Ax^2 + Ax + Bx + B + 1$$

Comparing the x^2 coefficients gives

$$3 = A$$

Comparing the constant terms gives

$$0 = B + 1$$

$$B = -1$$

Hence

$$\frac{3x^2 + 2x}{x + 1} = 3x - 1 + \frac{1}{x + 1}$$

Example 11.9

Express as partial fractions

$$\frac{x^4 + 2x^3 - 2x^2 + 4x - 1}{x^2 + 2x - 3}$$

Solution The degree of the numerator, n, is 4, the degree of the denominator, d, is 2, giving $n-d$ a value of 2. Thus the partial fractions contain a polynomial of degree 2, that is, a term of the form $Ax^2 + Bx + C$. The denominator is factorized to $(x+3)(x-1)$. Each of these two linear factors produces a partial fraction. So

$$\frac{x^4+2x^3-2x^2+4x-1}{x^2+2x-3} = \frac{x^4+2x^3-2x^2+4x-1}{(x+3)(x-1)}$$

$$= Ax^2 + Bx + C + \frac{D}{x+3} + \frac{E}{x-1}$$

Multiplying through by $(x+3)(x-1)$ yields

$$x^4+2x^3-2x^2+4x-1=(Ax^2+Bx+C)(x+3)(x-1)+D(x-1)+E(x+3)$$

Evaluating with $x=1$, we find

$$4=E(1+3)$$

$$E=1$$

Evaluating with $x=-3$, we find

$$-4=D(-3-1)$$

$$D=1$$

Thus

$$x^4+2x^3-2x^2+4x-1=(Ax^2+Bx+C)(x+3)(x-1)+(x-1)+(x+3)$$

$$x^4+2x^3-2x^2+2x-3=(Ax^2+Bx+C)(x+3)(x-1)$$

$$=(Ax^2+Bx+C)(x^2+2x-3)$$

$$=Ax^4+2Ax^3-3Ax^2+Bx^3$$

$$+2Bx^2-3Bx+Cx^2+2Cx-3C$$

$$=Ax^4+(2A+B)x^3+(2B+C-3A)x^2$$

$$+(2C-3B)x-3C$$

Comparing the x^4 coefficients gives

$$1=A$$

Comparing the constant terms gives

$$-3=-3C$$

$$C=1$$

Comparing the x^3 coefficients gives

$$2=2A+B$$

$$B=2-2A=0$$

Hence

$$\frac{x^4+2x^3-2x^2+4x-1}{x^2+2x-3}=x^2+1+\frac{1}{x+3}+\frac{1}{x-1}$$

Example 11.10

Express as partial fractions

$$\frac{4x^3+12x^2+13x+7}{4x^2+4x+1}$$

Solution   The fraction is improper, with $n-d=1$. Thus the partial fractions will include a polynomial of degree 1. The denominator has repeated linear factors of $(2x+1)^2$. So

$$\frac{4x^3+12x^2+13x+7}{4x^2+4x+1}=\frac{4x^3+12x^2+13x+7}{(2x+1)^2}$$

$$=Ax+B+\frac{C}{2x+1}+\frac{D}{(2x+1)^2}$$

By multiplying by $(2x+1)^2$, we obtain

$$4x^3+12x^2+13x+7=(Ax+B)(2x+1)^2+C(2x+1)+D$$

Evaluating with $x=-\frac{1}{2}$ produces

$$3=D$$

So

$$4x^3+12x^2+13x+7=(Ax+B)(2x+1)^2+C(2x+1)+3$$

$$4x^3+12x^2+13x+4=(Ax+B)(2x+1)^2+C(2x+1)$$

$$=(Ax+B)(4x^2+4x+1)+2Cx+C$$

$$=4Ax^3+4Ax^2+Ax+4Bx^2+4Bx+B+2Cx+C$$

$$=4Ax^3+(4A+4B)x^2+(A+4B+2C)x+B+C$$

Equating the x^3 coefficients gives

$$4=4A$$

$$A=1$$

Equating the x^2 coefficients gives

$$12=4A+4B$$

$$12=4+4B$$

$$B=2$$

Equating constant terms gives

$$4 = B + C$$

$$C = 2$$

Finally, we can write

$$\frac{4x^3 + 12x^2 + 13x + 7}{4x^2 + 4x + 1} = x + 2 + \frac{2}{2x+1} + \frac{3}{(2x+1)^2}$$

Example 11.11

Express as partial fractions

$$\frac{6x^3 + x^2 + 5x - 1}{x^3 + x}$$

Solution   The degrees of both numerator and denominator are 3, and so the fraction is improper. The partial fractions contain a polynomial of degree 0, that is, a constant. The denominator is factorized to $x(x^2 + 1)$. The linear factor x produces a partial fraction of the form $\dfrac{B}{x}$, and the quadratic factor $x^2 + 1$ produces a partial fraction of the form $\dfrac{Cx + D}{x^2 + 1}$. Hence

$$\frac{6x^3 + x^2 + 5x - 1}{x^3 + x} = \frac{6x^3 + x^2 + 5x - 1}{x(x^2 + 1)} = A + \frac{B}{x} + \frac{Cx + D}{x^2 + 1}$$

Multiplying through by $x(x^2 + 1)$ yields

$$6x^3 + x^2 + 5x - 1 = Ax(x^2 + 1) + B(x^2 + 1) + (Cx + D)x$$

Evaluating with $x = 0$ gives

$$-1 = B$$

Hence we have

$$6x^3 + x^2 + 5x - 1 = Ax(x^2 + 1) - (x^2 + 1) + (Cx + D)x$$

$$6x^3 + 2x^2 + 5x = Ax(x^2 + 1) + (Cx + D)x$$

$$= Ax^3 + Cx^2 + (A + D)x$$

Equating the x^3 coefficients gives

$$6 = A$$

Equating the x^2 coefficients gives

$$2 = C$$

Equating the x coefficient gives

$$5 = A + D$$

$$D = 5 - A = -1$$

Finally, we may write

$$\frac{6x^3 + x^2 + 5x - 1}{x^3 + x} = 6 - \frac{1}{x} + \frac{2x - 1}{x^2 + 1}$$

Self-assessment question 11.6

1. In what way do the partial fractions of an improper fraction differ from those of a proper fraction?

Exercise 11.6

1. Express as partial fractions

(a) $\dfrac{y^2 + 8y + 11}{y^2 + 4y + 4}$ (b) $\dfrac{z^3 + 3}{4z^2 + 4z + 1}$

(c) $\dfrac{x^2 + 8x + 12}{x^2 + 6x + 9}$ (d) $\dfrac{y^3 + y + 7}{y^2 + 2y + 1}$

(e) $\dfrac{2x^3 + 1}{x^2 + 1}$ (f) $\dfrac{x^4}{(x^2 + 1)(x^2 - 1)}$

(g) $\dfrac{3t^2 + t}{t^2 + 3t + 3}$ (h) $\dfrac{3t^2 + 12t - 13}{t^2 + 3t - 12}$

(i) $\dfrac{9t^2 + 12t + 5}{3t + 2}$ (j) $\dfrac{2x^3 + 12x^2 + 25x + 19}{x^2 + 5x + 6}$

Computer and calculator exercises 11

1. Use a computer algebra package to express the following as partial fractions:

(a) $\dfrac{2x - 4}{x(x - 1)(x - 3)}$ (b) $\dfrac{x^2 + x + 1}{(x - 1)^4(x + 2)}$

(c) $\dfrac{x(2x^2 - x + 3)}{(x^2 + 1)^2}$ (d) $\dfrac{x^4}{x^6 - 1}$

(e) $\dfrac{s - 2}{s^6(s^2 + 1)}$

2. Factorize $P(x) = x^5 - x^4 - 8x^3 - 8x^2 - x + 1$.

Express $\dfrac{x - 1}{P(x)}$ as partial fractions.

Review exercise 11

1 Express as partial fractions

(a) $\dfrac{s^2}{(s-1)(s-2)(s-3)}$

(b) $\dfrac{x}{(x+1)^2(x+2)}$

(i) $\dfrac{x^2-2x+1}{x^2-8x+16}$

(j) $\dfrac{6x+6}{x^2+3x}$

(c) $\dfrac{x^2+x-1}{x-1}$

(d) $\dfrac{s}{s(s-1)(s-2)}$

(k) $\dfrac{-4s+1}{2s^2+5s+2}$

(l) $\dfrac{x^2-x}{(x+1)(x^2+1)}$

(e) $\dfrac{s}{s(s-1)^2}$

(f) $\dfrac{x}{(x+1)(x^2+2)}$

(m) $\dfrac{x^4}{x^2+1}$

(n) $\dfrac{t^2+2t-1}{t+1}$

(g) $\dfrac{3-t^2}{(1+t)^2}$

(h) $\dfrac{y^3}{y^2-1}$

(o) $\dfrac{-x^2-2x-1}{1-x^2}$

Estimation

KEY POINTS

This chapter

- discusses situations in which a simplified calculation can be used

- analyses typical simplifications of several problems

- discusses the ways in which engineering variables are related in a formula

- illustrates quick methods of determining the effect of changing engineering variables

CONTENTS

12.1 Introduction

An engineer may have to carry out many calculations during a working day. Often only an approximate solution is required – usually when it is necessary to gain a feel for a situation. This is when the ability to estimate becomes vital. It is impossible to categorize all the situations in which estimation can be used. Fortunately, it is not necessary to do so, since the ability to estimate depends more on adopting the right attitude than on rigidly identifying certain types of problem. In this chapter we shall examine several different types of estimation. The aim is to give the reader confidence to identify appropriate situations in which a rough calculation can be used and to present some of the techniques needed to carry out such calculations. These calculations are often called 'back of the envelope' calculations precisely because they are simple enough to be worked out on the back of an envelope.

12.2 Simplifying a calculation

Suppose we are interested in the number of bricks required to construct a square building of side length 20 m and height 10 m, including foundations. The roof is to be flat. Let us assume that one square metre of wall requires 50 bricks. We could argue as follows. The area of one wall is the product of its length and its height; that is,

$$\text{area of one wall} = 20 \times 10 = 200 \text{ m}^2$$

With four walls, we have

$$\text{total area of walls} = 4 \times 200$$
$$= 800 \text{ m}^2$$

Given that 50 bricks have an area of 1 m^2, we shall require $800 \times 50 = 40\,000$ bricks in total.

In order to simplify the calculation, it is evident that a number of assumptions have been made. First, we ignored the presence of doors and windows in the building. Clearly, including these would have made the calculation much more complicated, since it would have been necessary to work out all of their areas, add these together and deduct them from the total area. This is unnecessary for a rough calculation because the effect of ignoring the doors and windows is to overestimate the number of bricks required, and so there is no danger of not ordering enough bricks. Also, casual observation of a building shows that doors and windows do not usually make up a large proportion of the total wall area, and so the error due to ignoring them will not be too large. Secondly, in carrying out our calculations we ignored the fact that when walls meet, the bricks interleave, and so slightly fewer bricks are needed than we estimated. However, some thought makes it clear that this is a very small difference and can be safely ignored in a rough calculation. This example illustrates several points concerning carrying out a rough calculation or estimation. These can be conveniently organized as the following checklist.

> 1. Before carrying out a rough calculation decide whether such a calculation is appropriate.

For example, it is not appropriate to do a rough calculation of the amount of radiation needed for cancer treatment. Such a calculation needs to be exact for the obvious reason that it is life-critical. It is useful to estimate the number of bricks needed to build a structure, since this can be used to gain a rough idea of such things as the cost, the amount of space the bricks will take up on site when stored before they are used or the amount of cement that will be required to lay the bricks. The key question to ask at this stage is whether a small error in the calculation is of any importance. If it is then it is not appropriate to do a rough calculation.

> 2. If a rough calculation is appropriate then decide what simplifications can be made without affecting the answer too much.

Returning to the example of the number of bricks needed to build a building, we see that ignoring the effect of the bricks interleaving at the corners is safe in a rough calculation. However, ignoring the spaces due to doors and windows is dependent upon the area that they take up. Some buildings have a large area of windows, and for these it may not be appropriate to ignore their effect. The key point at this stage is to try and make the calculation as simple as possible but not to ignore factors that are important.

> 3. If a simplification is made then arrange the calculation so that the estimate is safe.

For example, ignoring the saving on bricks due to the doors and windows is safe because it leads to an overestimate of the number of bricks required. If the estimate is to be used to cost a building then this means the cost will be overestimated, which is much better than it being underestimated. If the estimate is to be used to determine storage requirements then again an overestimate is better. The key phrase here is to 'err on the side of caution', which means making the calculation a safe estimate rather than a risky estimate.

We shall now illustrate these points with a number of examples.

Example 12.1

Decide whether or not a rough calculation is appropriate for the following cases:

(a)   calculating the amount of fuel needed by a space rocket making a journey to the Moon and back;

(b)   designing the shape of the fuselage of a jet aeroplane;

(c)   calculating the number of sandbags needed to distribute to a housing estate that is in danger of flooding due to bad weather;

(d)   calculating the amount of foundation stone needed to build a road.

Solution   (a)   This requires an exact calculation. The weight of fuel carried by a rocket is enormous, and carrying too much means that a bigger, more expensive rocket is needed. The consequences of not carrying enough fuel are obvious.

(b)   Designing the shape of an aircraft requires exact calculations. It is the sort of calculation that may require a great deal of engineering time together with a sophisticated computer.

(c)   This is a case where a rough calculation is appropriate. The calculation should err on the side of caution, since too many sandbags is not a major problem but too few is. Another consideration is that the calculation may need to be carried out quickly in order that the sandbags can be transported to the housing estate as rapidly as possible.

(d)   Again a rough calculation is appropriate. A typical simplification would be to ignore the effect of bends in the road.

Example 12.2

Suggest possible simplifications that could be made in the following cases in order to carry out a rough calculation.

(a)   A coal-fired power station is being designed, and it is decided to transport all of the coal by rail. In order to give the railway authorities time to plan their traffic, a rough estimate is needed of the daily consumption of coal and hence of the number of trains required to deliver the coal per day.

(b)   An engineer is designing a petrol station, and wishes to know how big to make the petrol storage tank.

(c)   A manufacturer requires to know the number of light bulbs consumed in domestic households per year in Britain in order to determine the potential market.

Solution   (a)   Consumption of coal at a power station varies from day to day and has seasonal fluctuations. Fortunately, power stations usually have plenty of space to store coal. This is useful because it allows the average consumption of coal to be delivered each day rather than the amount needed to take care of high demand. Once the average demand has been calculated, the number of railway wagons can be calculated by noting their carrying capacity. Hence, knowing how many railway wagons there are to a train, the required number of trains per day can be calculated.

(b) The main simplification here is not to worry about all the different cars likely to visit the petrol station, each with different-sized petrol tanks, but instead to estimate the typical amount of petrol purchased by a driver. This could be most easily achieved by observing how much petrol is purchased by drivers at an existing petrol station. The size of the tank needs to cater for the most extreme case that is likely to occur, and so any figures for the number of cars visiting the petrol station should take account of times such as bank holidays when people tend to use a lot of petrol. Again this sort of information can probably be obtained from observations made on existing petrol stations. However, it can be notoriously difficult to predict the demand for petrol, since it is very price-sensitive. Finally, account needs to be taken of how often the tank can be refilled conveniently by the petroleum company.

(c) This is potentially quite a complicated calculation, but several simplifications can be made. It is reasonable to assume that nearly every household in the country uses electricity. Clearly, some will not – but the error in assuming 100% usage is small. The life of a light bulb depends on several factors, such as the time it is switched on and the temperature of the room it is in. Here it is convenient to use an average light bulb life based on manufacturers' tests. An average number of light bulbs in use for a typical household would be required. Also, the total number of households in Britain would be needed. From this information, it would be possible to roughly estimate domestic light bulb usage in Britain.

Self-assessment questions 12.2

1. Explain what is meant by 'erring on the side of caution' when carrying out a rough calculation.

2. Explain why certain calculations need to be exact and a rough calculation is not appropriate.

Exercises 12.2

1. Decide whether or not a rough calculation is appropriate for the following situations, giving reasons for your answers:

   (a) calculating the amount of rubble and topsoil on a building site that needs to be cleared before building can take place;

   (b) calculating the stiffness of car springs to be incorporated into a new car design;

   (c) calculating the amount of waste material produced by a workshop during a day in order to design waste removal services;

   (d) calculating the size of a girder used in a road bridge;

   (e) calculating the amount of engine oil used by a fleet of buses based at a depot in order to determine oil storage requirements.

2. Suggest possible simplifications that could be made in the following situations in order to carry out a rough calculation (as with all exercises of this nature there is no single correct answer):

(a) calculating the number of tin cans available for recycling in Britain;

(b) calculating the amount of paint required to paint a steel bridge in order to prepare a financial budget;

(c) calculating the average energy use of a household in Britain prior to developing an energy conservation strategy.

12.3 Practical examples of estimation

In the previous section we examined when it was appropriate to carry out a rough calculation and how to simplify in order to make the calculation as simple as possible. It is not possible to cover all the different types of estimation that engineers carry out, nor is it necessary to do so. This section provides a number of examples to illustrate the process of estimation. It is important to realize that the ability to estimate is a skill gained with practice. Try carrying out rough estimates during your daily routines. Often the calculations are simple enough to be done in your head. Practise this every day and you will soon become a better engineer. One mechanical design engineer was once asked how many rough calculations he did in his head in a working day. His reply was 'about a hundred'; this itself was an estimate!

Example 12.3 Resistors in parallel

 When two resistors with resistances R_1 and R_2 are connected in parallel then the equivalent resistance R is given by

$$\frac{1}{R} = \frac{1}{R_1} + \frac{1}{R_2} \tag{12.1}$$

When electronic engineers are designing or analysing an electronic circuit, they often need to know the equivalent resistance of two resistors in parallel. It is not necessary to use Equation (12.1) each time. There are certain situations when a much simpler calculation is possible. Carry out the following.

(a) Devise a rule for calculating the resistance of two resistors connected in parallel when they both have the same resistance.

(b) Devise a rule for estimating the resistance of a network in which a large-value resistor and a small-value resistor are connected in parallel. Try out your rule for the following pairs of values:
(i) $1\,\text{k}\Omega$ and $10\,\text{k}\Omega$
(ii) $1\,\text{k}\Omega$ and $100\,\text{k}\Omega$
(iii) $2\,\text{k}\Omega$ and $5\,\text{k}\Omega$

Solution   (a)   We are given that $R_1 = R_2$, and so

$$\frac{1}{R} = \frac{1}{R_1} + \frac{1}{R_2}$$

$$= \frac{1}{R_1} + \frac{1}{R_1}$$

$$= \frac{2}{R_1}$$

Therefore

$$R = \tfrac{1}{2}R_1$$

Note that this is not an estimate but an exact calculation. No simplifications have been made at this stage. We see that the effect of putting the two resistors in parallel is to produce an equivalent resistance that is half that of each resistor. This is a convenient fact that is often used by engineers. It can be a useful way of getting a particular resistance value when you do not have it available in a single resistor. For example, suppose in a particular application we require a resistance of 16.5 Ω, but only standard 33 Ω resistors are available. Then by placing two of these in parallel, we can achieve the required 16.5 Ω resistance.

(b)   It is sometimes a good idea to 'play around' with a formula to see if any useful simplifications can be found. We are given

$$\frac{1}{R} = \frac{1}{R_1} + \frac{1}{R_2}$$

Let us rearrange this formula to calculate R. Thus

$$\frac{1}{R} = \frac{1}{R_1} + \frac{1}{R_2} = \frac{R_1 + R_2}{R_1 R_2}$$

Therefore

$$R = \frac{R_1 R_2}{R_1 + R_2}$$

Further, let us assume R_1 is the larger resistance and R_2 is the smaller. We see that in the formula for R the denominator is $R_1 + R_2$. Perhaps a simple estimate can be obtained by approximating this by R_1 because R_1 is very much greater than R_2, written $R_1 \gg R_2$. Our estimate R_{est} is therefore

$$R_{est} = \frac{R_1 R_2}{R_1} = R_2$$

This is interesting. In effect, by assuming $R_1 \gg R_2$, we can estimate the parallel network of resistors to have a resistance equal to that of the smaller resistor. Let us try out our estimate on the examples given.

(i) For $R_1 = 10\,\text{k}\Omega$, $R_2 = 1\,\text{k}\Omega$ the exact value is

$$R = \frac{10 \times 1}{10 + 1} = 0.91\,\text{k}\Omega$$

and the estimate is

$$R_{\text{est}} = R_2 = 1\,\text{k}\Omega$$

We see that for this case the estimate is in error by slightly less than 10%. For many of the sorts of calculations electronic engineers carry out this would be acceptable.

(ii) For $R_1 = 100\,\text{k}\Omega$, $R_2 = 1\,\text{k}\Omega$ the true value is

$$R = \frac{100 \times 1}{100 + 1} = 0.99\,\text{k}\Omega$$

and the estimate is

$$R_{\text{est}} = R_2 = 1\,\text{k}\Omega$$

We see that this is a very good estimate indeed.

(iii) For $R_1 = 5\,\text{k}\Omega$, $R_2 = 2\,\text{k}\Omega$ the true value is

$$R = \frac{2 \times 5}{2 + 5} = 1.43\,\text{k}\Omega$$

and the estimate is

$$R_{\text{est}} = 2\,\text{k}\Omega$$

We see that this is a very poor estimate. Clearly, the two resistors are too close in resistance to allow our simple estimate to be used.

On the basis of these calculations, we can conclude that if one resistor has a resistance of at least 10 times the other then, roughly, the resistance of the network is that of the smaller resistor. Electronic engineers often use this rule when designing electronic circuits.

Example 12.4 Punching sheets of metal

Many engineering products contain components that are punched out of sheet metal. It is important for a manufacturer to obtain as many components as possible out of a metal sheet in order to minimize costs. For complicated shapes the arrangement is done using a computer. Often, however, it is useful to get an estimate of how many components can be obtained from a sheet in order to get a rough idea of how many sheets will be needed. This rough estimate can be obtained by calculating a minimum-sized rectangle that encloses the shape and using this to calculate the approximate number of components that can be obtained from a single sheet. By using this method, estimate the number of components that can be obtained from a 2 m square sheet of metal for the following shapes, shown in (a) Figure 12.1   (b) Figure 12.2

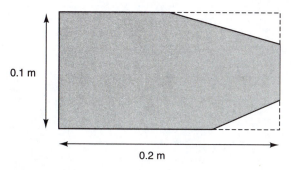

Figure 12.1   Shape for Example 12.4(a).

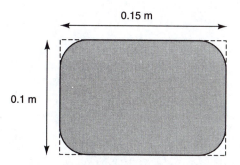

Figure 12.2   Shape for Example 12.4(b).

Solution   (a)   By examining Figure 12.1, we see the minimum rectangle that surrounds the shape is 0.2 m by 0.1 m. Therefore we can obtain approximately

$$\frac{2}{0.2} \times \frac{2}{0.1} = 10 \times 20 = 200 \text{ shapes}$$

(b)   Consider a strip of metal measuring 0.1 m by 2 m. Noting that $2/0.15 = 13.33$, we see that 13 of the complete shapes shown in Figure 12.2 can be cut from such a strip. The rectangular sheet, 2 m by 2 m, can be thought of as 20 strips, and so $13 \times 20 = 260$ shapes can be cut from it.

James Watt: the man who harnessed power

James Watt (see Figure 1) was born in 1736 at Greenock in Scotland. As a boy he was very interested in making models. For example, he made a miniature crane and a barrel organ. In 1753 his mother died and his father's business suffered a reverse. It became necessary for James to earn a living. He wanted to be a mathematical instrument maker. He was sent to his mother's kinsfolk in Glasgow to see if they could help him. Unfortunately, there was no work of this sort in Glasgow at the time since it was still mainly a university city. The best that could be obtained was to put him to work under an optician. James realized that learning a trade was not for him. He was introduced to Professor Robert Dick, who advised him to go to London, which was the centre for English instrument makers.

Figure 1 *James Watt: the man who harnessed power.*

In 1755 Watt set out to London on horseback since there was no coach running at that time. After a struggle, he found a master to work under named John Morgan. James's progress was extremely rapid. After a year, he had learnt enough to return home to Scotland and set up in business. He opened a shop in the College of Glasgow and was given the title of Mathematical Instrument Maker to the University. He set up a shop in the city two years later. At first, business was slow, but eventually it increased. In 1764 he started his experiments in steam. This was the year he got married. Watt was a modest and timid man, and so it was good for him to have someone to support him during this vital point in his life. The steam engine had already been invented, but it was inefficient. What was needed was a means of keeping the cylinder hot and yet condensing the steam.

Watt had the revolutionary idea of using a separate condenser. Watt described his thoughts to Robert Hart half a century later:

'It was in the Green of Glasgow. I had gone to take a walk in a fine Sabbath afternoon. I had entered the Green by the gate at the foot of Charlotte Street – had passed the old washing-house. I was thinking upon the engine at the time and had gone as far as the Herd's house when the idea came into my mind, that as steam was an elastic body it would rush into a vacuum, and if a communication was made between the cylinder and an exhausted vessel, it would rush into it, and might be there condensed without cooling the cylinder. I then saw that I must get quit of the condensed steam and injection water, if I used a jet as in Newcomen's engine. Two ways of doing this occurred to me. First the water might be run off by a descending pipe, if an offlet could be got at the depth of 35 or 36 feet, and any air might be extracted by a small pump; the second was to make the pump large enough to extract both water and air... I had not walked further than the Golf-house when the whole thing was arranged in my mind.'

Since it was the sabbath day, Watt had to wait until the following morning before building a model to prove his idea would work (see Figure 2). It did, and Watt went on, with his associates, to refine the steam engine into an extremely efficient and useful machine.

The industrial revolution was powered by steam. An efficient steam engine was required before steam engines could be used in bulk. James Watt provided this and by doing so started a revolution that transformed the Western world. H.W. Dickinson in his book, *James Watt: Craftsman and Engineer* (David and Charles, Newton Abbot, 1967; 1st edition Cambridge University Press, 1936) wrote:

'Glancing over the great changes that have taken place since Watt's day in every department of human activity, greater in extent than in any corresponding period of the world's history, we affirm that no one has made a greater individual contribution to these changes, for good or for ill, than has James Watt with his steam-engine.'

Figure 2 *Watt's original model of a surface condenser engine.*

Exercises 12.3

1.   When two capacitors with capacitances C_1 and C_2 are connected in series, the equivalent capacitance C is given by

$$\frac{1}{C} = \frac{1}{C_1} + \frac{1}{C_2}$$

(a)   Devise a rule for estimating the capacitance of two capacitors connected in series when they both have the same capacitance.

(b)   Devise a rule for estimating the capacitance of a network in which a large capacitor and a small capacitor are connected in series. Try out your rule for the following pairs of values:
(i) 1 μF and 50 μF
(ii) 1 μF and 10 μF

(iii) 3 μF and 4 μF
Comment on the results.

2.   When three resistors with resistances R_1, R_2 and R_3 are connected in parallel the equivalent resistance R is given by

$$\frac{1}{R} = \frac{1}{R_1} + \frac{1}{R_2} + \frac{1}{R_3}$$

Devise a rule for estimating the network resistance when one of the resistors has a much smaller resistance than the other two. Try out your rule for several resistor combinations and comment on the results.

3.   Estimate the number of components that can be obtained from a 3 m square sheet of metal for the shapes shown in Figure 12.3.

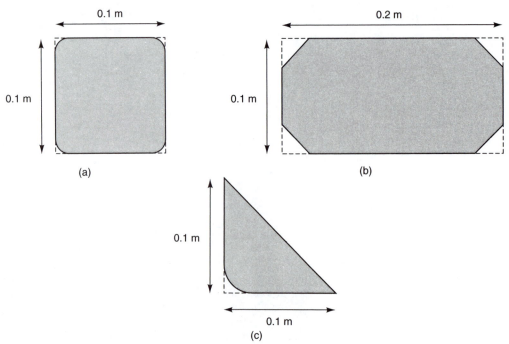

0.1 m

0.1 m

(a)

0.2 m

0.1 m

(b)

0.1 m

0.1 m

(c)

Figure 12.3   Shapes for Exercise 12.3 Q3.

12.4 Relationships between engineering variables

Engineers often need to know the effect that changing a variable has on other quantities, for example, the effect of doubling the current in a resistor, the effect of increasing the height of a chemical storage tank by 50% and the effect of halving the size of a girder. It is tedious to reach for a calculator every time such a calculation needs to be carried out, since a busy engineer does not have the time to do this. It is much better to be able to estimate the effect quickly by doing a mental calculation. This section will deal with the common types of these calculations.

12.4.1 Proportional relationships

Proportional relationships are the easiest to deal with. If two variables are directly proportional then doubling one variable leads to a doubling of the other variable, and increasing a variable by 50% leads to a 50% increase in the other. A few examples will illustrate these points. Before continuing, you may wish to revise your earlier work on proportion from Chapter 5.

Example 12.5 Ohm's law

Recall Ohm's law for a resistor, which is

$$V = IR$$

where V is the voltage across the resistor, I is the current through the resistor and R is the resistance of the resistor and is a constant. The variables V and I are directly proportional. Doubling the voltage across the resistor leads to a doubling of the current through the resistor. To see why this is so, suppose V is increased to $2V$; then the new current I_{new} is given by

$$2V = I_{new}R$$

$$I_{new} = \frac{2V}{R} = 2\left(\frac{V}{R}\right) = 2I$$

We see that the current through the resistor is doubled.

Sometimes one variable is inversely proportional to another. In such a case doubling one will lead to a halving of the other and so on. Consider the following example.

Example 12.6 Time taken for an object travelling at a constant velocity to cover a given distance

Recall that the distance s travelled by an object travelling at a constant velocity v is given by

$$s = vt$$

where t is the time for which the object has been travelling. This formula can be rearranged to make t the subject:

$$t = \frac{s}{v}$$

We see that t and v are inversely proportional; the greater the velocity, the shorter the time taken to travel the fixed distance s. For example, doubling the velocity will halve the time required to cover a fixed distance.

12.4.2 Square relationships

There are many examples where a variable is related to the square of another variable. It is possible to develop some simple rules for this case.

Example 12.7 Area of a circle

The area A of a circle is given by $A = \pi r^2$, where r is the length of the radius. If the radius is doubled, find the effect on the area.

Solution   We are given $A = \pi r^2$. If we double the radius r to $2r$ then the new area A_{new} is

$$A_{new} = \pi(2r)^2 = \pi(4r^2) = 4\pi r^2 = 4A$$

We see that when the radius is doubled, the area is increased by a factor of 4.

The previous example illustrates a general rule for use with square relationships. Doubling the value of a variable that is raised to the power of 2 leads to an increase by a factor of 4 in the related variable. This is a very useful rule, since it allows an engineer to calculate very quickly the effect of doubling a variable that is squared.

For example, if a square factory floor has sides of length x then its floor area A is given by the formula $A = x^2$. We can immediately see that doubling the side length leads to a fourfold increase in floor area.

Example 12.8 Kinetic energy of a moving car

A car of mass M and speed v has kinetic energy E given by

$$E = \tfrac{1}{2}Mv^2$$

(a)   Deduce the effect on the kinetic energy of doubling the speed of the car.
(b)   Deduce the effect on the kinetic energy of increasing the speed of the car by 5%.

Solution   (a)   If the speed of the car is doubled from v to $2v$ then its kinetic energy, E_{new} is given by

$$E_{\text{new}} = \tfrac{1}{2}M(2v)^2 = \tfrac{1}{2}M4v^2 = 4(\tfrac{1}{2}Mv^2) = 4E$$

We note that there is a fourfold increase in kinetic energy. This is the reason why fast-moving cars are potentially so much more dangerous in an accident. A tripling of speed leads to a ninefold increase in energy.

(b)   The speed is increased by 5% from v to $1.05v$. The new kinetic energy E_{new} is then given by

$$E_{\text{new}} = \tfrac{1}{2}M(1.05v)^2$$

$$= 1.1025\frac{Mv^2}{2}$$

$$= 1.1025E$$

So the kinetic energy is increased by 10.25% when the speed is increased by 5%.

Part (b) of the previous example illustrates another general rule for square relationships. If a variable that is raised to the power of 2 is only increased by a small percentage then the related variable increases by approximately twice that percentage. We saw that the speed of the car was increased by 5% and this led to an increase in kinetic energy of about 10%. This rule is reasonably good for increases in the squared variable of up to 10%.

12.4.3 General considerations

We have seen in earlier sections that it is possible to develop simple rules that enable quick mental calculations to be carried out without having to resort to a calculator each time. It is important not to see these rules as a mere list to be memorized. It is far better to develop your own rules and get into the habit of using them. You may also wish to develop rules for other common relationships, for example cubic and square root relationships. The important thing to remember is that in many calculations an exact answer is not necessary and an estimate is perfectly acceptable and, what is more important, is far quicker to carry out.

Self-assessment question 12.4

1.   Give reasons why an estimated calculation may be preferable to an exact calculation.

Exercises 12.4

1.   The energy E stored in a mass with moment of inertia J that is rotating with angular speed ω is given by

$$E = \tfrac{1}{2}J\omega^2$$

Note that J is a constant. Estimate the effect on the kinetic energy of increasing the angular speed by
(a) a factor of 2   (b) a factor of 3
(c) 3%   (d) 7%

2.   The pressure p on a submarine hull that is submerged to a depth h in sea water of density ρ is given by

$$p = \rho g h$$

Calculate the effect on the pressure on the submarine hull of
(a) increasing the depth of the submarine by 50%;
(b) halving the depth of the submarine.

3.   The power P dissipated in a resistor with resistance R and passing a current I is given by $P = I^2 R$.
(a) The current is doubled. What is the effect on the power?
(b) The resistance is halved. What is the effect on the power?
(c) The current and resistance are both doubled. What is the effect on the power?

Review exercises 12

1   Decide whether or not a rough calculation is appropriate for the following situations, giving reasons for your answers:
(a) calculating the quantity of a particular trace element to be included in a high-performance metal alloy;
(b) calculating the amount of gearbox oil used in an industrial complex in order to determine stock levels;
(c) calculating the current flow in a person when they accidentally touch the mains supply;
(d) calculating the amount of heat generated by an electronic circuit board in order to size the cooling fan needed.

2   Suggest possible simplifications that could be made in the following situations in order to carry out a rough calculation:
(a) calculating the fuel requirements of a fleet of lorries run by a haulage company in order to design the fuel storage depot;
(b) calculating the amount of newspaper that is available for recycling in Britain;
(c) calculating the emergency food requirements of a city hit by an earthquake;
(d) calculating the amount of paint needed to paint a ship.

3 The period T of a pendulum of length l is given by

$$T = 2\pi \sqrt{\frac{l}{g}}$$

where g is a constant known as the acceleration due to gravity. Calculate the effect on the period of the pendulum of
(a) quadrupling its length;
(b) doubling its length;
(c) increasing its length by 10%;
(d) hence deduce some simple rules for variables that have a square root relationship.

4 The volume V of a sphere of radius r is given by

$$V = \tfrac{4}{3}\pi r^3$$

Calculate the effect on the volume of the sphere of
(a) doubling its radius;
(b) tripling its radius;
(c) increasing its radius by 5%;
(d) hence deduce some simple rules for variables that have a cubic relationship.

13

Logarithms and Exponentials

KEY POINTS

This chapter

- defines the term 'logarithm'

- explains how to use a scientific calculator to calculate logarithms to various bases

- states the laws of logarithms and shows how to simplify expressions using the laws

- discusses the properties of the logarithmic graphs

- defines the exponential function

- relates the exponential function to the logarithm function

- explains how to solve, both algebraically and graphically, equations involving logarithmic and exponential expressions

CONTENTS

13.1 Introduction

Many physical processes are described using logarithmic and exponential functions. For example, the decay of radioactive material, the measurement of the intensity of sound and the growth of populations can all be modelled using logarithmic and exponential functions. In this chapter we examine these functions and draw their graphs. Laws are stated that allow expressions involving exponential and logarithmic terms to be simplified. Equations involving exponential and logarithmic expressions are solved.

13.2 Definition of a logarithm and common terminology

In the equation $16 = 2^4$ the **power** is 4 and the **base** is 2. Similarly, in the equation $25 = 5^2$ the power is 2 and the base is 5. An alternative name for power is **index**.

Example 13.1

State the base and the power for each of the following:
(a) $32 = 2^5$ (b) $64 = 2^6$ (c) $64 = 4^3$ (d) $64 = 8^2$

Solution

(a)   In the equation $32 = 2^5$ the base is 2 and the power is 5.
(b)   In the equation $64 = 2^6$ the base is 2 and the power is 6.
(c)   In the equation $64 = 4^3$ the base is 4 and the power is 3.
(d)   In the equation $64 = 8^2$ the base is 8 and the power is 2.

Note from (b), (c) and (d) that the same number can be expressed using various bases and powers.

We are now ready to introduce **logarithms**. Logarithms are an alternative way of writing expressions involving powers and bases. Logarithm is usually shortened to **log**.

Consider

$$16 = 2^4$$

We may express this as

$$\log_2 16 = 4$$

This is read as 'log to the base 2 of 16 equals 4'.

Example 13.2

Express $125 = 5^3$ using logarithms.

Solution
$$125 = 5^3$$

$$\log_5 125 = 3$$

We read this as 'log to the base 5 of 125 equals 3'.

From the previous example, we see that logarithms are simply powers.

Example 13.3

Express the following using logarithms.

(a) $32 = 2^5$ (b) $49 = 7^2$ (c) $9 = 3^2$ (d) $1000 = 10^3$ (e) $3 = 9^{\frac{1}{2}}$
(f) $0.01 = 10^{-2}$

Solution
(a) Since $32 = 2^5$, $\log_2 32 = 5$
(b) Since $49 = 7^2$, $\log_7 49 = 2$
(c) Since $9 = 3^2$, $\log_3 9 = 2$
(d) Since $1000 = 10^3$, $\log_{10} 1000 = 3$
(e) Since $3 = 9^{\frac{1}{2}}$, $\log_9 3 = \frac{1}{2}$
(f) Since $0.01 = 10^{-2}$, $\log_{10} 0.01 = -2$

In general,

$$\text{if} \quad a = b^c \quad \text{then} \quad \log_b a = c$$

The only restriction that is placed on the value of the base b is that it is a positive real number, excluding 1.

Self-assessment questions 13.2

1. Explain what is meant by the terms power, base, index and logarithm.

2. The logarithm of a number is always positive. True or false?

3. The logarithm of a number can never be less than -1. True or false?

4. If a number is doubled then the logarithm of the number is doubled. True or false?

Exercises 13.2

1. Rewrite the following equations using logarithms:
   (a) $100 = 10^2$ (b) $256 = 4^4$
   (c) $256 = 16^2$ (d) $10\,000 = 10^4$
   (e) $1728 = 12^3$ (f) $5^{-2} = \frac{1}{25}$
   (g) $3^{-1} = \frac{1}{3}$ (h) $\frac{1}{100} = 10^{-2}$
   (i) $27^{\frac{1}{3}} = 3$ (j) $27^{-\frac{1}{3}} = \frac{1}{3}$ (k) $10^1 = 10$
   (l) $10^0 = 1$ (m) $x^0 = 1$

2. Rewrite the following using powers:
   (a) $\log_2 32 = 5$ (b) $\log_3 27 = 3$
   (c) $\log_2 512 = 9$ (d) $\log_9 81 = 2$
   (e) $\log_{10} 1000 = 3$ (f) $\log_5 125 = 3$

   (g) $\log_6 36 = 2$ (h) $\log_2 128 = 7$

3. We may write $4 \times 16 = 64$. Also, we note that $4 = 2^2$, $16 = 2^4$, $64 = 2^6$. Carry out the following.
   (a) Calculate $\log_2 4$, $\log_2 16$, $\log_2 64$.
   (b) Can you see an equation connecting these three logarithms?

4. From the laws of indices, we know that $a^0 = 1$ for any number a. What can you deduce about the logarithm of 1 to any base?

13.3 Common bases and use of a calculator

In Section 13.2 we saw how logarithms to various bases were related to powers. Recall that if $a = b^c$ then $\log_b a = c$. The base of the logarithm is b. In practice, logarithms are calculated using only a few common bases. The commonly used bases are 10 and e. The letter e stands for an irrational number whose value is 2.718 281 .... This number has been found to occur in the description of many natural phenomena, for example radioactive decay.

Logarithms to the base 10 are usually denoted simply by 'log', the subscript being omitted. Logarithms to the base e are referred to as **natural logarithms** and are denoted by 'ln'. Buttons on scientific calculators are usually marked 'LOG' for logarithms to the base 10, and 'LN' or 'LN X' for natural logarithms.

Example 13.4

Use a scientific calculator to calculate (a) $\log_{10} 73$ (b) $\log_{10} 0.426$ (c) $\ln 5.64$ (d) $\ln 0.73$

Solution

Use your calculator to verify the following results: (a) $\log_{10} 73 = 1.8633$ (b) $\log_{10} 0.426 = -0.3706$ (c) $\ln 5.64 = 1.7299$ (d) $\ln 0.73 = -0.3147$

Example 13.5

Find the natural logarithms of (a) 10 (b) 65

Solution

Natural logarithms are logarithms to the base e. The 'LN' or 'LN X' button is used.
(a) $\ln 10 = 2.3026$ (b) $\ln 65 = 4.1744$

Example 13.6 *Velocity of a space rocket*

Consider a rocket travelling in outer space with velocity v_1. In order to increase the velocity of the rocket to a value of v_2, the jet engine is fired. It can be shown that

$$v_2 = v_1 + C \ln\left(\frac{M_1}{M_2}\right)$$

where M_1 is the mass of the rocket before the jet engine is fired, M_2 is the mass of the rocket after the jet engine is switched off and C is the effective velocity of the jet exhaust gases. Note that the rocket loses mass due to the burning of fuel to power the jet engine. This formula assumes that the effects of gravity and drag forces are negligible, and so is only valid in outer space. A more complicated formula is needed to deal with a space rocket taking off from a planet.

Let us now consider a simple example. A rocket is travelling in outer space with a velocity of 1000 m per second. The rocket has a mass of 10^5 kg. In order to increase the velocity of the rocket, the jet engine is fired for 40 s. While the engine is burning, the fuel is consumed at a rate of 250 kg per second. The effective velocity of the jet exhaust gases is 4000 m per second. Calculate the final velocity of the rocket.

Solution We are given that $v_1 = 1000$ m per second and $M_1 = 10^5$ kg. If the engine is fired for 40 s and the rate of fuel consumption is 250 kg per second then the total mass loss is $40 \times 250 = 10\,000 = 10^4$ kg. Therefore

$$M_2 = M_1 - 10^4 = 10^5 - 10^4 = 9 \times 10^4 \text{ kg}$$

Using the given formula, we can now calculate v_2. Thus

$$v_2 = v_1 + C \ln\left(\frac{M_1}{M_2}\right)$$

$$= 1000 + 4000 \ln\left(\frac{10^5}{9 \times 10^4}\right)$$

$$= 1000 + 4000 \ln \tfrac{10}{9}$$

$$= 1421 \text{ m per second}$$

The final velocity of the rocket is 1421 m per second.

Self-assessment questions 13.3

1. Which button on a calculator is used to calculate logarithms to the base 10?

2. Which button on a calculator is used to calculate logarithms to the base e?

3. What is the essential difference between $\log x$ and $\ln x$?

Exercises 13.3

1. Evaluate
   (a) ln 100   (b) $\log_{10} 10$ (c) ln e
   (d) $\log_{10} e$ (e) $\log_{10} 210$ (f) ln 37

2. Evaluate (a) log 3   (b) $\log \frac{1}{3}$ (c) ln 7
   (d) $\ln \frac{1}{7}$
   What do you notice about your answers to (a) and (b)? What do you notice about your answers to (c) and (d)?

3. (a) Calculate ln 6, ln 5, ln 30, $\ln \frac{6}{5}$ and $\ln \frac{5}{6}$.
   (b) Compare ln 30 with ln 6 + ln 5.
   (c) Compare ln 6 − ln 5 with $\ln \frac{6}{5}$.
   (d) Compare ln 5 − ln 6 with $\ln \frac{5}{6}$.

4. A rocket is travelling in outer space with a velocity of 500 m per second and has a mass of 8×10^4 kg. The jet engine is fired for 50 s in order to increase the velocity of the rocket. While the engine is firing, fuel is consumed at a rate of 150 kg per second. The effective velocity of the jet exhaust gases is 3000 m per second. By using the formula given in Example 13.6, calculate the final velocity of the rocket.

5. (a) Calculate $\log_{10} 2$, $\log_{10} 4$, $\log_{10} 8$, $\log_{10} 16$, $\log_{10} 32$.
   (b) Noting that $2 = 2^1$, $4 = 2^2$, $8 = 2^3$ and so on, try to deduce a rule connecting $\log_{10} 2^n$ to $\log_{10} 2$.

13.4 Laws of logarithms

There are various laws of logarithms. This section details the laws and shows how to use them to simplify expressions.

The first law of logarithms tells us how to add together two logarithms with the same base. Adding $\log A$ to $\log B$ results in $\log AB$.

$$\text{First law: } \log A + \log B = \log AB$$

The second law shows how to find the difference of two logarithms. Subtracting $\log B$ from $\log A$ results in $\log\left(\dfrac{A}{B}\right)$.

$$\text{Second law: } \log A - \log B = \log\left(\frac{A}{B}\right)$$

The third law relates $\log A$ to $\log A^n$. Note that we can write the power n as a coefficient of $\log A$.

$$\text{Third law: } \log A^n = n \log A$$

The fourth law is a statement that the logarithm of 1 to any base is 0.

Fourth law: $\log 1 = 0$

Finally, the fifth law states that the logarithm of a number to the same base is 1.

Fifth law: $\log_m m = 1$

These laws apply to logarithms of any base. However, the base must be kept the same throughout the application of a law. The following examples serve to illustrate the laws numerically.

Example 13.7

(a)   Use a calculator to find $\log_{10} 24$ and $\log_{10} 3$.
(b)   Use a calculator to find $\log_{10} 72$ and $\log_{10} 8$.
(c)   Verify that $\log_{10} 24 + \log_{10} 3 = \log_{10} (24 \times 3) = \log_{10} 72$.
(d)   Verify that $\log_{10} 24 - \log_{10} 3 = \log_{10} \frac{24}{3} = \log_{10} 8$.

Solution

(a)   Using a calculator, we find $\log_{10} 24 = 1.3802$ and $\log_{10} 3 = 0.4771$.
(b)   Using a calculator, we find $\log_{10} 72 = 1.8573$ and $\log_{10} 8 = 0.9031$
(c)   From (a), we have

$$\log_{10} 24 + \log_{10} 3 = 1.3802 + 0.4771 = 1.8573$$

From (b), we have

$$\log_{10} 72 = 1.8573$$

and so

$$\log_{10} 24 + \log_{10} 3 = \log_{10} 72$$

(d)   From (a), we have

$$\log_{10} 24 - \log_{10} 3 = 1.3802 - 0.4771 = 0.9031$$

From (b), we have

$$\log_{10} 8 = 0.9031$$

and so

$$\log_{10} 24 - \log_{10} 3 = \log_{10} 8$$

Part (c) of the above example illustrates the first law, that is, $\log A + \log B = \log AB$. Part (d) illustrates the second law, that is, $\log A - \log B = \log(A/B)$.

Example 13.8

    (a)   Use a calculator to evaluate $\log_{10} 64$ and $\log_{10} 4$.

    (b)   Verify that $3 \log_{10} 4 = \log_{10} 64$.

    (c)   By noting that 64 can be written as 4^3, verify that $3 \log_{10} 4$ is $\log_{10} 4^3$.

Solution   (a)   Using a calculator, we see $\log_{10} 64 = 1.806\,18$ and $\log_{10} 4 = 0.602\,06$.

    (b)   Using (a), we see that

$$3 \log_{10} 4 = 3(0.602\,06) = 1.806\,18$$

Also from (a), we see that

$$\log_{10} 64 = 1.806\,18$$

and so

$$3 \log_{10} 4 = \log_{10} 64$$

    (c)   From (b), we have $3 \log_{10} 4 = \log_{10} 64$. By writing 64 as 4^3, we see that

$$3 \log_{10} 4 = \log_{10} 64 = \log_{10} 4^3$$

as required.

Part (c) of the above example illustrates the third law, that is, $n \log A = \log A^n$.

The laws of logarithms can be used to simplify expressions involving logarithms.

Example 13.9

Use the laws of logarithms to simplify the following expressions to a single log term:

(a) $\log 6 - \log 2$ (b) $\log 12 + \log 3$ (c) $2 \log 4 - \log 2$

(d) $3 \log 6 - 2 \log 12 + \log 1$

Solution   (a)   We need to use the second law. This gives

$$\log 6 - \log 2 = \log\left(\frac{6}{2}\right) = \log 3$$

    (b)   We need to use the first law. This gives

$$\log 12 + \log 3 = \log(12 \times 3) = \log 36$$

(c)   We apply the third law to $2 \log 4$ to give

$$2 \log 4 - \log 2 = \log (4^2) - \log 2 = \log 16 - \log 2$$

We now apply the second law to obtain

$$\log 16 - \log 2 = \log \left(\frac{16}{2}\right) = \log 8$$

Hence

$$2 \log 4 - \log 2 = \log 8$$

(d)   Using the fourth law, the term $\log 1$ may be omitted, since its value is 0:

$$3 \log 6 - 2 \log 12 + \log 1 = 3 \log 6 - 2 \log 12$$

The third law is applied to both remaining terms:

$$3 \log 6 - 2 \log 12 = \log (6^3) - \log (12^2) = \log 216 - \log 144$$

Finally the second law is applied:

$$\log 216 - \log 144 = \log \left(\frac{216}{144}\right) = \log \left(\frac{3}{2}\right) = \log 1.5$$

So

$$3 \log 6 - 2 \log 12 + \log 1 = \log 1.5$$

Example 13.10

Simplify the following to single log expressions:
(a) $\ln 3x + \ln 2x$ (b) $\ln 6t - \ln 2t$ (c) $3 \ln y^2 - 2 \ln y^3$ (d) $\frac{1}{2} \ln z^2 + 3 \ln z$

Solution   (a)   Using the first law, we have

$$\ln 3x + \ln 2x = \ln (3x \cdot 2x) = \ln 6x^2$$

(b)   Using the second law,

$$\ln 6t - \ln 2t = \ln \left(\frac{6t}{2t}\right) = \ln 3$$

(c)   The third law is applied to both expressions to give

$$3 \ln y^2 - 2 \ln y^3 = \ln (y^2)^3 - \ln (y^3)^2 = \ln y^6 - \ln y^6 = 0$$

(d)   Consider the term $\frac{1}{2} \ln z^2$. Using the third law, we may write this as

$$\frac{1}{2} \ln z^2 = \ln (z^2)^{\frac{1}{2}} = \ln z$$

Also, $3 \ln z = \ln z^3$, using the third law. So

$$\frac{1}{2} \ln z^2 + 3 \ln z = \ln z + \ln z^3 = \ln (z \cdot z^3) = \ln z^4$$

Example 13.11

Simplify

(a) $3 \log y^2 - 4 \log y + 2 \log y^3$ (b) $2 \log abc + 3 \log a^2bc - 4 \log ab^2c$

Solution   (a)   We have

$$3 \log y^2 - 4 \log y + 2 \log y^3 = \log (y^2)^3 - \log y^4 + \log (y^3)^2$$

$$= \log y^6 - \log y^4 + \log y^6$$

$$= \log \left(\frac{y^6}{y^4}\right) + \log y^6$$

$$= \log y^2 + \log y^6$$

$$= \log (y^2 \cdot y^6)$$

$$= \log y^8$$

(b)   We have

$$2 \log abc + 3 \log a^2bc - 4 \log ab^2c = \log (abc)^2 + \log (a^2bc)^3 - \log (ab^2c)^4$$

$$= \log a^2b^2c^2 + \log a^6b^3c^3 - \log a^4b^8c^4$$

$$= \log (a^2b^2c^2 \cdot a^6b^3c^3) - \log a^4b^8c^4$$

$$= \log a^8b^5c^5 - \log a^4b^8c^4$$

$$= \log \left(\frac{a^8b^5c^5}{a^4b^8c^4}\right)$$

$$= \log \left(\frac{a^4c}{b^3}\right)$$

Example 13.12 Sound intensity

Sound is generated as a result of the movement of an object, for example, the vibration of a guitar string. Sound is transmitted away from a source by means of variations in the air pressure corresponding to sound waves. The strength or **intensity** of sound I is a measure of the power density of the sound waves and has units of watts per square metre. Table 13.1 gives the sound intensities of several typical sources. It is clear from the table that the range of sound intensities that people come into contact with is very large, varying from an intensity of 10^{-12} W m^{-2} for a just audible sound to 10 W m^{-2} for the noise from a jet aircraft. As a result of this, the human ear is adapted to cope with huge variations in sound intensity. It can be shown that the effect of doubling the intensity of a sound is to increase only slightly its loudness as perceived by the listener rather than double its loudness, which might naively be expected. In order to cope with this huge range of intensities, engineers prefer to use logarithms, the effect of which is to compress the measurement scale.

Table 13.1

Source	Intensity (W m^{-2})	Intensity level (dB)
Just audible sound	10^{-12}	0
Whisper	10^{-10}	20
Street traffic	10^{-5}	70
Thunder	10^{-1}	110
Jet aircraft (at 30 m)	10	130

This logarithmic measure of sound intensity level has units of **decibel** (dB) and is given by

$$\text{intensity level in dB} = 10 \log_{10}\left(\frac{I}{I_0}\right)$$

where I is the intensity of the source in W m^{-2}, and I_0 is the intensity of a just audible sound and has a value of $10^{-12}\,\text{W m}^{-2}$.

Returning to Table 13.1 we see that the intensity levels of the various sources have also been calculated in dB. Note that the choice of I_0 makes the just audible sound have an intensity level of 0 dB. This is easily seen by putting $I = I_0$ in the formula. We have

$$\text{intensity level in dB of a just audible sound} = 10 \log_{10}\left(\frac{10^{-12}}{10^{-12}}\right)$$

$$= 10 \log_{10} 1$$

$$= 10 \times 0 = 0\,\text{dB}$$

It is clear from examining this formula that the intensity level of a sound is a comparison with that of a just audible sound. Note that the dB values are more evenly distributed and have a smaller range. This illustrates the benefit of using a logarithmic measure of sound.

Self-assessment questions 13.4

1. State the laws of logarithms.

2. When applying any of the laws of logarithms, the base of the logarithms must be the same. True or false?

3. Explain how the laws of logarithms are related to the laws of indices.

Exercises 13.4

1. Simplify the following to single log terms:
   (a) $\log_{10} 10 + \log_{10} 5$ (b) $\log 10 - \log 5$
   (c) $\ln 5 - \ln 10$ (d) $2 \log 5 - \log 10$
   (e) $\ln 5 + 2 \ln 10$ (f) $3 \log_{10} 5 - 2 \log_{10} 10$
   (g) $\frac{1}{2} \ln 20 + \frac{1}{2} \ln 5$
   (h) $3 \log 30 + 2 \log 4 - 4 \log 5$
   (i) $2 \log 6 + 3 \log 10 - 6 \log 2$

2. Simplify to a single log term:
   (a) $\ln x + \ln y + \ln z$
   (b) $2 \log_{10} x + \log_{10} x^2 - \log_{10} x^3$
   (c) $2 \log x + 3 \log y + 4 \log z$
   (d) $\frac{1}{2} \ln x^2 + \frac{1}{3} \ln x^6 + \frac{1}{4} \ln x^8$
   (e) $3 \ln t^2 + 4 \ln t^3 - 6 \ln t$
   (f) $\frac{1}{2} \log_{10} 6x + \frac{1}{2} \log_{10} 24x$
   (g) $\log xyz - \log xy$ (h) $\frac{1}{3} \ln x^2 + \frac{2}{3} \ln x^3$
   (i) $3 \log xy + 2 \log xyz$ (j) $\ln\left(\dfrac{1}{z}\right) + \ln z$
   (k) $2 \ln\left(\dfrac{1}{z^2}\right) + 3 \ln z$

   (l) $4 \log\left(\dfrac{2}{t}\right) - 3 \log\left(\dfrac{2}{t^2}\right)$

   (m) $2 \log_{10} cd - 3 \log_{10}\left(\dfrac{c}{d}\right)$

   (n) $4 \log\left(\dfrac{a}{b}\right) - 3 \log\left(\dfrac{b}{a}\right)$

   (o) $3 \ln\left(\dfrac{a^2}{b}\right) - 2 \ln\left(\dfrac{b^2}{a^2}\right)$

3. (a) If $\log x + K = \log 3x$ for all values of x, find K.
   (b) Is it possible to find a constant p such that
   $$\log x + p = 3 \log x$$
   for all values of x?

4. Calculate the intensity level in decibels of a noise with an intensity of 10^{-9} W m^{-2}.

13.5 Logarithms to bases other than 10 and e

This section shows how to calculate logarithms to a base that is not 10 or e. We can use a scientific calculator to find the logarithm of a number. However, this limits us to logarithms to base 10 and base e. Occasionally we need to find logarithms to other bases. For example, logarithms to the base 2 are used in information technology. We need to be able to change the base of a logarithm. To do this, we use the formula

$$\log_a x = \frac{\log_b x}{\log_b a}$$

We can use this formula to calculate logarithms to base a, provided we can find logarithms to base b. In particular, we note that

$$\log_a x = \frac{\log_{10} x}{\log_{10} a}, \quad \text{by taking } b = 10$$

$$= \frac{\ln x}{\ln a}, \quad \text{by taking } b = e$$

Example 13.13

Find $\log_2 3$.

Solution   We let $a=2$, $b=10$ and $x=3$. This gives

$$\log_2 3 = \frac{\log_{10} 3}{\log_{10} 2} = \frac{0.477\,12}{0.301\,03} = 1.5850$$

Note that the same result is obtained if we use logarithms to the base e in our calculations.

$$\log_2 3 = \frac{\ln 3}{\ln 2} = \frac{1.098\,61}{0.693\,15} = 1.5850$$

Example 13.14

Find
(a) $\log_2 14$ (b) $\log_{20} 100$ (c) $\log_4 0.25$

Solution   (a) $$\log_2 14 = \frac{\log_{10} 14}{\log_{10} 2} = \frac{1.146\,13}{0.301\,03} = 3.8074$$

(b) $$\log_{20} 100 = \frac{\log_{10} 100}{\log_{10} 20} = \frac{2}{1.301\,03} = 1.5372$$

(c) $$\log_4 0.25 = \frac{\log_{10} 0.25}{\log_{10} 4} = \frac{-0.602\,06}{0.602\,06} = -1$$

Self-assessment question 13.5

1.   When changing the base of logarithms, the new base must always be less than the original base. True or false?

Exercises 13.5

1.   Evaluate (a) $\log_3 9$ (b) $\log_{12} 6$
     (c) $\log_4 4$ (d) $\log_8 30$ (e) $\log_2 16$
     (f) $\log_4 16$ (g) $\log_8 16$ (h) $\log_{100} 25$

2.   Find the value of a constant K such that

$$\ln x = K \log_{10} x$$

for all values of x.

3.   This exercise helps you derive the change-of-base formula

$$\log_a x = \frac{\log_b x}{\log_b a}$$

Consider a number x expressed as a power of a base a; that is, $x=a^n$.
(a)   By taking logs to base b of both sides of $x=a^n$, obtain an expression for n.
(b)   By taking logs to base a of both sides of $x=a^n$, obtain an expression for n.
(c)   Use (a) and (b) to derive the change-of-base formula.

13.6 The logarithm functions and their graphs

We are now ready to introduce the logarithm functions. They are defined as

$$y = \log_{10} x \quad \text{and} \quad y = \ln x$$

Recall that a function receives an input and produces a corresponding output. Consider the function $y = \log_{10} x$. If the input is, say, $x = 10$ then the output is $\log_{10} 10 = 1$. Hence

$$y(10) = \log_{10} 10 = 1$$

A scientific calculator can be used to find values of $\log_{10} x$ and $\ln x$ for various other values of x. Table 13.2 gives values of x and the corresponding values of $\log_{10} x$ and $\ln x$. The graphs of the functions $y = \log_{10} x$ and $\ln x$ are shown in Figure 13.1.

Table 13.2

x	0.1	0.5	1	2	5	10
$\log_{10} x$	-1	-0.3010	0	0.3010	0.6990	1
$\ln x$	-2.3026	-0.6931	0	0.6931	1.6094	2.3026

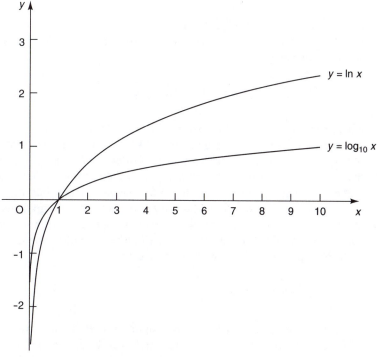

Figure 13.1   Graphs of $y = \log_{10} x$ and $\ln x$.

From the graphs, we note the following properties:

(a) As x increases, both functions increase. Indeed, by choosing x large enough, the values of $\log_{10} x$ and $\ln x$ can be made as large as desired. We write this as

$$\log_{10} x \to \infty \quad \text{as } x \to \infty$$

$$\ln x \to \infty \quad \text{as } x \to \infty$$

and read '$\log_{10} x$ and $\ln x$ approach infinity as x approaches infinity'.

(b) As x approaches 0, the function values become large and negative. This is expressed mathematically as

$$\log_{10} x \to -\infty \quad \text{as } x \to 0$$

$$\ln x \to -\infty \quad \text{as } x \to 0$$

This is read as '$\log_{10} x$ and $\ln x$ approach minus infinity as x approaches 0'.

(c) When $x = 1$, $\log_{10} x = \ln x = 0$.

(d) Both functions are strictly increasing; that is, as x increases, the functions increase.

(e) The logarithm functions are not defined when x is negative or zero.

Example 13.15 Electronic signal levels

Recall from Example 13.12 the formula for calculating the intensity level of a sound in decibels, namely

$$\text{intensity level in dB} = 10 \log_{10}\left(\frac{I}{I_0}\right)$$

where I is the intensity of the source in W m^{-2}, and I_0 is the intensity of a just audible sound and has a value of 10^{-12} W m^{-2}. Recall, also, that this measure of sound is a comparison with a sound of intensity I_0. Electronic engineers also make use of this formula, but in a slightly different form. They usually need to compare the difference in intensity of two signals rather than the intensity of a signal relative to a standard intensity. For example, a signal may have been increased in intensity as a result of being amplified by an amplifier. It is then convenient to calculate the gain of the amplifier by comparing the output signal with the input signal. The gain is defined as

$$\text{gain in dB} = 10 \log_{10}\left(\frac{P_2}{P_1}\right) \tag{13.1}$$

where P_2 is the power of the output signal and P_1 is the power of the input signal. As we are comparing two power levels, Equation (13.1) is known as a **power gain**. Note that if $P_2/P_1 > 1$ (that is, $P_2 > P_1$) then the gain is positive, but if

$P_2/P_1 < 1$ that is, $P_2 < P_1$ then the gain is negative. This is easily seen by examining the graph of the logarithm function shown in Figure 13.1. It can be shown that the overall gain due to an amplifier with several stages can be calculated by simply adding together the gains of the individual stages. This formula is more generally applicable and can be used to compare the intensity level of two signals. For example, it is quite common for engineers to talk about signals contaminated by noise as having a signal-to-noise ratio of, say, 20 dB. This simply means that the signal power is 20 dB larger than the noise power.

Electronic engineers sometimes prefer to work with the voltage level of a signal rather than its power level. Recall that

$$P = \frac{V^2}{R}$$

where P is power in W, V is voltage in V and R is resistance in Ω. If, for comparison, we assume that both signals are developed across $1\,\Omega$ resistors then we have

$$\text{gain in dB} = 10 \log_{10}\left(\frac{P_2}{P_1}\right)$$

$$= 10 \log_{10}\left(\frac{V_2^2}{V_1^2}\right)$$

$$= 10 \log_{10}\left(\frac{V_2}{V_1}\right)^2$$

$$= 10 \times 2 \times \log_{10}\left(\frac{V_2}{V_1}\right), \quad \text{by the laws of logarithms}$$

$$= 20 \log_{10}\left(\frac{V_2}{V_1}\right) \tag{13.2}$$

This formula is often used by electronic engineers. As we are comparing two voltage levels, Equation (13.2) is known as a **voltage gain**.

Self-assessment question 13.6

1. List properties that are common to both $y = \log x$ and $y = \ln x$. Can you think of any properties that distinguish $y = \log x$ and $y = \ln x$?

Exercises 13.6

1. It is impossible for $\ln x$ to exceed the value 10000. True or false?

2. An amplifier has an output signal of 100 mV for an input signal of 5 mV. Calculate the voltage gain of the amplifier in dB.

3.   A two-stage amplifier consists of a pre-amplifier and a power amplifier. Given the following information, calculate the voltage gain of the pre-amplifier, the power amplifier and the overall amplifier in dB:

pre-amplifier:	input signal 5 mV
	output signal 100 mV
power amplifier:	input signal 200 mV
	output signal 800 mV

Computer and calculator exercises 13.6

1.   (a)   Draw the graphs of $y = \log_{10} x$ and $y = \ln x$ for $0 < x \leqslant 10$.
   (b)   Measure the vertical separation between the two graphs for various values of x. What do you notice?

2.   Use a drawing package to sketch $y = \log_{10}(1/x)$ for $0 < x \leqslant 100$. Describe how this graph compares with that of $y = \log_{10} x$. Can you explain your observations using the laws of logarithms?

13.7 Exponential expressions

Exponential expressions are expressions of the form a^x where a is a constant. Recall that e is the irrational constant whose value is approximately 2.718. When $a = $ e, we obtain expressions of the form e^x. Such expressions obey the laws of indices, that is

$$e^x e^y = e^{x+y}$$

$$\frac{e^x}{e^y} = e^{x-y}$$

$$(e^x)^y = e^{xy}$$

$$\frac{1}{e^x} = e^{-x}, \qquad \frac{1}{e^{-x}} = e^x$$

$$e^0 = 1$$

A calculator can be used to calculate values of a^x for various values of a and x. Some calculators have a y^x button for this purpose. Most scientific calculators have 10^x and e^x buttons also. Make sure you can use these facilities on your calculator by checking the next example.

Example 13.16

Use a scientific calculator to evaluate (a) $(4.3)^{1.7}$ (b) $10^{-0.4}$ (c) $e^{2.5}$

Solution   (a)   Using a scientific calculator, we find

$$(4.3)^{1.7} = 11.9370$$

(b)   We use the 10^x button. We find

$$10^{-0.4} = 0.3981$$

(c)   We use the e^x button. We find

$$e^{2.5} = 12.1825$$

The following examples illustrate how to simplify exponential expressions.

Example 13.17

Simplify as much as possible (a) $e^x \cdot e^x$ (b) $e^x + e^x$ (c) $e^x(e^x + e^x)$

Solution   (a)   Using the laws of indices, we have

$$e^x \cdot e^x = e^{x+x} = e^{2x}$$

(b) $e^x + e^x = 2e^x$

(c) $e^x(e^x + e^x) = e^x(2e^x) = 2e^x e^x = 2e^{2x}$

Example 13.18

Simplify (a) $e^x \cdot e^y$ (b) $\dfrac{e^{3x}}{e^x}$ (c) $\dfrac{e^x}{e}$ (d) $(e^{2x})^3$

Solution   (a)   Using the laws of indices, we have $e^x \cdot e^y = e^{x+y}$.
           (b)   We have

$$\frac{e^{3x}}{e^x} = e^{3x-x} = e^{2x}$$

(c)   Recall that e is the same as e^1, and so

$$\frac{e^x}{e} = \frac{e^x}{e^1} = e^{x-1}$$

(d)   We use the law $(a^m)^n = a^{mn}$:

$$(e^{2x})^3 = e^{2x \times 3} = e^{6x}$$

Konstantin Tsiolkovsky and space flight

Konstantin Tsiolkovsky (see Figure 1) was born in the village of Izhevskoye in Russia in 1857. He had a fairly normal childhood, and engaged in all the usual activities that boys did at that time; he skated in winter, played games and flew kites.

Sadly, at the age of 10 he became seriously ill, and one of the complications was that he became almost totally deaf. The after effects of this illness continued until the age of 13. He later recalled:

'This three-year interval – through my lack of consciousness – was the saddest and darkest time of my life. I try to reconstruct it in my memory, but there is at present nothing that I can recall. There is simply nothing to mark that time. All that I can remember is skating, sleighing,....'

From the age of 13, he became very interested in inventing and built a range of engineering and scientific devices. For example, he built model steam engines, various types of windmills and a small lathe. He also began to study books on science from his father's library. Later he recalled:

'there were very few books, and I had no teachers at all, so I had to create and devise more than absorb and imbibe from others. There were no hints, no aid from anywhere; there was a great deal that I couldn't understand in these books, and I had to figure out everything by myself. In a word, then, the creative element, the element of self-development and originality was predominant.'

Tsiolkovsky continued this process of self-education throughout the rest of his life, and had very little formal education. In 1873 he was sent to Moscow by his father to continue his education. By studying on his own, he completed all of the secondary school course, and most of his university course. After passing the necessary examination, without attending lectures, he became a teacher and then began to spend most of his spare time on scientific investigations. He was interested in many areas of science and engineering, but his most famous achievements were in the areas of space flight and rocketry. This was amazing, given that much of his work was carried out in the 19th century. One of his most famous algebraic formulae is the one that allows the velocity of a rocket to be calculated as it burns fuel. It relates the velocity of a rocket at a particular time to the velocity with which the gas particles are expelled from the nozzle of the rocket, the mass of the rocket and the mass of the fuel that has been used. This is a difficult problem to analyse, because the mass of a rocket changes significantly as it burns fuel. The formula is now called Tsiolkovsky's formula (see Figure 2), and is

$$\frac{V}{V_1} = \ln\left(\frac{M_1 + M_2}{M_1 + M}\right)$$

where V is the velocity of the rocket, V_1 is the velocity of the exhaust gases relative to the rocket, M_1 is the mass of the rocket without fuel, M_2 is the mass of the rocket fuel before any is burnt and M is the mass of the unburnt fuel.

Let us consider the case when no fuel has been burnt. We have $M = M_2$, and Tsiolkovsky's formula becomes

$$V = V_1\ln\left(\frac{M_1 + M_2}{M_1 + M_2}\right) = V_1\ln 1 = V_1 \times 0 = 0$$

We see that the initial velocity of the rocket is zero, as we should expect.

Turning to the case when all the fuel has been burnt, we have $M = 0$, and then

$$V = V_1\ln\left(\frac{M_1 + M_2}{M_1}\right) = V_1\ln\left(1 + \frac{M_2}{M_1}\right)$$

This is the maximum velocity that the rocket can achieve. The important thing to note is that it shows that, given enough fuel, it is theoretically possible for the rocket to achieve any desired velocity. This in effect shows that space flight is theoretically possible and that a rocket can be built to attain a velocity greater than the escape velocity of the Earth.

Although many people had speculated on the possibility of space flight, it was Tsiolkovsky who first put it on a scientific basis. Events in the 20th century were to vividly demonstrate the validity of this formula, and it is a pity that Tsiolkovsky did not live to see the first manned space flight. In the 1920s his work began to receive the recognition it deserved. In 1929 Hermann Oberth, a German rocket engineer, wrote to Tsiolkovsky, saying:

> 'You have kindled a fire, and we shall not let it die out, but will bend every effort to make the greatest dream of mankind come true.'

Tsiolkovsky was not interested in the military uses of rockets. He saw them as a means by which man could expand to the planets and beyond. He wrote:

> 'Vehicles in revolution about the Earth and with all the accessories for the existence of intelligent beings may serve as a basis for the further expansion of humanity. People inhabiting the vicinity around the Earth in the form of a multitude of rings like those of Saturn... would increase 100- to 1000-fold the reserves of solar energy that are allotted to them on the surface of the Earth. Even so, man may not be satisfied, and from this conquered base he may extend his hands to capture the rest of the solar energy, which is two thousand million times greater than what the Earth gets.'

When evaluating his life, Tsiolkovsky wrote:

> 'The prime motive of my life is to do something useful for people, not to live my life purposelessly, but to advance humanity even the slightest bit. This is why I have interested myself in things that did not give me bread or strength. But I hope that my studies will, perhaps soon, but perhaps in the distant future, yield society mountains of grain and limitless power.'

Figure 1 *Portrait of Konstantin Tsiolkovsky.*

Figure 2 *A manuscript page (1897) showing Tsiolkovsky's formula of rocket motion.*

Exercises 13.7

1. Evaluate the following expressions using a scientific calculator:
(a) 5^7 (b) $8.1^{2.3}$ (c) $2^{-1.5}$
(d) $10^{1.8}$ (e) $10^{-2.3}$ (f) e^2
(g) $e^{-0.5}$

2. Use the laws of indices to simplify
(a) e^2e^4 (b) e^7e^5 (c) $e^{3.5}e^{2.3}$
(d) $e^{-2.5}e^6$ (e) $2e^43e^5$ (f) $5e^{-2}2e^{-3}$
(g) $3e^{-3}4e^{-1}$ (h) $\dfrac{e^3}{e^2}$ (i) $\dfrac{e^9}{e^2}$ (j) $\dfrac{6e^5}{2e^4}$
(k) $\dfrac{e^4}{e^8}$ (l) $\dfrac{e^{6.8}}{e^{-3}}$ (m) $\dfrac{16e^3}{12e^7}$
(n) $\dfrac{12e^{-2}}{30e^{-3}}$ (o) $\dfrac{e^2}{3}\times\dfrac{e^3}{2}$ (p) $3e^4\times\dfrac{2e^2}{e^6}$
(q) $\dfrac{8e^{-1.5}}{3e^2}\times\dfrac{6e^{2.5}}{e^{-2}}$ (r) $e^3\times3e^4\times e^{-3}\times2$
(s) $2e^2\times3e^3\times4e^{-4}$ (t) $\dfrac{2e^3\times5e^{-1}}{e^3\times4e^{-2}}$

3. Remove the brackets from the following expressions:
(a) $(e^3)^2$ (b) $(2e)^5$ (c) $(3e^{-1})^4$
(d) $(e^2e^3)^4$ (e) $(e^{-2}e^4)^{2.5}$ (f) $(e^5)^{-1}$
(g) $(e^{2.5})^{-2}$ (h) $(3e^2 4e^{-1})^2$ (i) $\left(\dfrac{e^4}{3}\right)^2$
(j) $\left(\dfrac{2e^3}{3e^2}\right)^{-1}$

4. Simplify the following expressions:
(a) $e^te^{2t}e^{3t}$ (b) $\dfrac{e^x}{2}+\dfrac{e^x}{3}$ (c) $3e^z\cdot4e^{4z}$
(d) $\dfrac{e^x}{e^{2x}}$ (e) $\dfrac{2e^{2t}}{4e^t}$ (f) $e^{2t}\cdot e^{-2t}$
(g) e^xe^{-2x} (h) $\dfrac{3e^{3x}}{6e^x}$ (i) $\dfrac{ee^x}{e^2}$

13.8 The exponential function

This section defines the exponential function and gives some graphical illustrations. The connection between the exponential function and the logarithm function is also explained.

> Exponential functions have the form
> $$y=a^x \qquad a>0$$
> where a is a constant.

13.8.1 Graph of exponential functions

Some exponential functions are shown in Figure 13.2 for $a=0.5$, $a=2$ and $a=3$. Values are listed in Table 13.3.

The most commonly used exponential function is $y=e^x$. This function is often referred to as **the** exponential function because of the way it dominates

Table 13.3

x	0.5^x	2^x	3^x
-3	8	0.125	0.037
-2	4	0.25	0.111
-1	2	0.5	0.333
0	1	1	1
1	0.5	2	3
2	0.25	4	9
3	0.125	8	27

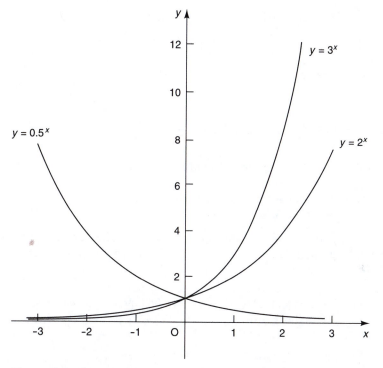

Figure 13.2   Some typical exponential functions.

engineering applications. Table 13.4 gives some values of e^x for various values of x. Figure 13.3 shows the function $y = e^x$. Some important properties of $y = e^x$ can be seen from the graph.

(a)   As x increases, e^x increases. We write $e^x \to \infty$ as $x \to \infty$ and say 'e^x approaches infinity as x approaches infinity'. This property is called **exponential growth**.

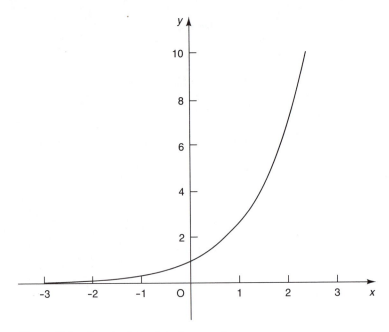

Figure 13.3   Graph of the function $y=e^x$.

(b)   We also note that e^x approaches 0 as x becomes large negatively; that is, $e^x \to 0$ as $x \to -\infty$.

(c)   The function is never negative.

(d)   When $x=0$, $e^x=1$.

Allied to $y=e^x$ is the function $y=e^{-x}$. Values are tabulated in Table 13.5 and a graph is shown in Figure 13.4. From the graph, we note some properties of the function.

(a)   As x becomes large negatively, the function increases without bound. We write this mathematically as $e^{-x} \to \infty$ as $x \to -\infty$.

Table 13.4

x	e^x
-3	0.050
-2	0.135
-1	0.368
0	1
1	2.718
2	7.389
3	20.086

Table 13.5

x	e^{-x}
-3	20.086
-2	7.389
-1	2.718
0	1
1	0.368
2	0.135
3	0.050

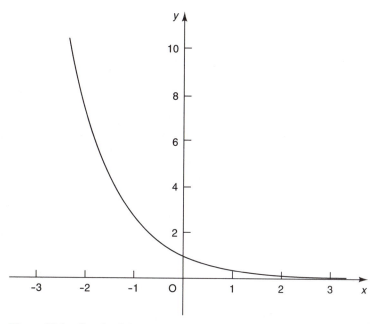

Figure 13.4   Graph of the function $y = e^{-x}$.

(b)   As x becomes large positively, the function approaches 0; that is, $e^{-x} \to 0$
       as $x \to \infty$. This property is called **exponential decay**.
(c)   The function is never negative.
(d)   When $x = 0$, e^{-x} is 1.

   The exponential function is useful for describing many different physical
phenomena, for example the discharge of a capacitor and growth of populations.

Example 13.19 The diode equation

A semiconductor diode at room temperature can be modelled by the equation

$$I = I_s(e^{40V} - 1)$$

where I is the current through the diode, V is the voltage across the diode, and
I_s is the reverse saturation current for the diode and is a constant. This is a
good model for germanium diodes, but only an approximate model for silicon
diodes. Figure 13.5 shows a graph of I against V for a typical diode. It is clear
from examining this graph that the diode conducts current more easily in one
direction than in the other. This is one of its main functions in an electronic
circuit. It is rather like a one-way valve in a pipe, for example a heart valve
that only allows blood to flow in one direction. Note from the graph that a
small current does flow when a negative voltage is applied across the diode. Its
size has been exaggerated in Figure 13.5 to make it stand out, and it is actually

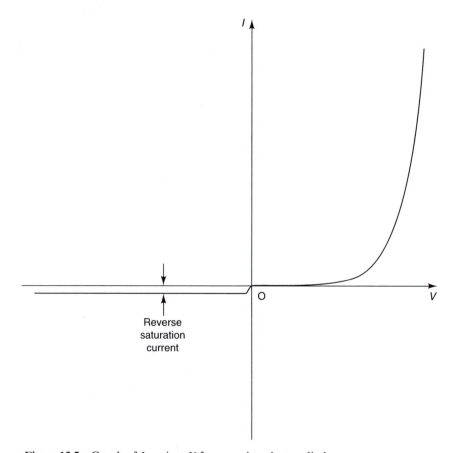

Figure 13.5 Graph of I against V for a semiconductor diode.

quite small for a real diode. When a positive voltage is applied to the diode then it is said to be forward-biased. Note that the forward characteristic is approximately the shape of an exponential function.

Self-assessment questions 13.8

1. Describe four essential properties of the function $y = e^x$.

2. Describe four essential properties of the function $y = e^{-x}$.

3. How are the functions $y = e^x$ and $y = e^{-x}$ related?

4. The function $y = e^{-x}$ is always negative. True or false?

Exercises 13.8

1.   A silicon diode has a reverse saturation current of 80 nA. Calculate the current flowing through the diode when a forward voltage of 0.25 V is applied to the diode.

2.   A germanium diode has a reverse saturation current of 5 μA. Sketch a graph of the current voltage characteristics of the diode.

Computer and calculator exercises 13.8

1.   (a)   Plot using the same axes $y = 3e^{2x}$ and $y = e^{-x}$ for $-2 \leqslant x \leqslant 2$.
   (b)   For what value of x is $3e^{2x} = e^{-x}$?

2.   (a)   Plot $y = \ln x$ and $y = e^{-x/4}$ for $0 < x \leqslant 3$.
   (b)   Find the value of x for which $\ln x = e^{-x/4}$.

13.9 Connection between e^x and ln x, and 10^x and log x

Recall that if $a = b^c$ then $\log_b a = c$. It follows immediately that

$$\text{if} \quad y = e^x \quad \text{then} \quad x = \ln y$$

This is an important result, and shows the connection between natural logarithms and the exponential function. The step of moving from $y = e^x$ to $x = \ln y$ can be explained using the laws of logarithms. Given

$$y = e^x$$

we can take the natural logarithm of both sides to obtain

$$\ln y = \ln (e^x)$$
$$= x \ln e, \quad \text{using the third law of logarithms}$$
$$= x, \quad \text{since } \ln e = 1$$

The step of moving from $y = e^x$ to $x = \ln y$ is known as 'taking natural logs of both sides of the equation'.

From this result, we can see that

$$\ln (e^x) = \ln y = x \tag{13.3}$$

and

$$e^{(\ln y)} = e^x = y \tag{13.4}$$

Equations 13.3 and 13.4 are illustrated in Figure 13.6.

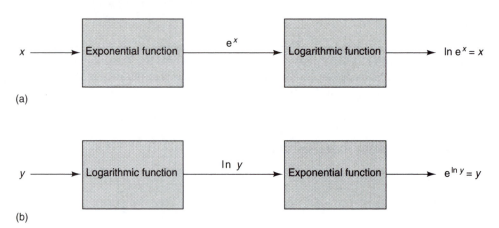

(a)

(b)

Figure 13.6 (a) The natural logarithm function is the inverse of the exponential function. (b) The exponential function is the inverse of the natural logarithm function.

In Figure 13.6(a) the input to the exponential function is x and the output is e^x. This forms the input to the natural logarithm function. The final output is x. Hence the natural logarithm function has undone the work of the exponential function; that is, the natural logarithm function, $\ln x$, is the inverse of the exponential function.

Similarly, in Figure 13.6(b) the input to the natural logarithm function is y and the output is $\ln y$. This forms the input to the exponential function. The final output is y. The exponential function has thus undone the work of the natural logarithm function; that is, the exponential function is the inverse of the natural logarithm function.

We can state a similar argument for 10^x and $\log_{10} x$. If $y = 10^x$ then taking logarithms to the base 10 yields $x = \log_{10} y$. Clearly, $\log_{10} y$ and 10^x are inverse functions in exactly the same way as e^x and $\ln x$ are.

Example 13.20
Find x given (a) $39 = e^x$ (b) $0.46 = e^x$ (c) $47 = 10^x$ (d) $0.32 = 10^x$

Solution (a) If $y = e^x$ then $x = \ln y$, and therefore

$$39 = e^x$$

$$x = \ln 39 = 3.6636$$

(b) If $y = e^x$ then $x = \ln y$, and so

$$0.46 = e^x$$

$$x = \ln 0.46 = -0.7765$$

(c)   If $y = 10^x$ then $x = \log_{10} y$, and so

$$47 = 10^x$$

$$x = \log_{10} 47 = 1.6721$$

(d)   If $y = 10^x$ then $x = \log_{10} y$, and so

$$0.32 = 10^x$$

$$x = \log_{10} 0.32 = -0.4949$$

Self-assessment questions 13.9

1.   State the inverse of the function $y = e^x$.

2.   State the inverse of the function $y = \ln x$.

3.   State the inverse of the function $y = \log x$.

4.   State the inverse of the function $y = 10^x$.

Exercises 13.9

1.   Find the value of t for which (a) $600 = e^t$
       (b) $0.8 = e^t$ (c) $25 = 10^t$ (d) $0.25 = 10^t$

2.   For which values of x is $e^x \geqslant 1\,000\,000$?

3.   For which values of x is $e^{-x} \leqslant 200\,000$?

4.   Find a value of x for which (a) $10^x = 3$
       (b) $10^x = \frac{1}{3}$
       What do you observe about your answers?
       Can you explain this using the laws of
       indices?

13.10 Solving equations involving logarithm and exponential expressions

This section explains how to solve algebraically equations involving logarithm and exponential expressions. It also shows how to interpret graphs in order to solve equations.

13.10.1 Algebraic solutions

We illustrate the methods of solution with examples.

Example 13.21

Solve (a) $2e^{2x}=100$ (b) $5e^{-x}=12$

Solution   (a) $2e^{2x}=100$

Dividing both sides of the equation by 2 gives

$$e^{2x}=50$$

Taking natural logs of both sides of the equation gives

$$2x=\ln 50=3.912\,02$$

and so

$$x=\frac{3.912\,02}{2}=1.9560$$

(b) $5e^{-x}=12$

Dividing the equation by 5 gives

$$e^{-x}=\frac{12}{5}=2.4$$

Taking natural logs gives

$$-x=\ln 2.4=0.8755$$

$$x=-0.8755$$

Example 13.22

Solve (a) $2\ln x=4.06$ (b) $\ln(2x)=1.36$

Solution   Recall that if $\ln x=y$ then $x=e^y$.
(a)   We have

$$2\ln x=4.06$$

$$\ln x=2.03$$

$$x=e^{2.03}=7.6141$$

(b)   We have

$$\ln(2x)=1.36$$

$$2x=e^{1.36}=3.8962$$

$$x=1.9481$$

Example 13.23

Solve (a) $e^{(x^2)} = 100$ (b) $3e^{\sqrt{x}} = 600$

Solution   (a)   We have

$$e^{(x^2)} = 100$$

Taking natural logs produces

$$x^2 = \ln 100 = 4.605$$

Taking the square root of both sides of the equation gives

$$x = \pm\sqrt{4.605} = \pm 2.146$$

(b)   We have

$$3e^{\sqrt{x}} = 600$$

Dividing the equation by 3 gives

$$e^{\sqrt{x}} = 200$$

Taking natural logs of both sides of the equation gives

$$\sqrt{x} = \ln 200 = 5.298$$

Squaring both sides of the equation gives

$$x = (5.298)^2 = 28.07$$

Example 13.24

Solve (a) $2 \cdot 10^x = 500$ (b) $5 \cdot 10^{-3x} = 600$

Solution   (a)   We have

$$2 \cdot 10^x = 500$$

$$10^x = 250$$

Taking logs to base 10 gives

$$x = \log_{10} 250 = 2.398$$

(b)   We have

$$5 \cdot 10^{-3x} = 600$$

$$10^{-3x} = 120$$

Taking logs of both sides produces

$$-3x = \log_{10} 120 = 2.0792$$

$$x = \frac{2.0792}{-3} = -0.693$$

Example 13.25

    Solve (a) $\log_{10} 3x = 1.5$ (b) $\log_{10}(3x - 6) = 0.76$

Solution   (a) $\log_{10} 3x = 1.5$

$$3x = 10^{1.5} = 31.623$$

$$x = 10.54$$

       (b) $\log_{10}(3x - 6) = 0.76$

$$3x - 6 = 10^{0.76} = 5.7544$$

$$3x = 11.7544$$

$$x = 3.918$$

13.10.2 Graphical solution

Some equations cannot be solved algebraically. In such cases we can find an approximate solution using graphs. Use of a graphics calculator or graph-plotting package is most useful, and saves the laborious task of plotting graphs by hand.

Example 13.26

    Find an approximate solution of $e^x + x = 5$.

Solution   We write the equation in the form

$$e^x + x - 5 = 0$$

Now we plot the function $y = e^x + x - 5$. Figure 13.7 shows the function plotted for values of x between 0 and 2. The value of the function is 0 where the graph cuts the x axis. The curve cuts the x axis at about $x = 1.3$. Hence the solution to $e^x + x = 5$ is approximately $x = 1.3$.

    This is an approximate solution. To improve the accuracy, we could redraw the graph for a smaller range of x values centred around $x = 1.3$. For example, we could now plot y for $1.2 \leqslant x \leqslant 1.4$. This allows a much finer scale to be used on each axis, and so greater accuracy can be achieved.

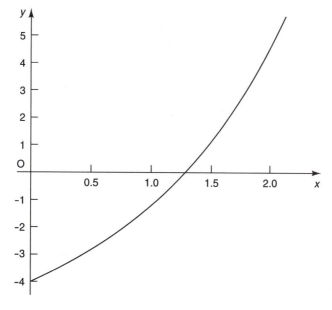

Figure 13.7 The function $y = e^x + x - 5$.

Example 13.27

Find an approximate solution of $5e^{-x} = \ln x$.

Solution The given equation is equivalent to

$$5e^{-x} - \ln x = 0$$

and hence we plot the function

$$y = 5e^{-x} - \ln x$$

We seek solutions of $5e^{-x} - \ln x = 0$ by looking where the graph cuts the x axis. This is shown in Figure 13.8 for $0 < x \leqslant 5$. The graph cuts the x axis at approximately $x = 2$. This accuracy may be improved by plotting the function again, this time for say $1.5 \leqslant x \leqslant 2.5$.

Self-assessment questions 13.10

1. Explain how the accuracy of a solution can be improved when solving equations graphically.

2. What advantages and disadvantages of the graphical method can you think of?

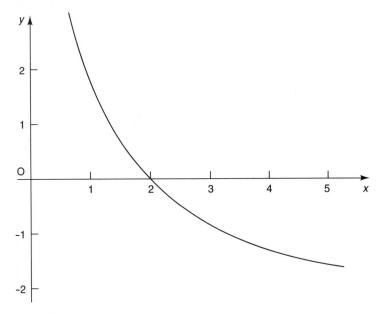

Figure 13.8 The function $y = 5e^{-x} - \ln x$.

Exercises 13.10

1. Solve algebraically (a) $e^x = 10$ (b) $3e^x = 6$
    (c) $10^x = 47$ (d) $10^{-x} = 1.3$
    (e) $4 \cdot 10^{-2x} = 20$ (f) $17e^{-x} = 23$
    (g) $e^{2x} = 3e^x$ (h) $10^{2x} = 16 \cdot 10^{-x}$
    (i) $e^{2x+3} = 20$ (j) $10^{-x+6} = 30$
    (k) $\dfrac{40}{10^x + 3} = 1.6$ (l) $\dfrac{1}{e^x + 1} = 0.3$
    (m) $\dfrac{2}{e^{-2x} - 1} = \dfrac{1}{3}$

2. Solve algebraically (a) $\log_{10} x = 1.6$
    (b) $\ln y = 3$ (c) $\ln(2t - 1) = 6$
    (d) $\log_{10}(\frac{1}{2}y) = 1.5$ (e) $3 \log_{10} b = 1.7$
    (f) $3 \log_{10} b^2 = 1.7$ (g) $2 \ln(3t - 5) = 4$
    (h) $2 \ln(1/t) = 6$ (i) $-\log_{10}(-x) = 2$
    (j) $\dfrac{2}{\ln(x^2 + 1)} = 1$ (k) $\log_{10}(3t^2 - 4) = 1.9$

Computer and calculator exercises 13.10

1. Use a graphics calculator or graph-plotting package to solve the following equations:
    (a) $e^x = 6x$ $0 \leqslant x \leqslant 4$
    (b) $\ln(x^3) = x$ $1 \leqslant x \leqslant 3$
    (c) $e^x = 5 - x^2$ $-2.5 \leqslant x \leqslant 2.5$
    (d) $e^{-x} = -x^2 + 2x + 3$ $-1.5 \leqslant x \leqslant 3.5$

2. Solve the following graphically:

    (a) $\log_{10} x = 7 - 2x$ (b) $2 \ln x = 10 - x$
    (c) $e^x = 10 - x - x^2$ (d) $10^x = 3 - 10x$

3. Use a graphics calculator to show that

    $$f(x) = e^x - x - 2 = 0$$

    has two roots – one positive and one negative – and locate them.

Review exercises 13

1 Simplify as much as possible (a) $\ln x + \ln y^2$
(b) $10^{2x} \cdot 10^x$ (c) $\log_{10} b^3 - \log_{10} b^2$
(d) $2 \ln y + \ln y^3$

(e) $\log AB + \log B^2 - \log \left(\dfrac{A}{B} \right)$ (f) $\dfrac{e^{2t}}{3e^t}$

(g) $\dfrac{e^{-n}}{e^{-2n}}$ (h) $3 \log xy - 2 \log y - 2 \log x$

(i) $\ln (abc)^2 - 2 \ln \left(\dfrac{b}{c} \right) + 3 \ln \left(\dfrac{ac}{b} \right)$

(j) $\dfrac{10^{x+y} 10^{-2y}}{10^{-x}}$ (k) $\dfrac{e^{2t}(e^{-t}+1)}{e^t}$

(l) $\ln \left(\dfrac{A}{B} \right) - \ln \left(\dfrac{B}{A} \right)$ (m) $\ln x + \log x$

2 Solve algebraically (a) $e^x = 12$

(b) $e^{3x} = 10$ (c) $e^{1/x} = 0.05$ (d) $\dfrac{2}{e^x} = 5$

(e) $e^{3x+2} = 1$ (f) $e^{5x-3} = 20$
(g) $e^{-2x+3} = 7$ (h) $e^{4-x} = 3e^x$
(i) $3e^{-x+2} = 5e^{x-1}$ (j) $e^{x^2} = 100$
(k) $e^{3x^2} = 250$ (l) $e^{2x^2-3} = 70$
(m) $3e^{2x^2-7} = 1000$

3 Solve algebraically (a) $\ln x = 1.96$
(b) $\log_{10} t = 2.3$ (c) $\ln (2b+1) = 1.6$

(d) $\log_{10} \left(\dfrac{3}{w} \right) = 0.75$

(e) $\log_{10} (2t+7) = 1.8$ (f) $\log_{10} 2t^2 = 2.1$
(g) $\ln (3h^2 - 5) = 2$ (h) $3 \ln (2t-1) = 1.8$
(i) $\log_{10} (t^5 + 1) = 2.5$
(j) $3 \log_{10} (3b^3 + 3) = 7$

(k) $\ln \left(\dfrac{4+b}{3+2b} \right) = 0.5$

14

Triangles and Their Solution

KEY POINTS

This chapter

● defines a right-angled triangle, an isosceles triangle, an equilateral triangle, and a scalene triangle

● states Pythagoras' theorem

● illustrates how to use Pythagoras' theorem to calculate the lengths of the sides of a right-angled triangle

● illustrates how to calculate angles and lengths of a given right-angled triangle

● states the sine rule and illustrates how it can be used to solve triangles

● states the cosine rule and illustrates how it can be used to solve triangles

CONTENTS

14.1 Introduction

Engineers are often required to calculate angles and lengths of the sides of a triangle. The sides of a triangle may represent various forces which are acting on a body. The force has a size and a direction, which are represented by the length and orientation of the side of a triangle.

When asked to solve a triangle, we must state the length of each of its sides and the size of each angle. We begin by looking at some special triangles, before moving on to the solution of right-angled triangles, which are relatively easy to solve. Then we introduce the sine and cosine rules, which allow any triangle to be solved.

14.2 Types of triangles

A triangle is a three-sided figure. The three angles inside the triangle always add up to 180°.

Sum of angles of any triangle $= 180°$

14.2.1 Right-angled triangle

An angle of 90° is often called a **right-angle**. A triangle containing a 90° angle is called a **right-angled triangle**. The side opposite the right-angle is called the **hypotenuse**. In Figure 14.1, AB is the hypotenuse.

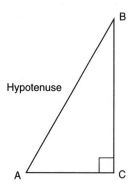

Figure 14.1   A right-angled triangle.

14.2.2 Equilateral triangle

An **equilateral triangle** has all sides of equal length, and all the angles are equal. Since the angles of a triangle sum to 180°, in an equilateral triangle all angles are 60°. Figure 14.2 shows an equilateral triangle.

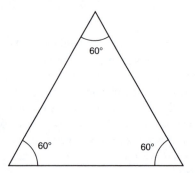

Figure 14.2   In an equilateral triangle all sides are equal and all angles are equal.

14.2.3 Isosceles triangle

A triangle in which two sides are of equal length is called an **isosceles triangle**. Figure 14.3 shows an isosceles triangle with sides AB and AC equal. The angles at B and C will also be equal. So an isosceles triangle has two equal sides and two equal angles.

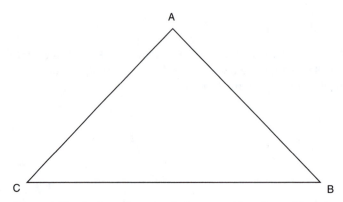

Figure 14.3   An isosceles triangle has two sides of equal length.

Example 14.1

ABC is a triangle with $AB = AC$. If $B = 40°$, find the other two angles.

Solution   Since $AB = AC$, ABC is an isosceles triangle, as shown in Figure 14.3. Now

$$B = C$$

and so $C = 40°$. The angles sum to $180°$, and so

$$B + C + A = 180°$$

$$40° + 40° + A = 180°$$

$$A = 100°$$

Example 14.2 Bridge lattices

Figure 14.4 shows two different bridge lattices, which form part of a larger bridge. For each case calculate the unknown angle θ.

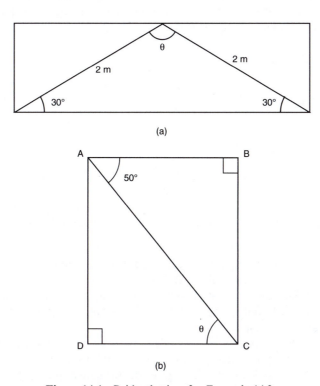

(a)

(b)

Figure 14.4   Bridge lattices for Example 14.2.

Solution   (a)   We see that the section contains an isosceles triangle, and two of the angles are 30°. Therefore

$$\theta = 180° - 30° - 30° = 120°$$

(b)   In the triangle ABC

$$\angle BCA = 180° - 90° - 50° = 40°$$

Since $\angle BCD = 90°$,

$$\theta + 40° = 90°$$

so that

$$\theta = 50°$$

14.2.4 Scalene triangle

A triangle in which all sides are of different lengths is called a **scalene triangle**. Note that all the angles are different too, although their sum is still 180°. A typical scalene triangle is shown in Figure 14.5.

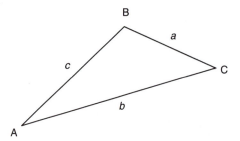

Figure 14.5   In a scalene triangle all the sides have different lengths.

In any triangle, the largest angle is opposite the longest side; the smallest angle is opposite the shortest side.

Recall that an angle between 0° and 90° is an **acute angle** and an angle between 90° and 180° is called an **obtuse angle**. Clearly, a triangle can have at most one obtuse angle.

We often shorten the angle notation. For example, for the triangle shown in Figure 14.5, rather than refer to $\angle ABC$, we often simply write B. The side opposite A is denoted a. In Figure 14.5 the side opposite A is BC and so $a = $ BC. Similarly, $b = $ AC and $c = $ AB. Solving a triangle requires finding A, B, C and a, b, c.

We say that AB and BC **include** B, or that B is included by AB and BC. Thus A is included by b and c, and C is included by a and b.

The triangle ABC is often written as $\triangle$ABC.

Self-assessment questions 14.2

1.   Define an equilateral triangle.

2.   Define an isosceles triangle.

3.   Define a scalene triangle.

4.   Can a triangle be both right-angled and isosceles?

5.   Can a triangle be both right-angled and scalene?

6.   All the angles of an isosceles triangle must be acute. True or false?

Exercises 14.2

1.   Calculate all the angles of a right-angled isosceles triangle.

2.   Calculate the unknown angle θ in the bridge lattices shown in Figure 14.6. The lattice forms part of a larger bridge.

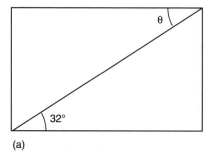

(a)

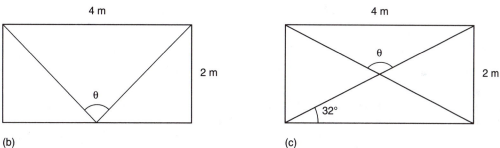

(b)   (c)

Figure 14.6   Bridge lattices for Exercise 14.2 Q2.

14.3 Pythagoras' theorem

Pythagoras' theorem applies only to right-angled triangles. Consider the right-angled triangle of Figure 14.7. The theorem states

$$c^2 = a^2 + b^2$$

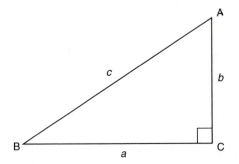

Figure 14.7   For a right-angled triangle $c^2 = a^2 + b^2$.

Example 14.3

A right-angled $\triangle ABC$ has a right-angle at C.
(a)   Calculate AB given AC = 3 cm and BC = 4 cm.
(b)   Calculate AC given AB = 13 m and BC = 12 m.

Solution   (a)   We are given $b = AC = 3$ and $a = BC = 4$.
From Pythagoras' theorem, we know that

$$c^2 = a^2 + b^2$$

and so

$$c^2 = 16 + 9 = 25$$

$$c = 5$$

The length of AB is 5 cm.
(b)   We are given $c = AB = 13$ and $a = BC = 12$. Pythagoras' theorem states

$$c^2 = a^2 + b^2$$

Substituting in the given values yields

$$13^2 = 12^2 + b^2$$

$$169 = 144 + b^2$$

$$b^2 = 25$$

$$b = 5$$

The length of AC is 5 m.

Example 14.4 Building frame

Figure 14.8 shows a metal building frame. Calculate the unknown length *l*.

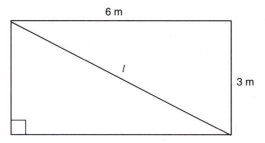

Figure 14.8   Building frame for Example 14.4.

Solution   The length l forms part of a right-angled triangle, and so, using Pythagoras' theorem, we have

$$l^2 = 3^2 + 6^2 = 9 + 36 = 45$$
$$l = \sqrt{45} = 6.708$$

The required length is 6.708 m.

Self-assessment questions 14.3

1. State Pythagoras' theorem in words.

2. Explain what is meant by the hypotenuse of a right-angled triangle.

Exercises 14.3

1. Consider a right-angled triangle ABC with a right-angle at C.
   (a) Calculate BC given $AB = 10$ m and $AC = 7$ m.
   (b) Calculate AB given $AC = 6$ mm and $BC = 9$ mm.
   (c) Calculate AC given $AB = 15$ cm and $BC = 12$ cm.

2. ABC is a right-angled isosceles triangle, as shown in Figure 14.9. If $AC = 3$ cm, calculate BC and AB.

3. In Figure 14.10 both ABC and ACD are right-angled isosceles triangles. If $AB = 2$ cm, calculate
   (a) BC   (b) AC   (c) CD   (d) AD

4. Calculate the unknown length l for each of the metal building frames shown in Figure 14.11.

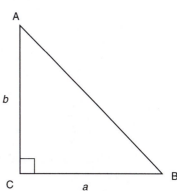

Figure 14.9   ABC is a right-angled isosceles triangle.

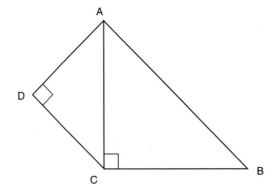

Figure 14.10   Figure for Exercise 14.3 Q3.

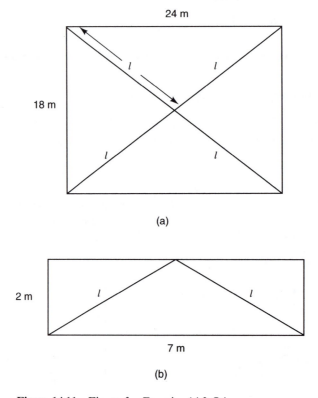

Figure 14.11   Figure for Exercise 14.3 Q4.

14.4 Solving right-angled triangles

Recall that when asked to solve a triangle, we must state all angles and the lengths of all sides. Knowing that a triangle is right-angled automatically tells us that one of the angles is 90°. To be able to solve the triangle, in addition we need to know either the length of one side and an angle, or the length of two sides. The following examples illustrate the technique.

Example 14.5
$\triangle ABC$ has a right-angle at C. Solve $\triangle ABC$ given (a) $AC=7$ cm and $B=40°$ (b) $AB=10$ cm and $BC=6$ cm

Solution   (a)   Figure 14.12 illustrates the situation. The sum of the angles of a triangle is 180°, so

$$A+B+C=180°$$

$$A+40°+90°=180°$$

$$A=50°$$

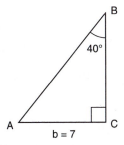

Figure 14.12 $\triangle ABC$ with $b=7$ and $B=40°$.

All the angles of the triangle are known. We now calculate the length of $a=BC$. We know

$$\tan B=\frac{b}{a}$$

$$a=\frac{b}{\tan B}=\frac{7}{\tan 40°}=\frac{7}{0.8391}=8.342$$

The length of BC is 8.342 cm. We now calculate $c=AB$. We know

$$\sin B=\frac{b}{c}$$

$$c=\frac{b}{\sin B}=\frac{7}{\sin 40°}=\frac{7}{0.6428}=10.890$$

The length of AB is 10.890 cm.
  The triangle is now solved:

$$a = BC = 8.342 \text{ cm}, \qquad A = 50°$$

$$b = AC = 7 \text{ cm}, \qquad B = 40°$$

$$c = AB = 10.890 \text{ cm}, \qquad C = 90°$$

Note that the longest side, AB, is opposite the largest angle, C, and that the shortest side, AC, is opposite the smallest angle, B.

(b)   Figure 14.13 illustrates the situation. We now calculate A. We know

$$\sin A = \frac{a}{c} = \frac{6}{10} = 0.6$$

$$A = \sin^{-1}(0.6) = 36.9°$$

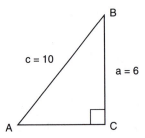

Figure 14.13 $\triangle ABC$ with $c = 10$ and $a = 6$.

We now calculate B. We have

$$\cos B = \frac{a}{c} = \frac{6}{10} = 0.6$$

$$B = \cos^{-1}(0.6) = 53.1°$$

We now calculate the length $b = AC$. Using Pythagoras' theorem, we have

$$c^2 = a^2 + b^2$$

$$10^2 = 6^2 + b^2$$

$$100 = 36 + b^2$$

$$b^2 = 64$$

$$b = 8$$

The triangle is now completely solved. In summary, we have

$$a = BC = 6 \text{ cm}, \qquad A = 36.9°$$

$$b = AC = 8 \text{ cm}, \qquad B = 53.1°$$

$$c = AB = 10 \text{ cm}, \qquad C = 90°$$

Example 14.6 Resultant of two forces at right-angles

Several forces acting on an object can be replaced by a single force that has the same effect on the object. Such a single force is called a **resultant force**, and can be considered as the sum of the original forces. Consider an object at A acted upon by forces AB and AC as shown in Figure 14.14. Force AB is 9 N, force AC is 5 N and they act at right-angles to one another. The resultant of forces AB and AC is the force AD. Calculate the size and direction of the resultant.

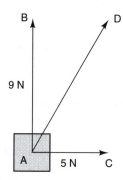

Figure 14.14   The resultant of forces AB and AC is the force AD.

Solution   Using Pythagoras' theorem, the size of the resultant, AD, is given by

$$(\text{size of resultant})^2 = 9^2 + 5^2$$

$$= 106$$

Hence

$$\text{size of resultant} = \sqrt{106} = 10.30$$

To calculate the direction of the resultant, we find $\angle DAC$:

$$\tan \angle DAC = \frac{9}{5} = 1.8$$

$$\angle DAC = \tan^{-1}(1.8) = 60.95°$$

The resultant is directed at 60.95° to the horizontal. Thus a force of 10.3 N directed at an angle of 60.95° to the horizontal has the same effect as the original forces.

Self-assessment questions 14.4

1. You wish to solve a right-angled $\triangle ABC$. You are told $A = 90°$. What additional information do you need in order to solve the triangle?

2. You are told $\triangle ABC$ is a right-angled isosceles triangle. What additional information do you need in order to solve the triangle?

Exercise 14.4

1. ABC is a right-angled triangle with $A = 90°$.
   Solve $\triangle ABC$ given (a) $C = 25°$, $AC = 6.2$ cm
   (b) $BC = 12$ cm, $AB = 9.5$ cm

   (c) $AB = 7.5$ m, $B = 42°$
   (d) $AC = 5.4$ cm, $AB = 7.3$ cm
   (e) $C = 37°$, $AB = 9.4$ cm

14.5 Solving equilateral and isosceles triangles

The methods of solving right-angled triangles can be applied to the solution of equilateral and isosceles triangles. Given any equilateral or isosceles triangle, it can be divided into two identical right-angled triangles as shown in Figure 14.15. In Figure 14.15(a) $\triangle ABC$ is an equilateral triangle; in Figure 14.15(b) $\triangle ABC$ is an isosceles triangle with $AB = AC$. The line AD divides each triangle into two identical right-angled triangles; that is, $\triangle ABD$ is identical to $\triangle ACD$. Note in particular that $BD = CD$ and $\angle BAD = \angle CAD$. When $\triangle ACD$ is solved, essentially $\triangle ABC$ is also solved.

Example 14.7

Solve $\triangle ABC$ given $AB = AC = 17$ mm and $B = 35°$.

Solution

Figure 14.16 illustrates the situation. Since $AB = AC$, $\triangle ABC$ is an isosceles triangle. The line AD divides the triangle into two identical right-angled triangles. We solve for BD in $\triangle ABD$:

$$\frac{BD}{AB} = \cos 35°$$

$$BD = AB \cos 35° = 17 \cos 35°$$

Since $BD = CD$,

$$BC = 2 \times BD = 2(17 \cos 35°) = 27.85$$

(a)

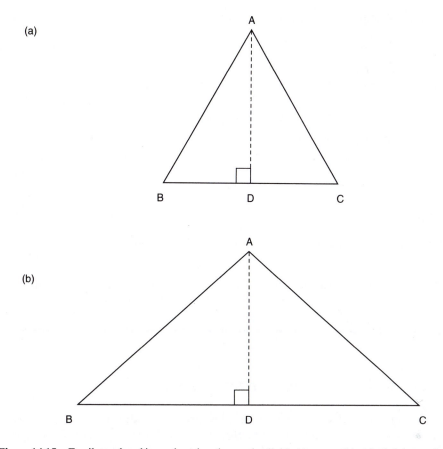

(b)

Figure 14.15   Equilateral and isosceles triangles can be divided into two identical right-angled triangles.

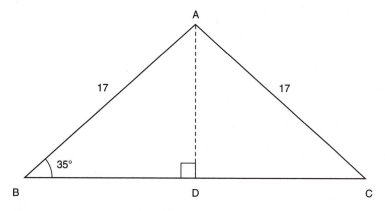

Figure 14.16   △ABC for Example 14.7.

Since △ABC is isosceles, $C = B = 35°$ and so

$$A = 180° - 35° - 35° = 110°$$

△ABC is now solved:

$$AB = 17 \text{ mm}, \qquad A = 110°$$

$$AC = 17 \text{ mm}, \qquad B = 35°$$

$$BC = 27.85 \text{ mm}, \qquad C = 35°$$

Exercises 14.5

1. ABC is an equilateral triangle of side 2 cm. Solve △ABC.

2. ABC is an isosceles triangle with $AB = BC = 21$ mm and $B = 50°$. Solve △ABC.

3. ABC is an isosceles triangle with $AB = AC$, $BC = 26$ mm and $A = 20°$. Solve △ABC.

4. ABC is an isosceles triangle with $AB = BC = 12$ cm and $A = 40°$. Solve △ABC.

14.6 Solving scalene triangles using the sine rule

Recall that a scalene triangle is one in which all the sides are of different lengths. To solve such a triangle, we use either the sine rule or the cosine rule, depending upon the information given. We examine each rule in turn.

Consider any △ABC. A typical triangle is shown in Figure 14.17. Recall the convention of labelling the side opposite A as a, the side opposite B as b

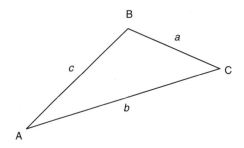

Figure 14.17   In a scalene triangle all the sides are of different lengths.

and the side opposite C as c. The sine rule states

in any $\triangle ABC$

$$\frac{a}{\sin A} = \frac{b}{\sin B} = \frac{c}{\sin C}$$

The sine rule can be used to solve any triangle provided we are given either

(i)   two angles and one side

or

(ii)   two sides and an angle that is not included between the given sides.

Example 14.8

Solve $\triangle ABC$ given $BC = 7$ cm, $A = 30°$ and $B = 110°$.

Solution   We are given one side and two angles and so the triangle can be solved using the sine rule. The angles of a triangle sum to 180°, and so $C = 180° - 30° - 110° = 40°$. We now apply the sine rule, noting that $BC = a$:

$$\frac{a}{\sin A} = \frac{b}{\sin B} = \frac{c}{\sin C}$$

Substituting in the known values, we obtain

$$\frac{7}{\sin 30°} = \frac{b}{\sin 110°} = \frac{c}{\sin 40°}$$

Solving for b, we obtain

$$b = \frac{7}{\sin 30°} \sin 110° = 13.16$$

and for c we obtain

$$c = \frac{7}{\sin 30°} \sin 40° = 9.00$$

Thus the solution to $\triangle ABC$ is

$$a = BC = 7 \text{ cm}, \qquad A = 30°$$
$$b = AC = 13.16 \text{ cm}, \qquad B = 110°$$
$$c = AB = 9.00 \text{ cm}, \qquad C = 40°$$

Note that the longest side is opposite the largest angle and the shortest side is opposite the smallest angle.

Example 14.9

Solve $\triangle ABC$ given $AB = 5$ cm, $BC = 9.8$ cm and $A = 53°$.

Solution The information given is illustrated in Figure 14.18. Since A is not included by the given sides, we have two sides and a non-included angle. Thus the sine rule can be used. Note that $AB = c$ and $BC = a$. The sine rule states

$$\frac{a}{\sin A} = \frac{b}{\sin B} = \frac{c}{\sin C}$$

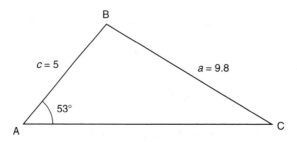

Figure 14.18 Two sides and a non-included angle are given in $\triangle ABC$.

Substituting in the given values produces

$$\frac{9.8}{\sin 53°} = \frac{b}{\sin B} = \frac{5}{\sin C}$$

We solve for $\sin C$:

$$\frac{9.8}{\sin 53°} = \frac{5}{\sin C}$$

$$\sin C = \frac{5 \sin 53°}{9.8} = 0.4075$$

Using a calculator and recalling that $\sin \theta = \sin (180° - \theta)$, we see that

$$C = \sin^{-1} (0.4075) = 24.05° \text{ or } 155.95°$$

Since $A = 53°$, it is impossible to have $C = 155.95°$ because the sum of the angles of a triangle cannot exceed $180°$, and so this value is ignored. Thus we have $C = 24.05°$.

We now calculate B:

$$B = 180° - A - C$$

$$= 180° - 53° - 24.05°$$

$$= 102.95°$$

Finally, we use the sine rule to calculate b:

$$\frac{9.8}{\sin 53°} = \frac{b}{\sin B}$$

$$b = \frac{9.8}{\sin 53°} \sin B$$

$$= \frac{9.8}{\sin 53°} \sin 102.95°$$

$$= 11.96$$

The solution is

$$a = BC = 9.8 \text{ cm}, \qquad A = 53°$$

$$b = AC = 11.96 \text{ cm}, \qquad B = 102.95°$$

$$c = AB = 5 \text{ cm}, \qquad C = 24.05°$$

Example 14.10

Solve $\triangle ABC$ given $B = 15°$, $AC = 5.1$ mm and $BC = 5.9$ mm.

Solution   We are given two sides and a non-included angle, and so the sine rule can be used. We are given $AC = b = 5.1$, $BC = a = 5.9$ and $B = 15°$; so

$$\frac{a}{\sin A} = \frac{b}{\sin B} = \frac{c}{\sin C}$$

$$\frac{5.9}{\sin A} = \frac{5.1}{\sin 15°} = \frac{c}{\sin C}$$

We solve for $\sin A$:

$$\sin A = \frac{5.9 \sin 15°}{5.1} = 0.2994$$

and hence

$$A = \sin^{-1}(0.2994) = 17.4° \text{ or } 162.6°$$

We have two choices for A, both of which are acceptable. Thus the triangle has two solutions.

Case 1: $A = 17.4°$. We calculate C:

$$C = 180° - 15° - 17.4° = 147.6°$$

Now we use the sine rule to calculate c:

$$\frac{5.1}{\sin 15°} = \frac{c}{\sin C} = \frac{c}{\sin 147.6°}$$

$$c = \frac{5.1}{\sin 15°} \sin 147.6° = 10.56$$

Case 2: $A = 162.6°$. We calculate C:

$$C = 180° - 15° - 162.6° = 2.4°$$

Now we use the sine rule to find c:

$$\frac{5.1}{\sin 15°} = \frac{c}{\sin C} = \frac{c}{\sin 2.4°}$$

$$c = \frac{5.1 \sin 2.4°}{\sin 15°} = 0.83$$

The two solutions are

$$a = BC = 5.9 \text{ mm}, \qquad A = 17.4°$$
$$b = AC = 5.1 \text{ mm}, \qquad B = 15°$$
$$c = AB = 10.56 \text{ mm}, \qquad C = 147.6°$$

and

$$a = BC = 5.9 \text{ mm}, \qquad A = 162.6°$$
$$b = AC = 5.1 \text{ mm}, \qquad B = 15°$$
$$c = AB = 0.83 \text{ mm}, \qquad C = 2.4°$$

Note that in both solutions the longest side is opposite the largest angle.

Example 14.11 Surveying

A surveyor wishes to calculate the distance to the peak of a hill at A. He uses a base line BC of length 100 m and establishes the angles B and C by means of a theodolite. The details are shown in Figure 14.19. Use this information to calculate the distance of A from B, that is, the length of AB.

Solution First, we need to calculate A. So

$$A = 180° - 87.2° - 87.3° = 5.50°$$

Now we make use of the sine rule. We have

$$\frac{a}{\sin A} = \frac{c}{\sin C}$$

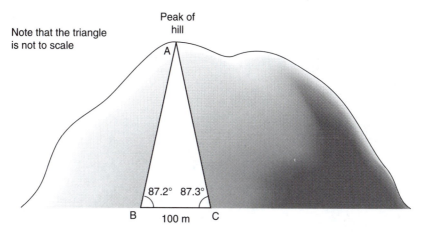

Figure 14.19 △ABC for Example 14.11.

$$c = \frac{a \sin C}{\sin A} = \frac{100 \sin 87.3°}{\sin 5.5°}$$

$$= 1042$$

The length of AB is 1042 m.

Example 14.12 Three forces in equilibrium

Several forces applied to an object can be replaced by a single force called the **resultant**. The forces may also have a resultant turning effect that would cause the object to rotate. Forces that have zero resultant and zero turning effect will not cause any change in the motion of the object to which they are applied. Such forces are said to be in **equilibrium**. Suppose the object is at O and is subjected to three forces *R*, *S* and *T* in equilibrium as shown in Figure 14.20. It can be shown that three forces in equilibrium can be represented in size and direction by the three sides of a triangle as shown in Figure 14.21. Suppose an object O of weight 50 N hangs in equilibrium supported by two strings as shown in Figure 14.22. Find the tension in each string.

Solution   The triangle corresponding to this situation is shown in Figure 14.23. Applying the sine rule, we find

$$\frac{T_1}{\sin 65°} = \frac{T_2}{\sin 60°} = \frac{50}{\sin 55°}$$

Then

$$T_1 = \frac{50}{\sin 55°} \sin 65° = 55.32 \text{ N}$$

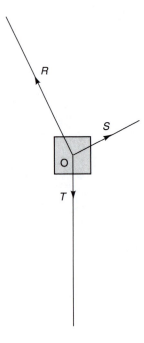

Figure 14.20   Forces *R*, *S* and *T* are in equilibrium.

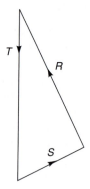

Figure 14.21   Three forces in equilibrium can be represented as the sides of a triangle.

$$T_2 = \frac{50}{\sin 55°} \sin 60° = 52.86 \text{ N}$$

The tensions in the strings are 55.32 N and 52.86 N.

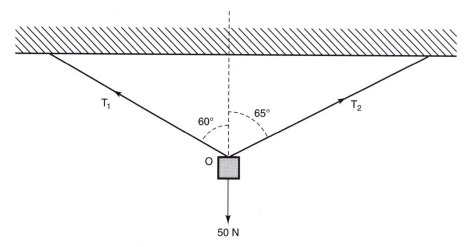

Figure 14.22 The object O is in equilibrium.

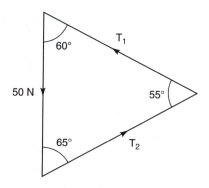

Figure 14.23 The three forces are in equilibrium.

Self-assessment questions 14.6

1. Under what conditions can the sine rule be used to solve a triangle?
2. Explain why two solutions are sometimes possible when solving a triangle using the sine rule.

Exercises 14.6

1. Solve $\triangle ABC$ using the sine rule when
   (a) $A=100°$, $B=17°$, $BC=15$ cm
   (b) $AB=10.6$ cm, $BC=7.9$ cm, $C=80°$
   (c) $B=40°$, $C=63°$, $AB=11$ cm

   (d) $AC=8.4$ cm, $BC=6.9$ cm, $A=20°$
   (e) $AC=10$ cm, $AB=17$ cm, $B=25°$
   (f) $B=20°$, $C=8°$, $AC=12$ cm

2. A triangle ABC is stated as having BC = 7.1 cm, AC = 6.3 cm and B = 93°. Explain why such a triangle cannot exist.

3. A surveyor wishes to calculate the distance to the peak of a hill at A. He carries out various measurements, which are detailed in Figure 14.24. Use this information to calculate the distance AB.

4. △ABC has AC = 5 cm, BC = 7 cm and C = 45°. Explain why △ABC cannot be solved using the sine rule.

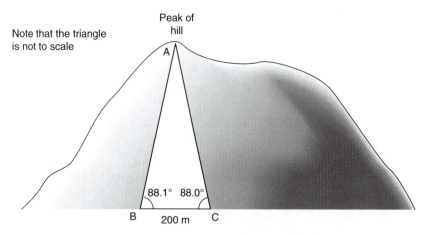

Note that the triangle is not to scale

Peak of hill

A

88.1°   88.0°

B   200 m   C

Figure 14.24 The triangle for Exercise 14.6 Q3.

14.7 Solving scalene triangles using the cosine rule

The cosine rule is used to solve triangles when we are given either

(i)   two sides and an included angle

or

(ii)   three sides.

The cosine rule states

in any △ABC

$$a^2 = b^2 + c^2 - 2bc \cos A$$

$$b^2 = a^2 + c^2 - 2ac \cos B$$

$$c^2 = a^2 + b^2 - 2ab \cos C$$

Example 14.13

Solve △ABC, given $A=40°$, AC$=5$ mm and AB$=9$ mm.

Solution The situation is shown in Figure 14.25. We are given two sides and the included angle, and so the cosine rule may be applied. We have $A=40°$, $b=5$ and $c=9$. We calculate a using the cosine rule:

$$a^2 = b^2 + c^2 - 2bc \cos A$$

$$= 5^2 + 9^2 - 2·5·9· \cos 40°$$

$$= 25 + 81 - 90 \cos 40° = 37.056$$

$$a = 6.0874$$

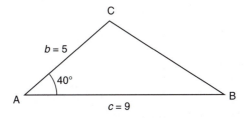

Figure 14.25 Two sides and an included angle are given in △ABC.

Having found a, we can now use the cosine rule again to find B. Note that at this stage the sine rule could equally well be used. We have

$$b^2 = a^2 + c^2 - 2ac \cos B$$

$$5^2 = (6.0874)^2 + 9^2 - 2(6.0874)(9) \cos B$$

$$25 = 118.056 - 109.573 \cos B$$

$$109.573 \cos B = 93.056$$

$$\cos B = \frac{93.056}{109.573} = 0.8493$$

$$B = \cos^{-1}(0.8493) = 31.9°$$

We now solve for C to obtain

$$C = 180° - A - B$$

$$= 180° - 40° - 31.9° = 108.1°$$

The solution is

$$a = BC = 6.087 \text{ mm}, \quad A = 40°$$

$$b=AC=5\,\text{mm}, \qquad B=31.9°$$

$$c=AB=9\,\text{mm}, \qquad C=108.1°$$

Example 14.14

An engineer is making a triangular component from a steel sheet with the dimensions shown in Figure 14.26. Calculate the angles and side lengths that are not given.

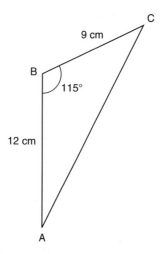

Figure 14.26   The triangular component for Example 14.14.

Solution   We are given two sides and the included angle, and so the cosine rule may be applied. We know $B=115°$, $a=BC=9$ cm and $c=AB=12$ cm. Using the cosine rule, we solve for b:

$$b^2 = a^2 + c^2 - 2ac \cos B$$

$$= 9^2 + 12^2 - 2 \cdot 9 \cdot 12 \cdot \cos 115°$$

$$= 316.2855$$

$$b = 17.7844$$

We now find A:

$$a^2 = b^2 + c^2 - 2bc \cos A$$

$$9^2 = (17.7844)^2 + 12^2 - 2(17.7844)(12) \cos A$$

$$81 = 316.29 + 144 - 426.83 \cos A$$

$$426.83 \cos A = 379.29$$

$$\cos A = \frac{379.29}{426.83} = 0.8886$$

$$A = \cos^{-1}(0.8886) = 27.3°$$

Finally, we find C:

$$C = 180° - A - B = 180° - 27.3° - 115° = 37.7°$$

The solution is

$$a = BC = 9 \text{ cm}, \qquad A = 27.3°$$
$$b = AC = 17.78 \text{ cm}, \qquad B = 115°$$
$$c = AB = 12 \text{ cm}, \qquad C = 37.7°$$

Example 14.15

A triangular component ABC has $AB = 17$ cm, $BC = 9$ cm and $AC = 15$ cm. Calculate A, B and C.

Solution

We are given three sides, and so the cosine rule can be used to find the three angles. We have $a = BC = 9$, $b = AC = 15$ and $c = AB = 17$. First, we find A:

$$a^2 = b^2 + c^2 - 2bc \cos A$$

$$9^2 = 15^2 + 17^2 - 2(15)(17) \cos A$$

$$81 = 225 + 289 - 510 \cos A$$

$$510 \cos A = 433$$

$$\cos A = \frac{433}{510} = 0.8490$$

$$A = \cos^{-1}(0.8490) = 31.9°$$

We solve for B to obtain

$$b^2 = a^2 + c^2 - 2ac \cos B$$

$$225 = 81 + 289 - 306 \cos B$$

$$306 \cos B = 145$$

$$\cos B = \frac{145}{306} = 0.4739$$

$$B = \cos^{-1}(0.4739) = 61.7°$$

Finally, we have

$$C = 180° - 31.9° - 61.7° = 86.4°$$

The solution is

$$a = BC = 9 \text{ cm}, \qquad A = 31.9°$$
$$b = AC = 15 \text{ cm}, \qquad B = 61.7°$$
$$c = AB = 17 \text{ cm}, \qquad C = 86.4°$$

Self-assessment question 14.7

1. State the conditions under which the cosine rule may be used to solve a triangle.

Exercises 14.7

1. Solve the following triangles:
   (a) $A = 67°$, $AC = 9$ cm, $AB = 14.3$ cm
   (b) $AC = 12$ mm, $AB = 15.5$ mm, $BC = 9.7$ mm
   (c) $B = 100°$, $BC = 10.6$ cm, $AB = 17.4$ cm
   (d) $C = 20°$, $AC = 5.5$ cm, $BC = 7.6$ cm
   (e) $AB = 10$ mm, $BC = 8$ mm, $AC = 11$ mm

2. The cosine rule states
   $$a^2 = b^2 + c^2 - 2bc \cos A$$
   If $A = 90°$, show that Pythagoras' theorem is obtained.

Review exercises 14

1 A rectangle ABCD has $AB = CD = 10$ cm and $AD = BC = 21$ cm. Calculate (a) AC (b) $\angle DAC$

2 Solve $\triangle ABC$, given
   (a) $AB = 17.5$ mm, $BC = 12$ mm, $AC = 9.7$ mm
   (b) $C = 59°$, $B = 48°$, $AB = 11$ cm
   (c) $BC = 15$ cm, $AC = 17$ cm, $A = 21°$
   (d) $AB = 31$ cm, $AC = 25$ cm, $A = 35°$
   (e) $AB = 30$ mm, $AC = 21$ mm, $BC = 26$ mm
   (f) $A = 30°$, $B = 35°$, $AB = 17.3$ cm
   (g) $AB = 16$ cm, $BC = 19$ cm, $A = 75°$
   (h) $AB = 19$ mm, $BC = 27$ mm, $B = 53°$

3 ABC is an equilateral triangle with sides of length 2 cm. D is the midpoint of BC and AD is perpendicular to BC.
   (a) Calculate CD and AD.
   (b) Show $\sin \angle CAD = \sin 30° = 0.5$.
   (c) Show that $\sin 60° = \frac{1}{2}\sqrt{3}$
   (d) Show that $\cos 30° = \frac{1}{2}\sqrt{3}$
   (e) Show that $\cos 60° = \frac{1}{2}$
   (f) Show that $\tan 30° = \sqrt{\frac{1}{3}}$
   (g) Show that $\tan 60° = \sqrt{3}$

15

Trigonometric Identities and Functions

KEY POINTS

This chapter

- develops some trigonometric identities

- states the commonly used trigonometric identities

- illustrates how these identities can be used to simplify trigonometric expressions

- introduces the trigonometric functions: $y = \sin \theta$, $y = \cos \theta$ and $y = \tan \theta$

- illustrates the graphs of the trigonometric functions

- tabulates values of $\sin \theta$, $\cos \theta$ and $\tan \theta$ for small values of θ

- introduces the small-angle approximations for $\sin \theta$, $\cos \theta$ and $\tan \theta$

CONTENTS

15.1 Introduction

This chapter develops the trigonometric work started in Chapter 10. Relationships, called identities, that connect the various trigonometric ratios are developed or stated. These identities may be used to simplify cumbersome expressions. The trigonometric functions are defined and their graphs plotted. Finally, simple approximations to the trigonometric functions close the chapter.

15.2 Trigonometric identities

An **identity** is a statement that two mathematical expressions are equal for all values of the variable they contain. Both an equation and an identity have a left-hand side and a right-hand side. With an equation, we try to find values of the variable, commonly x, so that the value of the left-hand side and the value of the right-hand side are equal. For example, given $2x+1=x+4$, we could try to find values of x so that $2x+1$ has the same value as $x+4$. Such values of x are known as solutions of the equation. With an identity, the values of the two sides are equal for all values of the variable. We do not have to solve anything. Identities are useful because they provide us with an alternative way of writing an expression. Trigonometry has an abundance of identities, and this section looks at some of the more common ones.

It is important to note that identities and equations are not the same, although they are often confused and the terms are sometimes used interchangeably.

15.2.1 Development of some common trigonometric identities

Figure 15.1 shows a right-angled triangle ABC. Recall that

$$\sin A = \frac{BC}{AC} = \frac{a}{b}, \quad \cos A = \frac{AB}{AC} = \frac{c}{b}$$

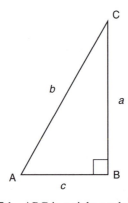

Figure 15.1   ABC is a right-angled triangle.

Squaring both sides gives

$$(\sin A)^2 = \frac{a^2}{b^2}, \quad (\cos A)^2 = \frac{c^2}{b^2}$$

If these are now added, we find

$$(\sin A)^2 + (\cos A)^2 = \frac{a^2}{b^2} + \frac{c^2}{b^2} = \frac{a^2 + c^2}{b^2}$$

From Pythagoras' theorem, we know that $b^2 = a^2 + c^2$, and so

$$(\sin A)^2 + (\cos A)^2 = \frac{b^2}{b^2} = 1$$

The term $(\sin A)^2$ is usually written as $\sin^2 A$, and $(\cos A)^2$ as $\cos^2 A$. Hence we have the identity

$$\sin^2 A + \cos^2 A = 1$$

It can be shown that this identity holds for any angle A, and not just angles less than 90°.

We return again to Figure 15.1. Recall that

$$\tan A = \frac{BC}{AB} = \frac{a}{c}$$

Now

$$\frac{\sin A}{\cos A} = \frac{a/b}{c/b} = \frac{a}{b} \times \frac{b}{c} = \frac{a}{c} = \tan A$$

and so

$$\frac{\sin A}{\cos A} = \tan A$$

This identity is true for angles of any size. Since $\tan A = \sin A / \cos A$ is an identity, $\tan A$ and $\sin A / \cos A$ have the same value for all values of A.

15.2.2 Further trigonometric identities

Table 15.1 lists some of the commonly used trigonometric identities. We use these identities in the following examples. Note that when asked to simplify an expression or to show that two expressions are equal, it is not usually obvious

Table 15.1

$$\sin^2 A + \cos^2 A = 1$$

$$\frac{\sin A}{\cos A} = \tan A$$

$$\sin(A+B) = \sin A \cos B + \sin B \cos A$$

$$\sin(A-B) = \sin A \cos B - \sin B \cos A$$

$$\cos(A+B) = \cos A \cos B - \sin A \sin B$$

$$\cos(A-B) = \cos A \cos B + \sin A \sin B$$

$$\tan(A+B) = \frac{\tan A + \tan B}{1 - \tan A \tan B}$$

$$\tan(A-B) = \frac{\tan A - \tan B}{1 + \tan A \tan B}$$

$$2 \sin A \cos B = \sin(A+B) + \sin(A-B)$$

$$2 \cos A \cos B = \cos(A+B) + \cos(A-B)$$

$$2 \sin A \sin B = \cos(A-B) - \cos(A+B)$$

$$\sin^2 A = \tfrac{1}{2}(1 - \cos 2A)$$

$$\cos^2 A = \tfrac{1}{2}(1 + \cos 2A)$$

which identities can be usefully employed. It is a matter of practice to know which ones to use.

Example 15.1

Show that $\sin 2A = 2 \sin A \cos A$.

Solution   We use the identity

$$\sin(A+B) = \sin A \cos B + \sin B \cos A$$

A special case of this identity occurs when $B = A$. We then have

$$\sin 2A = \sin(A+A) = \sin A \cos A + \sin A \cos A$$

that is

$$\sin 2A = 2 \sin A \cos A$$

Example 15.2

Show that $\dfrac{\sin^3 A}{\cos A} + \sin A \cos A$

is identical to $\tan A$.

Solution   By writing the expression with a common denominator, we obtain

$$\frac{\sin^3 A}{\cos A} + \sin A \cos A = \frac{\sin^3 A + \sin A \cos^2 A}{\cos A}$$

We note that $\sin A$ is a common factor of the numerator, which may be written as $\sin A (\sin^2 A + \cos^2 A)$. The identity $\sin^2 A + \cos^2 A = 1$ is now used to obtain

$$\frac{\sin^3 A + \sin A \cos^2 A}{\cos A} = \frac{\sin A (\sin^2 A + \cos^2 A)}{\cos A} = \frac{\sin A}{\cos A}$$

Finally we recall that $\sin A / \cos A = \tan A$. Hence

$$\frac{\sin^3 A}{\cos A} + \sin A \cos A = \tan A$$

Example 15.3

Show that (a) $\cos(-\theta) = \cos\theta$ (b) $\sin(-\theta) = -\sin\theta$

Solution   (a)   We use the identity for $\cos(A-B)$ with $A=0$ and $B=\theta$. Now

$$\cos(A-B) = \cos A \cos B + \sin A \sin B$$

With $A=0$ and $B=\theta$, this becomes

$$\cos(0-\theta) = \cos(-\theta) = \cos 0 \cos\theta + \sin 0 \sin\theta$$
$$= 1 \cos\theta + 0 \sin\theta$$
$$= \cos\theta$$

Hence

$$\cos(-\theta) = \cos\theta$$

(b)   We use the identity

$$\sin(A-B) = \sin A \cos B - \sin B \cos A$$

with $A=0$ and $B=\theta$. We then obtain

$$\sin(0-\theta) = \sin(-\theta) = \sin 0 \cos\theta - \sin\theta \cos 0$$
$$= 0 \cos\theta - \sin\theta \cdot 1$$
$$= -\sin\theta$$

Hence

$$\sin(-\theta) = -\sin\theta$$

Example 15.4

Use trigonometric identities to show that (a) $\sin(\theta + \frac{\pi}{2}) = \cos\theta$
(b) $\cos(\theta - \frac{\pi}{2}) = \sin\theta$

Solution   (a)   We use the identity for $\sin(A+B)$ with $A=\theta$ and $B=\frac{\pi}{2}$. Then

$$\sin(A+B) = \sin A \cos B + \sin B \cos A$$

Substituting in θ for A and $\frac{1}{2}\pi$ for B produces

$$\sin(\theta + \frac{\pi}{2}) = \sin\theta \cos\frac{1}{2}\pi + \sin\frac{1}{2}\pi \cos\theta$$

$$= \sin\theta\,(0) + (1)\cos\theta$$

$$= \cos\theta$$

Hence

$$\sin(\theta + \tfrac{1}{2}\pi) = \cos\theta$$

(b)   We use the identity for $\cos(A-B)$ with $A=\theta$ and $B=\frac{1}{2}\pi$. Now

$$\cos(A-B) = \cos A \cos B + \sin A \sin B$$

With $A=\theta$ and $B=\frac{1}{2}\pi$, we obtain

$$\cos(\theta - \tfrac{1}{2}\pi) = \cos\theta \cos\tfrac{1}{2}\pi + \sin\theta \sin\tfrac{1}{2}\pi$$

$$= \cos\theta\,(0) + \sin\theta\,(1)$$

$$= \sin\theta$$

Hence

$$\cos(\theta - \tfrac{1}{2}\pi) = \sin\theta$$

Self-assessment question 15.2

1.   Explain the difference between an equation and an identity.

Exercises 15.2

1.   From Table 15.1, we have

$$\tan(A+B) = \frac{\tan A + \tan B}{1 - \tan A \tan B}$$

Verify this identity for $A = 20°$ and $B = 30°$.

2.   Consider the identity

$$\cos(A-B) = \cos A \cos B$$
$$+ \sin A \sin B$$

Verify this identity with $A = 70°$ and $B = 20°$.

3.   Simplify $\sin A \cos A \tan A + \dfrac{\sin 2A}{2\tan A}$

4. (a) Show that $\cos 3\theta = \cos 2\theta \cos \theta - \sin 2\theta \sin \theta$.

   (b) Hence show that $\cos 3\theta = \cos^3 \theta - 3 \sin^2 \theta \cos \theta$.

5. Show that $\sin 3\theta = 3 \sin \theta \cos^2 \theta - \sin^3 \theta$.

6. Show that $R \cos(\theta + \alpha)$ can be written in the form

$$R \cos \theta \cos \alpha - R \sin \theta \sin \alpha$$

7. Show that $R \sin(\theta + \alpha)$ can be written in the form

$$R \sin \theta \cos \alpha + R \sin \alpha \cos \theta$$

8. Show, using trigonometric identities, that
   (a) $\sin(180° - \theta) = \sin \theta$
   (b) $\cos(180° - \theta) = -\cos \theta$
   (c) $\tan(180° - \theta) = -\tan \theta$
   (d) $\sin(180° + \theta) = -\sin \theta$
   (e) $\cos(180° + \theta) = -\cos \theta$
   (f) $\tan(180° + \theta) = \tan \theta$
   (g) $\sin(360° - \theta) = -\sin \theta$
   (h) $\cos(360° - \theta) = \cos \theta$
   (i) $\tan(360° - \theta) = -\tan \theta$

15.3 Trigonometric functions and their graphs

Having introduced the ratios $\sin \theta$, $\cos \theta$ and $\tan \theta$, we are ready to consider the functions $y = \sin \theta$, $y = \cos \theta$ and $y = \tan \theta$. The independent variable is θ, and for every value of θ the output $\sin \theta$, $\cos \theta$ or $\tan \theta$ can be found. The graphs of these functions are illustrated in this section.

15.3.1 The sine function, $y = \sin \theta$

We can plot the function $y = \sin \theta$ by drawing up a table of values, as for example, in Table 15.2. Plotting these values and joining them with a smooth curve produces the graph shown in Figure 15.2. It is possible to plot $y = \sin \theta$ using a graphics calculator or a graph-plotting package.

Table 15.2

θ	0°	30°	60°	90°	120°	150°	180°
$\sin \theta$	0	0.5	0.8660	1	0.8660	0.5	0

θ	210°	240°	270°	300°	330°	360°
$\sin \theta$	-0.5	-0.8660	-1	-0.8660	-0.5	0

Recall from Chapter 10 that adding or subtracting 360° to an angle does not alter the trigonometric ratio of that angle. Hence if we extend the values of θ below 0° and above 360°, the values of $\sin \theta$ are simply repeated. A graph of $y = \sin \theta$ for a larger domain of θ is shown in Figure 15.3. The pattern is repeated every 360°; that is,

$$\sin \theta = \sin(\theta + 360°) = \sin(\theta + 720°) = \sin(\theta + 1080°)$$

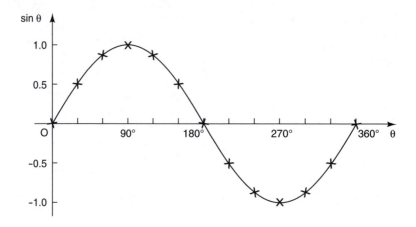

Figure 15.2   The function $y = \sin \theta$.

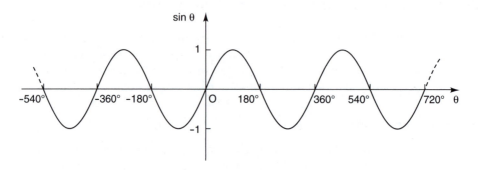

Figure 15.3   The values of $\sin \theta$ repeat every $360°$.

and so on. Similarly, subtracting $360°$ from an angle leaves the sine unchanged, so

$$\sin \theta = \sin (\theta - 360°) = \sin (\theta - 720°) = \sin (\theta - 1080°)$$

and so on. Hence $\sin \theta$ is a periodic function with period $360°$.

Note that the angle θ could be measured in radians. Recall that 2π radians and $360°$ are equal. Hence when measuring θ in radians, we have

$$\sin \theta = \sin (\theta + 2\pi) = \sin (\theta + 4\pi) = \sin (\theta + 6\pi)$$

and so on. Similarly,

$$\sin \theta = \sin (\theta - 2\pi) = \sin (\theta - 4\pi) = \sin (\theta - 6\pi)$$

and so on. Here the period of $\sin \theta$ is 2π.

15.3.2 The cosine function, $y = \cos \theta$

Table 15.3 gives values of θ and $\cos \theta$. Plotting these values and joining them with a smooth curve produces the graph of $y = \cos \theta$ shown in Figure 15.4.

Table 15.3

θ	0°	30°	60°	90°	120°	150°	180°
$\cos \theta$	1	0.8660	0.5	0	-0.5	-0.8660	-1

θ	210°	240°	270°	300°	330°	360°
$\cos \theta$	-0.8660	-0.5	0	0.5	0.8660	1

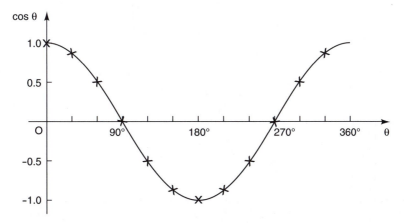

Figure 15.4 The function $y = \cos \theta$.

Extending the values of θ beyond 0° and 360° produces the graph shown in Figure 15.5, in which we see that the values are repeated every 360°; that is,

$$\cos \theta = \cos (\theta + 360°) = \cos (\theta + 720°) = \cos (\theta + 1080°)$$

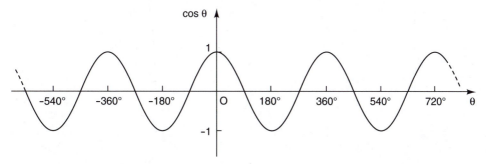

Figure 15.5 The values of $\cos \theta$ repeat every 360°.

and

$$\cos\theta = \cos(\theta - 360°) = \cos(\theta - 720°) = \cos(\theta - 1080°)$$

Note the similarity between $y = \sin\theta$ and $y = \cos\theta$. The two graphs are identical apart from the starting point.

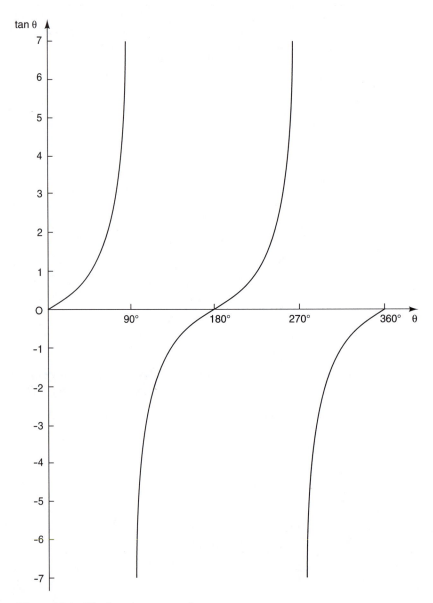

Figure 15.6 The function $y = \tan\theta$.

15.3.3 The tangent function, $y = \tan \theta$

By constructing a table of values and plotting points a graph of $y = \tan \theta$ may be drawn. Figure 15.6 shows a graph of $y = \tan \theta$ as θ varies from $0°$ to $360°$. Extending the values of θ produces Figure 15.7. Note that the pattern is repeated every $180°$. The values of $\tan \theta$ extend from minus infinity to plus infinity. The function has discontinuities at $\theta = 90°$, $270°$, $450°$,... and at $\theta = -90°$, $-270°$, $-450°$,...

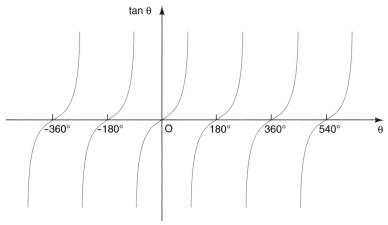

Figure 15.7   The values of $\tan \theta$ repeat every $180°$.

Self-assessment questions 15.3

1.   State two properties that are common to $y = \sin \theta$ and $y = \cos \theta$.

2.   State two properties of $y = \tan \theta$ that distinguish it from $y = \sin \theta$.

Computer and calculator exercises 15.3

1.   Plot a graph of $y = 2 \sin \theta$ for $0° \leqslant \theta \leqslant 360°$. What is the maximum value of y? At what value of θ does this maximum value occur?

2.   (a)   Plot a graph of $y = \cos 3\theta$ for $0° \leqslant \theta \leqslant 360°$. What is the maximum value of y? At what values of θ does this maximum value occur?

(b)   On the same axes plot a graph of $y = \cos \theta$. In what ways are $\cos 3\theta$ and $\cos \theta$ similar? In what ways are they different?

3.   On the same axes plot graphs of $y = \tan \theta$ and $y = \tan 2\theta$ for $0° \leqslant \theta \leqslant 360°$. In what ways are the graphs similar? In what ways are they different?

4.   On the same axes plot graphs of $y = \sin \theta$ and $y = \cos \theta$ for $0° \leqslant \theta \leqslant 360°$. Use your graphs to find approximate solutions to

$$\sin \theta = \cos \theta$$

15.4 The small-angle approximations

This section introduces the small-angle approximations for $\sin \theta$, $\cos \theta$ and $\tan \theta$. We aim to find simple functions of θ that approximate closely to the values of $\sin \theta$, $\cos \theta$ and $\tan \theta$. Table 15.4 gives values of θ in radians, with corresponding values of $\sin \theta$, $\cos \theta$ and $\tan \theta$. For angles up to 0.35 radians (about 20°), we see that $\sin \theta$, $\tan \theta$ and θ all have almost identical values. The values of $\cos \theta$ are very near to 1. Thus for small angles measured in radians we state the following approximations:

$$\sin \theta \approx \theta, \ \cos \theta \approx 1, \ \tan \theta \approx \theta$$

The symbol $\approx$ means 'is approximately equal to'. These approximations can be used to simplify equations so that approximate solutions can be found.

Table 15.4

θ (radians)	$\sin \theta$	$\cos \theta$	$\tan \theta$
0	0	1	0
0.05	0.05	1.00	0.05
0.10	0.10	1.00	0.10
0.15	0.15	0.99	0.15
0.20	0.20	0.98	0.20
0.25	0.25	0.97	0.26
0.30	0.30	0.96	0.31
0.35	0.34	0.94	0.37

Example 15.5

Solve $\sin \theta + \theta = 0.3$.

Solution   We know that for small angles measured in radians $\sin \theta \approx \theta$. Replacing $\sin \theta$ by θ gives

$$\theta + \theta \approx 0.3$$

$$2\theta \approx 0.3$$

$$\theta \approx 0.15$$

We can say that an approximate solution is $\theta = 0.15$ radians. Since $\sin \theta \approx \theta$ only for small angles, this method does not find large-value solutions.

Self-assessment question 15.4

1. Why is it useful to have approximations to the trigonometric functions?

Exercises 15.4

1. Use the small-angle approximations to find approximate solutions to
   (a) $3 \cos \theta + \theta = 3.2$ (b) $2\theta + \frac{1}{2} \sin \theta = 0.25$
   (c) $\theta - 2 \tan \theta = -0.15$
   (d) $\theta + \sin \theta + \cos \theta = 1.1$
   (e) $4 \tan \theta - \sin \theta = \frac{1}{3} \cos \theta$

2. Construct a table of small negative values of θ measured in radians, with corresponding values of $\sin \theta$, $\cos \theta$ and $\tan \theta$. Can you state the small-angle approximations for negative angles? How do they compare with those from Table 15.4?

Computer and calculator exercises 15

1. Use a graphics calculator/graph-plotting package to draw $y = 3 \tan \theta + \theta$ for θ between 0 and 2π radians. Hence find an approximate solution of

   $$3 \tan \theta + \theta = 1$$

2. Plot $y = \sin^2 \theta$ for $0° \leqslant \theta \leqslant 360°$. Hence find approximate solutions of

   $$\sin^2 \theta = 0.25$$

Review exercises 15

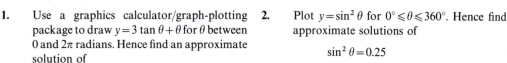

1 Simplify as much as possible
   (a) $\dfrac{\sin^4 A}{\cos^2 A} + \sin^2 A$ (b) $\left(\dfrac{\cos \theta \tan \theta}{\sin \theta}\right)^2$
   (c) $\dfrac{2 \sin A \cos A}{\cos 2A}$
   (d) $\cos (\pi - \theta) + \cos (\pi + \theta)$
   (e) $\cos (\pi + \theta) - \cos (\pi - \theta)$
   (f) $\sin (\theta + \frac{1}{2}\pi) + \sin(\frac{1}{2}\pi - \theta)$

2 Show that $(\sin \theta - \cos \theta)^2$ is identical to $1 - \sin 2\theta$.

3 Show that $\sin^4 \theta$ is identical to $(1 + \cos \theta)^2(1 - \cos \theta)^2$.

4 Find an approximate solution to

   $$3 \tan \theta + \theta = 1$$

   using the small-angle approximation.

5 Find approximate solutions to the following equations:
   (a) $\sin \theta + \cos \theta + 2 \tan \theta = 1.2$
   (b) $3 \sin \theta - \tan \theta = 1.25 - \cos \theta$
   (c) $\dfrac{1 + \sin \theta}{1 + \cos \theta} = 5 \tan \theta$

16

Engineering Waves

KEY POINTS

This chapter

- explains how sine and cosine functions are used in engineering to represent waves

- explains what is meant by the amplitude of a wave

- explains what is meant by the angular frequency of a wave

- explains what is meant by the period of a wave

- explains what is meant by the frequency of a wave

- explains what is meant by the phase angle and time displacement of a wave

- explains how the equation of a wave can be obtained by observing oscilloscope traces

- shows how to combine waves of the same frequency into a single sine or cosine wave

CONTENTS

16.1 Introduction

This chapter applies the trigonometric work from Chapters 10 and 15 to the analysis of engineering waves. The sine and cosine functions are used to describe waves, using various parameters such as amplitude, period and phase. The method of combining two waves together to form a single wave is described.

16.2 Waves that vary with time

In Chapter 15 we considered the functions $y = \sin \theta$ and $y = \cos \theta$. In these functions the independent variable is θ, which is usually measured in degrees or radians. In many engineering applications the independent variable is time t, usually measured in seconds. Consider for example, the function $y = \sin t$. The graph of this function is shown in Figure 16.1 for values of t between 0 and 2π seconds. If we extend the values of t below 0 and above 2π seconds, the values of $\sin t$ are simply repeated and the graph is as shown in Figure 16.2. This graph is sometimes referred to as a **sine wave** because it has the same shape as many of the waves found in nature. As t increases from 0 to 2π seconds, we see that one full cycle of the wave is completed. Similarly, the graph of the cosine function $y = \cos t$ is referred to as a **cosine wave**, and in 2π seconds one full cycle of this wave is completed.

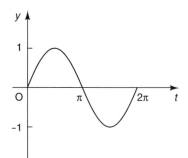

Figure 16.1   The graph of $y = \sin t$.

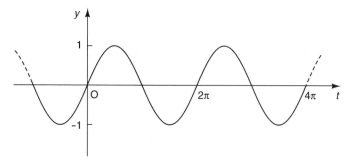

Figure 16.2   The sine wave repeats a cycle every 2π seconds.

16.2.1 The amplitude of a wave

Consider Figure 16.3, which shows a graph of $y = A \sin t$, where A is a constant. The quantity A is called the **amplitude** of the wave. For example, $y = 2 \sin t$ has an amplitude of 2. Note that $y = \sin t$ has an amplitude of 1. Similarly, $y = A \cos t$ has an amplitude of A.

> The amplitudes of both $y = A \sin t$ and $y = A \cos t$ are A.

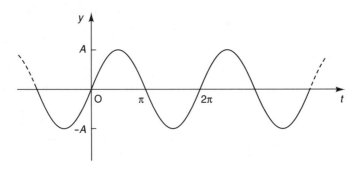

Figure 16.3 The graph of $y = A \sin t$.

Example 16.1

Determine the amplitudes of the sine waves shown in Figure 16.4 and state their equations.

Solution

Referring to Figure 16.4, we see that the amplitude of the wave marked (a) is 3. Therefore its equation is $y = 3 \sin t$. The amplitude of the wave marked (b) is 0.5. Its equation is $y = 0.5 \sin t$.

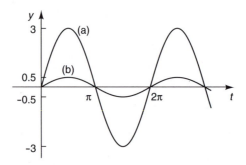

Figure 16.4 Sine waves for Example 16.1.

Example 16.2

State the amplitude of each of the following waves: (a) $y = 8 \sin t$

(b) $y = R \cos t$ (c) $y = \dfrac{\cos t}{3}$

Solution   (a)   The amplitude of $y = A \sin t$ is A, and so the amplitude of $y = 8 \sin t$ is 8.

(b)   The amplitude of $y = R \cos t$ is R.

(c)   We can rewrite $y = \dfrac{\cos t}{3}$ as $y = \frac{1}{3} \cos t$. We see that the amplitude is $\frac{1}{3}$.

16.2.2 The angular frequency of a wave

We now move on to look at waves such as $y = A \sin 2t$, $y = A \sin 3t$ and so on.

Consider the wave $y = A \sin \omega t$. We call ω the **angular frequency** of the wave. The angular frequency is measured in radians per second. For example $y = \sin 3t$ has an angular frequency of 3 radians per second. Note that the angular frequency of $y = \sin t$ is 1 radian per second. Since t is measured in seconds, ωt is measured in radians.

> The angular frequency of $y = A \sin \omega t$ is ω radians per second.

Example 16.3

State the angular frequency of each of the following waves: (a) $y = \sin 5t$

(b) $y = \sin 2\pi t$ (c) $y = \sin \dfrac{t}{10}$

Solution   (a)   Comparing the given wave with the general form $y = A \sin \omega t$, we see that $y = \sin 5t$ has an angular frequency of 5 radians per second.

(b)   Similarly, $y = \sin 2\pi t$ has an angular frequency of 2π radians per second.

(c)   The wave $y = \sin \dfrac{t}{10}$ can be written $y = \sin \frac{1}{10}t$, and so its angular frequency is $\frac{1}{10}$ radian per second.

Figure 16.5 shows the graph of $y = \sin \omega t$ for the particular case when $\omega = 2$, that is, $y = \sin 2t$:

when $t = 0$ seconds, note that $\omega t = 0$ radians

when $t = 1$ second, $\omega t = 2$ radians

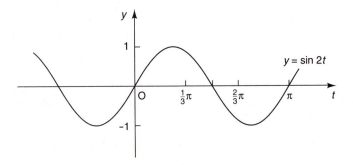

Figure 16.5 Graph of $y = \sin 2t$.

So, as t increases from 0 to 1 second, the angle ωt increases from 0 to 2 radians; that is, it increases at a rate of 2 radians per second. In this case the angular frequency is 2 radians per second. Note from the graph that one full cycle is completed in π seconds.

Figure 16.6 shows the graph of $y = \sin \omega t$ for the particular case when $\omega = 3$, that is, $y = \sin 3t$:

when $t = 0$ seconds, note that $\omega t = 0$ radians

when $t = 1$ second, $\omega t = 3$ radians

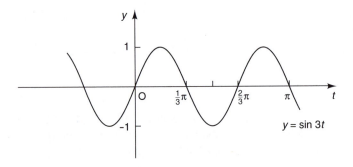

Figure 16.6 Graph of $y = \sin 3t$.

So, as t increases from 0 to 1 second, the angle ωt increases from 0 to 3 radians; that is, it increases at a rate of 3 radians per second. In this case the angular frequency is 3 radians per second. Note from the graph that one full cycle is completed in $\frac{2}{3}\pi$ seconds.

It is important to note that as the angular frequency is changed, so too is the time taken to complete a full cycle.

16.2.3 The period of a wave

Consider the wave $y = A \sin \omega t$:

when $t = 0$ seconds, $\omega t = 0$ radians

when $t = \dfrac{2\pi}{\omega}$ seconds, $\omega t = \omega \left(\dfrac{2\pi}{\omega} \right) = 2\pi$ radians

We can see that as t increases from 0 to $2\pi/\omega$ seconds, the angle ωt increases from 0 to 2π radians. We know that as the angle ωt increases by 2π radians, $A \sin \omega t$ completes a full cycle. Hence a full cycle is completed in $2\pi/\omega$ seconds. The time taken to complete a full cycle is called the **period** and is denoted by T.

> If $y = A \sin \omega t$ then the period $T = \dfrac{2\pi}{\omega}$

From Figure 16.1 we see that the period of $y = \sin t$ is 2π seconds. From Figure 16.5 we see that the period of $y = \sin 2t$ is $2\pi/2 = \pi$ seconds, and from Figure 16.6 that the period of $y = \sin 3t$ is $\frac{2}{3}\pi$ seconds.

Example 16.4

Find the period of each of the following sine waves: (a) $y = \sin 5t$
(b) $y = \sin 200\,\pi t$

Solution (a) Note that $y = \sin 5t$ has an angular frequency of $\omega = 5$ radians per second. Hence the period is $T = \frac{2}{5}\pi$ seconds.

(b) Note that $y = \sin 200\pi t$ has an angular frequency of $\omega = 200\pi$ radians per second. Hence the period is $T = 2\pi/200\pi = \frac{1}{100}$ seconds.

Example 16.5 Oscilloscope traces

Consider Figure 16.7, which shows two signals displayed on an oscilloscope. In each case identify the period, angular frequency and amplitude of the illustrated sine waves. Use this information to write down the equation of each wave.

Solution (a) The time taken to complete one cycle of the wave is $\frac{1}{3}\pi$ seconds, and so the period T is $\frac{1}{3}\pi$. Using the formula $T = 2\pi/\omega$, we have

$$\frac{\pi}{3} = \frac{2\pi}{\omega}$$

from which $\omega = 6$. The amplitude of the wave is 3, and so its equation is $y = 3 \sin 6t$.

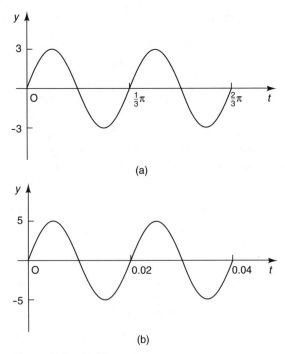

Figure 16.7 Oscilloscope traces.

(b)   The time taken to complete one cycle is 0.02 seconds, and so the period T is 0.02. Using the formula $T = 2\pi/\omega$, we have

$$0.02 = \frac{2\pi}{\omega}$$

from which

$$\omega = \frac{2\pi}{0.02} = 100\pi$$

The amplitude of the wave is 5, and so its equation is $y = 5 \sin 100\pi t$.

16.2.4 The frequency of a wave

Closely related to the period is the frequency of a wave. The **frequency** is the number of cycles completed in 1 second. Frequency is measured in units called hertz (Hz). One hertz is one cycle per second. We have seen that $y = A \sin \omega t$ takes

$$\frac{2\pi}{\omega} \text{ seconds to complete one cycle}$$

and so it will take

$$1 \text{ second to complete } \frac{\omega}{2\pi} \text{ cycles}$$

We use f as the symbol for frequency, and so

$$\text{frequency } f = \frac{\omega}{2\pi}$$

Note that, by rearrangement, we may write

$$\omega = 2\pi f$$

and so the wave $y = A \sin \omega t$ may also be written as $y = A \sin 2\pi f t$. From the definitions of period and frequency, we can see that

$$\text{period} = \frac{1}{\text{frequency}}$$

that is

$$T = \frac{1}{f}$$

We see that the period is the reciprocal of the frequency. Identical results apply for the wave $y = A \cos \omega t$.

Example 16.6

State the amplitude, angular frequency, period and frequency of the following waves: (a) $y = 3 \sin 2t$ (b) $y = 5 \cos 6t$ (c) $y = \frac{1}{2} \sin t$ (d) $y = \frac{2}{3} \cos \frac{1}{2} t$

Solution In each case we compare the given wave with the general form $y = A \sin \omega t$ or $y = A \cos \omega t$, as appropriate.

(a) The wave $y = 3 \sin 2t$ has amplitude 3. The angular frequency ω is 2 radians per second. The period is found from $T = 2\pi/\omega$, and is therefore $2\pi/2 = \pi$ seconds. The frequency is the reciprocal of the period; that is, $f = 1/\pi$ Hz.

(b) The amplitude is 5, and the angular frequency is 6 radians per second. The period is $2\pi/6 = \frac{1}{3}\pi$ seconds and the frequency is $3/\pi$ Hz.

(c) The amplitude is $\frac{1}{2}$, the angular frequency is 1 radian per second, the period is 2π seconds and the frequency is $1/2\pi$ Hz.

(d) The amplitude is $\frac{2}{3}$, the angular frequency is $\frac{1}{2}$ radian per second, the period is $2\pi/\frac{1}{2} = 4\pi$ seconds and the frequency is $1/4\pi$ Hz.

16.2.5 The phase and time displacement of a wave

We now consider how a sine wave can be shifted to the left or to the right by introducing the phase and time displacement of a wave. Figure 16.8 shows graphs of $y = \sin 2t$ and $y = \sin(2t+1)$. Starting at the peak of $\sin(2t+1)$ we need to travel to the right by 0.5 seconds to reach the peak of $\sin 2t$. As $\sin(2t+1)$ reaches its peak 0.5 seconds before $\sin 2t$, we say that $\sin(2t+1)$ **leads** $\sin 2t$ by 0.5 seconds and that the **time displacement** of $\sin(2t+1)$ is 0.5 seconds.

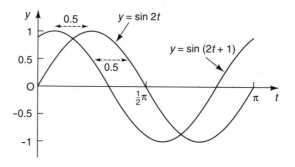

Figure 16.8   The waves $y = \sin 2t$ and $y = \sin(2t+1)$.

Example 16.7

Draw a graph of $y = \sin 3t$ and $y = \sin(3t-2)$. Measure the time displacement of $\sin(3t-2)$.

Solution   The waves are illustrated in Figure 16.9. Starting at the peak on $\sin(3t-2)$, we need to travel to the left by $\frac{2}{3}$ seconds to reach the peak of $\sin 3t$. The graph of $y = \sin(3t-2)$ reaches its peaks after $\sin 3t$, and so $\sin(3t-2)$ **lags** $\sin 3t$. The time displacement of $\sin(3t-2)$ is $-\frac{2}{3}$ seconds.

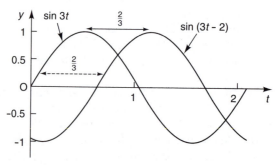

Figure 16.9   The waves $y = \sin 3t$ and $y = \sin(3t-2)$.

Note that $\sin(2t+1)$ can be written as $\sin 2(t+0.5)$, and the time displacement is 0.5 seconds. Also $\sin(3t-2)$ can be written as $\sin 3(t-\frac{2}{3})$, and the time displacement is $-\frac{2}{3}$ seconds. We are able to generalize the result. Consider the wave

$$y = \sin(\omega t + \alpha)$$

The angle α is known as the **phase angle** or simply as the **phase**.

The phase of $y = \sin(\omega t + \alpha)$ is α radians.

The wave may be written as

$$y = \sin\left[\omega\left(t + \frac{\alpha}{\omega}\right)\right]$$

We see that the time displacement of $\sin(\omega t + \alpha)$ is α/ω seconds.

The time displacement of $y = \sin(\omega t + \alpha)$ is α/ω seconds.

Amplitude has no effect on phase or time displacement, and so the results are identical for $y = A\sin(\omega t + \alpha)$. All the results for sine waves carry over unchanged to cosine waves. Hence the time displacement and phase of $y = A\cos(\omega t + \alpha)$ are α/ω seconds and α radians respectively.

Example 16.8

State the phase and time displacement of the following waves:
(a) $y = \sin(4t+3)$ (b) $y = 7\sin(4t+3)$ (c) $y = 2\sin(5t-3)$

Solution   (a)   For $y = \sin(4t+3)$, $\omega = 4$ and $\alpha = 3$, and so the phase is 3 radians and the time displacement is $\frac{3}{4}$ seconds.

(b)   For $y = 7\sin(4t+3)$ the amplitude, which is 7, does not affect the phase. Hence the phase is 3 radians and the time displacement is $\frac{3}{4}$ seconds.

(c)   For $y = 2\sin(5t-3)$, $\omega = 5$ and $\alpha = -3$, and so the phase is -3 radians and the time displacement is $-\frac{3}{5}$ seconds.

Example 16.9

State the phase and time displacement of (a) $\cos(3t-1)$ (b) $2\cos(5t+3)$

Solution   (a)   Here $\omega = 3$ and $\alpha = -1$, and so the phase is -1 radian and the time displacement is $-\frac{1}{3}$ seconds. Thus $\cos(3t-1)$ lags $\cos 3t$ by $\frac{1}{3}$ seconds.

(b)   The amplitude, 2, has no effect on the phase. Here $\omega = 5$ and $\alpha = 3$, giving a phase of 3 radians and a time displacement of $\frac{3}{5}$ second. Thus $2\cos(5t + 3)$ leads $2\cos 5t$ by $\frac{3}{5}$ seconds.

Example 16.10 Alternating current waveforms

Sine and cosine functions are often used to model alternating current (AC) waveforms. The equations for an AC current waveform are

$$I = I_m \sin(\omega t + \alpha) \quad \text{or} \quad I = I_m \cos(\omega t + \alpha)$$

where the amplitude I_m is the maximum current, ω is the angular frequency and α is the phase angle. We can use either the sine or the cosine function to model an AC waveform, depending on which is convenient. This is because either can be shifted along the time axis by adjusting the value of α.

Self-assessment questions 16.2

1.   Explain the term 'amplitude' of a wave.

2.   What is meant by the angular frequency of a wave?

3.   Explain the terms 'period' and 'frequency' of a wave.

4.   Explain the terms 'time displacement' and 'phase' of a wave.

Exercises 16.2

1.   Figure 16.10 shows some oscilloscope traces. In each case determine the period, angular frequency and amplitude. Hence state the equation of each wave.

2.   State the amplitude, angular frequency, frequency, time displacement and period of the following waves:
(a) $3\sin 6t$ (b) $4\cos 5t$ (c) $5\cos 4t$

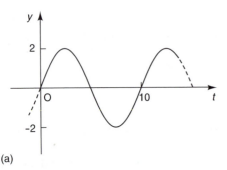

(a)

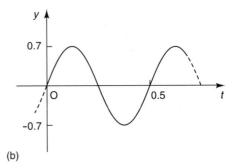

(b)

Figure 16.10   Waves for Exercise 16.2 Q1.

(d) $\frac{1}{6}\sin t$ (e) $6\sin\left(\dfrac{t}{6}\right)$ (f) $\frac{2}{3}\cos 7t$

(g) $0.6\cos\left(\dfrac{2t}{3}\right)$ (h) $\sin(t+2)$

(i) $\cos(2t-1)$

(a) $\sin(5t+3)$ (b) $\cos(t-3)$

(c) $\cos(2t+3)$ (d) $\sin\left(\dfrac{t}{2}+3\right)$

(e) $\cos\left(\dfrac{2t}{3}-3\right)$ (f) $\sin\left(\dfrac{2t-4}{7}\right)$

(g) $\cos\left(\dfrac{6t+5}{2}\right)$

3. State the phase and time displacement of the following waves:

Computer and calculator exercises 16.2

1. Use a graph-plotting package to draw $y=3\sin 4t$ and $y=3\sin(4t+5)$ for $0\leqslant t\leqslant 8$.
    (a) From your graphs, measure the amplitude of each wave.

    (b) Measure the horizontal distance between the waves. Hence state the time displacement of $3\sin(4t+5)$.

16.3 Combining waves of the same frequency

Recall that $y=A\sin(\omega t+\alpha)$ and $y=B\cos(\omega t+\beta)$ both have an angular frequency of ω radians per second and a frequency of $\omega/2\pi$ Hz. It is possible to add two waves of the same frequency to form a single sine or cosine wave. The resulting wave has the same frequency but a different amplitude and phase.

> If two waves of the same frequency are added, the result is a wave of the same frequency.

To find the amplitude and phase of the resulting wave, we make use of trigonometric identities. In particular, we need formulae for $\sin(A\pm B)$ and $\cos(A\pm B)$ and the identity $\cos^2\alpha+\sin^2\alpha=1$. The following examples illustrate how two waves are added.

Example 16.11

Express $3\sin 2t+4\cos 2t$ as a single wave of the form $A\sin(\omega t+\alpha)$.

Solution The angular frequency of $3\sin 2t$ is the same as that of $4\cos 2t$, namely 2 radians per second. When the waves are added, the angular frequency of the resulting wave will also be 2 radians per second. Let us write

$$3\sin 2t+4\cos 2t=A\sin(2t+\alpha)$$

so that the two waves on the left-hand side have been expressed as a single sine wave with unknown amplitude A and phase α. Note that the angular frequency of $A \sin(2t + \alpha)$ is also 2 radians per second. Using the trigonometric identity

$$\sin(A + B) = \sin A \cos B + \sin B \cos A$$

we can write

$$\sin(2t + \alpha) = \sin 2t \cos \alpha + \sin \alpha \cos 2t$$

and so

$$A \sin(2t + \alpha) = A(\sin 2t \cos \alpha + \sin \alpha \cos 2t)$$

Hence we have

$$3 \sin 2t + 4 \cos 2t = (A \cos \alpha) \sin 2t + (A \sin \alpha) \cos 2t$$

Looking at the $\sin 2t$ terms on the left- and right-hand sides, we see

$$3 = A \cos \alpha \tag{16.1}$$

Looking at the $\cos 2t$ terms on both sides we see

$$4 = A \sin \alpha \tag{16.2}$$

We need to solve Equations (16.1) and (16.2) in order to find A and α. To eliminate α, both equations are squared. This produces

$$9 = A^2 \cos^2 \alpha$$

$$16 = A^2 \sin^2 \alpha$$

These equations are added to give

$$25 = A^2 \cos^2 \alpha + A^2 \sin^2 \alpha$$

$$= A^2(\cos^2 \alpha + \sin^2 \alpha)$$

$$= A^2$$

where we have made use of the identity $\sin^2 \alpha + \cos^2 \alpha = 1$. Hence $A = 5$.

To eliminate A from Equations (16.1) and (16.2), we divide the latter by the former. This gives

$$\frac{4}{3} = \frac{A \sin \alpha}{A \cos \alpha} = \frac{\sin \alpha}{\cos \alpha} = \tan \alpha$$

From Equation (16.1), $\cos \alpha$ is positive, and from Equation 16.2, $\sin \alpha$ is positive, and so α is in the first quadrant. We require an angle α in the first quadrant such that $\tan \alpha = \frac{4}{3}$. Remembering that we are measuring angles in radians, and using a calculator, we obtain $\alpha = 0.927$. We may now state

$$3 \sin 2t + 4 \cos 2t = 5 \sin(2t + 0.927)$$

When the waves $3 \sin 2t$ and $4 \cos 2t$ are added, the result is a wave of amplitude 5, angular frequency 2 radians per second and a phase angle 0.927 radians. The time displacement is α/ω, that is, $0.927/2=0.464$ seconds.

Example 16.12

Express $2 \sin 5t - 4 \cos 5t$ in the form

$$A \cos (\omega t - \alpha)$$

where $\alpha > 0$.

Solution

The angular frequency of $2 \sin 5t$ and $4 \cos 5t$ is 5 radians per second, and so the angular frequency of the combined wave is also 5 radians per second, that is, $\omega = 5$. Using the trigonometric identity

$$\cos (A - B) = \cos A \cos B + \sin A \sin B$$

we see that

$$\cos (5t - \alpha) = \cos 5t \cos \alpha + \sin 5t \sin \alpha$$

and so

$$A \cos (5t - \alpha) = A \cos 5t \cos \alpha + A \sin 5t \sin \alpha$$

We require

$$2 \sin 5t - 4 \cos 5t = A \cos 5t \cos \alpha + A \sin 5t \sin \alpha$$
$$= (A \cos \alpha) \cos 5t + (A \sin \alpha) \sin 5t$$

Comparing the $\sin 5t$ terms on both sides, we see that

$$2 = A \sin \alpha \tag{16.3}$$

Comparing the $\cos 5t$ terms on both sides, we see that

$$-4 = A \cos \alpha \tag{16.4}$$

We need to solve Equations (16.3) and (16.4) to find A and α. First, we eliminate α and solve for A. Squaring Equations (16.3) and (16.4) results in

$$4 = A^2 \sin^2 \alpha$$
$$16 = A^2 \cos^2 \alpha$$

Adding these equations yields

$$20 = A^2 \sin^2 \alpha + A^2 \cos^2 \alpha$$
$$= A^2(\sin^2 \alpha + \cos^2 \alpha)$$
$$= A^2$$

and so

$$A=\sqrt{20}=4.472$$

In order to evaluate α, we now eliminate A from Equations (16.3) and (16.4). Dividing equation (16.3) by equation (16.4), we obtain

$$\frac{2}{-4}=\frac{A\sin\alpha}{A\cos\alpha}=\frac{\sin\alpha}{\cos\alpha}=\tan\alpha$$

that is,

$$\tan\alpha=-0.5$$

From Equation (16.3), $\sin\alpha$ is positive; from Equation (16.4), $\cos\alpha$ is negative, and so α must be an angle in the second quadrant. We also know that $\tan\alpha=-0.5$. Solving $\tan\alpha=-0.5$ with α in the second quadrant yields $\alpha=2.6779$. Thus

$$2\sin 5t-4\cos 5t=4.472\cos(5t-2.6779)$$

Note that the result of adding the waves $2\sin 5t$ and $-4\cos 5t$ is a wave of the same frequency having amplitude 4.472 and a phase angle of -2.6779 radians.

Self-assessment question 16.3

1.   Explain how sine and cosine waves of the same frequency can be combined into a single wave.

Exercises 16.3

1.   Express the following as single waves in the form $A\sin(\omega t-\alpha)$, $\alpha>0$:
(a) $4\sin 2t-2\cos 2t$
(b) $3\cos 4t+6\sin 4t$ (c) $\frac{3}{2}\sin t+\cos t$
(d) $4\cos\frac{1}{2}t+3\sin\frac{1}{2}t$

2.   Express the following as single waves in the form $A\sin(\omega t+\alpha)$, $\alpha>0$:
(a) $\sin t+\cos t$ (b) $3\sin 4t-5\cos 4t$
(c) $3\cos 2t-6\sin 2t$ (d) $\cos t+1.6\sin t$

3.   Express the following as single waves in the form $A\cos(\omega t+\alpha)$, $\alpha>0$:
(a) $\cos 3t-2\sin 3t$ (b) $6\sin t+2\cos t$
(c) $5\sin 2t-3\cos 2t$ (d) $\frac{7}{4}\sin 3t+\cos 3t$

4.   Express the following as single waves in the form $A\cos(\omega t-\alpha)$, $\alpha>0$:
(a) $7\sin 4t+2\cos 4t$ (b) $-\sin t-\cos t$
(c) $3\cos 5t-2\sin 5t$ (d) $\sin 3t-6\cos 3t$

Review exercises 16

1   State the amplitude, angular frequency, frequency, period, phase and time displacement of the following waves:

(a) $3\sin 2t$ (b) $4\cos t$ (c) $5\sin(3t-1)$

(d) $6\cos\left(\frac{t}{2}\right)$ (e) $3\sin\left(\frac{2t+3}{3}\right)$
(f) $\frac{2}{3}\sin(\frac{2}{3}t-1)$ (g) $\frac{3}{4}\cos(0.4t)$
(h) $\sin 100\pi t$

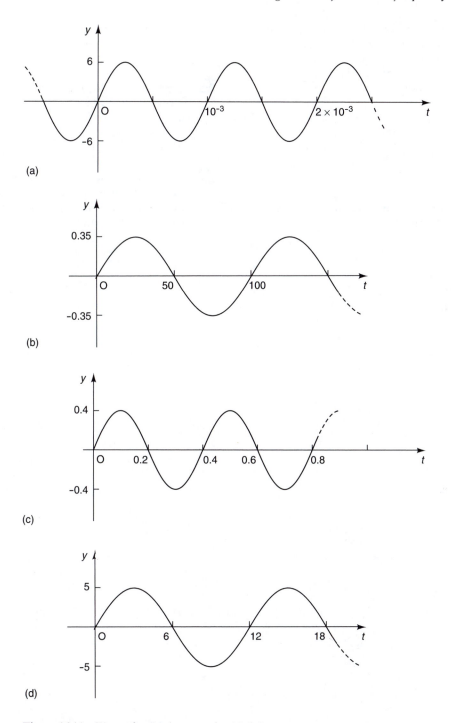

(a)

(b)

(c)

(d)

Figure 16.11 Waves for Review exercise 16 Q6.

2 Express the following in the form $A \cos(\omega t - \alpha)$, $\alpha > 0$:
(a) $3 \sin 3t - 2 \cos 3t$
(b) $0.7 \cos 5t + 0.9 \sin 5t$
(c) $\sin \pi t + 3 \cos \pi t$

3 Express the following in the form $A \sin(\omega t + \alpha)$, $\alpha > 0$:
(a) $2 \sin 2t - \cos 2t$ (b) $4 \cos 3t + \sin 3t$
(c) $\dfrac{\sin \pi t - 2 \cos \pi t}{2}$

4 Express the following in the form $A \sin(\omega t - \alpha)$, $\alpha > 0$:
(a) $5 \sin t + 3 \cos t$
(b) $-\cos 4t + \frac{1}{2} \sin 4t$
(c) $2 \cos 3t - \sin 3t$

5 Express the following in the form $A \cos(\omega t + \alpha)$, $\alpha > 0$:
(a) $2 \sin 5t + 3 \cos 5t$
(b) $\cos t - 3 \sin t$
(c) $4 \sin \frac{1}{2}t + \cos \frac{1}{2}t$

6 Figure 16.11 shows some oscilloscope traces. In each case determine the period, angular frequency and amplitude. Hence state the equation of each wave.

17

Geometry of the Straight Line

KEY POINTS

This chapter

- reviews the idea of the coordinates of a point

- explains the terms 'gradient' and 'vertical interception' of a straight line

- develops the equation of a straight line

- shows how to calculate the length of a line

- shows how to find the midpoint of a line

CONTENTS

17.1 Introduction

This chapter opens with a brief review of what is meant by the coordinates of a point. This is fundamental to the study of lines, their equations and graphs.

The gradient and vertical intercept of a straight line are introduced and used in the development of the equation of a line. Methods for calculating the length and midpoint of a line are described.

17.2 Cartesian coordinates

Engineers often need to solve problems involving lines, circles and other geometric figures. It is sometimes useful to employ **coordinate geometry** when tackling such problems.

A horizontal x axis and a vertical y axis intersect at the origin O. Each point in the (x, y) plane is identified by two numbers or **coordinates**. The x **coordinate** is the horizontal distance of the point from O, that is, the distance of the point from the y axis. The y **coordinate** is the vertical distance of the point from O, that is, the distance of the point from the x axis. When specifying the coordinates of a point, the x coordinate is always given first, then the y coordinate. Thus to locate the point (5, 3), we move 5 units in the x direction and 3 units in the y direction. This is shown in Figure 17.1.

The x and y coordinates are sometimes referred to as **cartesian coordinates**, after the French mathematician Descartes.

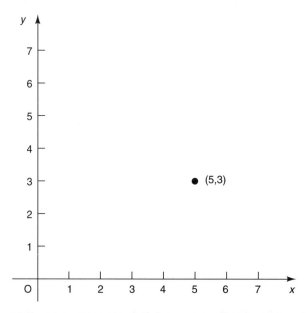

Figure 17.1   The x coordinate is 5, the y coordinate is 3.

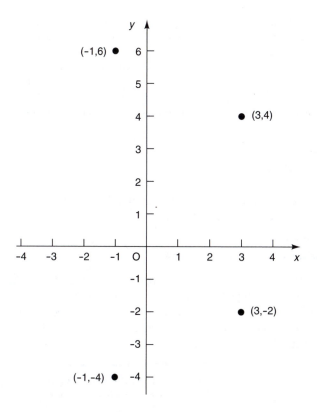

Figure 17.2 The points of Example 17.1.

Example 17.1
    Plot the points (3, 4), (− 1, 6), (− 1, − 4) and (3, − 2).

Solution   Figure 17.2 illustrates these points.

Self-assessment question 17.2

1.   Explain what is meant by the cartesian coordinates of a point.

Exercises 17.2

1.   State the coordinates of the origin O.

2.   (a)   On which axis does the point (0, 2) lie?
       (b)   On which axis does the point (2, 0) lie?

3. A straight line is drawn, passing through (0, 3) and parallel to the x axis. Which of the following points lie on the line?
(a) (3, 0)   (b) (−2, 3)   (c) (3, 2)
(d) (100, 3)

4. A line is drawn through (−6, 5), parallel to the y axis. Which of the following points lie on the line?
(a) (5, 5)   (b) (5, −6)   (c) (−6, 12)
(d) (0, 5)   (e) (−6, 0)

17.3 The straight line

This section develops the equation of a straight line. The terms 'gradient' and 'vertical intercept' are explained.

17.3.1 The gradient of a straight line

The **gradient** of a line measures the steepness of the line. In Figure 17.3 line 2 is steeper than line 1 and so has a greater gradient.

To treat gradient in a rigorous way we need to measure the change in the x and y coordinates of two points on the line. We measure change using the formula

> change = final value − initial value

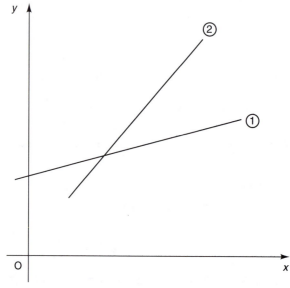

Figure 17.3 Line 2 is steeper than line 1.

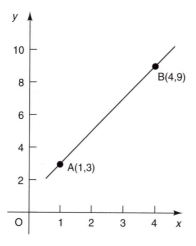

Figure 17.4 A straight line passes through A (1, 3) and B (4, 9).

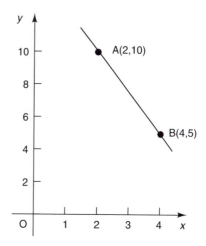

Figure 17.5 A straight line passes through A (2, 10) and B (4, 5).

Suppose A (1, 3) and B (4, 9) are two points on a line, as shown in Figure 17.4. We wish to measure the change in moving from A to B. The initial x value is 1; the final x value is 4. Hence the change in the x value is found as

change in x value $= 4 - 1 = 3$

Similarly, the change in the y value is $9 - 3 = 6$.

We are now ready to define the gradient of a line:

$$\text{gradient of a line} = \frac{\text{change in } y \text{ coordinate}}{\text{change in } x \text{ coordinate}}$$

To put it another way, the gradient of a line is the change in the vertical value for a unit increase in the horizontal value. The line shown in Figure 17.4 therefore has a gradient of $\frac{6}{3} = 2$. Hence an increase of 1 unit in x will create a change of 2 units in y.

Consider now the line shown in Figure 17.5. The initial point is (2, 10) and the final point is (4, 5). The change in the x coordinate is $4 - 2 = 2$. The change in the y coordinate is $5 - 10 = -5$. The gradient of the line is thus $-5/2 = -2.5$.

From Figure 17.4, we see that lines with positive gradients slope upwards; that is, as x increases, y increases. From Figure 17.5, we see that lines with negative gradients slope downwards; that is, as x increases, y decreases.

Example 17.2

Consider the lines shown in Figure 17.6. For each line state whether the gradient is positive or negative.

Solution   Line 1 has negative gradient.
Line 2 has positive gradient.
Line 3 has negative gradient.
Line 4 has positive gradient.
Line 5 has negative gradient.

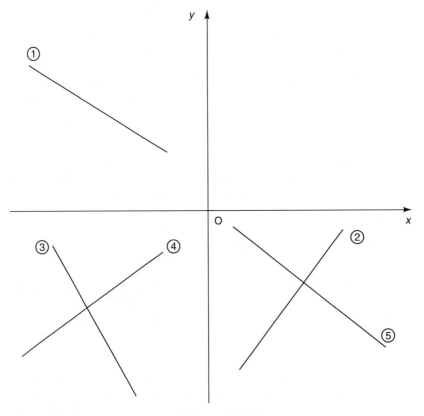

Figure 17.6   Lines for Example 17.2.

Example 17.3

Calculate the gradients of the lines shown in Figure 17.7.

Solution   Line A passes through $(-1, 6)$ and $(3, -4)$:

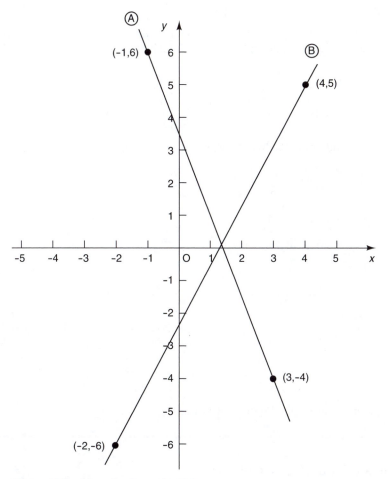

Figure 17.7 Lines for Example 17.3.

change in x coordinate $= 3 - (-1) = 4$

change in y coordinate $= -4 - 6 = -10$

gradient of line A $= \dfrac{-10}{4} = -2.5$

A gradient of -2.5 means that for every 1 unit increase in x the value of y decreases by 2.5.

We now consider line B, which passes through $(-2, -6)$ and $(4, 5)$:

change in x coordinate $= 4 - (-2) = 6$

change in y coordinate $= 5 - (-6) = 11$

gradient of line B $= \frac{11}{6}$

Example 17.4

A line passes through A (1, 7) and has a gradient of 5. Another point B lies on the line. The *x* coordinate of B is 3. Calculate the *y* coordinate of B.

Solution   We have

$$\text{gradient of line} = \frac{\text{change in } y \text{ value}}{\text{change in } x \text{ value}}$$

The gradient of the line is given as 5. The change in the *x* value is $3 - 1 = 2$. Hence

$$5 = \frac{\text{change in } y \text{ value}}{2}$$

and so

$$\text{change in } y \text{ value} = 10$$

Now

$$\text{change in } y \text{ value} = \text{final value} - \text{initial value}$$

and so

$$10 = \text{final value} - 7$$

from which

$$\text{final value} = 17$$

Hence the *y* coordinate of B is 17.

17.3.2 The equation of a straight line

Consider the straight line in Figure 17.8. The line cuts the *y* axis at A, a height *c* above the *x* axis. We say that the **vertical intercept** is *c*. Note that the coordinates of A are (0, *c*). Suppose that the gradient of the line is *m*. Let B (*x*, *y*) be any general point on the line. We seek an equation connecting *x* and *y*. The initial point is A and the final point is B. Thus the change in the *x* value is $x - 0 = x$. The change in the *y* value is $y - c$. Hence

$$\text{gradient} = \frac{\text{change in } y \text{ value}}{\text{change in } x \text{ value}}$$

$$m = \frac{y - c}{x}$$

$$mx = y - c$$

and so rearranging for *y* we obtain:

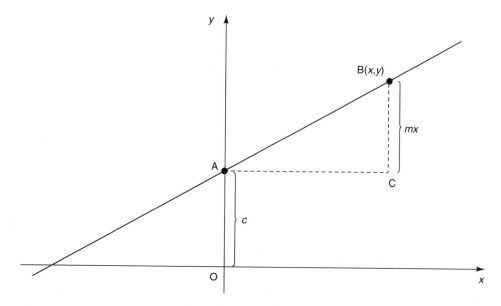

Figure 17.8 The vertical intercept is *c* and the gradient is *m*.

$$y = mx + c$$

This is the equation of a line with gradient *m* and vertical intercept *c*.

> The equation of a straight line is
>
> $$y = mx + c$$
>
> where *m* is the gradient and *c* is the vertical intercept.

Example 17.5

A straight line has gradient 3 and vertical intercept 4.
(a) State the equation of the line.
(b) Calculate where the line cuts the *x* axis.

Solution   (a)   gradient $= m = 3$

                  vertical intercept $= c = 4$

          The equation of the line is

$$y = mx + c$$
$$= 3x + 4$$

(b)   The value of y is zero on the x axis.

$$0 = 3x + 4$$

$$x = -\tfrac{4}{3}$$

The line cuts the x axis at $(-\tfrac{4}{3}, 0)$.

Example 17.6

A straight line passes through (5, 2) and has gradient -2. Find the equation of the line.

Solution   The equation of a line is

$$y = mx + c$$

We are told that the gradient m is -2, and so

$$y = -2x + c$$

The point (5, 2) lies on the line; that is, when $x = 5$, $y = 2$. Hence

$$2 = -2(5) + c$$

$$c = 12$$

and so

$$y = -2x + 12$$

Example 17.7 Car moving with constant acceleration

If a car moves with constant acceleration a then its velocity v is given by

$$v = u + at$$

where u is the initial velocity of the car and t is the time for which the car has been travelling. We see that this is the equation of a straight line with gradient a and vertical intercept u. Figure 17.9 is a graph of v against t.

Example 17.8

A high-performance car can accelerate from rest to 60 m.p.h. in 9 seconds. Given that 60 m.p.h. is equivalent to 26.8 m per second, calculate the acceleration of the car.

Solution   We know from Example 17.7 that $v = u + at$. In this example $u = 0$, since the car starts from rest. The velocity after 9 seconds is $v = 26.8$ m s^{-1}. Substituting these known values into the equation gives

$$26.8 = 0 + a(9)$$

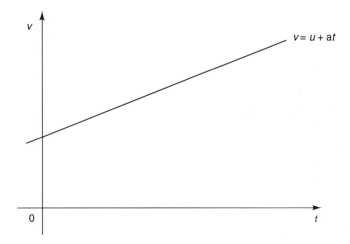

Figure 17.9 When acceleration is constant, the graph of v against t is a straight line.

$$a = \frac{26.8}{9} = 2.98$$

The acceleration of the car is 2.98 m s^{-2}.

Example 17.9

A straight line passes through two points A and B. The point A has coordinates (5, 4) and point B has coordinates (7, 12). Calculate the equation of the line.

Solution First, we calculate the gradient of the line:

$$\text{gradient } m = \frac{\text{change in } y \text{ coordinate}}{\text{change in } x \text{ coordinate}} = \frac{12 - 4}{7 - 5} = \frac{8}{2} = 4$$

The line has a gradient of 4, and so has an equation of the form

$$y = 4x + c$$

The line passes through (5, 4), that is, when $x = 5$, $y = 4$, and so

$$4 = 4(5) + c$$

$$c = 4 - 20 = -16$$

The equation of the line is $y = 4x - 16$.

When we wish to find the equation of a line given the coordinates of two points on the line, we can use the following method as an alternative.

Example 17.10

Find the equation of the line passing through $(6, -2)$ and $(1, 4)$.

Solution   The line has equation

$$y = mx + c$$

When $x = 6$, $y = -2$, and so

$$-2 = 6m + c \tag{17.1}$$

When $x = 1$, $y = 4$, and so

$$4 = m + c \tag{17.2}$$

Equations (17.1) and (17.2) are solved simultaneously for m and c. Equation (17.1) − Equation (17.2) yields

$$-6 = 5m$$

$$m = -\tfrac{6}{5}$$

From Equation (17.1),

$$c = -2 - 6m = -2 + \tfrac{36}{5} = \tfrac{26}{5}$$

The equation of the line is thus

$$y = -\tfrac{6}{5}x + \tfrac{26}{5}$$

Example 17.11

The equation of a line is given by $3x + 2y = 50$. State the gradient and vertical intercept of the line.

Solution   We rewrite $3x + 2y = 50$, so that y is the subject:

$$3x + 2y = 50$$

$$2y = -3x + 50$$

$$y = -\tfrac{3}{2}x + 25$$

The gradient is $-\tfrac{3}{2}$ and the vertical intercept is 25.

Self-assessment questions 17.3

1. Explain what is meant by the gradient of a straight line.
2. Explain what is meant by the vertical intercept of a straight line.

Exercises 17.3

1. State the gradient and vertical intercept of the following lines:
(a) $y=3x+2$ (b) $y=-2x-3$
(c) $y=\dfrac{x}{2}$ (d) $y=-\dfrac{2x}{3}+1$
(e) $2y=4x+1$ (f) $y+3x=9$
(g) $2x+4y-19=0$ (h) $\dfrac{y+x}{3}=2$
(i) $3x-4y-9=0$

2. Calculate the equations of the straight lines passing through the following pairs of points.
(a) $(1, 2)$ and $(3, 8)$ (b) $(0, 2)$ and $(3, 14)$
(c) $(-1, 2)$ and $(-2, 0)$ (d) $(0, 0)$ and $(-2, 3)$
(e) $(6, -1)$ and $(-2, 3)$

3. Calculate the equation of the straight line
(a) passing through $(1, 2)$ and with a gradient of 2
(b) passing through $(2, 4)$ and with a gradient of -1
(c) passing through $(-1, 3)$ and with a gradient of 0.5
(d) passing through $(0, -2)$ and with a gradient of $-\frac{2}{3}$
(e) passing through $(-1, -1)$ and with a gradient of -1

4. Calculate where the following lines cut the x axis:
(a) $y=3x-9$ (b) $y=2x+8$

(c) $y=-2x+4$ (d) $y=\dfrac{x}{2}+1$
(e) $y=-\dfrac{x}{2}+6$

5. Determine whether or not the given point lies on the line:
(a) line $y=3x+1$, point $(2, 7)$
(b) line $y=-2x+6$, point $(6, 0)$
(c) line $y=\dfrac{x}{2}-1$, point $(6, 2)$
(d) line $y=-\dfrac{2x}{3}+4$, point $(6, 1)$
(e) line $y=10x-90$, point $(10, 10)$

6. The line $y=2x+3$ meets the line $y=3x-1$. Calculate the coordinates of the point where the lines meet.

7. A line is parallel to the x axis and passes through the point $(10, 3)$. Calculate the equation of the line.

8. Calculate the equation of the line
(a) passing through $(-1, 4)$ with gradient 3
(b) passing through $(4, -2)$ with gradient -2
(c) passing through the origin with gradient 3
(d) passing through the origin with gradient k

17.4 Distance between two points on a line

Suppose we have two points A $(5, 9)$ and B $(7, 15)$, and we wish to find the distance between them. Figure 17.10 illustrates the situation. We note that the x coordinate of C is the same as the x coordinate of B, namely 7. The y coordinate of C is the same as the y coordinate of A, namely 9. Hence C has coordinates $(7, 9)$. Since AC is in the x direction, the distance AC is $7-5=2$. Similarly, since BC is in the y direction, the distance BC is $15-9=6$.

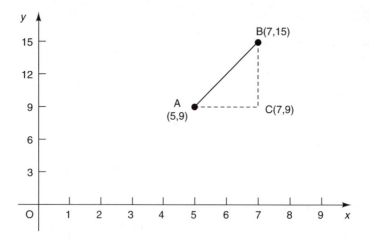

Figure 17.10   The distance from A to B is found using Pythagoras' theorem.

We now apply Pythagoras' theorem to calculate the length of AB:

$$(AB)^2 = (AC)^2 + (BC)^2$$
$$= 2^2 + 6^2$$
$$= 40$$
$$AB = \sqrt{40} = 6.3246$$

In general, if A and B have coordinates (x_1, y_1) and (x_2, y_2) then the distance AC is $(x_2 - x_1)$ and BC is $(y_2 - y_1)$, as shown in Figure 17.11. Hence,

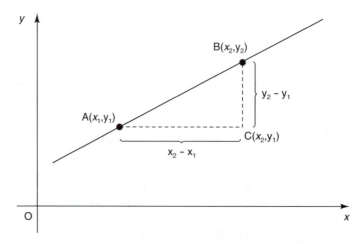

Figure 17.11   The distance from A to B is $\sqrt{(x_2 - x_1)^2 + (y_2 - y_1)^2}$.

from Pythagoras' theorem,

$$(AB)^2 = (x_2 - x_1)^2 + (y_2 - y_1)^2$$
$$AB = \sqrt{(x_2 - x_1)^2 + (y_2 - y_1)^2}$$

The distance from (x_1, y_1) to (x_2, y_2) is given by

$$\text{distance} = \sqrt{(x_2 - x_1)^2 + (y_2 - y_1)^2}$$

Example 17.12

Find the distance from (3, 7) to (6, 12).

Solution   We have $x_1 = 3$, $y_1 = 7$; $x_2 = 6$, $y_2 = 12$. Hence

$$\text{distance} = \sqrt{(6-3)^2 + (12-7)^2}$$
$$= \sqrt{3^2 + 5^2} = \sqrt{34} = 5.8310$$

Example 17.13

Calculate the distance from $(-1, 5)$ to $(-3, -2)$.

Solution   We have $x_1 = -1$, $y_1 = 5$; $x_2 = -3$, $y_2 = -2$. Hence

$$\text{distance} = \sqrt{(-3-(-1))^2 + (-2-5)^2}$$
$$= \sqrt{(-2)^2 + (-7)^2}$$
$$= \sqrt{53} = 7.2801$$

Self-assessment question 17.4

1.   Explain how Pythagoras' theorem is used to calculate the length of the line joining the point A (x_1, y_1) to B (x_2, y_2).

Exercise 17.4

1.   Calculate the distance between the following pairs of points:
(a) (3, 2) and (5, 9)   (b) (5, 0) and (0, 5)
(c) (4, -2) and (6, 1)
(d) (3, -2) and (-2, -3)
(e) (1, 4) and (2, 6)   (f) (0, 0) and (x, y)
(g) (x_1, y_1) and (x_2, y_2)

17.5 Midpoint of a line

Consider the straight line joining A (x_1, y_1) and B (x_2, y_2). We wish to find the coordinates of its midpoint. The situation is shown in Figure 17.12. The coordinates of C are (x_2, y_1). The distance AC is $x_2 - x_1$. Consider E, the midpoint of AC. The x coordinate of E is

$$x_1 + \tfrac{1}{2}AC = x_1 + \tfrac{1}{2}(x_2 - x_1)$$

$$= \frac{x_1 + x_2}{2}$$

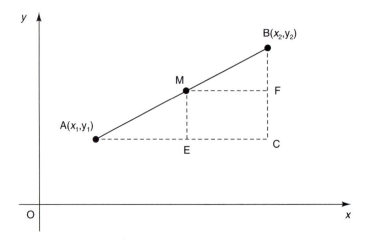

Figure 17.12 M is the midpoint of AB.

Thus the x coordinate of M is $\dfrac{x_1 + x_2}{2}$.

By a similar analysis, the y coordinate of M is $\dfrac{y_1 + y_2}{2}$. Thus the coordinates of M are $\left(\dfrac{x_1 + x_2}{2}, \dfrac{y_1 + y_2}{2} \right)$.

The midpoint of the line joining (x_1, y_1) and (x_2, y_2) is $\left(\dfrac{x_1 + x_2}{2}, \dfrac{y_1 + y_2}{2} \right)$.

Example 17.14

Find the midpoint of the line joining (a) $(-1, 6)$ and $(3, -2)$ (b) $(3, 6)$ and $(5, 17)$

Solution   (a)   Here $x_1 = -1$, $y_1 = 6$; $x_2 = 3$, $y_2 = -2$. So

$$\frac{x_1 + x_2}{2} = \frac{2}{2} = 1, \qquad \frac{y_1 + y_2}{2} = \frac{4}{2} = 2$$

The midpoint is (1, 2).

(b)   Here $x_1 = 3$, $y_1 = 6$; $x_2 = 5$, $y_2 = 17$. So

$$\frac{x_1 + x_2}{2} = 4, \qquad \frac{y_1 + y_2}{2} = 11.5$$

The midpoint is (4, 11.5).

Self-assessment question 17.5

1.   Explain how to calculate the midpoint of the line joining A (x_1, y_1) to B (x_2, y_2).

Exercise 17.5

1.   Calculate the coordinates of the midpoint of the line from

(a) (3, 2) to (5, 8)   (b) (2, 6) to (3, 4)

(c) (−1, 0) to (0, 1)   (d) (0, 0) to (a, b)

(e) (a,b) to (c,d)

Review exercises 17

1   A line passes through (3, 4) and has a gradient of 2.
(a) Calculate where the line cuts the x axis.
(b) Calculate where the line cuts the y axis.

2   Calculate the equations of the following lines:
(a) the line passing through (3, 6) and (5, 10)
(b) the line passing through (−1, 4) and (−5, 6)
(c) the line passing through (−4, −6) and (−2, 0)
(d) the line passing through (1, −3) and (−1, −3)
(e) the line with a gradient of 3 and passing through (−1, 4)
(f) the line with a gradient of 1 and passing through (0, 0)
(g) the line with a gradient of −2 and passing through (−1, −6)

3   Calculate the gradient, vertical intercept and intercept with the x axis of the following lines:
(a) $y = 2x - 1$ (b) $y = -\frac{1}{2}x + 3$
(c) $2y - x = 6$ (d) $3x - 4y + 16 = 0$
(e) $2x + 7y = 16$

4   Calculate the distance between the following pairs of points:
(a) (9, 3) and (10, 5)   (b) (7, 6) and (5, 12)
(c) (−1, 0) and (5, −3)
(d) the origin and (a, b)

5   The lines

$$2x + 3y = 0$$

$$x - 6y - 3 = 0$$

intersect at P. Find the cartesian coordinates of P.

The Channel Tunnel: one of the greatest civil engineering projects of the 20th century

The Channel Tunnel Act received Royal Assent on 23 July 1987. The Channel Tunnel Group and France Manche SA, the consortia awarded the concession to construct and operate the tunnel between England and France under the English Channel, were given the final go-ahead. The project was to construct two tunnels each 7.6 m in diameter together with a 4.8 m diameter service tunnel (see Figure 1). The service tunnel was to be linked by passages to the main tunnels at 375 m intervals.

The engineering task was enormous. In the early stages of the project surveyors used the Navstar Global Positioning System to fix the position of the tunnels. The Navstar system comprises 21 satellites orbiting the Earth at a height of some 10 898 nautical miles. The satellites are financed by the American Department of Defense but their navigation signals are available to anyone who has a receiver. Such a receiver on the ground picks up signals from four or more of the satellites, and sophisticated software determines its three-dimensional position coordinates: longitude, latitude and altitude.

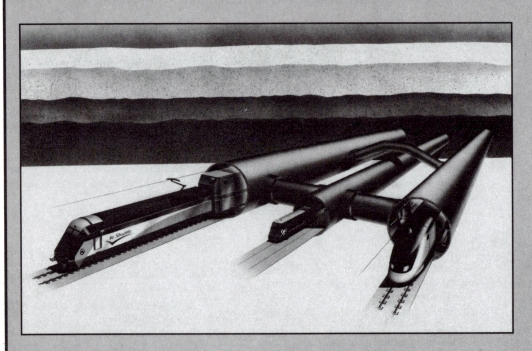

Figure 1 *Two tunnels each 7.6 m in diameter together with a 4.8 m diameter service tunnel.*

The system is so accurate that very precise determination of position is possible. Forty metres below the sea bed, massive tunnel boring machines slowly cut their way through the chalk, each one computer-controlled by a laser guidance system. With this system, minute corrections could be made to the tunnel alignment as boring proceeded. Teams of engineers drilled from both the French and English ends of the Tunnel, and in the final stages two tunnel boring machines came to a halt about 100 m apart. Slowly, a 40 mm diameter hole was bored through the final chalk barrier to make the first undersea contact between England and Europe. The British machine was then entombed forever while the French one was dismantled and returned to France. The final few metres were finished off using power tools and shovels! The final location check showed that the alignment error between the two tunnels was only a few centimetres - a remarkable testimony to engineering precision. For further reference detailing the work see *Tunnelling the Channel* by Derek Wilson (Century Publications, 1991).

(a)

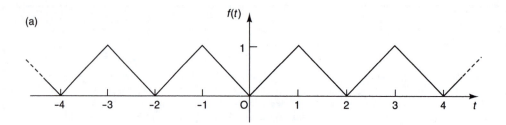

(b)

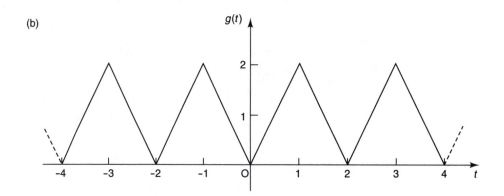

(c)

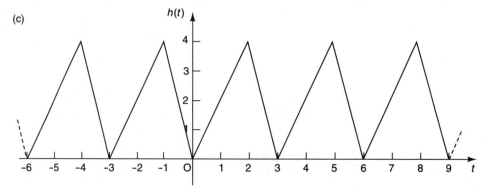

Figure 17.13 Three periodic waveforms.

6 Calculate the coordinates of the midpoint of the line joining

(a) (3, 7) to (6, 10)
(b) (−2, 1) to (0, 4)
(c) (0, 3) to (−3, 0)

7 Figure 17.13 shows three periodic waveforms taken from the display of an oscilloscope. Define each of the waveforms mathematically and state its period.

The Conic Sections

CONTENTS

18.1 Introduction

The circle, the ellipse, the parabola and the hyperbola are four curves that collectively are known as the **conic sections**. This chapter states the equations and main properties of the conic sections, as well as illustrating their curves. These equations are used extensively in describing planetary motion and are also useful in many mechanical applications.

Polar coordinates are introduced. These are a way of describing the position of a point, just as cartesian coordinates do. The link between the two coordinate systems is explained so that polar coordinates can be converted to cartesian coordinates and vice versa.

Finally, the equations of the conic sections are given an alternative form, known as a parametric form. The parametric form of the equation of a circle utilizes polar coordinates.

18.2 Slicing a double cone

Each of the conic sections can be obtained by slicing a double cone by a plane, as shown in Figure 18.1. As the angle at which the plane slicing the double cone changes, different curves are generated. For example, if the double cone is sliced parallel to its base, a circle is obtained. Note that the hyperbola is in two parts.

18.3 The circle and the ellipse

Many engineering components have a circular cross-section, for example pistons, connecting rods and pipes. However, some components have elliptical cross-sections, for example cams in certain mechanisms. Coordinate measuring machines can measure how accurately such components fit to an idealized shape. If the component is within a specified tolerance, it is accepted; otherwise it is rejected. This section studies the equations of the circle and the ellipse.

18.3.1 The circle

Figure 18.2 shows a circle with centre C (a, b) and radius r. We wish to find the equation of this circle. Let P (x, y) be a typical point on the circle. We consider the right-angled triangle CPQ as shown. The coordinates of Q are (x, b). The distance CQ is $(x - a)$. The distance PQ is $(y - b)$. The distance CP is r, the radius of the circle. Using Pythagoras' theorem, we see that

$$(CQ)^2 + (PQ)^2 = (CP)^2$$

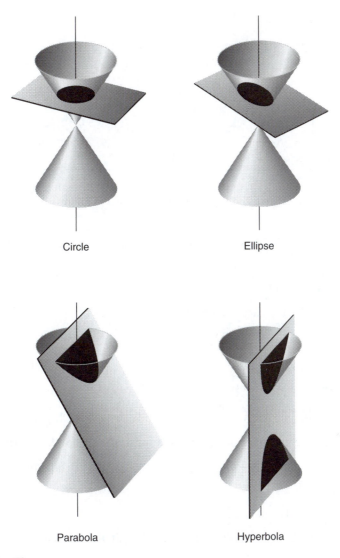

Circle

Ellipse

Parabola

Hyperbola

Figure 18.1 The conic sections are obtained by slicing a double cone at different angles.

and so

$$(x-a)^2 + (y-b)^2 = r^2$$

This is the equation of a circle, centre (a, b) and radius r.

> The equation of a circle with centre (a, b) and radius r is
> $$(x-a)^2 + (y-b)^2 = r^2$$

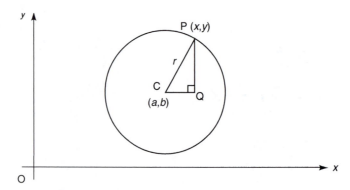

Figure 18.2   A circle, centre (a, b) and radius r.

The equation may be written as

$$x^2 - 2ax + a^2 + y^2 - 2by + b^2 = r^2$$

or more commonly as

$$x^2 + y^2 - 2ax - 2by + a^2 + b^2 - r^2 = 0$$

Example 18.1

Describe the circle whose equation is

$$(x-4)^2 + (y+3)^2 = 49$$

Solution   When we compare the given equation with the standard form

$$(x-a)^2 + (y-b)^2 = r^2$$

we see that

$$a = 4, \quad b = -3, \quad r = 7$$

Hence the centre of the circle is at $(4, -3)$ and the radius is 7.

Example 18.2

Find the equation of the circle, centre $(3, 4)$ and radius 2.

Solution   Here $r = 2$, $a = 3$ and $b = 4$. The equation of the circle is

$$(x-3)^2 + (y-4)^2 = 4$$

Using the form $x^2 + y^2 - 2ax - 2by + a^2 + b^2 - r^2 = 0$, this may be written as

$$x^2 + y^2 - 6x - 8y + 21 = 0$$

Example 18.3

Find the equation of the circle, centre (1, 3) and passing through (2, 5).

Solution We need to calculate the radius of the circle, r. This is the distance from the centre (1, 3) to the point (2, 5). Recall the formula for the distance between two points:

$$r = \sqrt{(x_2 - x_1)^2 + (y_2 - y_1)^2}$$

So we have

$$r = \sqrt{(2-1)^2 + (5-3)^2} = \sqrt{5}$$

from which $r^2 = 5$. The equation of the circle may now be stated

$$(x-1)^2 + (y-3)^2 = 5$$

Example 18.4

Find the radius and centre of the circle given by

$$x^2 + y^2 + 4x - 2y + 1 = 0$$

Solution We compare the equation given to the standard form

$$x^2 + y^2 - 2ax - 2by + a^2 + b^2 - r^2 = 0$$

By comparing the x coefficients, we see that

$$-2a = 4$$
$$a = -2$$

By comparing the y coefficients, we see that

$$-2b = -2$$
$$b = 1$$

By comparing the constant terms, we have

$$a^2 + b^2 - r^2 = 1$$
$$4 + 1 - r^2 = 1$$
$$r^2 = 4$$
$$r = 2$$

Recall that the centre of the circle is (a, b). Hence the centre of the circle is $(-2, 1)$. The radius is 2.

Example 18.5

A circle has centre $(-1, 1)$ and radius 3.

(a) Calculate where the circle cuts the x axis.

(b) Calculate where the circle cuts the y axis.

Solution   (a)   The equation of the circle is

$$(x+1)^2+(y-1)^2=9$$

Where the circle cuts the x axis, the y coordinate is 0. Hence

$$(x+1)^2+(-1)^2=9$$

$$(x+1)^2=8$$

$$x+1=\pm\sqrt{8}$$

$$x=-1\pm\sqrt{8}$$

Hence there are two solutions: $x=1.828$ and $x=-3.828$. The circle cuts the x axis at $(1.828, 0)$ and $(-3.828, 0)$.

(b)   At points where the circle cuts the y axis the x coordinate is 0. Hence

$$1+(y-1)^2=9$$

$$(y-1)^2=8$$

$$y-1=\pm\sqrt{8}$$

$$y=1\pm\sqrt{8}$$

Hence there are two solutions: $y=3.828$ and $y=-1.828$. The circle cuts the y axis at $(0, 3.828)$ and $(0, -1.828)$.

18.3.2 The ellipse

The simplest form of an equation of an ellipse is

$$\frac{x^2}{a^2}+\frac{y^2}{b^2}=1$$

Figure 18.3 shows such an ellipse.

An ellipse may be thought of as a circle that has been stretched or flattened. The amount of stretching depends upon the constants a and b. Figure 18.4 shows two ellipses with the equation

$$\frac{x^2}{a^2}+\frac{y^2}{b^2}=1$$

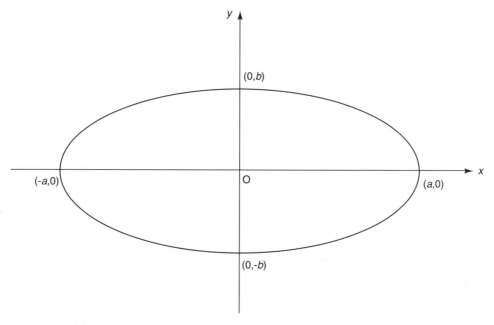

Figure 18.3 The ellipse $\dfrac{x^2}{a^2} + \dfrac{y^2}{b^2} = 1$.

Ellipse 1 has $a=1$, $b=2$, that is

$$x^2 + \frac{y^2}{4} = 1$$

and ellipse 2 has $a=1$, $b=\frac{1}{2}$, that is

$$x^2 + 4y^2 = 1$$

We see that the shape of the ellipse is changed by changing the value of b. Similarly, changing the value of a also changes the shape of an ellipse.

Example 18.6

Find where the ellipse

$$\frac{x^2}{4} + y^2 = 1$$

cuts (a) the x axis   (b) the y axis.

Solution   (a)   On the x axis the y coordinate is 0. Hence the equation becomes

$$\frac{x^2}{4} = 1$$

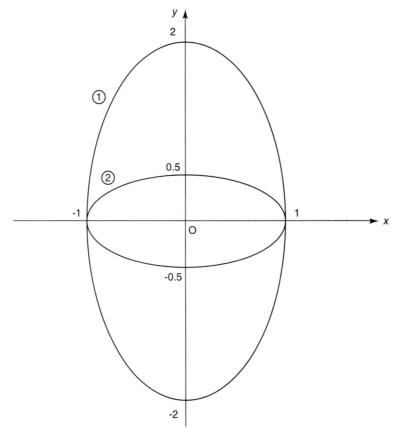

Figure 18.4   Changing the value of *b* changes the shape of the ellipse.

$$x^2 = 4$$
$$x = 2, -2$$

The ellipse cuts the *x* axis at (2, 0) and (−2, 0).

(b)   On the *y* axis the *x* coordinate is 0. Hence the equation becomes

$$y^2 = 1$$

$$y = 1, -1$$

The ellipse cuts the *y* axis at (0, 1) and (0, −1).

Self-assessment question 18.3

1.   Describe the circumstances under which an ellipse will become a circle.

Exercises 18.3

1. Calculate the equations of the following circles:

(a) centre $(3, -1)$, radius 2
(b) centre $(-1, 0.5)$, radius 5
(c) centre $(0, 0)$, radius 7
(d) centre $(0, 0)$, radius r
(e) centre $(-1, 1)$, passing through $(2, 3)$
(f) centre $(3, 3)$ and passing through $(4, 5)$

2. Calculate where the following circles cut the x and y axes:

(a) $x^2 + y^2 - 2x - 2y - 2 = 0$
(b) $x^2 + y^2 = 100$
(c) $x^2 + y^2 + 4x + 2y = 20$

3. An ellipse

$$x^2 + \frac{y^2}{4} = 1$$

intersects the line $y = x$. Calculate the points of intersection.

4. An ellipse

$$\frac{x^2}{4} + \frac{y^2}{6} = 1$$

intersects the line $y = 2x$. Find the points of intersection.

5. Calculate the equations of the following circles:

(a) radius 9, centre $(3, 2)$
(b) centre $(4, 1)$ and intersecting the x axis at $(6, 0)$

6. State the centre and radius of the following circles:

(a) $(x - 7)^2 + (y + 4)^2 = 36$
(b) $x^2 + y^2 + 6x + 2y + 6 = 0$
(c) $2x^2 + 2y^2 + 2x - 2y - 1 = 0$

Computer and calculator exercises 18.3

Hint. When plotting $x^2 + y^2 = r^2$ on a graphics calculator, we need to rearrange the equation in the form

$$y = +\sqrt{r^2 - x^2}, \qquad y = -\sqrt{r^2 - x^2}$$

Both curves must be plotted on the same axes. This will produce a circle. However, on some graphics calculators the different x and y scales may give the appearance that it is an ellipse and not a circle.

1. Draw the circles $x^2 + y^2 = 9$ and $(x - 4)^2 + (y - 2)^2 = 4$. State the points of intersection of the circles.

2. Draw the circle $x^2 + y^2 = 20$ and the ellipse $x^2 + 4y^2 = 36$. Find their points of intersection.

3. Plot the ellipse

$$\frac{x^2}{a^2} + y^2 = 4$$

for (a) $a = 1$ (b) $a = 4$ (c) $a = 0.5$
Comment upon the effect of changing the value of a.

4. The equation of an ellipse is

$$\frac{x^2}{a^2} + \frac{y^2}{b^2} = 1$$

Investigate the effect of
(a) changing the value of a for b fixed;
(b) changing the value of b for a fixed.

Satellites in Orbit: global communication and navigation

An immense amount of international communication, including sound, pictures and computer programs now takes place via satellite. Information in digital form is transmitted to a satellite, which then relays it to a receiving station, where it is processed into an understandable form. Weather prediction uses information gathered by satellite and aircraft, and ships commonly use satellites to aid navigation. The Landsat satellites launched in the 1970s were specifically designed to help study the Earth's resources, and information gathered has been used for forest inventories, mineral exploitation and monitoring environmental impact.

Communication satellites commonly use microwave frequencies of 1-30 GHz. These signals are not appreciably deflected by the Earth's atmosphere. To build receiving and transmitting towers large enough to enable transatlantic communication is not feasible. Orbiting satellites, however, overcome this problem.

A satellite that is always above the same point of the Earth's surface is said to be in a **geosynchronous orbit**. A system of three satellites in geosynchronous orbit can cover almost all of the Earth's surface. The global coverage overcomes the problem of installation and maintenance of systems in remote areas, oceans and areas of harsh climate. The science-fiction writer Arthur C. Clarke first envisaged such a system of geosynchronous satellites in 1945: by 1969, it was a reality. A circular orbit 35 768 km above the Earth is geosynchronous. Communication satellites have increased in size and mass as the launch vehicle capability has increased. The development of the Space Shuttle has dramatically increased this capacity. To put a satellite successfully into orbit requires knowledge of the equations that describe the motion of planets and the forces that act upon them. Newton's three laws of motion are used extensively. The first law states that a body remains at rest or continues in uniform motion in the absence of forces. The second law states that the force on a body equals the rate of change of its linear momentum. The third law states that the forces of action and reaction are equal in size and opposite in direction. In addition, the law of gravitational attraction is used. This states that the force of attraction between two bodies is proportional to the product of their masses and inversely proportional to the square of their distance apart. With Newton's laws and the gravitational law as building blocks, it is possible to develop the equations that describe the orbits of planets and satellites. These orbits are ellipses.

The 21 Navstar satellites were launched to provide a worldwide navigation system. A receiver on the Earth picks up a signal from at least four of the satellites. The distance from the receiver to each satellite is estimated using the time required for a signal to travel from a satellite to the receiver. As the speed of a signal is so high, 3×10^8 m per second, the timing devices must be extremely accurate. The satellites carry atomic clocks that are accurate to within 1 second every 160 000 years. From these distances, the position of the receiver is uniquely determined to within 100 feet.

For further reference see *Design of geosynchronous spacecraft* by B.N. Agrawal, and *The Navstar global positioning system* by T. Logsdon.

Figure 1 *Seasat: A satellite studying the oceans.*

18.4 The parabola and hyperbola

18.4.1 The parabola

The equation of a parabola comes in a variety of forms. Traditionally, it is given as

$$y^2 = 4ax$$

Figure 18.5 shows this curve for $a = 1, 3, -1, -3$. Note that the curve is not defined for negative values of x when $a > 0$ and not defined for positive values of x when $a < 0$.

The parabola may be reoriented by writing the equation in the less familiar form

$$y = ax^2 + bx + c$$

In this form we recognize that the right-hand side is a polynomial of degree 2. Figure 18.6 illustrates a typical parabola, $y = x^2 + x + 7$.

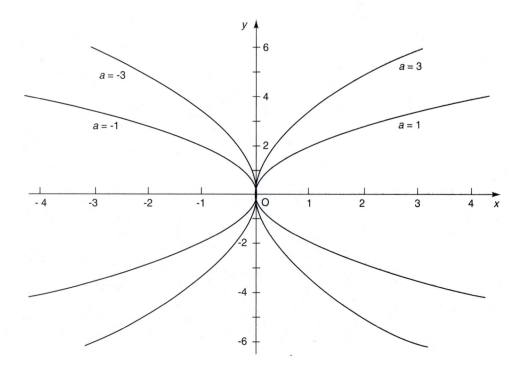

Figure 18.5 The parabola $y^2 = 4ax$ for $a = 1, 3, -1, -3$.

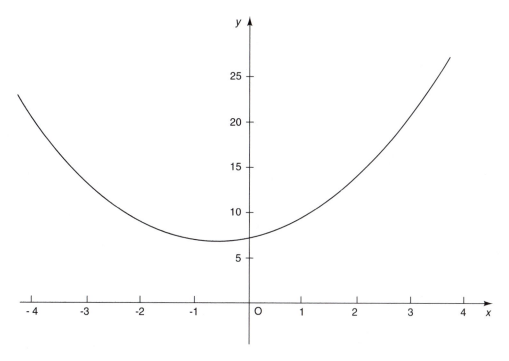

Figure 18.6 The parabola $y=x^2+x+7$.

Example 18.7

(a) Draw the parabola $y=2x^2-3x-4$ for $-3\leqslant x\leqslant4$.

(b) Use your graph to find approximate solutions to

$$2x^2-3x-4=0$$

Solution (a) A table of values is drawn up as in Table 18.1. Points are plotted and joined by a smooth curve to produce the curve shown in Figure 18.7.

Table 18.1

x	-3	-2	-1	0	1	2	3	4
y	23	10	1	-4	-5	-2	5	16

(b) When the parabola cuts the x axis, $y=0$; that is,

$$2x^2-3x-4=0$$

The x values found from the graph are $x=-0.9$ and $x=2.4$. These are approximate solutions to the equation.

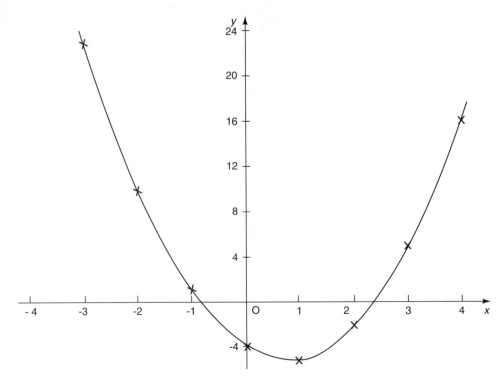

Figure 18.7   The parabola $y = 2x^2 - 3x - 4$.

18.4.2 The hyperbola

The **hyperbola** has an equation of the form

$$\frac{x^2}{a^2} - \frac{y^2}{b^2} = 1$$

A typical hyperbola is shown in Figure 18.8. When $a = b$, the equation simplifies to

$$x^2 - y^2 = a^2$$

Such a hyperbola is known as a **rectangular hyperbola**.

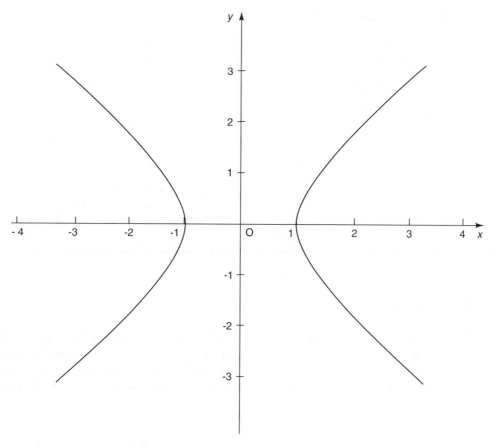

Figure 18.8 A typical hyperbola.

Self-assessment questions 18.4

1. State one property of a circle that distinguishes it from a hyperbola and a parabola.
2. State one property of a hyperbola that distinguishes it from a parabola.

Exercises 18.4

1. Calculate where the parabola $y = x^2 - 4x + 1$ cuts (a) the x axis (b) the y axis

2. The rectangular hyperbola $x^2 - y^2 = 2$ intersects the circle $x^2 + y^2 = 16$. Calculate the points of intersection.

Computer and calculator exercises 18.4

1.   A parabola has the equation

$$y = ax^2 + bx + c$$

Investigate the effect of
(a) changing the value of a, keeping b and c fixed;
(b) changing the value of b, keeping a and c fixed;
(c) changing the value of c, keeping a and b fixed.

2.   Plot the hyperbola

$$\frac{x^2}{a^2} - \frac{y^2}{b^2} = 1$$

for (a) $a=2$, $b=1$ (b) $a=1$, $b=2$
(c) $a=b=1$ (d) $a=b=2$

3.   The equation of a hyperbola is

$$\frac{x^2}{a^2} - \frac{y^2}{b^2} = 1$$

Investigate the effect of
(a) changing the value of a for fixed b;
(b) changing the value of b for fixed a.

18.5 Polar coordinates

This section introduces polar coordinates. It shows how to convert from cartesian coordinates to polar coordinates and vice versa. Consider Figure 18.9 which shows a point P in the (x, y) plane. We can describe the position of P by stating its x and y coordinates, that is, its cartesian coordinates. An alternative

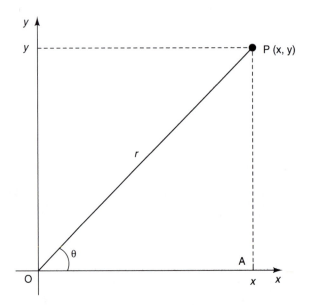

Figure 18.9   The polar coordinates of P are r and θ.

way of describing the position of P is to give the angle θ and the distance r. The values r and θ are called the **polar coordinates** of P.

The angle θ is the angle between the positive x axis and the arm OP. We use the usual convention that positive angles are measured anticlockwise and negative angles are measured clockwise. The angle is measured so that $-\pi < \theta \leqslant \pi$ using radians, or $-180° < \theta \leqslant 180°$ using degrees.

The length of the arm OP is denoted by r. It is always positive. When writing the polar coordinates of a point, the value of r is given first. For example, we write $3 \angle 2$ to mean $r = 3$ and $\theta = 2$ radians, and $2 \angle 30°$ to mean $r = 2$ and $\theta = 30°$. If the degree symbol $°$ is not used, the angle is taken to be in radians.

The polar coordinates of a point P are $r \angle \theta$. The value of r is the distance OP. The angle θ is the angle between the positive x axis and OP.

Example 18.8

Plot the points whose polar coordinates are (a) $2 \angle 90°$ (b) $2 \angle 146°$ (c) $3 \angle -\frac{1}{2}\pi$ (d) $2 \angle -1.75$

Solution   Figure 18.10 shows the positions of the four points.

18.5.1 Conversion between cartesian and polar coordinates

Consider the point P as shown in Figure 18.9. It has cartesian coordinates (x, y) and polar coordinates $r \angle \theta$. In $\triangle$OAP, OA $= x$ and AP $= y$. Now

$$\cos\theta = \frac{\text{OA}}{\text{OP}} = \frac{x}{r}$$

and so

$$x = r\cos\theta$$

Also

$$\sin\theta = \frac{\text{AP}}{\text{OP}} = \frac{y}{r}$$

and so

$$y = r\sin\theta$$

Hence, given polar coordinates $r \angle \theta$, we can calculate cartesian coordinates

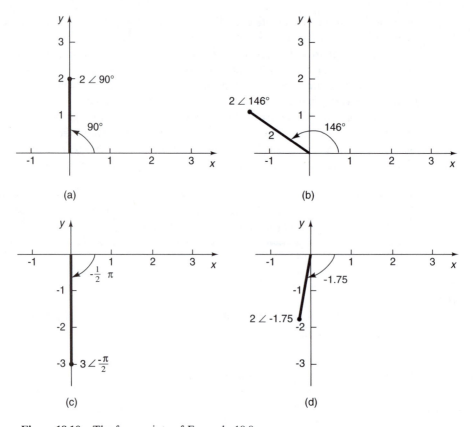

Figure 18.10 The four points of Example 18.8.

(x, y) by using

$$x = r \cos \theta, \quad y = r \sin \theta$$

We can rearrange these equations to make r and θ subject.

$$x = r \cos \theta \tag{18.1}$$

$$y = r \sin \theta \tag{18.2}$$

Squaring Equations (18.1) and (18.2), we obtain

$$x^2 = r^2 \cos^2 \theta \tag{18.3}$$

$$y^2 = r^2 \sin^2 \theta \tag{18.4}$$

Adding Equations (18.3) and (18.4) yields

$$x^2 + y^2 = r^2 \cos^2 \theta + r^2 \sin^2 \theta = r^2(\cos^2 \theta + \sin^2 \theta) = r^2$$

using the trigonometric identity $\sin^2\theta + \cos^2\theta = 1$. Hence

$$r = \sqrt{x^2 + y^2}$$

By dividing Equation (18.2) by Equation (18.1), we obtain

$$\frac{y}{x} = \frac{r\sin\theta}{r\cos\theta} = \tan\theta$$

so that

$$\theta = \tan^{-1}\left(\frac{y}{x}\right)$$

When calculating θ, care must be taken to ensure it is located in the correct quadrant.

Hence, given cartesian coordinates (x, y), we can calculate polar coordinates $r\angle\theta$ using

$$r = \sqrt{x^2 + y^2}, \qquad \theta = \tan^{-1}\left(\frac{y}{x}\right)$$

Example 18.9

Calculate the cartesian coordinates of the following points:
(a) $3\angle 2$ (b) $4\angle 0.7$ (c) $1\angle 180°$

Solution   Note that the angles in (a) and (b) are given in radians. We use $x = r\cos\theta$, $y = r\sin\theta$.

(a)   We have $r = 3$, $\theta = 2$, and so

$$x = 3\cos 2 = -1.25, \qquad y = 3\sin 2 = 2.73$$

The cartesian coordinates are $(-1.25, 2.73)$.

(b)   Here $r = 4$, $\theta = 0.7$, and so

$$x = 4\cos 0.7 = 3.06, \qquad y = 4\sin 0.7 = 2.58$$

The cartesian coordinates are $(3.06, 2.58)$.

(c)   Here $r = 1$, $\theta = 180°$, and so

$$x = 1\cos 180° = -1, \qquad y = 1\sin 180° = 0$$

The cartesian coordinates are $(-1, 0)$.

Example 18.10

Calculate the polar coordinates of the following points: (a) $(3, 4)$ (b) $(-2, 1)$
(c) $(-2, -3)$

Solution (a) Here $x=3$, $y=4$, and so

$$r=\sqrt{3^2+4^2}=\sqrt{25}=5$$

Now

$$\theta=\tan^{-1}\left(\frac{y}{x}\right)=\tan^{-1}\tfrac{4}{3}$$

Since $\tan\theta$ is positive, θ could be in the first or third quadrant. At this point, we note that (3, 4) is in the first quadrant, and so θ will be in the range 0 to $\frac{1}{2}\pi$. Using a calculator, we have

$$\theta=\tan^{-1}\tfrac{4}{3}=0.927$$

The required polar coordinates are $5\angle0.927$.

(b) Here $x=-2$, $y=1$, and so

$$r=\sqrt{(-2)^2+1^2}=\sqrt{5}$$

Also

$$\theta=\tan^{-1}\left(\frac{y}{x}\right)=\tan^{-1}\left(\frac{1}{-2}\right)=\tan^{-1}(-0.5)$$

Since $\tan\theta$ is negative, θ is in either the second or the fourth quadrant. The point $(-2, 1)$ is in the second quadrant, and so θ will be in the range $\frac{1}{2}\pi$ to π. Using a calculator and taking care to use the appropriate quadrant, we find

$$\theta=\tan^{-1}(-0.5)=2.678$$

The polar coordinates are $\sqrt{5}\angle2.678$.

(c) Here $x=-2$, $y=-3$, and so

$$r=\sqrt{(-2)^2+(-3)^2}=\sqrt{13}$$

Also

$$\theta=\tan^{-1}\left(\frac{y}{x}\right)=\tan^{-1}\left(\frac{-3}{-2}\right)=\tan^{-1}1.5$$

Since $\tan\theta$ is positive, θ is in the first or third quadrant. The point $(-2, -3)$ is in the third quadrant. Hence θ is in the third quadrant. We find

$$\tan^{-1}1.5=4.124$$

However, it is usual to express θ as an angle between $-\pi$ and π. Hence

$$\theta=4.124-2\pi=-2.159$$

Figure 18.11 illustrates the situation. The polar coordinates are $\sqrt{13}\angle-2.159$.

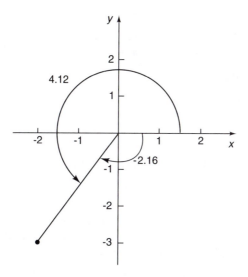

Figure 18.11 θ is measured anticlockwise.

Self-assessment questions 18.5

1. Describe what is meant by polar coordinates.

2. What is the usual range when giving values of θ in polar coordinates?

Exercises 18.5

1. Determine the cartesian coordinates of the following points:
    (a) $5\angle 1$ (b) $1\angle 5$ (c) $2\angle -1$
    (d) $3\angle 2.6$ (e) $4\angle -2$ (f) $3\angle 40°$
    (g) $10\angle 100°$ (h) $3\angle 70°$ (i) $4\angle -30°$

2. In which quadrants do the following points lie?
    (a) $(3, -2)$ (b) $5\angle -1$ (c) $(-2, 3)$

    (d) $200\angle 1.7$ (e) $0.16\angle -2$
    (f) $(-1, 10\,000)$

3. Determine the polar coordinates of the following points:
    (a) $(3, 7)$ (b) $(-2, 5)$ (c) $(-4, -2)$
    (d) $(7, -4)$ (e) $(3, 0)$ (f) $(-3, 0)$
    (g) $(0, 3)$ (h) $(0, -3)$

18.6 Parametric form of the equations of the conics

Curves are often expressed in the form $y = f(x)$. For every value of x, the corresponding y value can be found and the point (x, y) plotted. Sometimes it is useful to express the x and y coordinates in terms of a third variable, known as a **parameter**. Commonly we use t or θ to denote a parameter. Thus the coordinates (x, y) of the points on a curve may be expressed in the form

$$x = f(t), \quad y = g(t)$$

These are **parametric equations**.

Example 18.11

Sketch the curve defined parametrically by

$$x = t^2, \quad y = 2t \quad 0 \leqslant t \leqslant 5$$

Solution

We calculate x and y values for various values of the parameter t. These are shown in Table 18.2. Plotting the points (x, y) and joining by a smooth curve produces the graph shown in Figure 18.12.

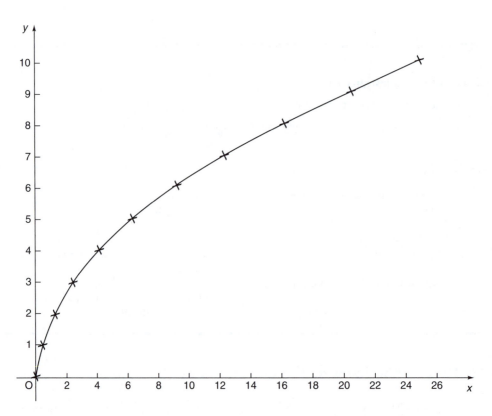

Figure 18.12   Each point on the curve corresponds to a value of the parameter t.

Table 18.2

t	0	0.5	1	1.5	2	2.5	3	3.5	4	4.5	5
x	0	0.25	1	2.25	4	6.25	9	12.25	16	20.25	25
y	0	1	2	3	4	5	6	7	8	9	10

By eliminating t, it is possible to write down the equation of the curve in terms of x and y.

Example 18.12

A curve is defined parametrically by

$$x = t^2, \qquad y = 2t$$

Express the equation of the curve using only x and y.

Solution

The parameter t must be eliminated. Now

$$y = 2t$$

and so

$$t = \tfrac{1}{2}y$$

Hence

$$x = t^2 = (\tfrac{1}{2}y)^2 = \tfrac{1}{4}y^2$$

and finally

$$y^2 = 4x$$

We recognize this as the equation of a parabola.

It is possible to write the equations of all the conic sections in parametric form.

18.6.1 Parametric equation of a circle

Consider a circle, centre O, radius r. The equation of such a circle is

$$x^2 + y^2 = r^2$$

This equation may be expressed parametrically as

$$x = r \cos \theta, \quad y = r \sin \theta \quad 0° \leqslant \theta \leqslant 360°$$

Note that r is a constant and the parameter is θ. Clearly the parametric form uses the polar coordinates of points on the circle. For every value of θ the point $(r \cos \theta, r \sin \theta)$ lies on the circle. As θ varies from $0°$ to $360°$, all the points on the circle are generated.

Let us now consider a circle with centre (a, b), radius r. The equation of this circle is

$$(x - a)^2 + (y - b)^2 = r^2$$

In parametric form this is

$$x = a + r \cos \theta, \quad y = b + r \sin \theta \quad 0° \leqslant \theta \leqslant 360°$$

The parametric form of a circle, centre (a, b), radius r is

$$x = a + r \cos \theta, \quad y = b + r \sin \theta \quad 0° \leqslant \theta \leqslant 360°$$

18.6.2 Parametric form of an ellipse

The equation of an ellipse is

$$\frac{x^2}{a^2} + \frac{y^2}{b^2} = 1$$

Parametrically, the equation is expressed as

$$x = a \cos \theta, \quad y = b \sin \theta \quad 0° \leqslant \theta \leqslant 360°$$

As θ varies, every point on the ellipse is generated.

18.6.3 Parametric form of a parabola

The equation of a parabola $y^2 = 4ax$ is expressed parametrically as

$$x = at^2, \quad y = 2at \quad -\infty < t < \infty$$

18.6.4 Parametric form of a hyperbola

The equation of a hyperbola is

$$\frac{x^2}{a^2} - \frac{y^2}{b^2} = 1$$

The parametric form of this equation is

$$x = a \sec \theta, \qquad y = b \tan \theta \qquad 0° \leqslant \theta \leqslant 360°$$

Recall that $\sec \theta = 1/\cos \theta$. As θ varies, every point on the hyperbola is generated.

Self-assessment question 18.6

1.   Explain what is meant by a parameter. Which symbols are commonly used to represent a parameter?

Exercises 18.6

1.   Verify that the point $(a \cos \theta, b \sin \theta)$ lies on the ellipse

$$\frac{x^2}{a^2} + \frac{y^2}{b^2} = 1$$

2.   Verify that the point $(a \sec \theta, b \tan \theta)$ lies on the hyperbola

$$\frac{x^2}{a^2} - \frac{y^2}{b^2} = 1$$

Computer and calculator exercises 18.6

Some graphics calculators and graph-plotting packages are able to plot graphs described in parametric form. Check whether your calculator or package can do this.

1.   Plot $x = 5t^2$, $y = 10t$ for $t = -3$ to $t = 3$. Describe the curve that is generated.

2.   Plot $x = 3 \cos \theta$, $y = 3 \sin \theta$ for $0° \leqslant \theta \leqslant 90°$. Describe the curve that is generated.

3.   Plot $x = 2 \sec \theta$, $y = \tan \theta$ for $0° \leqslant \theta \leqslant 360°$. Describe the curve that is generated.

4.   Plot the curve defined parametrically by

$$x = 1 + t^2, \qquad y = 2 + t \qquad |t| \leqslant 5$$

Is the curve a conic section? If so, which one?

5.   (a)   Plot $x = \frac{1}{3}t$, $y = t^2$ for $t = 0$ to $t = 3$. Is the curve a conic section? If so, which one?

(b)   Plot $x = a + \frac{1}{3}t$, $y = t^2$ for $t = 0$ to $t = 3$ for various values of a. What do you notice?

(c)   Plot $x = \frac{1}{3}t$, $y = b + t^2$ for $t = 0$ to $t = 3$ for various values of b. What do you notice?

6.   The curve with parametric equations

$$x = \theta - \sin \theta$$

$$y = 1 - \cos \theta$$

where θ is measured in radians is called a **cycloid**. This is the path traced out by a point on the circumference of a wheel rolling along the ground. Use a graphics calculator to plot the graph of this cycloid for $0 \leqslant \theta \leqslant 3\pi$.

Review exercises 18

1 Calculate the centre and radius of the following circles:
(a) $x^2 + y^2 - 4x - 4y + 2 = 0$
(b) $x^2 + y^2 + 4x - 6y + 1 = 0$
(c) $x^2 + y^2 = 10$ (d) $x^2 + y^2 + 6x - 3 = 0$

2 A circle has centre (2, 1) and radius 4.
(a) State the equation of the circle.
(b) Calculate where the circle cuts the x axis.
(c) Calculate where the circle cuts the y axis.

3 Calculate the polar coordinates of the following points:
(a) (1, 1)   (b) (1, −1)   (c) (−2, 2)
(d) (0, 3)   (e) (−2, 0)

4 Calculate the cartesian coordinates of the following points:
(a) $3\angle 1$ (b) $2\angle -30°$ (c) $4\angle 1.5$

(d) $4\angle -1.5°$ (e) $\frac{1}{2}\angle 120°$

5 Find the points of intersection of the line $y = -x + 2$ and the circle $x^2 + y^2 = 9$.

6 (a)   Plot the rectangular hyperbola
$$x^2 - y^2 = 1$$

(b)   Find the points of intersection of the rectangular hyperbola with the circle $x^2 + y^2 = 3$.

(c)   Find the points of intersection of the rectangular hyperbola with the ellipse
$$x^2 + \frac{y^2}{2} = 1$$

7 Determine the points of intersection of the line $y = x$ and the hyperbola $x^2 - \frac{y^2}{4} = 1$

19

Mensuration

KEY POINTS

This chapter

- states formulae for calculating the area of many common shapes: a rectangle, a triangle, a circle, a trapezium, a parallelogram, a sector and a segment

- states formulae for calculating the surface area of a sphere, a cylinder and a cone

- states formulae for calculating the volume of a sphere, a cylinder, a cone and a cuboid

CONTENTS

19.1 Introduction

Mensuration is the study of areas of surfaces and volumes of bodies. Section 19.2 looks at several shapes and shows how to calculate their area. Sections 19.3 and 19.4 examine the surface areas and volumes of some three-dimensional bodies.

19.2 Areas of common shapes

In this section we calculate the areas of various common shapes.

Square

> **A square with sides of length l has area**
>
> $$A = l^2$$

Example 19.1

Calculate the area of a square with sides of length 3 m.

Solution Here $l = 3$, and so

$$A = l^2 = 3^2 = 9$$

The area of the square is 9 m^2.

Rectangle

> **A rectangle with length l and width w has area**
>
> $$A = lw$$

Example 19.2

Calculate the area of a rectangle that measures 9 m by 7 m.

Solution Here $l = 9$ and $w = 7$, and hence

$$A = lw = (9)(7) = 63$$

The area of the rectangle is 63 m^2.

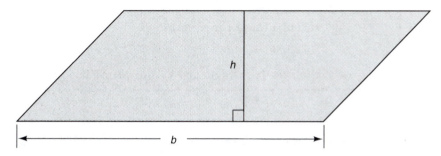

Figure 19.1 The area of a parallelogram is bh.

Parallelogram

Figure 19.1 shows a typical parallelogram.

> The area of a parallelogram is given by
>
> $A = bh$
>
> where b is the base length and h is the perpendicular height.

Example 19.3

Calculate the area of a parallelogram with base length 12 cm and height 6 cm.

Solution Here $b = 12$ and $h = 6$, and so

$$A = bh = (12)(6) = 72$$

The area of the parallelogram is 72 cm^2.

Triangle

Figure 19.2 shows a triangle.

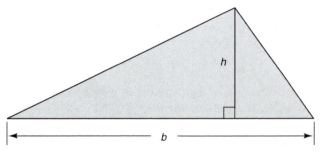

Figure 19.2 The area of a triangle is $\frac{1}{2}bh$.

> The area of a triangle is given by
>
> $A = \frac{1}{2}bh$
>
> where b is the base length and h is the perpendicular height.

Example 19.4

 A rectangular piece of metal ABCD, as shown in Figure 19.3, measures 90 mm by 70 mm. Calculate the area of the triangle ABC.

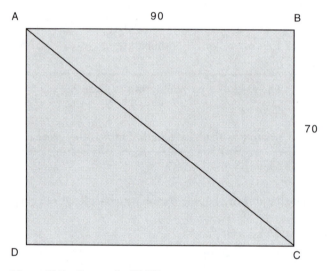

Figure 19.3   Rectangle ABCD.

Solution   The base length AB = 90 mm. The perpendicular height is BC = 70 mm. The area, A, is then

$A = \frac{1}{2}90(70) = 3150$

The area of △ABC is 3150 mm².

Example 19.5

 Figure 19.4 shows a triangular piece of plastic, ABC, with $A = 25°$, AB = 17 cm and AC = 36 cm. Calculate the area of △ABC.

Solution   The base of the triangle is AC, and this has length 36 cm. The height of the triangle is BD. To calculate BD, we consider the right-angled triangle ABD:

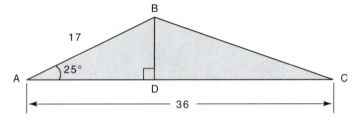

Figure 19.4   △ABC for Example 19.5.

$$\frac{BD}{AB} = \sin A$$

$$BD = AB \sin A = 17 \sin 25°$$

The area of triangle ABC is

$$A = \tfrac{1}{2}bh = \tfrac{1}{2}(36)(17 \sin 25°) = 129.3$$

The area of the triangle is 129.3 cm².

The previous example illustrates a general formula for the area of a triangle. If we know the length of two sides, say $c = AB$ and $b = AC$, and the included angle A, then the area is given by

$$\text{area} = \tfrac{1}{2}bc \sin A$$

Similarly, the area is also found from $\tfrac{1}{2}ac \sin B$ and $\tfrac{1}{2}ab \sin C$.

> **The area of a triangle is given by**
>
> $$\text{area} = \tfrac{1}{2}bc \sin A = \tfrac{1}{2}ac \sin B = \tfrac{1}{2}ab \sin C$$

In addition, the area of a triangle may also be calculated from knowledge of the lengths of the sides. Suppose the lengths are known to be a, b and c. Then the area A is given by

> $$A = \sqrt{s(s-a)(s-b)(s-c)}$$
>
> where $s = \tfrac{1}{2}(a+b+c)$.

Example 19.6

A triangular metal template ABC has dimensions $c = AB = 9$ cm, $b = AC = 12$ cm and $a = BC = 11$ cm. Calculate the area of △ABC.

Solution   Here $a+b+c=9+12+11=32$ cm. Thus $s=32/2=16$ cm. The area A is now found:

$$A = \sqrt{s(s-a)(s-b)(s-c)}$$

$$= \sqrt{16(16-11)(16-12)(16-9)}$$

$$= \sqrt{(16)(5)(4)(7)}$$

$$= 47.3$$

The area of the triangle is 47.3 cm².

Circle

A circle with radius r and diameter $d=2r$ has area A given by

$$A = \pi r^2 = \tfrac{1}{4}\pi d^2$$

Example 19.7

A circular disc of aluminium has radius 60 mm. Calculate its area.

Solution   Here we have $r=60$, so

$$A = \pi r^2 = \pi(60)^2 = 11\,309.7$$

The area of the circle is 11 309.7 mm².

Example 19.8

A ring is made by cutting away a circle of radius 3 cm from a circle of radius 7 cm. Calculate the area of the ring.

Solution   area of circle of radius 7 cm $= \pi r^2 = \pi(7^2) = 49\pi$

area of circle of radius 3 cm $= \pi r^2 = \pi(3^2) = 9\pi$

area of ring $= 49\pi - 9\pi = 125.7$

The ring has an area of 125.7 cm².

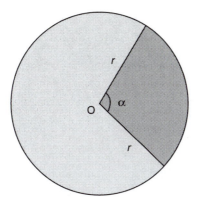

Figure 19.5 The deeply shaded area is the sector of a circle.

Sector of a circle

Figure 19.5 shows the sector of a circle, centre O. The radius of the circle is r and the angle subtended at the centre is α radians. The area of the sector depends upon r and α:

> area of sector $= \frac{1}{2}\alpha r^2$,   with α in radians

Example 19.9

A circular plastic plate has a radius of 50 mm. A sector subtends an angle of 1.7 radians at the centre. Calculate the area of the sector.

Solution   Here we have $r = 50$ and $\alpha = 1.7$. Hence

$$\text{area of sector} = \tfrac{1}{2}\alpha r^2 = \tfrac{1}{2} \times 1.7 \times 50^2 = 2125$$

The area of the sector is 2125 mm^2.

Segment of a circle

A circle, centre O, has a radius r. Referring to Figure 19.6, the straight line joining A and B is called a **chord**. The chord AB subtends an angle α radians at O. The deeply shaded area in Figure 19.6 is known as a **segment**. Note that it is distinct from a sector. The area of a segment is given by

> area of segment $= \frac{1}{2}r^2(\alpha - \sin\alpha)$,   with α in radians

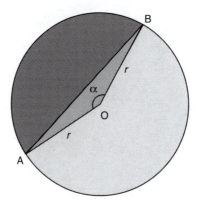

Figure 19.6   The deeply shaded area is a segment.

Example 19.10

A circle has a radius of 12 cm. The chord AB subtends an angle of 70° at the centre. Calculate the area of the segment.

Solution   To use the formula, angles must first be expressed in radians. The angle at the centre is $70° = \frac{70}{180}\pi = 0.3889\pi = 1.2217$ radians. The area of the segment is

$$A = \tfrac{1}{2}r^2(\alpha - \sin \alpha)$$

$$= \tfrac{1}{2} \times 12^2(1.2217 - \sin 1.2217)$$

$$= 20.31$$

The area of the segment is 20.31 cm².

Trapezium

A trapezium is a four-sided figure with one pair of sides parallel. Figure 19.7 shows a trapezium ABCD with AB and CD parallel. If *x* and *y* are the lengths

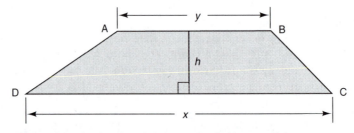

Figure 19.7   A trapezium has one pair of sides parallel.

of the parallel sides and the perpendicular height is h then the area is given as follows:

$$\text{area of trapezium} = \tfrac{1}{2}(x+y)h$$

The area of a trapezium is half the sum of the parallel sides multiplied by the perpendicular distance between them.

Example 19.11

In Figure 19.7, $AB = 6$ cm, $DC = 20$ cm and the height h is 7 cm. Calculate the area of the trapezium.

Solution   Here we have $x = 20$, $y = 6$, $h = 7$. The area A is therefore

$$A = \tfrac{1}{2}(20+6)7 = 91$$

The area of the trapezium is 91 cm^2.

Self-assessment questions 19.2

1. State the formula for the area of (a) a square   (b) a rectangle   (c) a trapezium   (d) a circle (e) a triangle   (f) a sector of a circle   (g) a segment of a circle

2. What is the distinction between a chord AB and an arc AB?

3. What is the distinction between a segment of a circle and a sector of a circle?

Exercises 19.2

1. Calculate the area of a circular disc of steel whose radius is (a) 160 cm   (b) 5 m

2. Calculate the area of a circle of radius 3 m. If the radius is doubled to 6 m, what is the area of the new circle?

3. A square piece of plastic has sides of length 200 mm. Calculate the radius of a circle with the same area as the square.

4. A square has side l m. Calculate the radius of a circle with the same area as the square.

5. An equilateral triangle is attached to one side of a square. All sides are of length 12 cm.

The situation is shown in Figure 19.8. Calculate the area of the figure.

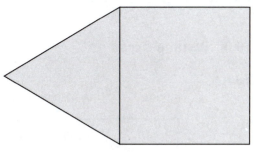

Figure 19.8   An equilateral triangle is attached to a square.

6. A circle, centre O, has a radius of 8 cm. Arc AB subtends an angle of 50° at O. The situation is shown in Figure 19.9.
   (a) Calculate the area of the circle.
   (b) Calculate the area of the sector AOBCA.
   (c) Calculate the area of the sector AOBDA.
   (d) Calculate the area of the segment ABC.

7. In Figure 19.10, ABCDE is a trapezium and ABCD is a parallelogram, with AB = 12 cm, CE = 20 cm and perpendicular height 7 cm.
   (a) Calculate the area of the trapezium ABCDE.
   (b) Calculate the area of △ADE.
   (c) Calculate the area of △ABD.
   (d) Calculate the area of △ABC.

8. Calculate the area of an equilateral triangle of side *l* cm.

9. Calculate the area of △ABC, given
   (a) $a = 12$ cm, $b = 17$ cm, $c = 20$ cm
   (b) $a = b = c = 15$ cm
   (c) $a = 10$ cm, $b = 15$ cm, $\angle C = 37°$
   (d) $b = 6$ cm, $c = 10$ cm, $\angle A = 120°$

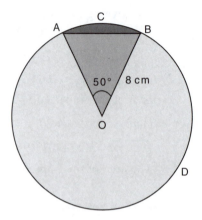

Figure 19.9   Figure for Exercise 19.2 Q6.

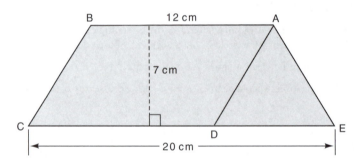

Figure 19.10   The trapezium for Exercise 19.2 Q7.

19.3 Surface areas

Sphere

Figure 19.11 shows a sphere of radius *r*.

> A sphere of radius *r* has surface area *A* given by
>
> $$A = 4\pi r^2$$

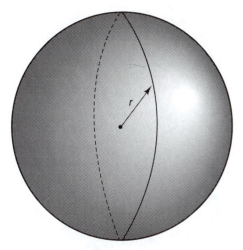

Figure 19.11 A sphere of radius r.

Example 19.12

Calculate the surface area of a sphere of radius 2 m.

Solution

$$A = 4\pi r^2 = 4\pi(2^2) = 16\pi = 50.27$$

The surface area is 50.27 m².

Cylinder

An open-ended cylinder is shown in Figure 19.12. It has radius r and length l. The curved surface can be thought of as a rectangular sheet that has been bent so that a pair of edges meet. The length of the rectangular sheet is l and the width equals the circumference, that is, $2\pi r$. The curved surface area S is therefore

$$S = (\text{circumference}) \times (\text{length})$$

$$= 2\pi r l = \pi d l$$

where the diameter $d = 2r$. If the cylinder has ends then these areas must be included. Each end is a circle, and so has an area of πr^2. We summarize as follows:

> surface area of cylinder with no ends $= 2\pi r l = \pi d l$
>
> surface area of cylinder with one end $= 2\pi r l + \pi r^2 = \pi d l + \frac{1}{4}\pi d^2$
>
> surface area of cylinder with two ends $= 2\pi r l + 2\pi r^2 = \pi d l + \frac{1}{2}\pi d^2$

Figure 19.12 An open-ended cylinder of radius *r* and length *l*.

Example 19.13

An open-ended cylindrical copper tube has a radius of 40 mm and a length of 750 mm. Calculate the surface area of the tube.

Solution   Here $r = 40$ and $l = 750$, so the surface area is given by

$$\text{area} = 2\pi rl$$
$$= 2\pi(40)(750)$$
$$= 188\,496$$

The surface area of the tube is $188\,496$ mm^2.

Example 19.14

A lorry has a closed cylindrical chemical storage tank of length 4 m and diameter 1.5 m. These are external dimensions. Calculate the surface area of the tank.

Solution   The storage tank is a cylinder with two ends. We have $d = 1.5$ and $l = 4$. Hence the area is found as

$$\text{area} = \pi dl + \tfrac{1}{2}\pi d^2$$
$$= \pi(1.5)(4) + \tfrac{1}{2}\pi(1.5)^2$$
$$= 6\pi + 1.125\pi$$
$$= 7.125\pi = 22.38$$

The surface area of the tank is 22.38 m^2.

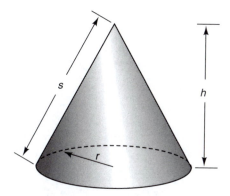

Figure 19.13 A cone of base radius r, perpendicular height h and slant height s.

Cone

A typical cone is illustrated in Figure 19.13. The radius of the base is r, the perpendicular height is h and the slant height is s. The base of the cone is a circle, and so has an area of πr^2. The curved surface area is πrs.

---

area of base of cone $= \pi r^2$

curved surface area $= \pi rs$

Example 19.15

A cone has a base radius of 10 cm and a perpendicular height of 15 cm. Calculate the curved surface area of the cone.

Solution

Here $r = 10$ and $h = 15$. The slant height s is found using Pythagoras' theorem. Figure 19.14 illustrates the relevant right-angled triangle. Using Pythagoras' theorem, we have

$$s^2 = h^2 + r^2$$

$$= 15^2 + 10^2$$

$$= 325$$

$$s = 18.03$$

The curved surface area is

$$\text{area} = \pi rs = \pi(10)(18.03) = 566.4$$

The curved surface area is 566.4 cm^2.

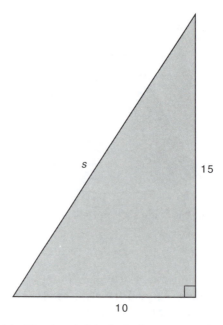

Figure 19.14 The slant height is the hypotenuse of the triangle.

Self-assessment questions 19.3

1. State the formula for the surface area of a sphere.

2. State the formula for the surface area of an open-ended cylinder.

3. State the formula for the curved surface area of a cone.

Exercises 19.3

1. A cylindrical copper tube with one closed end has a length of 5 m and an external diameter of 12 cm. Calculate the external surface area of the tube.

2. Calculate the surface area of a sphere whose diameter is 14 cm.

3. An open-ended steel pipe has a length of 4 m and an external diameter of 300 mm. Calculate the external surface area of the pipe.

4. A cone has base radius of 0.5 m and a perpendicular height of 3.2 m. Calculate the curved surface area.

5. Calculate the surface area of a sphere of radius 0.7 m.

6. A cylindrical cup has a radius of 45 mm and a height of 150 mm. Calculate the external surface area of the cup.

19.4 Volumes of common bodies

This section examines the volumes of some common bodies.

Sphere

> A sphere of radius r and diameter $d = 2r$ has volume V given by
>
> $$V = \tfrac{4}{3}\pi r^3 = \tfrac{1}{6}\pi d^3$$

Example 19.16

A ball bearing has a diameter of 20 mm. Calculate the volume of steel in the bearing.

Solution   Here $d = 20$, and so the volume is found as

$$V = \tfrac{1}{6}\pi d^3 = \tfrac{1}{6}\pi(20)^3 = \frac{8000}{6}\pi = \frac{4000}{3}\pi = 4189$$

The volume of steel is 4189 mm³.

Cylinder

> A cylinder of radius r, diameter $d = 2r$ and height h has a volume
>
> $$V = \pi r^2 h = \tfrac{1}{4}\pi d^2 h$$

Example 19.17

A cylinder has radius 9 cm and height 20 cm. Calculate the volume of the cylinder.

Solution   Here $r = 9$, $h = 20$, and so

$$V = \pi r^2 h = \pi(9^2)(20) = \pi(81)(20) = 5089$$

The volume of the cylinder is 5089 cm³.

Example 19.18

A cylindrical chemical storage tank has an internal length of 6 m and an internal diameter of 2 m. Calculate the storage capacity of the tank.

Solution   Here $h=6$ and $d=2$. The volume is found as

$$V = \tfrac{1}{4}\pi d^2 h = \tfrac{1}{4}\pi 2^2 6 = 6\pi = 18.85$$

The storage capacity of the tank is 18.85 m³.

Cone

Figure 19.13 shows a typical cone, with perpendicular height h and radius r.

> **A cone of base radius r and perpendicular height h has volume**
>
> $$V = \tfrac{1}{3}\pi r^2 h$$

Example 19.19

A cone has a base radius of 11 cm and a height of 15 cm. Calculate the volume of the cone.

Solution   Here we are given that $r=11$ and $h=15$. The value of V can be found from the formula

$$V = \tfrac{1}{3}\pi r^2 h = \tfrac{1}{3}\pi (11^2)(15) = 1900.7$$

The volume of the cone is 1900.7 cm³.

Example 19.20

A conical feed to a process line is shown in Figure 19.15. The feed supplies flour to a dough mixing line. Calculate the maximum volume of flour that the feed can hold.

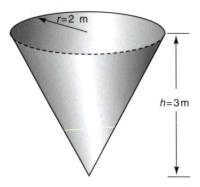

Figure 19.15   A conical feed as described in Example 19.20.

Solution   We have $h=3$ and $r=2$, and so the volume of the feed is

$$V = \tfrac{1}{3}\pi r^2 h = \tfrac{1}{3}\pi 2^2 3 = 4\pi = 12.57$$

The maximum volume of flour in the feed is 12.57 m^3.

Cuboid

A **cuboid** is a rectangular solid. Figure 19.16 shows a typical cuboid.

> **The volume of a cuboid is given by**
>
> $$V = (\text{width})(\text{height})(\text{length}) = whl$$

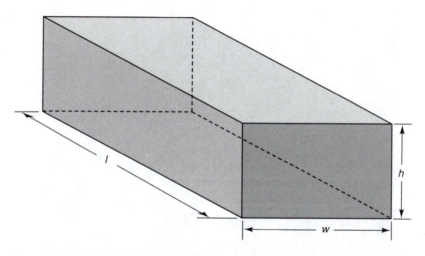

Figure 19.16   A cuboid, width w, height h and length l has volume whl.

Example 19.21

A steel ingot in the shape of a cuboid has length 3 m, width 0.5 m and height 0.45 m. Calculate the volume of steel in the ingot.

Solution   We have $w=0.5$, $h=0.45$ and $l=3$. Thus

$$V = whl = (0.5)(0.45)(3) = 0.675$$

The volume of steel in the ingot is 0.675 m^3.

Computer-controlled coordinate measuring machines

Over the past decade, the use of computers in manufacturing industry has steadily increased, and has been well publicized, for example, in the case of robot welders in car plants. One aspect of computerization that has not received much public attention is the use of computer-assisted or computer-controlled coordinate measuring machines (CMM).

The combination of computers and CMMs has revolutionized quality control in metalworking. A CMM consists of a means of moving a probe within a three-dimensional rectangular coordinate system. In some CMMs the probe is attached to rollers that have an electric current passing through them. A deflection of the probe, made when contact with the manufactured component is established, causes a change of resistance, which is used to cause a trigger signal. This enables the spatial coordinates of selected contact points to be accurately recorded.

Repeatable measurements of 0.0005 mm can be achieved with such systems. Noncontact CMMs use lasers to detect faults on castings, to check surface conditions and tool wear. Often different inspection requirements necessitate the use of a variety of probes. Sophisticated CMMs have the ability to change the probe automatically. Some CMMs are suitable for inspecting components during manufacture, others are used for post-process inspection. To avoid errors due to vibration CMMs require special foundations, often several metres deep. The figures show some probes in current use.

Figure 1 *A three-dimensional probe head.*

A probe can be manually operated with spatial coordinates fed into a computer by an operator (computer-assisted CMM) or have all movement, data recording and data processing controlled by computer (directly computer-controlled CMM).

Various physical arrangements for CMMs have evolved, for example bridge, cantilever, column, gantry and horizontal. Each arrangement has certain advantages that suit it to a particular need; for example, the gantry arrangement is suitable for large components and comprises a vertical column with probe attachments moving on a mutually perpendicular beam (*y* axis), which in turn is fixed to two columns that move along base rails (*x* axis).

The advantages of CMMs over traditional manual measuring techniques are flexibility, speed of measurement and improved accuracy. Some measuring tasks can be performed by a CMM in a few seconds rather than the several minutes needed for manual inspection. In addition, a CMM can be used in hazardous environments unsafe for a human operator.

For further reference see *Total Quality Control and JIT Management in CIM* by P. Ranky, and the *Journal of Quality Assurance*, vol. II, September 1985, pp. 74–5.

Figure 2 *Measuring the dimensions of a gear.*

Self-assessment questions 19.4

1.   The volume of a cuboid may be calculated using

$$\text{volume} = \text{cross-sectional area} \times \text{length}$$

True or false?

2.   The volume of a cylinder may be calculated using

$$\text{volume} = \text{cross-sectional area} \times \text{length}$$

True or false?

Exercises 19.4

1.   Calculate the volume of a sphere whose radius is (a) 3 cm   (b) 17 cm (c) 21 mm   (d) 2.3 m

2.   A cylinder has height 1.3 m and radius 250 mm. Calculate the volume of the cylinder.

3.   A cylinder has length 2 m and radius 4 cm. Calculate the volume of the cylinder.

4.   Calculate the volume of a cone whose base radius is 9 cm and whose perpendicular height is 12 cm.

5.   The slant height of a cone is equal to its base diameter. If the perpendicular height is 350 mm, calculate the volume of the cone.

6.   A cuboid has a square cross-section and its length is three times its width. Calculate the volume of the cuboid if its height is 5 cm.

7.   A cube is a special case of a cuboid, with all the edges of equal length, that is, height, width and length are equal. If the length of one side of a cube is 0.14 m, calculate its volume.

8.   A spherical ball-bearing has diameter of 45 mm. Calculate the volume of steel in the bearing.

9.   A spherical chemical storage tank has an internal diameter of 6.4 m. Calculate the storage capacity of the tank.

10.   A cone has perpendicular height 1 m and base radius 0.12 m.
    (a) Calculate the area of the base of the cone.
    (b) Calculate the volume of the cone.

Review exercises 19

1   Calculate the volume of a sphere whose radius is (a) 0.7 cm   (b) 1.2 m   (c) 19 mm

2   An equilateral triangle has sides of length 25 cm. Calculate the area of the triangle.

3   A circle of radius r has the same area as a sphere of radius R. Deduce a relationship connecting r and R.

4   The diameter of a cylinder is a quarter of its

height. If the radius is 0.24 m, calculate the volume of the cylinder.

5 A sphere has radius 8 cm. The base radius of a cone is also 8 cm, and the sphere and cone have equal volumes. Calculate the height of the cone.

6 Figure 19.17 shows a circle inside a square. The side of the square is 10 cm. Calculate the area of the deeply shaded portion.

7 A solid steel cylinder of length 3 m and diameter 0.2 m is melted down to form 100 equal solid steel spheres. Calculate the radius of the spheres.

8 A cone and a sphere have the same volume. If the radius of the sphere is 9 cm and the base radius of the cone is 12 cm, calculate the height of the cone.

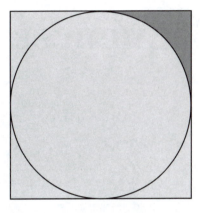

Figure 19.17   Figure for Review exercise 19 Q6.

20

Sequences and Series

KEY POINTS

This chapter

● explains the meaning of the term 'sequence' and introduces the symbols commonly used to denote a sequence

● explains the meaning of the terms 'arithmetic progression' and 'geometric progression' and shows how the terms of such progressions are calculated

● explains the meaning of the term 'series' and introduces the 'sigma' notation

● introduces arithmetic series and geometric series, and the formulae for their sums

● explains how to expand expressions of the form $(a+b)^n$ using Pascal's triangle

● introduces the binomial theorem

CONTENTS

20.1 Introduction

Sequences are important in engineering mathematics because they can be used to describe discrete time signals. These are signals that have a nonzero value only at certain specific and usually equally spaced instants in time. For example, in digital computers calculations are carried out at fixed intervals of time governed by an electronic clock. Series arise when the terms of a sequence are added. They are important because the solutions of some mathematical problems can be expressed as series.

20.2 Sequences

A **sequence** is a set of numbers written down in a specific order. For example,

1, 3, 5, 7, 9

$-1, -2, -3, -4, -5, -6, -7, -8$

are both sequences.

Each number in the sequence is called a **term** of the sequence. The number of terms in the first sequence given above is five, and the number of terms in the second sequence is eight. We can use fractions as terms of a sequence, for example

$1, \frac{1}{2}, \frac{1}{4}$ and   0.1, 0.2, 0.3, 0.4, 0.5

It does not matter if some of the terms are the same, for example 1, 0, 1, 0, 1, 0 is a sequence, despite the repetition. Sometimes we use the symbol '...' to indicate that the sequence continues. For example, the sequence 1, 2, 3,..., 20 is the sequence of integers from 1 to 20 inclusive. All of the sequences given above have a finite number of terms. They are known as **finite sequences**. Some sequences, though, go on forever, and these are called **infinite sequences**. To indicate that a sequence might go on forever, we can use the '...' notation. So, when we write

1, 3, 5, 7, 9,...

it can be assumed that this sequence continues indefinitely.

Very often you will be able to spot a rule that allows you to find the next term in a sequence. For example 1, 3, 5, 7, 9,... is a sequence of odd integers. The next term is 11. The next term in the sequence $1, \frac{1}{2}, \frac{1}{4},...$ is probably $\frac{1}{8}$.

20.2.1 Notation used for sequences

It is necessary to introduce a notation for handling sequences. Consider again the sequence of odd integers, 1, 3, 5, 7, 9,.... Suppose we let $f[1]$ stand for the first term, $f[2]$ stand for the second term, and so on. That is,

$f[1] = 1, \quad f[2] = 3, \quad f[3] = 5, \quad f[4] = 7, \quad ...$

A general term in the sequence is then referred to as $f[k]$, where k is 1, 2, 3, ... In some applications an alternative notation, using a subscript, is used. We write

$$f_1 = 1, \quad f_2 = 3, \quad f_3 = 5, \quad \text{and so on}$$

Suppose we were asked to write down the 1000th term in the sequence 1, 3, 5, 7, 9, ... Clearly it would be very laborious to write out all the terms up to the 1000th one. If we could deduce a formula from which the 1000th term could be found, this would be advantageous. From inspection of the sequence 1, 3, 5, 7, 9, ..., we find that

$$f[k] = 2k - 1$$

For example,

$$f[1] = 2(1) - 1 = 1 \quad \text{and} \quad f[5] = 2(5) - 1 = 9$$

and so on. It follows from this that

$$f[1000] = 2(1000) - 1 = 1999$$

20.2.2 Sequences as functions

We have seen that the sequence of odd integers 1, 3, 5, 7, 9, ... can be expressed using the formula

$$f[k] = 2k - 1, \quad \text{with } k = 1, 2, 3, \ldots$$

We can think of $f[k]$ as the function block shown in Figure 20.1. In this example the values allowed as input to the function are 1, 2, 3, ... Thus $f[k]$ can be thought of as a function whose domain is the set of positive integers. The range of this function is the set of terms in the sequence. The independent variable is k.

Functions were introduced in Chapter 6. Previously the domain was all values in a given interval. When dealing with sequences, the domain is restricted

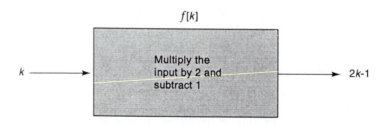

Figure 20.1   A sequence can be generated from a function block.

to integer values. It is in order to emphasize this important difference that we use the square-bracket notation for sequences.

In general, any sequence can be regarded as the output from a function whose input is a subset of the integers.

Example 20.1

The function f whose domain is the set of positive integers between 1 and 5 is defined by

$$f[k] = 3^k \qquad k = 1, 2, 3, 4, 5$$

Write down the range of this function.

Solution   If $f[k] = 3^k$ then

$$f[1] = 3^1 = 3, \qquad f[2] = 3^2 = 9, \qquad f[3] = 3^3 = 27,$$
$$f[4] = 3^4 = 81, \qquad f[5] = 3^5 = 243$$

The range of $f[k]$ is the set $\{3, 9, 27, 81, 243\}$. We see that the range of the function is the set of terms in the sequence 3, 9, 27, 81, 243.

Of course it is not necessary to always use f as the symbol for a function. This will be apparent from what follows.

20.2.3 Graphing a sequence

We can plot a graph of a sequence. Consider the following example.

Example 20.2

Plot a graph of the sequence

$$f[k] = 2k \qquad k = 0, 1, 2, 3, \ldots, 8$$

Solution   Using the definition $f[k] = 2k$, we calculate the terms of the sequence as follows (note that the domain given in this example is the set of integers between 0 and 8):

$$f[0] = 2(0) = 0, \qquad f[1] = 2(1) = 2, \qquad \text{and so on up to} \qquad f[8] = 16$$

Hence the sequence is 0, 2, 4, 6, 8, ..., 16. When plotting the graph, note that k is the independent variable and so the k axis is horizontal.

The graph of $f[k] = 2k$ is shown in Figure 20.2. Instead of a continuous line, the graph of a sequence is made up of distinct points.

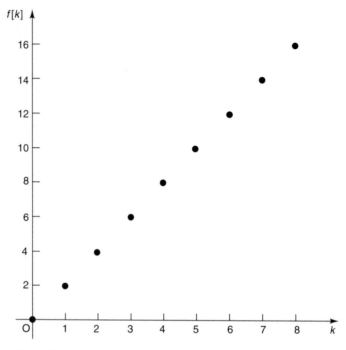

Figure 20.2 Graph of the sequence $f[k] = 2k$, $k = 0, 1, 2, \ldots, 8$.

Self-assessment questions 20.2

1. Explain what is meant by a sequence.

2. Write down a finite sequence and plot its graph.

3. Explain the distinction between an infinite sequence and a finite sequence.

Exercises 20.2

1. Write down the first five terms of the sequences given by the following, paying particular attention to the domain given:
   (a) $f[k] = 3k$, $k = 0, 1, 2, \ldots$
   (b) $g[k] = k/2$, $k = 1, 2, \ldots$
   (c) $h[k] = 3k + 2$, $k = 0, 1, 2, \ldots$
   (d) $x[k] = k^2$, $k = 0, 1, 2, \ldots$
   (e) $x[k] = k^2$, $k = 1, 2, \ldots$
   (f) $x[k] = 4^k$, $k = 2, 3, 4, \ldots$
   (g) $y[k] = 4^{k-1}$, $k = 0, 1, 2, 3, \ldots$

   (h) $f[n] = 3(n + 1)$, $n = 0, 1, 2, 3, \ldots$
   (i) $f[k] = 2k^2 + 3k - 1$, $k = 0, 1, 2, 3, \ldots$

2. A sequence is defined by $x[n] = 3n^2 + 2n - 1$, for $n = -3, -2, \ldots, 2, 3$. Write down the terms of this sequence.

3. The function g is defined by $g[k] = 5k + 2$, $k = 0, 1, 2, \ldots$. Identify the dependent and independent variables.

4. Plot a graph of the sequence defined in Q2.

5. Graph the sequence defined by

$$f[k] = \begin{cases} 1 & k=1 \text{ and } 3 \\ 4 & k=2 \\ 0 & \text{otherwise} \end{cases}$$

where $k \in \mathbb{N}$

6. A sequence is given by

$$1, \tfrac{1}{2}, \tfrac{1}{3}, \tfrac{1}{4}, \tfrac{1}{5}, \ldots$$

Find a suitable function and domain to represent this sequence.

20.3 Sequences arising from sampling

Consider the continuous function $f(t)=t^2$, for $0 \leqslant t \leqslant 5$. This function is shown in Figure 20.3. Suppose we take measurements or **samples** of this function when $t=0, 0.5, 1, 1.5, \ldots$. Doing so, we obtain the values $f(0)=0$, $f(0.5)=0.5^2=0.25$, $f(1)=1$, $f(1.5)=1.5^2=2.25$, and so on. In this way, we can produce the sequence

$$0, \quad 0.25, \quad 1, \quad 2.25, \quad \ldots$$

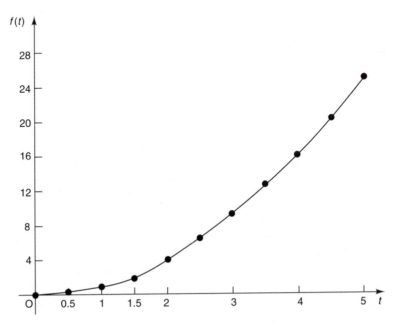

Figure 20.3 The function $f(t)=t^2$, $0 \leqslant t \leqslant 5$.

We see that a sequence can arise by **sampling** a function. When the independent variable t represents time, the time interval between samples is called the **sample interval**, and is often given the symbol T. In this example $T=0.5$. The samples are taken when $t=0.5k$ for $k=0, 1, 2, \ldots$, so that we can write the sampled sequence as

$$f[k]=(0.5k)^2 \qquad k=0, 1, 2, \ldots$$

Example 20.3

The continuous function $f(t)=\cos \frac{1}{2}t$, is sampled when $t=0, \pi, 2\pi, 3\pi, \ldots$.
(a) Identify the sample interval T.
(b) Write down the first five terms of the sampled sequence $f[k]$.

Solution   (a)   Samples are taken at intervals of π; that is, $T=\pi$.
(b)   Noting that $t=k\pi$, the sampled sequence is given by $f[k]=\cos \frac{1}{2}k\pi$, for $k=0, 1, 2, \ldots$. Evaluating this sequence we find that the first five terms are

$$\cos 0, \quad \cos \tfrac{1}{2}\pi, \quad \cos \pi, \quad \cos \tfrac{3}{2}\pi, \quad \cos 2\pi$$

that is

$$1, \quad 0, \quad -1, \quad 0, \quad 1$$

Self-assessment question 20.3

1.   Explain how a sequence can arise from sampling a continuous function.

Exercises 20.3

1.   The function $f(t)=t^3$, $t\geqslant 0$, is sampled at $t=0, 1, 2, \ldots$. Write down the first five terms of the sampled sequence.

2.   The function $f(t)=3\cos 100\pi t$ is sampled when $t=0, 0.01, 0.02, \ldots$. Identify the sample interval and find the first five terms of the sampled sequence.

3.   The function $g(t)=e^t$ is sampled at $t=0, 1, 2, \ldots$. Find the first five terms of the sampled sequence.

4.   The function $g(t)=\sin 2t$, $t\geqslant 0$, is sampled at $t=0, \frac{1}{4}\pi, \frac{1}{2}\pi, \frac{3}{4}\pi, \ldots$.
(a) State the sampling interval.
(b) Write down the first six terms of the sequence.
(c) State $g[k]$.

5.   The function $f(t)=t\cos t$, $t\geqslant 0$, is sampled every $\frac{1}{10}$ s, starting at $t=0$. Write down
(a) $f[0]$ (b) $f[1]$ (c) $f[k]$

20.4 Arithmetic progressions

One particularly simple way of forming a sequence is to calculate each new term by adding a fixed amount to the previous term. For example, suppose the first term of a sequence is 1, and we find the second, third and fourth terms by repeatedly adding on 5. We obtain the sequence

1, 6, 11, 16

Such a sequence is known as an **arithmetic progression**. The fixed amount that is added each time is called the **common difference**.

Example 20.4

Write down the first six terms of the arithmetic progression that has first term 3 and common difference 2.

Solution

The second term is found by adding the common difference 2 to the first term 3; that is, the second term is 5. To find the third term, the common difference is added to the second term, and so on. Proceeding in this way, we obtain the sequence

3, 5, 7, 9, 11, 13

Example 20.5

Write down the first six terms of the arithmetic progression that has first term 2 and common difference -3.

Solution

The second term is found by adding the common difference -3 to the first term 2; that is, the second term is $2+(-3)=-1$. To find the third term, the common difference is added to the second term, and so on. Proceeding in this way, we obtain the sequence

2, -1, -4, -7, -10, -13

Example 20.6

Write down the first term and the common difference of the arithmetic progression

$\frac{1}{2}, \frac{5}{2}, \frac{9}{2}, \frac{13}{2}, \ldots$

Solution

The first term is $\frac{1}{2}$. To find the common difference, we must ask what amount must be added to $\frac{1}{2}$ to give $\frac{5}{2}$. Clearly, this amount is 2. Alternatively, the common difference can be found by calculating the difference between any term and its predecessor; for example,

$$\tfrac{13}{2} - \tfrac{9}{2} = \tfrac{4}{2} = 2$$

We shall now introduce a general notation that can be used to describe any arithmetic progression. Suppose the first term is a and the common difference is d. The second term is found by adding the common difference to the first term; that is, the second term is $a+d$. The third term is found by adding the common difference to the second term; that is, $(a+d)+d=a+2d$; and so on. Hence we note the following result.

> An arithmetic progression can be written
>
> $$a, \quad a+d, \quad a+2d, \quad a+3d, \quad a+4d, \quad \ldots$$

Note that

> the first term is a
> the second term is $a+d$
> the third term is $a+2d$
> the fourth term is $a+3d$

and so on. This leads to the following formula for the nth term.

> The nth term of an arithmetic progression is given by $a+(n-1)d$.

Example 20.7

Find the 23rd term of the arithmetic progression with first term 2 and common difference 7.

Solution   Using the formula

$$n\text{th term} = a+(n-1)d$$

with $a=2$, $n=23$ and $d=7$, we find

$$23\text{rd term} = 2+(23-1)7 = 2+(22)(7) = 156$$

so that the 23rd term of the arithmetic progression is 156.

Self-assessment questions 20.4

1. What is meant by an arithmetic progression?

2. Give an example of one sequence that is an arithmetic progression and one that is not.

Exercises 20.4

1. Write down the first 5 terms of the arithmetic progression with common difference 4 and first term 1.

2. Write down the first 6 terms of the arithmetic progression with common difference 3 and first term 2.

3. Write down the first 5 terms of the arithmetic progression with common difference -3 and first term 7.

4. Write down the first term and common difference of the arithmetic progression 19, 16, 13, 10, ...

5. Write down the 17th and 111th terms of the arithmetic progressions (a) 7, 5, 3, ... (b) 4, 0, -4, ...

6. What is the behaviour of an arithmetic progression that has a common difference 0?

7. The first term of an arithmetic progression is 3. The 20th term is 74. Find the common difference.

8. The 13th term of an arithmetic progression is 10 and the 25th term is 20:
   (a) calculate the common difference;
   (b) calculate the first term;
   (c) calculate the 17th term.

20.5 Geometric progressions

Another simple way of forming a sequence is to calculate each new term by multiplying the previous term by a fixed amount. For example, suppose the first term of a sequence is 1, and we find the second, third and fourth terms by repeatedly multiplying by 3. We obtain the sequence

1, 3, 9, 27

Such a sequence is known as a **geometric progression**. The fixed amount by which each term is multiplied is called the **common ratio**.

Example 20.8

Write down the first six terms of the geometric progression with first term 3 and common ratio 2.

Solution   The first term is 3. The second term is found by multiplying the first by the common ratio, 2. That is, the second term is 6. The third term is found by multiplying the second by the common ratio. So the third term is 12. Continuing in this way, we find 3, 6, 12, 24, 48, 96.

Example 20.9

A geometric progression is given by 1, $\frac{1}{2}$, $\frac{1}{4}$, ... What is its common ratio?

Solution   The first term is 1. We must ask by what must the first term be multiplied to give the second. The answer is clearly $\frac{1}{2}$. The common ratio is therefore $\frac{1}{2}$.

We shall now introduce a general notation that can be used to describe any geometric progression. Suppose the first term is *a* and the common ratio is *r*. The second term is found by multiplying the first term by the common ratio; that is, the second term is *ar*. The third term is found by multiplying the second term by the common ratio; that is, $(ar) \times r = ar^2$; and so on. Hence we have the following result.

> A geometric progression can be written
>
> $$a, \quad ar, \quad ar^2, \quad ar^3, \quad ar^4, \quad ...$$

Note that

the first term is *a*
the second term is *ar*
the third term is ar^2
the fourth term is ar^3

and so on. This leads us to the following formula for the *n*th term.

> The *n*th term of a geometric progression is given by ar^{n-1}.

Example 20.10

Find the 7th term of a geometric progression with first term 2 and common ratio of 3.

Solution   Here $a=2$, $r=3$ and we require the 7th term, so that $n=7$. Hence, using the formula nth term $=ar^{n-1}$, we have

$$\text{7th term} = 2(3^{7-1}) = 2(3^6) = 1458$$

Example 20.11

Find the 8th term in the geometric progression that has first term 2 and common ratio $-\frac{1}{2}$.

Solution   Note that in this example the common ratio is negative. Using the formula nth term $=ar^{n-1}$, with $a=2$, $r=-\frac{1}{2}$ and $n=8$, we find

$$\text{8th term} = (2)(-\tfrac{1}{2})^7 = -0.015\,625$$

Self-assessment questions 20.5

1. Explain what is meant by a geometric progression.

2. Give one example of a sequence that is a geometric progression and one that is not.

Exercises 20.5

1. A geometric progression has first term 4 and common ratio 2. Find (a) the 5th term (b) the 11th term

2. A geometric progression has first term 2 and common ratio $\frac{1}{3}$. Write down the first 6 terms.

3. A geometric progression is given by 2, -1, $\frac{1}{2}$, $-\frac{1}{4}$, ... What is its common ratio?

4. If the common ratio of a geometric progression lies between 0 and 1, what happens to the terms of the sequence as more and more are calculated? What about when the ratio lies between -1 and 0?

5. What can you say about the terms of a geometric progression when the common ratio is 1?

6. The 19th term of a geometric progression is 6 and the 20th term is 7. Calculate the 21st term.

7. The 4th term of a geometric progression equals the 6th term. What can you deduce about the common ratio?

20.6 The limit of a sequence

We have seen that some sequences continue instead of stopping after a finite number of terms, and these sequences are called infinite sequences. Sometimes the terms of an infinite sequence get closer and closer to a fixed value. For example, the sequence $x[k] = 1/k$, $k = 1, 2, 3, \ldots$, is given by

$$1, \tfrac{1}{2}, \tfrac{1}{3}, \tfrac{1}{4}, \ldots$$

We see that the terms are getting smaller and smaller. If we continue on forever then eventually these terms will approach zero. This happens as k gets larger and larger and approaches infinity. We say that 'as k tends to infinity, the **limit** of the sequence is zero'. We write this concisely as

$$\lim_{k \to \infty} \frac{1}{k} = 0$$

When a sequence possesses a limit, it is said to **converge**. Consider another example. Suppose a sequence is given by

$$y[k] = \sqrt{2 + \frac{1}{k}} \qquad k = 1, 2, 3, \ldots$$

Then as k tends to infinity, the quantity $1/k$ tends to zero, and so $y[k]$ tends to $\sqrt{2}$. We write this as

$$\lim_{k \to \infty} y[k] = \sqrt{2}$$

However, not all sequences possess a limit. Consider the sequence defined by $x[k] = 3k - 2$, $k = 1, 2, 3, \ldots$. Writing out the first few terms, we obtain the sequence

$$1, 4, 7, 10, \ldots$$

As k gets larger and larger, so too do the terms of the sequence. In fact, as k tends to infinity, the value $x[k]$ tends to infinity as well. This sequence does not possess a limit. When a sequence does not possess a limit, it is said to **diverge**.

Example 20.12

(a)   Write down the first 5 terms of the sequence $x[k] = 2 + 1/k^2$, $k = 1, 2, 3, \ldots$.
(b)   Find, if possible, the limit of this sequence as k tends to infinity.

Solution   (a)   The first five terms of the sequence

$$x[k] = 2 + \frac{1}{k^2} \qquad k = 1, 2, 3, \ldots$$

are given by

$$x[1] = 2 + \frac{1}{1} = 3, \quad x[2] = 2 + \frac{1}{4} = 2\tfrac{1}{4}, \quad x[3] = 2 + \frac{1}{9} = 2\tfrac{1}{9},$$

$$x[4]=2+\frac{1}{16}=2\tfrac{1}{16}, \quad x[5]=2+\frac{1}{25}=2\tfrac{1}{25}$$

and so the first five terms are

$$3, \quad 2\tfrac{1}{4}, \quad 2\tfrac{1}{9}, \quad 2\tfrac{1}{16}, \quad 2\tfrac{1}{25}$$

(b) As more and more terms are included, we see that $x[k]$ approaches 2, because the quantity $1/k^2$ becomes smaller and smaller. We write

$$\lim_{k\to\infty}\left(2+\frac{1}{k^2}\right)=2$$

The sequence converges to the limit 2.

Example 20.13

(a) Write down the first six terms of the sequence $x[n]=(-1)^n$, $n=1, 2, 3, \dots$.
(b) Find, if it exists, the limit of $x[n]$ as n tends to infinity.

Solution (a) The first six terms of $x[n]=(-1)^n$, $n=1, 2, 3, \dots$, are

$$(-1)^1, \quad (-1)^2, \quad (-1)^3, \quad (-1)^4, \quad (-1)^5, \quad (-1)^6$$

that is

$$-1, 1, -1, 1, -1, 1$$

(b) As more and more terms are taken, we see that the sequence alternates between -1 and 1. However far we go along the sequence, we never actually approach a fixed value. So, this sequence has no limit, that is, it diverges.

Self-assessment questions 20.6

1. Explain what is meant by the limit of a sequence.
2. Explain what is meant by (a) a convergent sequence and (b) a divergent sequence.

Exercises 20.6

1. Find the following limits if they exist:

(a) $\lim_{k\to\infty} (k+1)$ (b) $\lim_{k\to\infty}\left(\dfrac{2}{k^2}\right)$

(c) $\lim_{k\to\infty} k$ (d) $\lim_{k\to\infty} (-\tfrac{1}{2})^k$

(e) $\lim_{k\to\infty}\left(2-\dfrac{2}{k}\right)$ (f) $\lim_{k\to\infty} 3^k$

(g) $\lim_{n\to\infty} \cos n\pi$ (h) $\lim_{n\to\infty} \sin n\pi$

(i) $\lim_{k\to\infty} (\tfrac{9}{10})^k$

2. Give an example of a divergent sequence whose terms do not get larger and larger in size.

20.7 Series

If the terms of a sequence are added, the result is known as a **series**. For example, if we add the terms of the sequence 1, 2, 4, 7, 11, we obtain the series

$$1+2+4+7+11$$

Example 20.14

Write down the series formed from the terms of the sequence $1, \frac{1}{2}, \frac{1}{4}, \frac{1}{8}$, and find its sum.

Solution   The series is formed by adding the terms of the sequence, that is,

$$1+\tfrac{1}{2}+\tfrac{1}{4}+\tfrac{1}{8}$$

The sum of this series is

$$1+\tfrac{1}{2}+\tfrac{1}{4}+\tfrac{1}{8} = \frac{8+4+2+1}{8} = \tfrac{15}{8}$$

20.7.1 The sigma notation for a series

Consider the sequence of ten terms $x[k]$, for $k = 1, 2, \ldots, 10$, given by

$$x[1], \quad x[2], \quad x[3], \quad \ldots, \quad x[9], \quad x[10]$$

The series formed from the terms of this sequence is

$$x[1]+x[2]+x[3]+ \ldots +x[9]+x[10]$$

An abbreviated notation for this sum is

$$\sum_{k=1}^{k=10} x[k], \quad \text{or simply} \quad \sum_{k=1}^{10} x[k]$$

where the Greek capital letter sigma Σ stands for the sum of all the values of $x[k]$ as k ranges from 1 to 10. Note that the lower-most and upper-most values of k are written at the top and bottom of the sigma sign.

Example 20.15

If $x[k] = 3^k$ find $\sum_{k=1}^{k=4} x[k]$.

Solution $\sum_{k=1}^{k=4} x[k]$ stands for

$$\sum_{k=1}^{k=4} x[k] = x[1]+x[2]+x[3]+x[4]$$

But $x[k]=3^k$, and thus $x[1]=3^1$, $x[2]=3^2$, and so on. Therefore

$$\sum_{k=1}^{k=4} x[k] = 3^1 + 3^2 + 3^3 + 3^4 = 3 + 9 + 27 + 81 = 120$$

If the sequence is infinite then the series will comprise an infinite number of terms. In such a case we write

$$\sum_{k=1}^{\infty} x[k]$$

20.7.2 The sums of some important series

Sum of the first n whole numbers

The sum of the first seven whole numbers is

$$1+2+3+4+5+6+7=28$$

that is,

$$\sum_{k=1}^{7} k = 28$$

Similarly, the sum of the first n whole numbers is given by

$$1+2+3+4+ \dots +n$$

In sigma notation we can write this series as

$$\sum_{k=1}^{n} k$$

There is a formula for finding this sum for any value of n:

$$\sum_{k=1}^{n} k = \tfrac{1}{2}n(n+1)$$

Example 20.16

Find the sum of the first 100 whole numbers.

Solution   Using the formula with $n=100$, we find

$$\sum_{k=1}^{100} k = \tfrac{1}{2} \times 100(101) = 5050$$

Sum of the squares of the first n whole numbers

The sum of the squares of the first five whole numbers is

$$1^2 + 2^2 + 3^2 + 4^2 + 5^2 = 55$$

that is,

$$\sum_{k=1}^{5} k^2 = 55$$

Similarly, the sum of the squares of the first n whole numbers is given by

$$1^2 + 2^2 + 3^2 + 4^2 + \ldots + n^2$$

In sigma notation we can write this series as

$$\sum_{k=1}^{n} k^2$$

There is a formula for finding this sum for any value of n:

$$\sum_{k=1}^{n} k^2 = \tfrac{1}{6}n(n+1)(2n+1)$$

Example 20.17

Find the sum of the squares of the first 10 whole numbers; that is, $1^2 + 2^2 + 3^2 + \ldots + 9^2 + 10^2$.

Solution   Using the formula with $n = 10$, we find

$$\sum_{k=1}^{10} k^2 = \tfrac{1}{6} \times 10(11)(21) = 385$$

Self-assessment questions 20.7

1. Explain the meaning of the word 'series'. How is a series different from a sequence?

2. Explain what is meant by each of the following:

    (a) $\displaystyle\sum_{k=1}^{5} x[k]$ (b) $\displaystyle\sum_{k=1}^{n} x[k]$ (c) $\displaystyle\sum_{k=1}^{\infty} x[k]$

Exercises 20.7

1. Given $x[n] = 1 + n^2$, $n = 1, 2, 3, \ldots$, find

   (a) $\displaystyle\sum_{n=1}^{4} x[n]$ (b) $\displaystyle\sum_{n=4}^{7} x[n]$ (c) $\displaystyle\sum_{k=1}^{4} x[k]$

2. Given $y[k] = 3k + 2$, $k = 0, 1, 2, \ldots$, find

   (a) $\displaystyle\sum_{k=0}^{3} y[k]$ (b) $\displaystyle\sum_{k=4}^{7} y[k]$ (c) $\displaystyle\sum_{k=0}^{7} y[k]$

   Show that $\displaystyle\sum_{k=0}^{3} y[k] + \sum_{k=4}^{7} y[k] = \sum_{k=0}^{7} y[k]$

3. Find

   (a) the sum of the first 20 positive integers

   (b) the sum of the squares of the first 20 positive integers

---

20.8 Arithmetic series

If the terms of an arithmetic progression are added, the result is known as an **arithmetic series**.

Example 20.18

(a) Write down the first five terms of the arithmetic progression with first term 3 and common difference 5.

(b) Write down the corresponding arithmetic series and find its sum.

Solution

(a) The first five terms of the arithmetic progression with first term 3 and common difference 5 are

$$3, 8, 13, 18, 23$$

(b) The corresponding series is found by adding the terms of the progression; that is,

$$3 + 8 + 13 + 18 + 23$$

The sum of this series is 65.

Example 20.19

(a) Write down the first six terms of the arithmetic progression with first term 1 and common difference -2.

(b) Write down the corresponding arithmetic series, and find its sum.

Solution

(a) The first six terms of the arithmetic progression with first term 1 and common difference -2 are

$$1, -1, -3, -5, -7, -9$$

(b) The corresponding arithmetic series is found by adding the terms of the progression; that is,

$$1+(-1)+(-3)+(-5)+(-7)+(-9)$$

The sum of this series is -24.

If the series contains a large number of terms then finding its sum by directly adding all the terms will be laborious. Fortunately, there is a formula that enables us to find the sum of an arithmetic series.

> The sum of the first n terms of an arithmetic series with first term a and common difference d is
>
> $$S_n = \tfrac{1}{2}n[2a+(n-1)d]$$

Example 20.20

(a) Find the sum of the first six terms of the arithmetic series with first term 3 and common difference 4.

(b) Find the sum of the first 200 terms of this series.

Solution (a) Using the formula $S_n = \tfrac{1}{2}n[2a+(n-1)d]$ with $a=3$, $n=6$ and $d=4$, we find

$$S_6 = \tfrac{6}{2}[(2)(3)+(6-1)(4)] = 3(6+20) = 78$$

(b) Using the formula with $n=200$, we find

$$S_{200} = \tfrac{200}{2}[(2)(3)+(199)(4)] = 100(6+796) = 100(802) = 80\,200$$

Self-assessment question 20.8

1. Explain what is meant by an arithmetic series.

Exercises 20.8

1. Find the sum of the first 10 terms of the arithmetic series with first term 3 and common difference 5.

2. Find the sum of the first 12 terms of the arithmetic series with first term -2 and common difference -3.

3. An arithmetic series has first term 5 and common difference 1.3. Find the smallest value of n for which $S_n > 900$. That is, find the smallest number of terms required such that their sum is greater than 900.

4.   An arithmetic series of n terms has first term a and common difference d. Show that

$$S_n = \frac{n[\text{first term} + n\text{th term}]}{2}$$

5.   An arithmetic progression has first term 50,

and the sum of the first 100 terms is 0. Calculate the 100th term of the sequence.

6.   The sum of the first 4 terms of an arithmetic progression is 22 and the sum of the first 6 terms is 51. Calculate the sum of the first 10 terms of the progression.

20.9 Geometric series

If the terms of a geometric progression are added, the result is known as a **geometric series**.

Example 20.21

(a)   Write down the first five terms of the geometric progression with first term 3 and common ratio 2.

(b)   Write down the corresponding geometric series, and find its sum.

Solution   (a)   The first five terms of the geometric progression with first term 3 and common ratio 2 are

3, 6, 12, 24, 48

(b)   The corresponding geometric series is found by adding the terms of the progression; that is,

$$3 + 6 + 12 + 24 + 48$$

The sum of this series is 93.

If the geometric series contains a large number of terms then finding its sum by directly adding all the terms will be laborious. Fortunately, there is a formula for finding the sum of a geometric series.

The sum of the first n terms of a geometric series with first term a and common ratio r is given by

$$S_n = \frac{a(1 - r^n)}{1 - r}, \quad r \neq 1$$

Note that this formula does not apply if $r = 1$.

Example 20.22

Find the sum of the first 12 terms of the geometric series with first term 5 and common ratio 0.9.

Solution   Using the formula $S_n = a(1-r^n)/(1-r)$ with $a=5$, $r=0.9$ and $n=12$, we find

$$S_{12} = \frac{5(1-0.9^{12})}{1-0.9} = \frac{5(1-0.2824)}{0.1} = 35.8785$$

Self-assessment questions 20.9

1.   Explain what is meant by a geometric series.

2.   The sum of a geometric series is given by $S_n = a(1-r^n)/(1-r)$. This formula cannot be used when $r=1$, since this would require division by 0. Try to find another formula that would give the sum of n terms when $r=1$.

Exercises 20.9

1.   Find the sum of the first 6 terms of the geometric series with first term 3 and common ratio 2.

2.   Find the sum of the first 12 terms of the geometric series with first term -2 and common ratio -2.

3.   Find the sum of the first 9 terms of the geometric series with first term 5 and common ratio -1.

4.   A geometric series has first term 1 and common ratio $\frac{1}{3}$. Find the sum of the first 1000 terms.

20.10 Infinite series

When the terms of an infinite sequence are added, we obtain an **infinite series**. It may seem strange to try to add together an infinite number of terms, but under some circumstances their sum is finite and can be found. For example, consider the infinite series formed from the terms of the sequence $x[k]=1/2^k$, $k=0, 1, 2, \ldots$; that is,

$$1 + \tfrac{1}{2} + \tfrac{1}{4} + \tfrac{1}{8} + \tfrac{1}{16} + \ldots$$

We can calculate the sum of n terms, S_n, for various values of n. For example, the sum of just the first term is

$$S_1 = 1$$

The sum of the first two terms is

$$S_2 = 1 + \tfrac{1}{2} = 1.5$$

Similarly, $S_3 = 1.75$, $S_4 = 1.875, \ldots, S_{10} = 1.9980$. As we calculate S_n for larger and larger values of n, we note that S_n gets nearer and nearer to 2. We write S_∞ to stand for the sum of an infinite number of terms. Then

$$S_\infty = \sum_{k=0}^{\infty} \frac{1}{2^k} = 2$$

In general, it is difficult to determine whether or not an infinite series has a finite sum. Even if it has, this sum may be difficult to evaluate. However, for an infinite geometric series a formula is available for calculating the sum provided the common ratio lies between -1 and 1.

The sum of an infinite number of terms of a geometric series is

$$S_\infty = \frac{a}{1-r}, \quad \text{provided that } -1 < r < 1$$

Example 20.23

Find, if possible, the sum of the infinite geometric series with first term 2 and common ratio $\tfrac{1}{2}$.

Solution   Because the common ratio lies between -1 and 1, we can apply the formula $S_\infty = a/(1-r)$, with $a = 2$ and $r = \tfrac{1}{2}$. That is,

$$S_\infty = \frac{2}{1 - \tfrac{1}{2}} = \frac{2}{\tfrac{1}{2}} = 4$$

Example 20.24

Find, if possible, the sum of the infinite geometric series with first term 2 and common ratio 3.

Solution   Because the common ratio is greater than 1, we cannot apply the formula. The geometric series is given by $2 + 6 + 18 + 54 + \ldots$. Because successive terms are increasing, it is clear that no finite sum exists. Hence the series diverges.

Self-assessment question 20.10

1. Under what circumstances is it possible to find the sum of an infinite geometric series?

Exercises 20.10

1. Find, if possible, the sums of the infinite geometric series with first term and common ratio given by
    (a) $a=1, r=9$ (b) $a=2, r=\frac{1}{4}$
    (c) $a=-1, r=-0.5$ (d) $a=17, r=0$
    (e) $a=2, r=0.9$ (f) $a=-2, r=4$
    (g) $a=5, r=1$ (h) $a=7, r=-\frac{1}{3}$

2. Find the sums of the following infinite geometric series:
    (a) $1+\frac{1}{4}+\frac{1}{16}+\ldots$ (b) $2+\frac{2}{3}+\frac{2}{9}+\ldots$
    (c) $2-\frac{2}{3}+\frac{2}{9}-\ldots$ (d) $24+6+\frac{3}{2}+\ldots$

3. Find the sum to infinity of the geometric series with first term 17 and common ratio 0.8.

20.11 Pascal's triangle and the binomial theorem

In mathematical work it is frequently necessary to expand expressions of the form $(a+b)^2$ or $(a+b)^3$. The expansion of an expression of the form $(a+b)^n$ is called a **binomial expansion**. When the power involved is small, it may be possible to expand the brackets by directly multiplying them out. Thus

$$(a+b)^2 =(a+b)(a+b)=a^2 + ab + ba + b^2 = a^2 + 2ab + b^2$$

and

$$(a+b)^3 =(a+b)^2(a+b)$$
$$=(a^2 + 2ab + b^2)(a+b)$$
$$=a^3 + 2a^2b + b^2a + a^2b + 2ab^2 + b^3$$
$$=a^3 + 3a^2b + 3ab^2 + b^3$$

However, when the power is larger than 3, it becomes extremely tedious to directly multiply out the brackets. An alternative technique is to make use of **Pascal's triangle**.

20.11.1 Pascal's triangle

Consider the triangle of numbers shown in Figure 20.4, where every entry is obtained by adding the two entries on either side in the preceding row, always starting and finishing a row with a 1.

Figure 20.4 The first four rows of Pascal's triangle.

Example 20.25

Complete the next two rows in Pascal's triangle.

Solution Each additional row is started and finished with a 1. Every other entry is obtained by adding the entries on either side in the row before. Thus, for the next two rows, we obtain the triangle shown in Figure 20.5.

Figure 20.5 Pascal's triangle for Example 20.25.

Example 20.26

Find the row in Pascal's triangle starting with 1   7.

Solution In order to reach the row beginning 1   7, two additional rows must be added to Figure 20.5. We obtain the triangle shown in Figure 20.6.

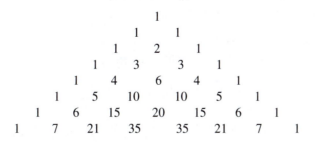

Figure 20.6 The first eight rows of Pascal's triangle.

20.11.2 Using Pascal's triangle to expand $(a+b)^n$

Let us compare the third and fourth rows of Pascal's triangle with the expansions of $(a+b)^2$ and $(a+b)^3$:

$$(a+b)^2 = a^2 + 2ab + b^2$$

$$(a+b)^3 = a^3 + 3a^2b + 3ab^2 + b^3$$

It is clear that the triangle gives the coefficients in the expansions. Furthermore, the terms in these expansions are composed of decreasing powers of a and increasing powers of b. If we want to expand $(a+b)^4$, the row in the triangle beginning 1   4 will provide the necessary coefficients. These are 1   4   6   4   1. We must simply insert the appropriate powers of a and b, starting with the highest power of a, namely a^4. Thus

$$(a+b)^4 = a^4 + 4a^3b + 6a^2b^2 + 4ab^3 + b^4$$

Example 20.27

Use Pascal's triangle to expand $(a+b)^5$.

Solution   We look to the row commencing 1   5, that is 1   5   10   10   5   1, because $a+b$ is raised to the power 5. This row provides the necessary coefficients. We must now insert the appropriate powers of a and b, starting with the highest power of a, namely a^5. We find

$$(a+b)^5 = a^5 + 5a^4b + 10a^3b^2 + 10a^2b^3 + 5ab^4 + b^5$$

20.11.3 The binomial theorem

When we wish to expand an expression such as $(a+b)^n$, the method of Pascal's triangle becomes extremely laborious when n is large. This is when the **binomial theorem** is useful. This states that

when n is a positive integer

$$(a+b)^n = a^n + na^{n-1}b + \frac{n(n-1)}{2!}a^{n-2}b^2$$

$$+ \frac{n(n-1)(n-2)}{3!}a^{n-3}b^3 + \ldots + b^n$$

This is a sum of a finite number of terms. Recall the factorial notation discussed in Chapter 1:

$$2! = 2 \times 1, \quad 3! = 3 \times 2 \times 1, \quad 4! = 4 \times 3 \times 2 \times 1, \quad \text{and so on}$$

The binomial theorem is often quoted for the particular case when $a = 1$ and $b = x$. In this case we obtain

$$(1+x)^n = 1 + nx + \frac{n(n-1)}{2!}x^2 + \frac{n(n-1)(n-2)}{3!}x^3 + \ldots + x^n$$

Example 20.28

Use the binomial theorem to expand $(1+x)^4$.

Solution   Using the previous formula with $n=4$, we find

$$(1+x)^4 = 1 + 4x + \frac{(4)(3)}{2!}x^2 + \frac{(4)(3)(2)}{3!}x^3 + \frac{(4)(3)(2)(1)}{4!}x^4$$

$$= 1 + 4x + 6x^2 + 4x^3 + x^4$$

The binomial theorem can also be applied when n is not a positive integer, provided that $-1 < x < 1$. However, when n is not a positive integer, the series is infinite, and in this case we have

$$(1+x)^n = 1 + nx + \frac{n(n-1)}{2!}x^2 + \frac{n(n-1)(n-2)}{3!}x^3 + \ldots \qquad -1 < x < 1$$

Example 20.29

Use the binomial theorem to find the first four terms in the expansion of $(1+x)^{\frac{1}{2}}$.

Solution   We use the binomial theorem with $n = \frac{1}{2}$. This gives

$$(1+x)^{\frac{1}{2}} = 1 + \frac{1}{2}x + \frac{\frac{1}{2}(\frac{1}{2}-1)}{2!}x^2 + \frac{\frac{1}{2}(\frac{1}{2}-1)(\frac{1}{2}-2)}{3!}x^3 + \ldots$$

$$= 1 + \frac{x}{2} - \frac{x^2}{8} + \frac{x^3}{16} + \ldots$$

Provided $-1<x<1$, the left- and right-hand sides are equal. If this is the case we say that the expansion is **valid**. When $x\leqslant-1$ or $x\geqslant1$, the two sides are different, and the expansion is **invalid** and should not be used.

Self-assessment question 20.11

1. Describe the rule for constructing a new row of Pascal's triangle.

Exercises 20.11

1. Use Pascal's triangle to expand $(a+b)^7$.

2. (a) Use Pascal's triangle to expand $(a+b)^4$.
   (b) By letting $a=1$ and $b=3y$, find the expansion of $(1+3y)^4$.

3. Expand $(a+b)^3$ using Pascal's triangle. By letting $a=3x$ and $b=4y$, find the expansion of $(3x+4y)^3$.

4. (a) Use the binomial theorem to find the expansion of $(1+y)^5$.

   (b) By replacing y by $4x$, find the expansion of $(1+4x)^5$.

5. Use the binomial theorem to find the expansion of $(2x+y)^4$.

6. Obtain the first five terms in the expansion of $(1+x)^{-2}$. State the range of values of x for which your expansion is valid.

7. Obtain the first four terms in the expansion of $(1+\frac{1}{2}x)^{-3}$. State the range of values of x for which your expansion is valid.

Review exercises 20

1 Write down and graph the first five terms of the sequences $x[k]$ defined by
(a) $x[k]=3^k$, $k=0$, 1, 2, ...
(b) $x[k]=k^3$, $k=0$, 1, 2, ...
(c) $x[k]=(-2)^k$, $k=0$, 1, 2, ...
(d) $x[k]=5$, for all natural numbers k

2 State whether the following sequences are arithmetic, geometric or neither:
(a) 1, -1, -3, -5,...   (b) 4, 2, 1, 0.5,...
(c) 6, 7, 8, 9, ...   (d) 4, 5, 7, 10, ...
(e) 1, 0.1, 0.01, 0.001, ...
(f) 1, -1, 1, -1, 1, ...   (g) 1, 1, 1, 1, ...

3 State which of the sequences in Q2 are convergent.

4 Use the binomial theorem to find the first four terms in the expansion of $(1+3x)^{-3}$. State the range of values of x for which the expansion is valid.

5 An arithmetic progression has first term -3 and common difference 4. State
(a) the 10th term   (b) the 300th term

6 An arithmetic series has first term 4 and common difference $\frac{1}{2}$. Find the sum of
(a) the first 20 terms
(b) the first 100 terms

7 A geometric progression has first term -2 and common ratio $-\frac{3}{4}$. State the 20th term.

8 A geometric series has first term equal to 3 and a common ratio of 1.5. Calculate the sum of
(a) the first 10 terms
(b) the first 50 terms
(c) the 30th to the 49th term inclusive

9 If $x[k] = 2 - k$, evaluate
(a) $\sum_{k=1}^{4} x[k]$ (b) $\sum_{k=3}^{10} x[k]$

10 The function $x(t) = 3t^2 + t + 1, t \geq 0$ is sampled at $t = 0, 0.5, 1, 1.5, \ldots$
(a) State the sampling interval.
(b) Write down the first five terms of the sampled sequence $x[k]$.

11 Find (a) $\sum_{k=1}^{32} k^2$ (b) $\sum_{k=1}^{18} k$

21

Change and Rates of Change

KEY POINTS

This chapter

- explains the term 'change' and shows how to calculate it

- explains the term 'average rate of change' of a function and shows how to calculate this

- shows how the concept of average rate of change of a function across an interval is extended to give the rate of change of a function at a specific point

- introduces the technique known as 'differentiation from first principles'

- shows how the technique can be applied to find the rates of change of functions such as x^2 and $1/x$

- explains the term 'derivative' and introduces the notations

$$\frac{\delta y}{\delta x}, \quad \frac{\mathrm{d}y}{\mathrm{d}x}, \quad y'(x)$$

CONTENTS

21.1 Introduction

It is often important to know how quickly an engineering quantity is changing. For example, the rate at which the temperature of an industrial chemical is rising in a tank, the rate at which the speed of a car is increasing or decreasing, and the rate at which the charge on a capacitor is changing, are all of interest to engineers. The technique for analysing the rate at which a function is changing is known as differentiation. We shall develop this topic in this chapter.

21.2 Changes in the value of a function

Recall from Chapter 17 that the change in the value of a function is measured by the following formula:

> change = final value − initial value

A positive change indicates that the value of the function has increased. A negative change indicates that the value of the function has decreased. Consider the following example.

Example 21.1 Volume of liquid in a chemical storage tank

Figure 21.1 shows the variation in the volume of liquid in a chemical storage tank which is part of an oil refinery. The results are taken over a 24 h period. These plots are common in the process industries and are usually obtained by means of a chart recorder. Calculate the change in the volume of liquid in the tank during the following time intervals:
(a) 03 30 h to 07 15 h   (b) 09 00 h to 18 00 h

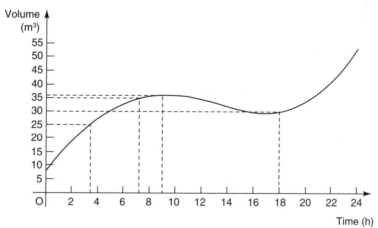

Figure 21.1   Volume of liquid in the storage tank.

Solution   (a)   At 03 30 the volume is 25 m³, while at 07 15 the volume is 35 m³. Therefore the change in volume is

$$\text{final value} - \text{initial value} = 35 - 25$$

$$= 10 \text{ m}^3$$

This corresponds to an increase in the volume of liquid in the tank.

       (b)   At 09 00 the volume is 36 m³, while at 18 00 the volume is 30 m³. Therefore the change in volume is $30 - 36 = -6$ m³. This corresponds to a decrease in volume of liquid in the tank.

Example 21.2

If the argument t of the function $f(t) = 4t^2$ changes from 5 to 8, calculate the change in the value of the function.

Solution   When $t = 5$, $f(t) = 4(5^2) = 100$. When $t = 8$, $f(t) = 4(8^2) = 256$. The change in the value of the function is given by

$$\text{change} = \text{final value} - \text{initial value} = 256 - 100 = 156$$

We see that over the interval [5, 8] the change in the value of the function is 156.

Self-assessment questions 21.2

1. Explain how the change in the value of a function is calculated.

2. Initially a variable has a value of -7. Finally it has a value of -13. Is this a positive or negative change?

3. At time $t = 0$ a variable has a value -13. At time $t = 3$ seconds it has a value -2. Is this change positive or negative?

4. The value of an engineering variable is measured at different times. The change in its value from time $t = 0$ to $t = 1$ seconds is found to be 0. Does this mean that the variable is constant over the time interval [0, 1]? Give reasons for your answer.

5. The value of an engineering quantity is measured at $t = 0$ and found to be 32. At time $t = 1200$ seconds its value is 164. Does this mean that the value of the quantity was always increasing over the time interval [0, 1200]?

Exercises 21.2

1. The temperature of the liquid in a vessel is measured at time $t = 0$ and found to be 28 °C. Five seconds later, the temperature is found to be 12 °C. Calculate the temperature change.

2. The potential difference across an electrical component is measured at $t = 5$ s and found to be 35 V. Fifteen seconds later, the measured potential difference is 14 V. Calculate the change in potential difference.

3. A particle changes its velocity from 22 m per second to 38 m per second. Find the change in velocity.

4. If the argument x of the function $y(x)=3x^5$ changes from -3 to -2, calculate the change in the value of the function.

5. Calculate the change in the value of the function $y(x)=\sin x$ as x changes from 0 to $\frac{1}{2}\pi$, where x is in radians.

6. Figure 21.2 shows the variation in the volume of sulphuric acid in a storage tank that forms part of a production line in a factory. The results are taken over a 24 h

period. Calculate the change in the volume of the acid in the tank during the following time intervals:
(a) 02 00 h to 06 00 h
(b) 00 00 h to 24 00 h
(c) 09 00 h to 16 00 h
(d) 14 00 h to 19 00 h

7. The height y of a projectile fired from the origin with speed V at an angle of $45°$ to the horizontal is given by $y=x-gx^2/V^2$, where x is the horizontal distance travelled. The constant g is known as the acceleration due to gravity and is approximately $10\ \mathrm{m\ s^{-2}}$. If the projectile is fired at 20 m per second, calculate the change in y as x changes from 1 to 2.

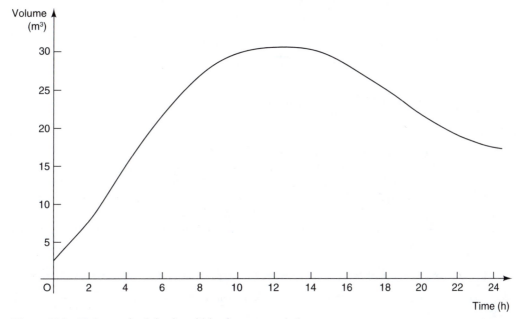

Figure 21.2 Volume of sulphuric acid in the storage tank.

21.3 The average rate of change of a function

Consider the graph of the function $y(x)=x^2$ as shown in Figure 21.3. The point A has coordinates $(1, 1)$ and the point B has coordinates $(2, 4)$. As x increases from 1 to 2, the value of y increases by 3, from 1 to 4. On the other hand, the

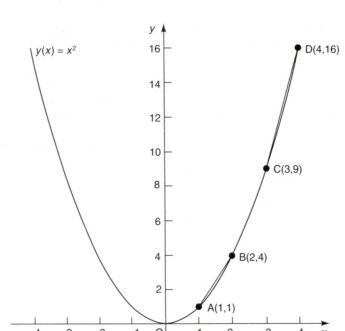

Figure 21.3   Graph of the function $y(x) = x^2$.

point C has coordinates (3, 9), and the point D has coordinates (4, 16). As x increases from 3 to 4, the value of y increases by 7, from 9 to 16.

The increase in y over the interval [3, 4] is larger than the increase in y over the interval [1, 2]. It is clear that the rate at which the value of y changes depends upon the interval over which we measure.

We calculate the **average rate of change** of y across an interval from the formula

$$\text{average rate of change of a function across an interval} = \frac{\text{change in } y}{\text{change in } x}$$

Consider the interval [1, 2]. When $x = 1$, $y = 1$ and when $x = 2$, $y = 4$, we have

$$\text{average rate of change of } y \text{ across } [1, 2] = \frac{\text{change in } y}{\text{change in } x} = \frac{4 - 1}{2 - 1} = 3$$

We say that 'the average rate of change of y with respect to x across the interval [1, 2] is 3'. The phrase 'with respect to' is often abbreviated to w.r.t.

The straight line joining A and B is known as a **chord**. Using our knowledge of straight lines from Chapter 17, we can calculate the gradient of this chord from

$$\text{gradient of chord AB} = \frac{\text{change in } y}{\text{change in } x} = \frac{4-1}{2-1} = 3$$

We see that the gradient of the chord AB measures the average rate of change of y across the interval $[1, 2]$.

Let us now calculate the average rate of change of the same function across the interval $[3, 4]$. When $x=3$, $y=9$. When $x=4$, $y=16$.

$$\text{Average rate of change of } y \text{ across } [3, 4] = \frac{\text{change in } y}{\text{change in } x} = \frac{16-9}{4-3} = 7$$

The average rate of change of y w.r.t. x across $[3, 4]$ is 7. This is also the gradient of the chord CD.

The fact that the average rate of change over the interval $[3, 4]$ is 7 shows that the value of y is increasing at a faster rate than over the interval $[1, 2]$ where the average rate of change is 3.

In a similar way, we find that over the interval $[-3, -2]$,

$$\text{average rate of change of } y \text{ across } [-3, -2] = \frac{\text{change in } y}{\text{change in } x}$$

$$= \frac{4-9}{-2-(-3)} = -5$$

The average rate of change is negative, and this reflects the fact that the value of y is decreasing over the interval $[-3, -2]$, as can be seen from Figure 21.3.

Example 21.3

(a) Sketch a graph of the function $y(x) = 3x^3$ for values of x between 0 and 2.
(b) Calculate the average rate of change of this function over the intervals
(i) $[1, 2]$ (ii) $[1, 1.5]$ (iii) $[1, 1.25]$

Solution

(a) A graph of the function $y = 3x^3$ is shown in Figure 21.4.
(b) (i) Consider the interval $[1, 2]$. When $x=1$, $y=3(1)^3 = 3$, and when $x=2$, $y=3(2)^3 = 24$. As x changes from 1 to 2, the value of y changes from 3 to 24. Therefore

$$\text{average rate of change of } y \text{ across } [1, 2] = \frac{\text{change in } y}{\text{change in } x}$$

$$= \frac{24-3}{2-1} = 21$$

This is the gradient of the chord AB in Figure 21.4.

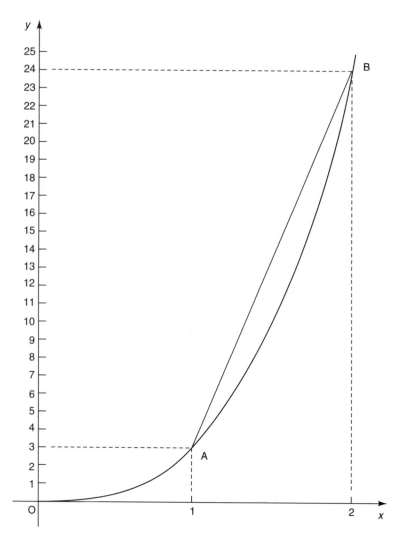

Figure 21.4   Graph of the function $y = 3x^3$.

(ii)   Now consider the interval $[1, 1.5]$. When $x = 1.5$, $y = 3(1.5)^3 = 10.125$. As x changes from 1 to 1.5, the value of y changes from 3 to 10.125. Therefore

$$\text{average rate of change of } y \text{ across } [1, 1.5] = \frac{\text{change in } y}{\text{change in } x}$$

$$= \frac{10.125 - 3}{1.5 - 1} = 14.25$$

(iii)   Consider the interval [1, 1.25]. When $x=1.25$, $y=3(1.25)^3=5.8594$. Therefore, as x changes from 1 to 1.25, the value of y changes from 3 to 5.8594. So

$$\text{average rate of change of } y \text{ across } [1, 1.25] = \frac{\text{change in } y}{\text{change in } x}$$

$$= \frac{5.8594-3}{1.25-1}$$

$$= 11.44$$

Now let us extend what we have seen to the case of a more general function $y(x)$ as shown in Figure 21.5. The point A has coordinates $(x_1, y(x_1))$,

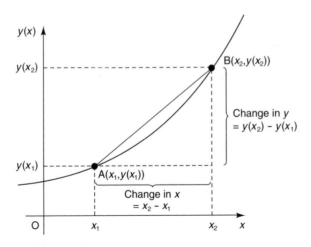

Figure 21.5   A general function $y(x)$.

and the point B has coordinates $(x_2, y(x_2))$. Therefore, as x changes from x_1 to x_2, the value of y changes from $y(x_1)$ to $y(x_2)$. The function changes by an amount $y(x_2)-y(x_1)$ over the interval $[x_1, x_2]$. The average rate of change of the function over the interval is given by

$$\text{average rate of change} = \frac{\text{change in } y}{\text{change in } x} = \frac{y(x_2)-y(x_1)}{x_2-x_1}$$

This is also the gradient of the chord **AB**.

Recall from Chapter 17 that the equation of a straight line is $y = mx + c$, and that m is the gradient of the line.

Example 21.4

The gradient of the line $y = \frac{1}{2}x - 3$ is $\frac{1}{2}$. Verify that the average rate of change of y across the interval $[4, 7]$ is also $\frac{1}{2}$.

Solution Using the equation of the straight line $y = \frac{1}{2}x - 3$, we find that when $x = 4$, $y = -1$. When $x = 7$, we find that $y = \frac{1}{2}$. Therefore

$$\text{average rate of change of } y \text{ across } [4, 7] = \frac{\text{change in } y}{\text{change in } x}$$

$$= \frac{\frac{1}{2} - (-1)}{7 - 4}$$

$$= \frac{\frac{3}{2}}{3}$$

$$= \frac{1}{2}$$

We see that the average rate of change of y is the same as the gradient of the line.

The result obtained in the previous example is true for any linear function. When $y(x)$ is a linear function, the average rate of change of y w.r.t. x is the same across any interval, and is equal to the gradient of the line.

Example 21.5 Furnace temperature

The temperature T of a furnace increases linearly with time t, and is given by $T = 0.2t + 20$.

(a) Sketch a graph of T against t for $t = 0$ to $t = 400$, given that T has units of degrees Celsius (°C) and t has units of seconds (s).

(b) Calculate the average rate of change of temperature across the interval $[100, 200]$, and confirm that this is the same as the gradient of the graph.

Solution (a) The graph of $T = 0.2t + 20$ is shown in Figure 21.6.

(b) Examining Figure 21.6, we see that when $t = 100$ seconds, $T = 40\,°C$, and that when $t = 200$ seconds, $T = 60\,°C$. Therefore,

$$\text{average rate of change of } T \text{ across } [100, 200] = \frac{\text{change in } T}{\text{change in } t}$$

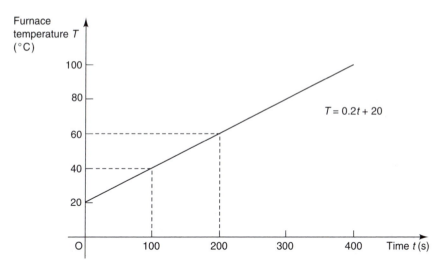

Figure 21.6 Graph of furnace temperature against time.

$$= \frac{60-40}{200-100}$$

$$= \frac{20}{100}$$

$$= 0.2 \,^{\circ}\text{C per second}$$

Therefore T increases at a rate of $0.2\,^{\circ}\text{C}$ for every one second increase in time. Note that this is the same as the gradient obtained by inspection of the equation $T(t) = 0.2t + 20$.

Example 21.6 Displacement–time graph: constant velocity

 If an object such as a car moves along in a straight line, we can measure its displacement s from some fixed point at any instant in time t. A graph can be plotted showing displacement against time, such as that shown in Figure 21.7. This is called a **displacement–time** graph. In this particular example the graph is linear. We see that the displacement in the first second is $5 - 3 = 2$ m. The average rate of change of s across the interval $[0, 1]$ is therefore 2 m per second. The object has travelled 2 m in 1 second, and so the velocity of the object is the same as the average rate of change of s. When the velocity is constant, equal distances will always be covered in equal intervals of time, and so the graph will be a straight line.

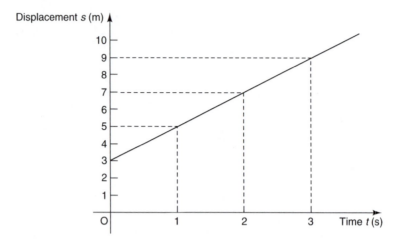

Figure 21.7   Displacement–time graph.

Example 21.7 *Velocity–time graph: constant acceleration*

 If an object moves in a straight line with constant acceleration and at $t=0$ has velocity $1\,\mathrm{m\,s^{-1}}$, while at $t=3$ s its velocity is $10\,\mathrm{m\,s^{-1}}$, its velocity has increased by $9\,\mathrm{m\,s^{-1}}$ in 3 s. The average rate of change of velocity across $[0,3]$ is $\frac{9}{3}=3$ metres per second per second, written $3\,\mathrm{m\,s^{-2}}$. This is the object's acceleration. The graph of velocity against time is shown in Figure 21.8. When the acceleration of an object is constant, the velocity–time graph will be linear because the velocity is increasing at a steady rate. A constant acceleration is often referred to as a **uniform** acceleration.

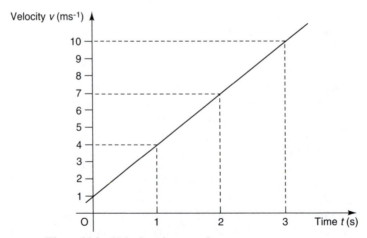

Figure 21.8   Velocity–time graph.

Example 21.8 Pumped hydro power station

A pumped hydro power station is one that can store energy as well as generate it. It is used to accommodate the variations in electrical demand that occur throughout a 24 h period. During the day, at peak demand times, electricity is generated, while at night, when demand is low, energy is stored. A typical scheme is shown in Figure 21.9. The main piece of equipment is a reversible pump–turbine. At night, excess electricity is used to pump water from the lower reservoir to the upper reservoir, thus refilling the latter and so storing energy. At peak times, water is allowed to flow from the upper to the lower reservoir, thus generating electricity. A meter records the volume of water passing from the upper to the lower reservoir. Figure 21.10 shows a meter reading for a typical 24 h period. For convenience, the meter is assumed to be reset to 0 at the start of the measuring period at midnight. Notice from the figure that, during the early hours of the day, water is flowing from the lower to the upper reservoir. Calculate the following:

(a)   the average flow rate from the upper to the lower reservoir between 00 00 h and 06 00 h;

(b)   the average flow rate from the upper to the lower reservoir between 06 00 h and 20 00 h;

(c)   the average flow rate from the upper to the lower reservoir between 01 00 h and 10 00 h.

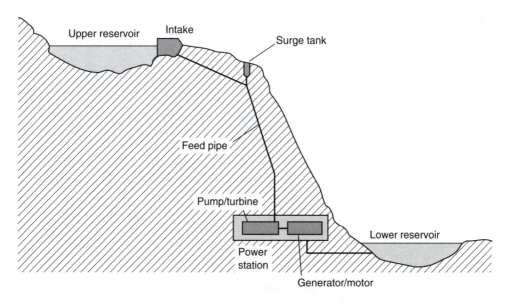

Figure 21.9   Schematic diagram of the pumped hydro power station.

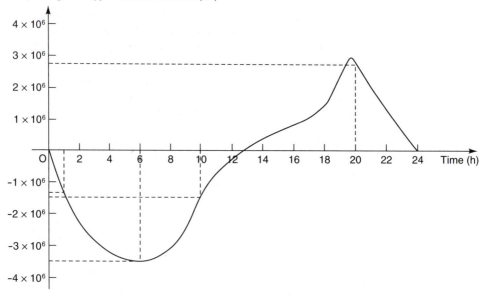

Water passing from upper to lower reservoir (m³)

Figure 21.10 Water passing from upper reservoir to lower reservoir during a 24 h period.

Solution (a) At 00 00 h the volume of water having passed through the meter is 0 m³, while at 06 00 h this volume is -3.5×10^6 m³. Therefore

$$\text{average flow rate} = \frac{-3.5 \times 10^6 - 0}{6 - 0} = -5.8 \times 10^5 \text{ m}^3 \text{ h}^{-1}$$

The negative sign means that there is net flow from the lower to the upper reservoir.

(b) At 06 00 h the volume of water having passed through the meter is -3.5×10^6 m³, while at 20 00 h this volume is 2.8×10^6 m³. Therefore

$$\text{average flow rate} = \frac{2.8 \times 10^6 - (-3.5 \times 10^6)}{20 - 6} = 4.5 \times 10^5 \text{ m}^3 \text{ h}^{-1}$$

This means that there is net flow from the upper to the lower reservoir.

(c) At 01 00 h the volume of water having passed through the meter is -1.3×10^6 m³, while at 10 00 h this volume is -1.5×10^6 m³. Therefore

$$\text{average flow rate} = \frac{-1.5 \times 10^6 - (-1.3 \times 10^6)}{10 - 1}$$

$$= -2.2 \times 10^4 \text{ m}^3 \text{ h}^{-1}$$

The negative sign means that there is net flow from the lower to the upper reservoir.

Self-assessment questions 21.3

1.   If a particle moves in a straight line with constant acceleration, and a velocity–time graph is drawn, is it true that the gradient of the graph represents the acceleration of the object?

2.   A displacement–time graph is found to be linear with gradient -2. Is it true that the vehicle is slowing down?

3.   A graph is plotted of the displacement s m of a vehicle against time t s, and is found to be a straight line with gradient 3. Is it true that the vehicle is accelerating?

4.   The average rate of change of a function $y(x)$ from $x=a$ to $x=b$ is -2. Does this mean that $y(x)$ is decreasing across the interval from a to b?

Exercises 21.3

1.   A straight line passes through the point A with coordinates (1, 2) and also through the point B with coordinates $(-2, -34)$. Find the equation of the straight line joining A and B and determine the rate of change of y with respect to x.

2.   Figure 21.11 shows a velocity–time graph for a car on a journey between two points. Calculate the acceleration of the car during the different parts of the journey.

3.   Calculate the average rate of change of the function $y(x)=6x^2$ over the intervals (a) [2, 4] and (b) [4, 6]. Comment upon your results.

4.   Sketch a graph of the function $y(x)=\dfrac{1}{x}$. Calculate its average rate of change over the intervals [1, 2] and $[-2, -1]$. Comment upon your results.

5.   A car travels 156 miles during a 6 hour journey. Calculate the average speed over the journey.

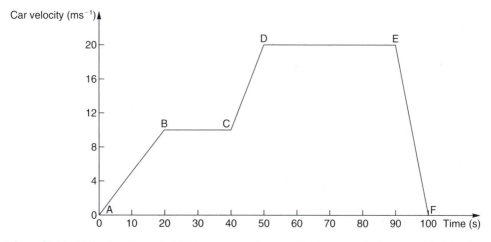

Figure 21.11   Velocity–time graph for a car on a journey between two points.

6.   Consider the graph of the function $y(x)=2x^5$.   **7.**   The average rate of change of a function $y(x)$
The point A on the curve has coordinates
from $x=2$ to $x=5$ is -3. If $y(5)=16$, what
(2, 64). The point B on the curve has
is $y(2)$?
coordinates (4, 2048). Calculate the gradient
of the chord AB.

21.4 Estimating the rate of change of a function at a point

Consider the graph of $y(x)$ in Figure 21.12 and in particular the chord AB.
Suppose that the chord AB is extended as a straight line on both sides as shown.
Now suppose that the point B is moved along the curve closer and closer to
A until both points eventually coincide. Figure 21.12 shows what happens to the
extended chord as B gets closer and closer to A. The extended chord eventually
becomes a **tangent** to the curve at A when B coincides with A. This is the
straight line that just touches the curve at A.

We can calculate the gradient of the chord AB and this gives the average
rate of change of y w.r.t. x over the interval from A to B. As B gets closer and
closer to A, the gradient of the extended chord AB approaches the gradient of
the tangent to the curve at A. So we can identify the gradient of the tangent
at A as **the rate of change of the function $y(x)$ at the point** A. Alternatively, we
say that the **instantaneous rate of change** of $y(x)$ at A equals the gradient of the
tangent at A.

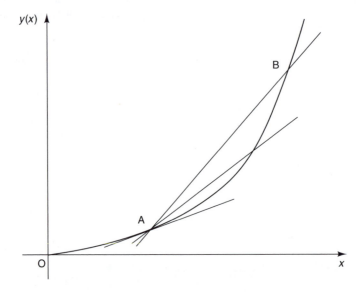

Figure 21.12   The gradient of the extended chord AB is an estimate of the rate of change of $y(x)$ at A.

We also define the **gradient of the curve** at the point A to be the gradient of the tangent at that point. Note that the gradient of a curve varies from point to point, and may be positive, negative or zero.

Consider the following example.

Example 21.9

(a) Sketch a graph of the function $y(x)=4x^2-2x$ for values of x between 0 and 2.
(b) Calculate the average rate of change of the function over the interval $[1,2]$.
(c) Calculate the average rate of change of the function over the interval $[1, 1.5]$.
(d) Draw in by eye a tangent to the curve at A where $x=1$ and measure its gradient.

Solution   (a)   A graph of the function $y(x)=4x^2-2x$ is shown in Figure 21.13.

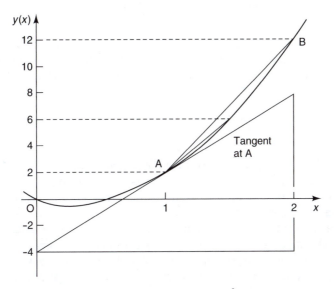

Figure 21.13   Graph of the function $y=4x^2-2x$.

(b)   When $x=1$ the function has the value $y(1)=4-2=2$. When $x=2$, the function has the value $y(2)=16-4=12$. Therefore

$$\text{average rate of change of } y \text{ over } [1,2]=\frac{\text{change in } y}{\text{change in } x}=\frac{12-2}{2-1}$$

$$=\frac{10}{1}=10$$

This is the gradient of the chord AB in Figure 21.13.

(c) When $x = 1.5$, the function has the value $y(1.5) = 4(1.5)^2 - 2(1.5) = 6$. Therefore

$$\text{average rate of change of } y \text{ over } [1, 1.5] = \frac{\text{change in } y}{\text{change in } x} = \frac{6-2}{1.5-1}$$

$$= \frac{4}{0.5} = 8$$

(d) The tangent at A is drawn in by eye as shown in Figure 21.13. The gradient of the tangent at $x = 1$ is found by measuring the changes in x and y between two points on the tangent. For example, estimating from the graph, we find

$$\frac{\text{change in } y}{\text{change in } x} = \frac{8-(-4)}{2-0} = \frac{12}{2} = 6$$

The answers to (b), (c) and (d) are all estimates of the rate of change of $y(x)$ at $x = 1$. Clearly, as B moves nearer to A along the curve, we obtain better approximations to the gradient of the tangent at A.

If several people were to estimate the gradient of the tangent drawn in by eye, a variety of estimates would be obtained, depending upon how well they had drawn the graph and how careful they were at measuring the required distances. What is required is a mathematical technique that will remove the guesswork and estimation, and allow the gradient of the tangent at a point to be calculated exactly. This is done by considering the gradient of the extended chord as B gets closer and closer to A. This requires a knowledge of limits. This topic has already been covered in Chapter 8 and if necessary should be revised before proceeding further.

Self-assessment questions 21.4

1. What is the distinction between a tangent and a chord?

2. Explain the distinction between average rate of change across an interval and rate of change at a point.

Exercises 21.4

1. (a) Sketch a graph of the function $y(x) = 3x^2 + x$ for values of x between 0 and 4.
   (b) Calculate the average rate of change of y across
   (i) the interval $[1, 2]$
   (ii) the interval $[1, 1.5]$
   (c) Draw in the tangent at the point where $x = 1$ and measure its gradient.

2. (a) Sketch a graph of the function $y(x) = \dfrac{1}{x}$ for values of x between -4 and -1.
   (b) Calculate the rate of change of $y(x)$ across the intervals $[-3, -2]$, and $[-3, -2.5]$.
   (c) Draw in the tangent at $x = -3$ and measure its gradient.

21.5 Differentiation from first principles

Consider the function $y(x)$ shown in Figure 21.14 and suppose we wish to find the rate of change of the function at the point A. This point has coordinates $(x, y(x))$. To estimate the rate of change at A, we shall use the technique of the previous sections, which is to choose a nearby point B on the curve and construct the chord AB as shown. Our experience tells us that B should be chosen as close as possible to A in order that the gradient of the chord AB is an accurate estimate of the gradient of the tangent at A. This means that the x coordinate of point B should be only slightly larger than that at A. We therefore let the

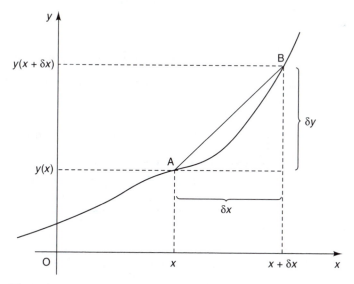

Figure 21.14 A general function $y(x)$.

x coordinate at B be $x+\delta x$ where δx stands for a small change or **increment** in x. We write the y coordinate of B as $y+\delta y$, where δy stands for a small change in y. Because B lies on the curve $y(x)$, the y coordinate at B can also be written as $y(x+\delta x)$. We can now write down the gradient of the chord AB:

$$\text{the gradient of the chord AB} = \frac{\text{change in } y}{\text{change in } x} = \frac{\delta y}{\delta x} = \frac{y(x+\delta x)-y(x)}{\delta x}$$

This quantity is the gradient of the chord AB and is an estimate of the rate of change of y at A. We can improve the estimate by moving B closer and closer to A, as before. In fact, we eventually want to move B so close to A that δx approaches 0. We write this as $\delta x \to 0$. In other words, we must evaluate the quantity

$$\frac{\delta y}{\delta x} \quad \text{in the limit as } \delta x \to 0$$

In mathematical notation, this is written as

$$\lim_{\delta x \to 0} \frac{\delta y}{\delta x}, \quad \text{or} \quad \lim_{\delta x \to 0} \frac{y(x+\delta x)-y(x)}{\delta x}$$

This quantity defines the gradient of the tangent to $y(x)$ at the point A, and thus gives the **rate of change of** $y(x)$ **at** A, that is, the **instantaneous rate of change** at A. These quantities may look daunting when first met, but we shall soon see that in many cases they can be quickly calculated and simplified. An important notation must be introduced that helps to abbreviate this formula. We write

$$\frac{dy}{dx} \quad \text{to stand for} \quad \lim_{\delta x \to 0} \frac{y(x+\delta x)-y(x)}{\delta x}$$

The quantity $\frac{dy}{dx}$ or (dy/dx) is read 'dy by dx'. It is important to realize that $\frac{dy}{dx}$ is not a fraction but the notation, or the symbol, used to stand for the limit of the fraction $\frac{\delta y}{\delta x}$ as $\delta x \to 0$.

Thus the rate of change of the function $y(x)$ is given by

$$\text{rate of change of } y(x) = \frac{dy}{dx} = \lim_{\delta x \to 0} \frac{y(x+\delta x)-y(x)}{\delta x}$$

The quantity $\frac{dy}{dx}$ is also known as the **derivative** of y with respect to x. It is often abbreviated to $y'(x)$ or simply y', pronounced 'y prime' or 'y dash'.

The quantity $y'(1)$ means the derivative evaluated when $x=1$. Thus $y'(1)$ is the gradient of the tangent to $y(x)$ where $x=1$. It is also the instantaneous rate of change of $y(x)$ at the point where $x=1$.

Similarly, $y'(-5)$ means the derivative evaluated when $x=-5$. This is the gradient of the tangent where $x=-5$. It is also the instantaneous rate of change of $y(x)$ where $x=-5$.

The process of finding the derivative of a function is called **differentiation**, and when this is done by means of the formula just introduced, it is called **differentiation from first principles**.

Example 21.10

(a) Find the rate of change of the function $y(x)=x^2$ at an arbitrary point A $(x, y(x))$ using differentiation from first principles.

(b) Use your result to find the rate of change of the function when $x=3$ and when $x=-4$.

Solution

(a) A graph of $y(x)=x^2$ is shown in Figure 21.15. The point A is shown, as is the nearby point B. The x coordinate at A is x. Because the x coordinate at B is only slightly larger, we write this as $x+\delta x$ as shown. Because $y(x)=x^2$, the y coordinate at A is x^2, while the y coordinate at B is $(x+\delta x)^2$. Then the gradient of the chord AB is

$$\text{gradient of chord AB} = \frac{\text{change in } y}{\text{change in } x}$$

$$= \frac{(x+\delta x)^2 - x^2}{\delta x}$$

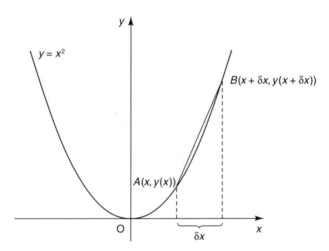

Figure 21.15   Graph of the function $y=x^2$.

$$= \frac{(x^2 + 2x\,\delta x + (\delta x)^2) - x^2}{\delta x}$$

$$= \frac{2x\,\delta x + (\delta x)^2}{\delta x}$$

$$= \frac{\delta x\,(2x + \delta x)}{\delta x}$$

$$= 2x + \delta x$$

This quantity, being the gradient of the chord AB, is only an estimate of the rate of change of $y(x)$ at A. We can improve it by moving B closer and closer to A, as before. Mathematically, this means that δx approaches 0, written $\delta x \to 0$. We must evaluate $2x + \delta x$ as $\delta x \to 0$, that is,

$$\lim_{\delta x \to 0} (2x + \delta x) = 2x$$

We have succeeded in finding the rate of change of the function $y(x)$. In summary,

$$\text{when} \quad y(x) = x^2, \quad \frac{dy}{dx} = 2x$$

(b)   Because dy/dx tells us the rate of change of the function at any point, we can use it to evaluate the rate of change at specific points. So, when $x = 3$, $dy/dx = 6$, that is, the rate of change of the function is 6. Thus, at $x = 3$, the function is increasing at a rate of 6 units per unit increase in x.
        When $x = -4$, $dy/dx = -8$, and so the rate of change of the function is -8. Thus at $x = -4$ the function is decreasing at a rate of 8 units per unit increase in x.

Note from the previous example that when dy/dx is positive, the function increases as x increases. When dy/dx is negative, the function decreases as x increases. The precise value of dy/dx at a point tells us how quickly the function is increasing or decreasing there.

Example 21.11

(a)   Use differentiation from first principles to find the derivative of the function $y(x) = \dfrac{1}{x}$

(b)   Hence find $y'(2)$ and $y'(-3)$.

Solution (a)   We use the formula

$$\text{derivative of } y(x) = \frac{dy}{dx} = \lim_{\delta x \to 0} \frac{y(x + \delta x) - y(x)}{\delta x}$$

We are given that $y(x) = \frac{1}{x}$ and so $y(x + \delta x) = \frac{1}{x + \delta x}$. Therefore

$$\frac{y(x + \delta x) - y(x)}{\delta x} = \frac{\frac{1}{x + \delta x} - \frac{1}{x}}{\delta x}$$

$$= \frac{\frac{x - (x + \delta x)}{(x + \delta x)x}}{\delta x}$$

$$= \frac{-\delta x}{(x + \delta x)x} \bigg/ \delta x$$

$$= \frac{-\delta x}{(x + \delta x)x \, \delta x}$$

$$= \frac{-1}{x^2 + x \, \delta x}$$

We must now let $\delta x \to 0$ to find the derivative. That is,

$$\lim_{\delta x \to 0} \frac{-1}{x^2 + x \, \delta x} = -\frac{1}{x^2}$$

In summary

$$\text{when } y(x) = \frac{1}{x}, \quad \frac{dy}{dx} = -\frac{1}{x^2}$$

(b)   Because dy/dx tells us the rate of change of the function at any point, we can use it to evaluate the rate of change at specific points. In particular, $y'(2) = -1/2^2 = -\frac{1}{4}$. This means that when $x = 2$, y is decreasing at a rate of $\frac{1}{4}$ unit per unit increase in x. Similarly, $y'(-3) = -1/(-3)^2 = -\frac{1}{9}$, so that when $x = -3$, y is decreasing at a rate of $\frac{1}{9}$ unit per unit increase in x.

Self-assessment questions 21.5

1.   Explain the term 'small increment in x'.

2.   What is the distinction between $\frac{\delta y}{\delta x}$ and $\frac{dy}{dx}$?

3.   Explain what is meant by $y'(x)$.

Exercises 21.5

1.   Consider the function $y(x)=3x^2$.

  (a) Find $\dfrac{dy}{dx}$ from first principles.

  (b) Hence find $y'(3)$.

2.   Find the rate of change of $y(x)=7x^2$ at the point where $x=2$ using differentiation from first principles.

3.   Find the rate of change of $y(x)=2x^3$ at an arbitrary point x using differentiation from first principles.

4.   Find the gradient of the tangent of $f(t)=\dfrac{3}{t}$ at the point where $t=1$, using differentiation from first principles.

5.   Use differentiation from first principles to find y' when $y(x)=7$.

Review exercises 21

1   An object changes its velocity from $10\ \mathrm{m\,s^{-1}}$ to $20\ \mathrm{m\,s^{-1}}$ in $15\ \mathrm{s}$. Find the average acceleration in that time.

2   Find the average rate of change of $y(x)=7x^2$ over the interval $[0, 3]$.

3   Differentiate from first principles
  (a) $y(x)=3x^2+100$ (b) $y(x)=5x^3$
  (c) $y(x)=x^2+x$

4   Calculate the average rate of change of $y(x)=e^x$ across the interval $[-1, 1]$.

5   The function $y(x)$ is given by $y(x)=3x^2+4$.
  (a) Calculate y' using differentiation from first principles.
  (b) Find the gradient of the tangent to $y(x)$ when $x=1$.
  (c) Hence find the equation of the tangent to $y(x)$ at $x=1$.

6   Find the rate of change of $y(x)=x-x^2$ at
  (a) $x=1$ (b) $x=0$ (c) $x=-1$

22

Some Rules of Differentiation

KEY POINTS

This chapter

- gives several rules and formulae that enable the derivatives of a variety of functions to be found

- explains the meaning of the terms 'turning point', 'local maximum point' and 'local minimum point' of a function, and shows how differentiation can be used to find such points

- shows how to calculate the second derivative of a function, and illustrates its application to identifying turning points

---

CONTENTS

22.1 Introduction

This chapter continues the development of differentiation started in Chapter 21. A number of rules are given that allow the derivatives of a variety of functions to be found, thereby avoiding the need to differentiate from first principles. Then the application of differentiation to finding maximum and minimum points is introduced.

22.2 Derivative of x^n

A rule exists for differentiating any function of the form $y(x)=x^n$, for example $y(x)=x^7$ or $y(x)=x^{-2}$. This rule states

$$\text{if} \quad y(x)=x^n \quad \text{then} \quad \frac{dy}{dx}=nx^{n-1}$$

The rule applies whether n is positive, negative or fractional. Consider the following examples.

Example 22.1

If $y(x)=x^8$ find $\frac{dy}{dx}$.

Solution   Using the rule with $n=8$, we find

$$\frac{dy}{dx}=8x^{8-1}=8x^7$$

Example 22.2

(a)   If $y(x)=x$, find $\frac{dy}{dx}$.

(b)   If $y(x)=\frac{1}{x}$, find $\frac{dy}{dx}$.

Solution   (a)   Note that $y(x)=x$ can be written as $y(x)=x^1$, and so using the rule with $n=1$ gives

$$\frac{dy}{dx}=1x^{1-1}=1x^0=1$$

(b)   Note that $y(x) = \dfrac{1}{x}$ can be written as $y(x) = x^{-1}$. So using the rule with $n = -1$ gives

$$\frac{dy}{dx} = -1x^{-2} = -\frac{1}{x^2}$$

Example 22.3

Find $\dfrac{dy}{dx}$ if $y(x) = x^{\frac{3}{2}}$.

Solution   Using the rule with $n = \frac{3}{2}$ gives $\dfrac{dy}{dx} = \frac{3}{2}x^{\frac{1}{2}}$.

Example 22.4

Use the rule to differentiate the following functions:

(a) $y(x) = x^{13}$ (b) $y(x) = x^{-3}$ (c) $y(x) = x^{\frac{1}{2}}$ (d) $y(x) = \sqrt{x}$ (e) $y(x) = x^{\frac{2}{3}}$

(f) $y(x) = \dfrac{1}{x^7}$

Solution   (a)   Using the rule with $n = 13$, we find $y' = 13x^{12}$.

(b)   Using the rule with $n = -3$, we find $y' = -3x^{-4} = -\dfrac{3}{x^4}$.

(c)   Using the rule with $n = \frac{1}{2}$, we find $y' = \frac{1}{2}x^{-\frac{1}{2}}$.

(d)   Since $\sqrt{x} = x^{\frac{1}{2}}$, we can use the result of (c) to give $y' = \frac{1}{2}x^{-\frac{1}{2}}$.

(e)   Using the rule with $n = \frac{2}{3}$, we find $y' = \frac{2}{3}x^{-\frac{1}{3}}$.

(f)   Note that $\dfrac{1}{x^7} = x^{-7}$. Therefore, using the rule with $n = -7$, we find

$$y' = -7x^{-8} \text{ or } -\frac{7}{x^8}.$$

Self-assessment question 22.2

1.   State the rule for differentiating functions of the form $y(x) = x^n$.

Exercises 22.2

1. Use the rule to differentiate the following functions:

   (a) $y(x) = x^{16}$ (b) $y(x) = \dfrac{1}{x^2}$

   (c) $y(x) = x^{\frac{1}{3}}$ (d) $y(x) = x^{-2}$

   (e) $y(x) = \sqrt[3]{x}$ (f) $y(x) = x^{\frac{7}{2}}$

2. Use the rule to differentiate $y(x) = 1$. (Hint: Recall that $x^0 = 1$.)

22.3 Differentiation as an operator

When we are given a function $y(x)$ and are asked to find dy/dx, we are being instructed to carry out an operation on the function $y(x)$. In this case the operation is that of differentiation. A notation for this operation exists, and is helpful in later work.

$\dfrac{d}{dx}$ stands for the instruction 'differentiate with respect to x'

We call $\dfrac{d}{dx}$ a **differential operator**. For example,

$$\frac{d}{dx}(x^5)$$

is an instruction to differentiate x^5 w.r.t. x; that is,

$$\frac{d}{dx}(x^5) = 5x^4$$

Example 22.5

Find $\dfrac{d}{dx}(x^{-2})$.

Solution   We are required to differentiate x^{-2} with respect to x. Using the rule that if $y(x) = x^n$ then $dy/dx = nx^{n-1}$, with $n = -2$, we find

$$\frac{d}{dx}(x^{-2}) = -2x^{-3} = -\frac{2}{x^3}$$

Exercise 22.3

1. Find

    (a) $\dfrac{d}{dx}(x^8)$ (b) $\dfrac{d}{dx}(x^{-8})$ (c) $\dfrac{d}{dx}\left(\dfrac{1}{x^8}\right)$

    (d) $\dfrac{d}{dx}(x^{-\frac{1}{4}})$ (e) $\dfrac{d}{dx}(x^9)$ (f) $\dfrac{d}{dx}(x^{-9})$

    (g) $\dfrac{d}{dx}(x^{1.5})$

22.4 Independent variables other than x

Recall from Chapter 6 that, given a function $y(x)$, x is the independent variable and y is the dependent variable, because the value of y depends upon the value chosen for x. In order to solve a variety of engineering problems, it is essential to be able to carry out differentiation using variables other than y and x. In many engineering problems the independent variable is time t, and a 'dot' notation is often used to denote differentiation. Thus if v is a function of t then the derivative of v with respect to t is written as $\dot{v}$, pronounced 'v dot'.

Example 22.6

Modify the formula of Section 22.2 to the case when the independent variable is t. Hence find dy/dt when $y(t) = t^4$.

Solution   Modifying the formula so that the independent variable is t, we find

$$\text{if}\quad y(t) = t^n \quad\text{then}\quad \frac{dy}{dt} = nt^{n-1}$$

Hence if $y(t) = t^4$, $\dfrac{dy}{dt} = \dfrac{d}{dt}(t^4) = 4t^{4-1} = 4t^3$. Equivalently, we could write

$$\dot{y} = 4t^3$$

Example 22.7 Electric current in a wire

Electric current in a wire is a flow of charged particles. The electric current i can be found from the formula

$$i = \frac{dq}{dt}$$

where q is electric charge and t is time. In effect, the current is a measure of the rate at which charge passes a particular point. Note that to relate the

Michael Faraday: the father of electrical engineering

Michael Faraday (see Figure 1) was born at Newington, near London, in 1791. At the age of 12 he went to work for a bookseller and became apprenticed as a bookbinder. However, not content to just bind books, he also spent a large amount of time reading their contents. He was especially interested in scientific works. One of the customers was so impressed by the amount that Faraday had taught himself that he obtained tickets for him to go to some public lectures to be given by Sir Humphry Davy at the Royal Institution. Faraday took detailed notes of these lectures and afterwards he copied them out neatly and bound them with a title page which he dedicated to Davy. He then presented them to Davy as a gift. Davy arranged to see Faraday and was so impressed by his character that he offered him a job as a laboratory assistant at the Royal Institution.

Faraday spent many years at the Royal Institution, and during this time he laid the foundations of the electrical engineering industry. His most important discovery was the link between electricity and magnetism. He found that if a coil of wire was moved in a magnetic field then this motion caused an electric current to flow in the coil. This is the basis of the electrical generator. The amount of current that was produced depended on the rate of change of the magnetic field within the coil. Differentiation is concerned with the rate of change of quantities, and so is particularly useful for describing Faraday's work in a precise manner. For example, the relationship between the voltage v induced in a coil and the current i passing through the coil is given by

$$v = L\left(\frac{\mathrm{d}i}{\mathrm{d}t}\right)$$

where L is the inductance of the coil.

Faraday was a modest man who enjoyed his work. Hermann Ludwig Ferdinand von Helmholtz, in a letter to his wife, wrote that Faraday was a simple, gentle man and was as modest as a child. Also, he was most obliging to Helmholtz and showed him all there was to be seen in his laboratory.

Faraday was an experimental scientist, and so some justification is needed for including him in a book on engineering mathematics. Although mathematical symbolism did not feature heavily in Faraday's work, James Clark Maxwell, who used a lot of mathematics in his work, commented that Faraday's method of conceiving of phenomena was a mathematical one even though he did not make use of the conventional approach of mathematical symbols.

Maxwell used Faraday's work as a basis for describing electromagnetic phenomena in the form of a concise set of mathematical equations. He too was a great man. Although it was to be many years before the full potential of electricity was realized, Faraday is rightly acknowledged by modern electrical engineers as one of the most important founders of their discipline. In fact, the Institution of Electrical Engineers still holds an annual Faraday Lecture. The lecture tours the country, with the aim of interesting young people in the subject of electrical engineering. Faraday must have had some glimpse of the potential of electricity. One story is that he was once asked by a politician what use were his discoveries, to which he replied that he did not know but one day the politicians would be able to tax them.

Faraday died in 1867 at the age of 76. Figure 2 shows a cartoon published in 1881 which provides a vivid insight into the way that the potential of electricity was beginning to be understood. It depicts three energy sources: coal, steam and infant electricity. It provides an unwitting tribute to Faraday's genius.

One of the secrets of Faraday's success was his love of his work rather than the superficial rewards of status and power. He thought that the reason people were not more successful in experimental work was because they were more interested in the external rewards from that work than the work itself. The world's praise was more important to them than the joy of being a successful experimental philosopher.

Figure 1 *Michael Faraday: the father of electrical engineering.*

Figure 2 *Punch cartoon published 1881.*

quantities current and charge, we need to use a derivative. We could have written dq/dt as $\dot{q}$.

Example 22.8 Acceleration, velocity and displacement of an object

Consider Figure 22.1, which shows an object moving in a straight line. The displacement s of the object is measured from an origin which corresponds to the position of the object when time t was zero. The velocity v of the object can be obtained using the formula

$$v = \frac{ds}{dt}$$

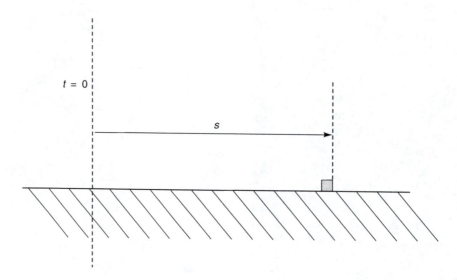

Figure 22.1   The object has travelled a distance s.

In other words, the instantaneous rate of change of displacement w.r.t. time corresponds to the velocity of the object. The acceleration a of the object can be obtained using the formula

$$a = \frac{dv}{dt}$$

In other words, the instantaneous rate of change of velocity w.r.t. time corresponds to the acceleration of the object.

Example 22.9 Voltage across a coil

The voltage v across a coil with inductance L is related to the current i flowing through the coil by

$$v = L\frac{di}{dt}$$

This relationship follows from Faraday's law of electromagnetic induction, which states that the voltage induced in a coil is proportional to the rate of change of magnetic flux in the coil. A changing magnetic flux in a coil corresponds to a changing current in the coil. Note that if the current is changing rapidly, that is, di/dt is large, then the induced voltage is large.

Self-assessment question 22.4

1. Explain the meaning of the notation $\dot{y}$.

Exercises 22.4

1. Find the derivatives of the following functions:

   (a) $f(t) = t^{14}$ (b) $g(z) = \dfrac{1}{z}$ (c) $F(s) = s^{-\frac{1}{2}}$

   (d) $f(t) = \sqrt{t}$

2. Find

   (a) $\dfrac{d}{dt}(t^{\frac{1}{3}})$ (b) $\dfrac{d}{ds}(s^{-3})$

3. Differentiate $f(t) = t^{17}$.

4. Differentiate $F(s) = \dfrac{1}{s^2}$.

5. Differentiate (a) $h(t) = t^{1.7}$

   (b) $A(n) = \dfrac{1}{n^4}$ (c) $x(y) = y^4$

22.5 Using a table of derivatives

Table 22.1 contains a list of common functions and their derivatives. Once the basic principles of differentiation have been mastered, it is quicker to use such a table, rather than differentiating from first principles each time. Repeated use of the table quickly leads to the more common results being memorized. Note that when differentiating trigonometric functions, the angle x must always be measured in radians.

Table 22.1 Table of derivatives.

$y(x)$	$\dfrac{dy}{dx}$	
constant, k	0	
x^n	nx^{n-1}	
$\sin x$	$\cos x$	
$\sin kx$	$k\cos kx$	
$\cos x$	$-\sin x$	x is in radians
$\cos kx$	$-k\sin kx$	k is a constant
$\tan x$	$\sec^2 x$	
$\tan kx$	$k\sec^2 kx$	
e^x	e^x	
e^{kx}	ke^{kx}	k is a constant
$\ln x$	$\dfrac{1}{x}$	
$\ln kx$	$\dfrac{1}{x}$	k is a constant

Example 22.10

Use Table 22.1 to find $\dfrac{dy}{dx}$ when (a) $y(x)=e^{2x}$ (b) $y(x)=e^{-x}$

Solution   From the table, we note that if $y(x)=e^{kx}$ then $\dfrac{dy}{dx}=ke^{kx}$.

(a)   Using this result with $k=2$ means that if $y(x)=e^{2x}$ then $\dfrac{dy}{dx}=2e^{2x}$.

(b)   Using the same result with $k=-1$ means that if $y(x)=e^{-x}$ then $\dfrac{dy}{dx}=$
$-1e^{-x}=-e^{-x}$.

Example 22.11

(a)   Find the derivative of $y(t)=\sin 3t$, and hence deduce the rate of change
of y when $t=4$.

(b)   Find the derivative of $y(x)=e^{-5x}$, and hence deduce the gradient of the
tangent to the curve $y=e^{-5x}$ when $x=0.2$.

(c)   Find the derivative of $y(t)=\ln 5t$.

Solution   (a)   The table is written with independent variable x. However, it can still be
used when other independent variables are involved. From the table, we

deduce that if $y(t)=\sin kt$ then $\dfrac{dy}{dt}=k\cos kt$. Therefore, with $k=3$, we find

$$\text{if}\quad y(t)=\sin 3t\quad\text{then}\quad\frac{dy}{dt}=3\cos 3t$$

When $t=4$ we find $\dfrac{dy}{dt}=3\cos 12=2.53$. Hence when $t=4$, the rate of change of y is 2.53.

(b) From the table, if $y(x)=e^{kx}$ then $\dfrac{dy}{dx}=ke^{kx}$. Therefore when $k=-5$, we have

$$\text{if}\quad y(x)=e^{-5x}\quad\text{then}\quad\frac{dy}{dx}=-5e^{-5x}$$

When $x=0.2$, we find $\dfrac{dy}{dx}=-5e^{-1}=-1.839$. This is the gradient of the tangent to the curve when $x=0.2$.

(c) From the table we deduce that if $y(t)=\ln kt$ then $\dfrac{dy}{dt}=\dfrac{1}{t}$. Therefore when $k=5$, we have

$$\text{if}\quad y(t)=\ln 5t\quad\text{then}\quad\frac{dy}{dt}=\frac{1}{t}$$

Exercises 22.5

1. Use the table of derivatives to differentiate the following functions:
   (a) $f(x)=\cos 3x$ (b) $f(x)=x^3$
   (c) $f(x)=\tan 3x$ (d) $f(x)=\ln 2x$
   (e) $f(x)=e^x$ (f) $f(x)=e^{3x}$
   (g) $f(x)=e^{-2x}$ (h) $f(x)=e^{-4x}$

2. Find the rate of change of the function $f(x)=e^{3x}$ at the point where $x=2$.

3. Find the rate of change of the function $f(t)=\sin t$ at the time when $t=1$.

4. Calculate the gradient of the tangent to the curve $y=x^7$ when $x=-1$.

5. Calculate the values of t for which $f(t)=\sin t$ has a derivative equal to zero.

22.6 Differentiation of $k\,f(x)$ and $f(x)\pm g(x)$

22.6.1 Derivative of $k\,f(x)$, where k is a constant

When the derivative of a function is already known, it is a simple matter to find the derivative of a constant multiple of that function. This

involves using the rule

$$\frac{d}{dx}[kf(x)] = k\frac{df}{dx}$$

That is, the derivative of k times a function is simply k times the derivative of the function. Consider the following example.

Example 22.12

The derivative of x^5 is $5x^4$. Find the derivatives of
(a) $3x^5$ (b) $-17x^5$ (c) $0.01x^5$

Solution   (a)   We have

$$\frac{d}{dx}(3x^5) = 3\frac{d}{dx}(x^5) = 3(5x^4) = 15x^4$$

(b)

$$\frac{d}{dx}(-17x^5) = -17\frac{d}{dx}(x^5) = -17(5x^4) = -85x^4$$

(c)

$$\frac{d}{dx}(0.01x^5) = 0.01\frac{d}{dx}(x^5) = 0.01(5x^4) = 0.05x^4$$

Example 22.13

Find the derivatives of
(a) $f(x) = -\dfrac{2x}{3}$ (b) $g(x) = \dfrac{4x^4}{7}$

Solution   (a)

$$\frac{df}{dx} = \frac{d}{dx}\left(-\frac{2x}{3}\right) = -\frac{2}{3}\frac{d}{dx}(x) = -\frac{2}{3}(1) = -\frac{2}{3}$$

(b)

$$\frac{dg}{dx} = \frac{d}{dx}\left(\frac{4x^4}{7}\right) = \frac{4}{7}\frac{d}{dx}(x^4) = \frac{4}{7}(4x^3) = \frac{16x^3}{7}$$

Example 22.14 Flow of water into a tank

Consider Figure 22.2, which shows a pipe feeding water into a tank. The tank has cross-sectional area A m², the volume of water in the tank is V m³ and the height of the water is h m. The flow rate of water into the tank is q m³ s⁻¹. The rate of increase of the volume of water in the tank is equal to the flow rate of

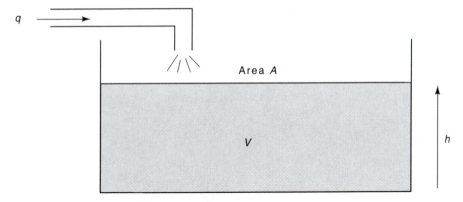

Figure 22.2   Flow of water into a tank.

water into the tank, and so we can write

$$q = \frac{dV}{dt}$$

If the tank has vertical sides then its cross-sectional area is always constant. We can write

$$q = \frac{dV}{dt} = \frac{d(Ah)}{dt} = A\frac{dh}{dt}$$

Therefore the rate at which the height of water increases is proportional to the flow rate of water into the tank.

22.6.2 Derivatives of the sum and difference of two functions

It is frequently necessary to differentiate the sum or difference of two functions, for example functions such as

$$x^2 + \sin x, \quad 3x^3 + 2x^2, \quad 7x^5 - \frac{1}{x^3}$$

To differentiate such functions is straightforward: we simply differentiate each of the terms. Mathematically, if f and g are functions of x then

$$\frac{d}{dx}[f(x)+g(x)] = \frac{df}{dx} + \frac{dg}{dx}$$

and

$$\frac{d}{dx}[f(x)-g(x)] = \frac{df}{dx} - \frac{dg}{dx}$$

Example 22.15

Find $\dfrac{dy}{dx}$ when

(a) $y = x^2 + x$ (b) $y = 0.5x^2 - 16$

Solution   (a)   We have already established in Section 22.2 that

$$\frac{d}{dx}(x^2) = 2x, \quad \frac{d}{dx}(x) = 1$$

Therefore

$$\frac{d}{dx}(x^2 + x) = \frac{d}{dx}(x^2) + \frac{d}{dx}(x)$$

$$= 2x + 1$$

That is,

$$\frac{dy}{dx} = 2x + 1$$

(b) $\dfrac{d}{dx}(0.5x^2 - 16) = \dfrac{d}{dx}(0.5x^2) - \dfrac{d}{dx}(16)$

$$= x - 0$$

$$= x$$

Example 22.16 Particle moving in a straight line

The distance *s* travelled by a particle moving in a straight line is given by

$$s = 8t + 3t^2 + t^3$$

where *t* is the time for which the particle has been moving. Derive expressions for the speed *v* and acceleration *a* of the particle.

Solution   The speed of the particle is given by $v = \dfrac{ds}{dt}$, and so

$$v = \frac{ds}{dt} = 8 + 3(2t) + 3t^2 = 8 + 6t + 3t^2$$

The acceleration of the particle is given by $a = \dfrac{dv}{dt}$, and so

$$a = \frac{dv}{dt} = 0 + 6 + 3(2t) = 6 + 6t$$

Self-assessment question 22.6

1. State the rules for differentiating $kf(x)$ and $f(x) \pm g(x)$.

Exercises 22.6

1. Differentiate the following functions:
   (a) $y(x) = 4x^3 + 2x^2$ (b) $f(x) = 3 - 2x$
   (c) $f(x) = 9x^{\frac{1}{2}} - 3x^4$

2. Differentiate the following functions:
   (a) $y(t) = 3t^2 - 3t^4 + 7t$ (b) $f(t) = \sin t + e^t$
   (c) $f(t) = 4 \sin t$

3. Find the derivatives of the following functions:
   (a) $f(x) = e^x + \cos x$
   (b) $f(t) = 6t^3 - 3t^2 + 7$
   (c) $g(t) = \sin t + 4t^2$
   (d) $f(s) = s^{\frac{1}{2}} - \sqrt{s}$
   (e) $f(x) = 3 \sin x - \cos x$
   (f) $f(x) = \dfrac{e^x - 2 \sin x}{3}$

4. Find $f'(1)$ where $f(x)$ is given by
   (a) $e^x - x$ (b) $3 + \cos x$
   (c) $3x - \dfrac{3}{x} + 3e^x + 3 \sin x$ (d) $\dfrac{1}{2x} + 1$
   (e) $\ln x$ (f) $3 \ln x + x$

5. By using the law of logarithms $\ln x^n = n \ln x$, find the derivative of
   (a) $f(x) = \ln x^2$ (b) $f(x) = \ln \sqrt{x}$

6. Differentiate (a) $3 \ln x$ (b) $\ln (3x^2)$
   (c) $\ln (2x^3)$

7. Find the rates of change of the following functions when $x = 2$:
   (a) $y = e^{0.3x}$ (b) $y = 3e^x$ (c) $y = \dfrac{3}{e^x}$
   (d) $y = \dfrac{1}{3e^x}$ (e) $y = \dfrac{1}{e^{3x}}$

22.7 The product rule

Suppose that $u(x)$ and $v(x)$ are two functions of x, and we wish to differentiate the product $y = u(x)v(x)$. Then the **product rule** is given by the following formula:

$$\text{if} \quad y = u(x)v(x) \quad \text{then} \quad \frac{dy}{dx} = u\frac{dv}{dx} + v\frac{du}{dx}$$

In words, we have '$\dfrac{dy}{dx}$ equals the first function times the derivative of the second plus the second function times the derivative of the first'.

Consider the following examples.

Example 22.17

If $y = xe^{2x}$, find $\dfrac{dy}{dx}$.

Solution   The function y is clearly a product of the two functions x and e^{2x}. In order to apply the product rule, let

$$u(x) = x \quad \text{and} \quad v(x) = e^{2x}$$

We shall require the derivatives of these two functions; these are

$$\frac{du}{dx} = 1 \quad \text{and} \quad \frac{dv}{dx} = 2e^{2x}$$

Applying the product rule, we find

$$\frac{dy}{dx} = u\frac{dv}{dx} + v\frac{du}{dx}$$

$$= x(2e^{2x}) + e^{2x}(1)$$

$$= 2xe^{2x} + e^{2x}$$

$$= e^{2x}(2x + 1)$$

Example 22.18

If $f(t) = t^{\frac{1}{2}}\cos t$, find $\dfrac{df}{dt}$.

Solution   We are required to find the derivative of the product of the functions $t^{\frac{1}{2}}$ and $\cos t$. Let

$$u(t) = t^{\frac{1}{2}} \quad \text{and} \quad v(t) = \cos t$$

Then

$$\frac{du}{dt} = \tfrac{1}{2}t^{-\frac{1}{2}} \quad \text{and} \quad \frac{dv}{dt} = -\sin t$$

Applying the product rule, we find

$$\frac{df}{dt} = u\frac{dv}{dt} + v\frac{du}{dt}$$

$$= t^{\frac{1}{2}}(-\sin t) + (\cos t)(\tfrac{1}{2}t^{-\frac{1}{2}})$$

$$= -t^{\frac{1}{2}}\sin t + \tfrac{1}{2}t^{-\frac{1}{2}}\cos t$$

Self-assessment questions 22.7

1. Describe what is meant by the product of two functions.

2. State the product rule for differentiation.

3. For which of the following functions might it be appropriate to use the product rule for differentiation. Given reasons for your decisions.
   (a) $f(x) = (x^2 + 3x - 7)e^{-5x}$ (b) $f(x) = e^{\sin x}$ (c) $f(x) = (\sin x)(\ln x)$
   (d) $f(x) = \sin(\ln x)$ (e) $f(x) = x^{-\frac{1}{3}}$ (f) $f(x) = 150e^{-x}$ (g) $f(x) = 3\pi$

Exercises 22.7

1. Use the product rule to find the derivative of each of the following functions:
   (a) $f(x) = x^3 \tan x$
   (b) $f(x) = (x^2 + x)\cos 2x$
   (c) $f(x) = e^{-x}\cos 2x$
   (d) $f(x) = \dfrac{1}{x}\cos x$
   (e) $f(x) = x^{\frac{1}{2}}\sin 3x$
   (f) $f(t) = t\cos 3t$
   (g) $f(x) = \sqrt{x}\ln x$

2. Use the product rule to find y' when
   (a) $y = (3x^3 + 2x^2)\cos x$

   (b) $y = 4(\cos 2x)e^{2x}$
   (c) $y = \dfrac{1}{x^2}\cos x$
   (d) $y = x^{\frac{3}{2}}\tan 3x$

3. Find a value of t for which the rate of change of $y = te^{-t}$ equals zero.

4. Find the rate of change of $y = t^2e^{2t}$ when $t = 0.5$.

5. Find the rate of change of $y = t\cos t$ when $t = 1$. Remember to set your calculator to radian mode.

22.8 The quotient rule

Suppose that $u(x)$ and $v(x)$ are two functions of x, and we wish to differentiate the quotient function $y(x) = \dfrac{u(x)}{v(x)}$. The **quotient rule** is given by the following formula:

$$\text{if } y = \frac{u(x)}{v(x)} \quad \text{then} \quad \frac{dy}{dx} = \frac{v\dfrac{du}{dx} - u\dfrac{dv}{dx}}{v^2}$$

Consider the following examples.

Example 22.19

If $y(x) = \dfrac{\sin x}{x}$, find $\dfrac{dy}{dx}$.

Solution   We are required to find the derivative of the quotient of the functions $\sin x$ and x. In order to apply the quotient rule let

$$u(x) = \sin x \quad \text{and} \quad v(x) = x$$

We shall require the derivatives of these functions:

$$\frac{du}{dx} = \cos x \quad \text{and} \quad \frac{dv}{dx} = 1$$

Then, from the quotient rule, we find

$$\frac{dy}{dx} = \frac{v\dfrac{du}{dx} - u\dfrac{dv}{dx}}{v^2}$$

$$= \frac{x \cos x - (\sin x) \cdot 1}{x^2}$$

$$= \frac{x \cos x - \sin x}{x^2}$$

Example 22.20

If $y(t) = \dfrac{e^{-t}}{t^2 + 2}$, find $\dfrac{dy}{dt}$.

Solution   We are required to find the derivative of the quotient of the functions e^{-t} and $t^2 + 2$. So, let

$$u(t) = e^{-t} \quad \text{and} \quad v(t) = t^2 + 2$$

The derivatives of these functions are given by

$$\frac{du}{dt} = -e^{-t} \quad \text{and} \quad \frac{dv}{dt} = 2t$$

Then, applying the quotient rule, we find

$$\frac{dy}{dt} = \frac{v\dfrac{du}{dt} - u\dfrac{dv}{dt}}{v^2}$$

$$= \frac{(t^2+2)(-e^{-t}) - e^{-t}(2t)}{(t^2+2)^2}$$

$$= \frac{-e^{-t}(t^2+2t+2)}{(t^2+2)^2}$$

Self-assessment questions 22.8

1.   Describe what is meant by the quotient of two functions.

2.   State the quotient rule for differentiation.

Exercises 22.8

1.   Use the quotient rule to differentiate the following functions:

   (a) $\dfrac{x^2}{x^2+1}$ (b) $\dfrac{\sin 2x}{3x}$ (c) $\dfrac{4t^2+1}{e^t}$

   (d) $\dfrac{\cos 2t}{2t+1}$ (e) $\dfrac{x}{e^x+1}$ (f) $\dfrac{\ln x}{x}$

   (g) $\dfrac{x}{\ln x}$ (h) $\dfrac{2x}{e^{-x}+1} + \dfrac{e^{-x}+1}{2x}$

   (i) $\dfrac{4-t}{4+t}$ (j) $\dfrac{3z^2-z}{z+1}$

2.   Use the quotient rule to find y' when

   (a) $y = \dfrac{\tan x}{2x}$ (b) $y = \dfrac{3x^2}{\sin x}$

   (c) $y = \dfrac{x^2+3x+2}{e^x}$

3.   Use the product or quotient rule as appropriate to find f' when

   (a) $f = \sqrt{x}\cos x$ (b) $f = (\cos x)e^{-x}$

   (c) $f = \dfrac{\cos x}{e^x}$ (d) $f = t \ln t$

   (e) $f = \dfrac{1}{x+1}\sin x$

4.   Use the quotient rule to differentiate

   (a) $f(x) = \dfrac{3x+2}{3x-2}$ (b) $f(x) = \dfrac{2x^2+3x+1}{3x-3}$

5.   Differentiate

   (a) $g(x) = \dfrac{x+2}{x+2}$ (b) $f(x) = \dfrac{3x+6}{x+2}$

   (c) $f(t) = \dfrac{t^2-1}{t+1}$

6. Recall that $\tan x = \dfrac{\sin x}{\cos x}$. Use the quotient rule to show that

$$\frac{d}{dx}(\tan x) = \sec^2 x$$

7. Differentiate

(a) $\dfrac{x \sin x}{e^x}$

(b) $\dfrac{e^x \sin x}{x}$

(c) $\dfrac{x e^x}{\sin x}$

(d) $\dfrac{x}{e^x \sin x}$

(e) $\dfrac{e^x}{x \sin x}$

(f) $\dfrac{\sin x}{x e^x}$

22.9 The chain rule

Suppose we are given a function $y(x)$, where the variable x is itself a function of another variable, t say. We say that y is a **function of a function**. For example, suppose that

$$y(x) = x^3 \quad \text{and} \quad x(t) = \sin t$$

Then we can write

$$y = x^3 = (\sin t)^3, \quad \text{which can be written as } \sin^3 t$$

The **chain rule** states

$$\frac{dy}{dt} = \frac{dy}{dx} \times \frac{dx}{dt}$$

The application of the chain rule to finding the derivative of $\sin^3 t$ is shown in the next example.

Example 22.21

If $y(t) = \sin^3 t$ find $\dfrac{dy}{dt}$.

Solution To emphasize that $y(t) = \sin^3 t$ is a function of a function, we note that

$$y(t) = (\sin t)^3$$

We then make the substitution $x = \sin t$, so that

$$y = (\sin t)^3 = x^3$$

We have

$$y = x^3, \quad \text{and so} \quad \frac{dy}{dx} = 3x^2$$

Also
$$x = \sin t, \quad \text{and so} \quad \frac{dx}{dt} = \cos t$$

Then using the chain rule we can write
$$\frac{dy}{dt} = \frac{dy}{dx} \times \frac{dx}{dt}$$
$$= 3x^2 \times \cos t$$
$$= 3(\sin t)^2 \cos t \quad (\text{since } x = \sin t)$$
$$= 3 \sin^2 t \cos t$$

Thus the derivative of $\sin^3 t$ has been found to be $3 \sin^2 t \cos t$.

Example 22.22

Use the chain rule to find $\dfrac{dy}{dt}$ when $y = \sqrt{3t+1}$.

Solution Note that $y = \sqrt{3t+1}$ is a function, namely the square root, of another function, $3t+1$. Suppose we make a substitution and let $x = 3t+1$. Then
$$y = \sqrt{3t+1} \quad \text{can be written as} \quad y = \sqrt{x}$$

To use the chain rule, we require $\dfrac{dy}{dx}$ and $\dfrac{dx}{dt}$. These are found as follows:

$$\text{if} \quad y = \sqrt{x} = x^{\frac{1}{2}} \quad \text{then} \quad \frac{dy}{dx} = \tfrac{1}{2} x^{-\frac{1}{2}}$$

$$\text{if} \quad x = 3t+1 \quad \text{then} \quad \frac{dx}{dt} = 3$$

Then, from the chain rule,
$$\frac{dy}{dt} = \frac{dy}{dx} \times \frac{dx}{dt}$$
$$= \tfrac{1}{2} x^{-\frac{1}{2}} \times 3$$
$$= \tfrac{3}{2}(3t+1)^{-\frac{1}{2}}$$

Example 22.23

By making the substitution $u = x^2 + 3x$, use the chain rule to find $\dfrac{dy}{dx}$ when $y = \ln (x^2 + 3x)$.

Solution   Letting $u = x^2 + 3x$, we find

$$y = \ln u, \quad \text{so that} \quad \frac{dy}{du} = \frac{1}{u}$$

In this example the chain rule becomes

$$\frac{dy}{dx} = \frac{dy}{du} \times \frac{du}{dx}$$

From $u = x^2 + 3x$, we see that $\dfrac{du}{dx} = 2x + 3$. Applying the chain rule, we find

$$\frac{dy}{dx} = \frac{1}{u} \times (2x + 3) = \frac{2x + 3}{x^2 + 3x}$$

Thus the derivative of $\ln(x^2 + 3x)$ is $\dfrac{2x + 3}{x^2 + 3x}$. Note also that the numerator is the derivative of the denominator.

The result of the previous example can be generalized to any function of the form $y = \ln f(x)$. We have

$$\text{if} \quad y = \ln f(x) \quad \text{then} \quad \frac{dy}{dx} = \frac{f'(x)}{f(x)}$$

Self-assessment questions 22.9

1.   Decide which rule – product, quotient or chain – it might be appropriate to use in order to differentiate the following functions:

   (a) $e^x \ln x$ (b) $\sin \sqrt{x}$ (c) $\dfrac{1}{x^2} \ln x$ (d) $\sin \dfrac{1}{x}$ (e) $\dfrac{1}{x+1}$

2.   State the chain rule for differentiation.

Exercises 22.9

1. Find $\dfrac{dy}{dx}$ when

(a) $y=\cos 4x^2$ (b) $y=\cos(2x+1)$

(c) $y=\sin(x^2+3x+4)$ (d) $y=\tan\dfrac{1}{x}$

(e) $y=e^{x^2}$ (f) $y=e^{x^2+3x+2}$

2. Find $\dfrac{df}{dt}$ when

(a) $f=(6t+1)^5$ (b) $f=(4t+3)^{-2}$
(c) $f=\tan e^t$ (d) $f=\sqrt{t+1}$
(e) $f=\sin(3t^2+2t)$

3. If $z=6t^2$ and $y=3z-7$, find

(a) $\dfrac{dz}{dt}$ (b) $\dfrac{dy}{dz}$ (c) $\dfrac{dy}{dt}$

4. Find $y'(2)$ where $y(t)$ is

(a) $\dfrac{1}{(1+t)^2}$ (b) $(3+2t)^{1.6}$

(c) $\sin(2+t^2)$ (d) e^{t^2} (e) e^{t^2+1}

(f) $\sqrt{t^2+1}$ (g) $\sqrt{3t^2+1}$

5. Find $\dfrac{dy}{dx}$ when

(a) $y=\ln(3x^2+3x+9)$
(b) $y=\ln(t^7+4t)$ (c) $\ln(\sin t)$

22.10 Maxima and minima

22.10.1 Turning points

Consider the graph of the function $y(x)$ shown in Figure 22.3. Both B and D are important points on this curve. The curve is rising as we move from A to B. In other words, the gradient of the curve here is positive. As we move from B to C, the curve is falling. In other words, the gradient of the curve is negative

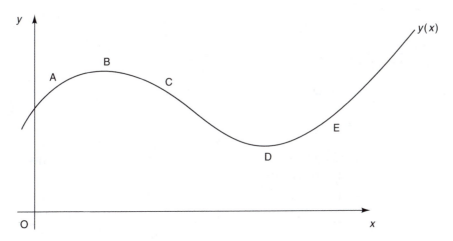

Figure 22.3 Graph of $y(x)$.

here. For that part of the curve near to B, B is the highest point. We call point B a **local maximum**. A tangent drawn at B is parallel to the *x* axis; that is, it has zero gradient. Hence the gradient of the curve at B is zero. Clearly, as we pass through the point B from left to right, the gradient of the curve is first positive, then zero, then negative.

The curve is falling as we move from C to D. The gradient of the curve here is negative. As we move from D to E, the curve is rising. The gradient of the curve is positive here. For that part of the curve near to D, D is the lowest point. We call D a **local minimum**. A tangent drawn at D is parallel to the *x* axis, and hence the gradient at D is also zero. As we pass through the point D from left to right, the gradient of the curve is first negative, then zero, then positive.

It is common practice to drop the word 'local' and to refer to each of B and D as simply a maximum or a minimum. Note that the plural of maximum is **maxima**, and the plural of minimum is **minima**. Together, points such as B and D are called **turning points** because the gradient of the curve changes from positive to negative, or vice versa. At such points the tangent to the graph is parallel to the *x* axis, that is, its gradient is zero. Furthermore, because the gradient of the tangent at B and at D is zero, the derivative d*y*/d*x* will be zero too. We can use this as a means of finding local maxima and minima.

At maximum and minimum points, the gradient of the curve is zero; that is,

$$\frac{dy}{dx} = 0 \quad \text{at a turning point}$$

Consider the following example.

Example 22.24

Find the turning points of $y = x^2 + 4x + 8$.

Solution   By differentiating $y = x^2 + 4x + 8$, we obtain

$$\frac{dy}{dx} = 2x + 4$$

At a turning point, d*y*/d*x* = 0, and so we can locate any turning points by setting $2x + 4$ equal to 0. That is,

$$2x + 4 = 0$$
$$2x = -4$$
$$x = -2$$

We conclude that there is a turning point when $x=-2$. We can find the y coordinate of this turning point from the equation $y=x^2+4x+8$; that is, $y(-2)=(-2)^2+4(-2)+8=4$. So the point with coordinates $(-2, 4)$ is a turning point. A further investigation is needed to decide whether $(-2, 4)$ is a maximum or minimum. The graph of the function $y=x^2+4x+8$ is shown in Figure 22.4, where the turning point, in this case a minimum, can be clearly seen.

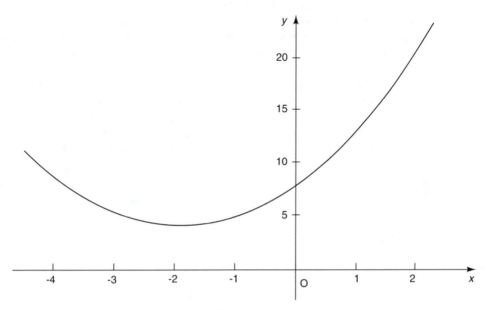

Figure 22.4   Graph of the function $y=x^2+4x+8$.

Note that solving $dy/dx=0$ only locates the position of a turning point. It does not tell us whether a turning point is a maximum or a minimum point. To distinguish between maximum and minimum points without drawing a graph, we can find the gradient of the function on either side of the turning point.

Example 22.25

Find the turning point of the function $y=-x^2+2x-1$ and determine its type.

Solution   By differentiating $y=-x^2+2x-1$ we obtain

$$\frac{dy}{dx}=-2x+2$$

At a turning point, $dy/dx=0$ and so we can locate any turning points by setting $-2x+2$ equal to 0. That is,

$$-2x+2=0$$

$$2x=2$$

$$x=1$$

We conclude that there is a turning point when $x=1$. We can find the y coordinate of this turning point from the equation $y=-x^2+2x-1$; that is, $y(1)=-1^2+2(1)-1=0$. So the point with coordinates $(1, 0)$ is a turning point. To determine the type of turning point, we look at the gradient on either side of the point $x=1$.

A little to the left of $x=1$, say at $x=0$, we find that $dy/dx=-2(0)+2=2$.

A little to the right, say at $x=2$, we find that $dy/dx=-2(2)+2=-2$. Consequently, as we move through the turning point from left to right, the gradient of the curve changes from positive to negative. This behaviour is depicted in Figure 22.5. This reveals that the turning point must be a maximum. Finally, the graph of the function $y=-x^2+2x-1$ is shown in Figure 22.6.

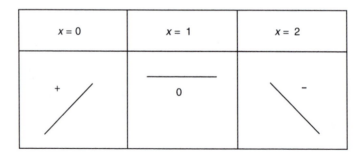

Figure 22.5   In moving from left to right, the gradient changes from positive, through zero, to negative.

Example 22.26

Find the turning points of $y=2x^3-3x^2-12x$, and determine whether they are local maxima or minima.

Solution   Differentiating $y=2x^3-3x^2-12x$, we find

$$\frac{dy}{dx}=6x^2-6x-12$$

At turning points, $dy/dx=0$, so we can locate turning points by solving the equation

$$6x^2-6x-12=0$$

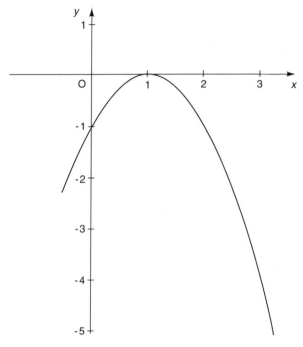

Figure 22.6 Graph of $y = -x^2 + 2x - 1$.

To do this, we take out a factor of 6 and then factorize the quadratic. Thus

$$6(x^2 - x - 2) = 0$$

$$6(x + 1)(x - 2) = 0$$

that is,

$$x = -1 \quad \text{and} \quad x = 2$$

so that turning points will occur when $x = -1$ and when $x = 2$. This function has two turning points. The y coordinates are found from the equation $y = 2x^3 - 3x^2 - 12x$. It is easy to verify that the turning points occur at $(-1, 7)$ and $(2, -20)$.

To determine the type of turning point, we consider the gradient at either side of each point.

For the turning point at $(-1, 7)$, a little to its left, at $x = -2$, we find $dy/dx = 6(-2)^2 - 6(-2) - 12 = 24$, which is positive. A little to the right, at $x = 0$, we find $dy/dx = -12$, which is negative. This is shown in Figure 22.7(a). As we move through the turning point from left to right, the gradient changes from positive to negative, and so the turning point is a maximum.

For the turning point at $(2, -20)$, a little to its left, at $x = 1$, we find $dy/dx = 6(1)^2 - 6(1) - 12 = -12$, which is negative. A little to the right, at $x = 3$,

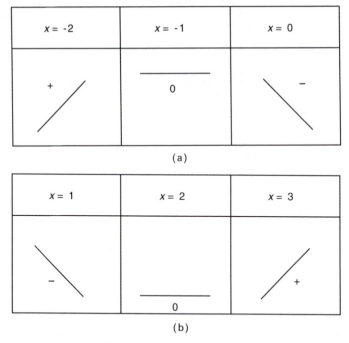

Figure 22.7 (a) In moving from left to right, the gradient changes from positive, through zero, to negative. (b) In moving from left to right, the gradient changes from negative, through zero, to positive.

we find $dy/dx=6(3)^2-6(3)-12=24$, which is positive. This is shown in Figure 22.7(b). As we move through the turning point $(2, -20)$ from left to right, the gradient changes from negative to positive, and so the turning point is a minimum. A graph of the function is shown in Figure 22.8, from which these turning points can be clearly seen.

We now summarize the behaviour of the gradient close to a maximum or minimum point.

> If to the left of the turning point the gradient is positive and to the right the gradient is negative, the point is a maximum.
> If to the left of the turning point the gradient is negative and to the right the gradient is positive, the point is a minimum.

22.10.2 Points of inflection

Sometimes, when trying to find the type of a turning point, the signs of the gradients on both the left and the right of the point where $dy/dx=0$ turn out

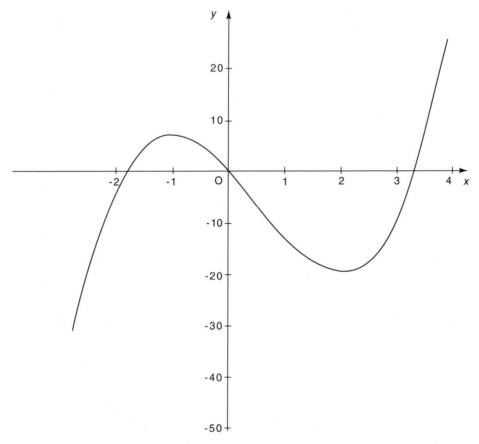

Figure 22.8 The function $y = 2x^3 - 3x^2 - 12x$.

to be the same. This situation is shown in Figure 22.9. When this occurs, the point in question is neither a maximum nor a minimum point, but instead is known as a **point of inflection**.

Example 22.27

(a) Determine any points at which the gradient of $y = x^3 - 3x^2 + 3x - 1$ is zero.

(b) Determine whether such points are maxima, minima or points of inflection.

Solution   (a) By differentiating, we find $y' = 3x^2 - 6x + 3$. Setting $y' = 0$ in order to find where the gradient is zero, we obtain

$$3x^2 - 6x + 3 = 0$$

$$3(x^2 - 2x + 1) = 0$$

$$3(x - 1)(x - 1) = 0$$

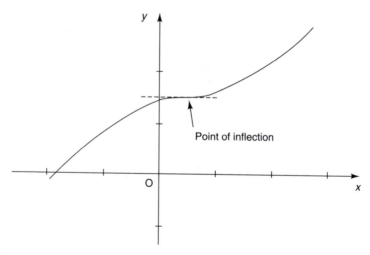

Figure 22.9   The signs of the gradient on both sides of a point of inflection are the same.

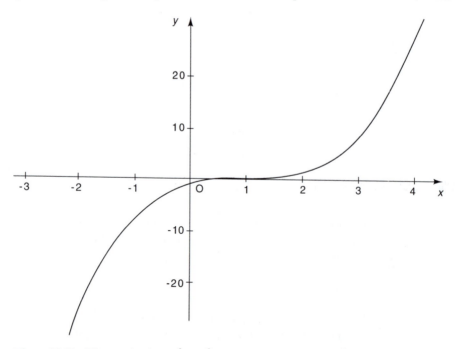

Figure 22.10   The graph of $y = x^3 - 3x^2 + 3x - 1$ has a point of inflection at (1, 0).

so that the gradient is zero only when $x = 1$. The y coordinate when $x = 1$ is 0, and so at the point (1, 0) the gradient is zero.

(b)   A little to the left of $x = 1$, say at $x = 0$, we find that $y' = 3$, that is, positive. A little to the right of $x = 1$, say at $x = 2$, we find that $y' = 3(2)^2 - 6(2) + 3 = 3$, which is also positive. Therefore the point $x=1$ must be a point of inflection. The graph of $y=x^3-3x^2+3x-1$ is shown in Figure 22.10, from which the point of inflection can be seen.

Other, more complicated, points of inflection can occur, but these are beyond the scope of this book.

Self-assessment questions 22.10

1. Explain the distinction between a maximum point and a minimum point.

2. Explain why we sometimes refer to these as 'local'.

3. Describe the mathematical test used to locate a turning point.

Exercises 22.10

1. Copy the graphs shown in Figure 22.11. On each graph mark the sign of dy/dx next to the graph, showing the points where dy/dx$=0$.

2. Determine the positions and types of the turning points of the functions
   (a) $y=x^2$ (b) $y=-x^2$

3. Determine the positions and types of the turning points of the functions
   (a) $f=t^3-6t^2+9t+2$ (b) $f=t^3-6t^2$
   (c) $f=t^3$

4. The function $f(t)=3t^2+\dfrac{k}{t}$ has gradient 4 when $t=1$.
   (a) Calculate the value of k.
   (b) Determine the positions and types of the turning points of $f(t)$.

5. Find the turning points of the function $y=2x^3+9x^2-24x+3$.

6. Find the maxima and minima of $f(t)=2t^3-15t^2+24t$.

22.11 The second derivative

22.11.1 Finding second derivatives

The derivative dy/dx of a function $y(x)$ is more correctly referred to as the **first derivative**. The **second derivative** of y is found by differentiating the first derivative:

$$\text{second derivative of } y(x) = \frac{d}{dx}\left(\frac{dy}{dx}\right)$$

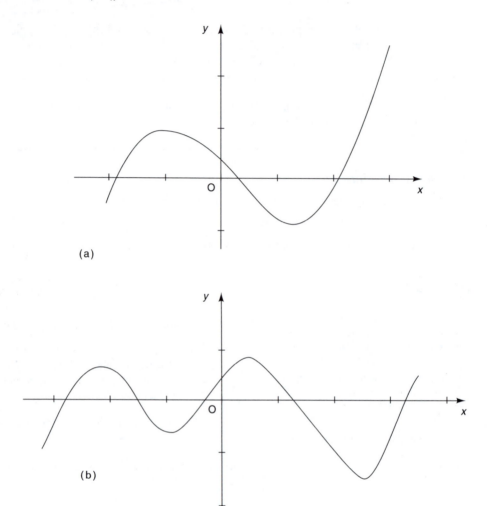

(a)

(b)

Figure 22.11   Graphs for Exercises 22.10, Q1.

This is written $\dfrac{d^2y}{dx^2}$ (or d^2y/dx^2), or simply y''. If f is a function of t, we often

use the dot notation to denote its second derivative and write $\ddot{f}$ for $\dfrac{d^2f}{dt^2}$. We

refer to $\ddot{f}$ as 'f double dot'.

Example 22.28

    Find the first and second derivatives of $y = x^3 + 2x^2$.

Solution   By differentiating $y = x^3 + 2x^2$, we find $dy/dx = 3x^2 + 4x$. The second derivative is found by differentiating the first derivative. That is,

$$\frac{d^2y}{dx^2} = \frac{d}{dx}\left(\frac{dy}{dx}\right) = 6x + 4$$

We could write this as $y' = 3x^2 + 4x$ and $y'' = 6x + 4$.

Example 22.29

Find the first and second derivatives of $f(t) = e^{3t} + \sin 2t$.

Solution   Using Table 22.1, we find

$$\frac{df}{dt} = 3e^{3t} + 2 \cos 2t$$

$$\frac{d^2f}{dt^2} = \frac{d}{dt}\left(\frac{df}{dt}\right) = 9e^{3t} - 4 \sin 2t$$

We could write $\dot{f} = 3e^{3t} + 2 \cos 2t$ and $\ddot{f} = 9e^{3t} - 4 \sin 2t$.

22.11.2 Use of second derivatives to find the type of a turning point

A simple test exists for finding the type of a turning point using the second derivative. Suppose we know the location of a turning point but not its type. We find d^2y/dx^2 and evaluate it at the turning point.

If $\dfrac{d^2y}{dx^2} > 0$ at the turning point, the point is a minimum.

If $\dfrac{d^2y}{dx^2} < 0$ at the turning point, the point is a maximum.

If $\dfrac{d^2y}{dx^2} = 0$ at the turning point, this test does not give us any useful information.

The test using d^2y/dx^2 is an alternative to examining the sign of dy/dx on both sides of the turning point.

Example 22.30

Find the turning points of the function $y = 4x^3 + 15x^2 - 18x + 6$, and determine their type.

Solution   By differentiation, we find

$$\frac{dy}{dx} = 12x^2 + 30x - 18 \quad \text{and} \quad \frac{d^2y}{dx^2} = 24x + 30$$

To find the turning points, we set $dy/dx = 0$; that is,

$$12x^2 + 30x - 18 = 0$$

$$6(2x^2 + 5x - 3) = 0$$

$$6(2x - 1)(x + 3) = 0$$

so that turning points occur at $x = -3$ and $x = \frac{1}{2}$.

To determine the type of turning point that exists at $x = -3$, we evaluate d^2y/dx^2 there. At $x = -3$,

$$\frac{d^2y}{dx^2} = 24(-3) + 30 = -42$$

which is negative. The turning point is therefore a maximum.

At $x = \frac{1}{2}$,

$$\frac{d^2y}{dx^2} = 24(0.5) + 30 = 42$$

which is positive. The turning point is therefore a minimum.

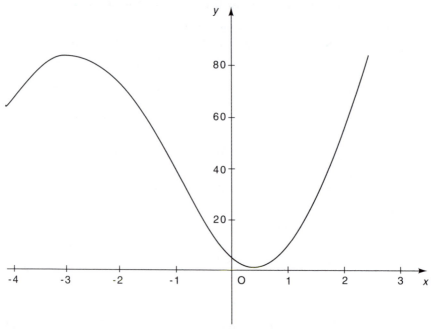

Figure 22.12   The function $y = 4x^3 + 15x^2 - 18x + 6$.

The y coordinates of the turning points can be found by substituting the x values into the equation $y=4x^3+15x^2-18x+6$. Doing this, we find that $(-3, 87)$ is the maximum point and $(\frac{1}{2}, \frac{5}{4})$ is the minimum point. A graph of the function is shown in Figure 22.12.

Self-assessment questions 22.11

1. How is the second derivative of a function $y(x)$ found? Describe the notation used for the second derivative of $y(x)$.

2. State the second-derivative test for finding the type of a turning point.

3. What can you conclude about the type of a turning point if d^2y/dx^2 is zero there?

Exercises 22.11

1. Find $\dfrac{d^2y}{dx^2}$ when

(a) $y=x^3+3x^2+1$ (b) $y=2$
(c) $y=3x+7$ (d) $y=x^6-7x^3$
(e) $y=9x^4$ (f) $y=7x^2+7x+7$

2. Find y'' when

(a) $y=\dfrac{1}{x}$ (b) $y=\dfrac{1}{x^2}$ (c) $y=\dfrac{3}{x}$

(d) $y=\sin x$ (e) $y=\sin x+\cos x$

(f) $y=3x^6+\dfrac{1}{x}$ (g) $y=\tan x$

(h) $y=\tan 2x$ (i) $y=e^x$ (j) $y=e^{2x}$
(k) $y=e^{kx}$ (l) $y=\ln x$

3. Find the second derivative of each of the following:

(a) $y=te^t$ (b) $y=x\sin x$ (c) $y=\dfrac{1}{t+1}$

(d) $y=\sqrt{t+1}$ (e) $y=\sin kx$ (f) $y=\dfrac{e^x}{x}$

4. Determine the type of each turning point of the following functions:
(a) $y=12x^4-3x^2$ (b) $y=3x^3$

5. If $y=xe^{-x}$ find $\dfrac{dy}{dx}$ and $\dfrac{d^2y}{dx^2}$. Show that
$$\dfrac{d^2y}{dx^2}+2\dfrac{dy}{dx}+y=0$$

Computer and calculator exercises 22.11

1. Plot a graph of $f(x)=x^3e^x$ for values of x between -5 and 5. By zooming in on any maxima or minima, determine their location.

2. Plot a graph of $f(x)=\dfrac{\ln x}{x^2}$ for $0<x\leqslant 5$. Locate the position of any local maxima or minima in this interval.

Review exercises 22

1 Find the gradient of the tangent to $f(x) = -x^3$ at $x = 3$.

2 Write down the derivative of
(a) $f(x) = x^{11}$ (b) $f(x) = 3$
(c) $f(x) = \dfrac{4}{x^5}$ (d) $f(x) = -x^{-5}$
(e) $g(t) = -3\sqrt{t}$ (f) $h(x) = \sqrt{7x^3}$
(g) $A(x) = 10x^{3.5}$

3 Find f' when
(a) $f(x) = e^{-5x}$ (b) $f(x) = 3\sin 2x$
(c) $f(x) = 6x^{-7}$ (d) $f(x) = \frac{1}{2}\ln 3x$
(e) $g(t) = -3\sqrt{t}$ (f) $h(x) = \sqrt{7x^3}$

4 Differentiate $f(x) = (1 + 2x)^2$.

5 Differentiate
(a) $f(t) = t^2 - 2t^3$ (b) $g(t) = 5e^{4t}$
(c) $f(z) = \cos 4z + \sin 4z$
(d) $f(t) = \tan 3t$

6 Show that if $y = e^{5x}$ then $\dfrac{dy}{dx} = 5y$.

7 Find values of x where $\dfrac{dy}{dx} = 0$ when $y = x^3 - 5x^2 + 8x - 4$.

8 The pressure P of a gas at constant temperature is inversely proportional to its volume V, that is, $P = \dfrac{K}{V}$, where K is a constant. Find $\dfrac{dP}{dV}$.

9 The period T of a simple pendulum of length l is given by
$$T = 2\pi\sqrt{\dfrac{l}{g}}$$
where g is a constant called the acceleration due to gravity. Find $\dfrac{dT}{dl}$.

10 The voltage v of a device is related to the current i through the device by $v = k(1 + \alpha i + \beta i^2)$, where k, α and β are constants.
Find $\dfrac{dv}{di}$.

11 Find the slopes of the following curves at the given points:
(a) $y = 3x^2$ at $x = 2$
(b) $4y = x^2$ at $x = -1$
(c) $y = \dfrac{3}{x^4}$ at $x = 0.5$
(d) $y = \sqrt{x}$ at $x = 1$

12 If $f(x) = x^2 - 2x + 3$, find a value of x where the gradient of the graph is zero.

13 The average rate of change of a function $f(x)$ from $x = a$ to $x = b$ is 4. If $f(b) = 14$, what is $f(a)$?

14 The height h of the liquid in a chemical storage tank is given by
$$h = 80(1 - e^{-t/50})$$
where t is the time from which the tank starts to be filled. The units of h are m and the units of time are s. Calculate the speed, in m per second, at which the liquid level rises in the tank.

15 A projectile is launched at an angle θ from the horizontal as shown in Figure 22.13. If air resistance is neglected then the formula for the vertical displacement y of the projectile at time t after launch is given by
$$y = (V_0 \sin \theta)t - \frac{1}{2}gt^2$$
where V_0 is the initial speed of the projectile and g is a constant called the acceleration due to gravity. Derive expressions for
(a) the vertical velocity of the projectile,
(b) the vertical acceleration of the projectile, and comment upon the results.

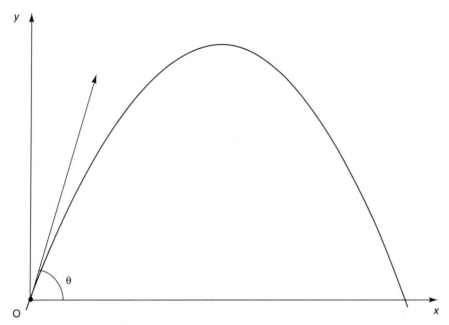

Figure 22.13   Projectile launched at an angle θ.

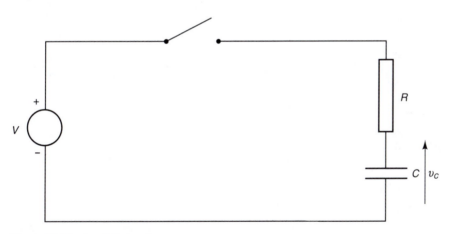

Figure 22.14   An RC circuit.

16   Figure 22.14 shows a capacitor and resistor in series, connected to a DC voltage source of voltage V by means of a switch. Initially, the capacitor has no charge. At time $t=0$, the switch is closed. The voltage v_C across the capacitor is given by

$$v_C = V(1-e^{-t/RC})$$

Carry out the following.

(a)   For a capacitor, the voltage v_C across it and the current i through it are related by

$$i = C\frac{dv_C}{dt}$$

Use this formula to calculate the current flowing through the capacitor in the circuit of Figure 22.14.

(b)   Using Kirchhoff's voltage law, confirm the result of (a) by calculating the current flow through the resistor.

17   Find the position of any local maxima of the function $f(x) = 12 \sin x + 5 \cos x$ in the interval $0 \leqslant x \leqslant \frac{1}{2}\pi$.

18   Find the local maximum and minimum points of $f(x) = x^2 e^x$.

19   Find the types of the turning points of the functions

(a) $f(x) = x(x-2)^3$ (b) $f(x) = \dfrac{x+9}{x^2}$

20   Consider the curve $y = x^2 + 3x - 7$. A point P with x coordinate 2 lies on the curve.
(a) Find the y coordinate of P.
(b) Find the gradient of the tangent to the curve at P.
(c) Determine the equation of the tangent to the curve at P.

21   Find the first derivative of each of the following functions:
(a) $h(t) = t^4 e^{2t}$ (b) $g(t) = (t^4 + 1)e^{2t}$
(c) $m(x) = \sin x \cos x$
(d) $y(t) = (1 + \sin t)^5$ (e) $q(r) = \dfrac{r^2 + 1}{r + 1}$
(f) $A(c) = \ln(3c^2 + 2)$

Integration – the Reverse of Differentiation

KEY POINTS

This chapter

- explains what is meant by an indefinite integral

- explains the term 'integrand'

- explains what is meant by a constant of integration

- provides a table of indefinite integrals

- gives rules for finding integrals of sums, differences and constant multiples of functions

CONTENTS

23.1 Introduction

In the previous two chapters we explained some of the techniques and applications of differentiation. We showed how, given a function, we can determine its derivative. In some applications we know the derivative but do not know the function from which it was derived. This is when we need knowledge of indefinite integrals, which are used to reverse the process of differentiation.

23.2 Indefinite integrals

Suppose we ask what functions we must differentiate to obtain $2x$. From our knowledge of differentiation, we know that differentiating x^2 with respect to x yields $2x$. There are also other functions that we can differentiate to obtain $2x$. Differentiating any of

$$x^2 + 7, \quad x^2 - 3, \quad x^2 + 0.5$$

yields $2x$. In fact, any function of the form $x^2 + c$, where c is a constant, will be the answer to our question. So, starting with the function $2x$, we can reverse the process of differentiation to obtain $x^2 + c$. This reverse process of differentiation is called **indefinite integration** and is given the symbol $\int$. The $\int$ is known as an **integral sign**. Accompanying the integral sign is always a term of the form dx, which indicates the independent variable involved, in this case x. So we write

$$\int 2x \, dx = x^2 + c$$

The constant c is called the **constant of integration**, and it must always be included when finding an indefinite integral.

Example 23.1

(a)   State the derivative of x^3.
(b)   Hence find the indefinite integral of $3x^2$.

Solution   (a)   From our knowledge of differentiation, the derivative of x^3 is $3x^2$.
(b)   Indefinite integration reverses the process of differentiation, and so we write

$$\int 3x^2 \, dx = x^3 + c$$

We always include the additional constant of integration when finding indefinite integrals. Note that our answer can be checked by differentiating $x^3 + c$ to obtain $3x^2$.

More generally, we have the following relationship between derivatives and indefinite integrals:

$$\text{if} \quad \frac{d}{dx}[F(x)] = f(x) \quad \text{then} \quad \int f(x)\,dx = F(x) + c$$

In the expression $\int f(x)\,dx$, the function $f(x)$ is referred to as the **integrand**. When we have calculated $\int f(x)\,dx$, we say $f(x)$ has been **integrated with respect to** x to yield $F(x) + c$.

Example 23.2

In the expression $\int 3t^2\,dt$, state (a) the independent variable (b) the integrand

Solution (a) The independent variable is t. We can tell this by observing the term dt.
            (b) The integrand is $3t^2$.

Example 23.3

(a) State the derivative of $\sin x$.
(b) Find the indefinite integral of $f(x) = \cos x$.

Solution (a) From our knowledge of differentiation, we have $\dfrac{d}{dx}(\sin x) = \cos x$.

(b) The indefinite integral of $\cos x$ is given by

$$\int \cos x\,dx = \sin x + c$$

It is possible to find indefinite integrals by referring to a table such as Table 23.1.

Note that when calculating indefinite integrals using the table, a constant of integration should always be added. It is important to note that when dealing with the trigonometric functions, the variable x must be measured in radians and not degrees. The modulus sign is inserted in the last entry in the table to remind us that only logarithms of positive numbers can be found.

Table 23.1

Function $f(x)$	Indefinite integral $\int f(x)\,dx$	
constant, k	kx	
x	$\dfrac{x^2}{2}$	
x^2	$\dfrac{x^3}{3}$	
x^n	$\dfrac{x^{n+1}}{n+1}$	$n \neq -1$
$\sin x$	$-\cos x$	
$\cos x$	$\sin x$	
$\sin kx$	$\dfrac{-\cos kx}{k}$	
$\cos kx$	$\dfrac{\sin kx}{k}$	
$\tan kx$	$\dfrac{1}{k}\ln|\sec kx|$	
e^x	e^x	
e^{-x}	$-e^{-x}$	
e^{kx}	$\dfrac{e^{kx}}{k}$	
$x^{-1} = \dfrac{1}{x}$	$\ln|x|$	

Example 23.4

Consider the indefinite integral $\int 3\,dx$.
(a)   State the independent variable.
(b)   State the integrand.
(c)   Use Table 23.1 to find $\int 3\,dx$.

Solution   (a)   The independent variable is x. We can tell this by inspecting the term dx.
(b)   The integrand is 3.
(c)   The first entry in Table 23.1 states that $\int k\,dx = kx + c$. With $k = 3$, we find

$$\int 3\,dx = 3x + c$$

Note that we can check our answer by differentiating it. We find

$$\frac{d}{dx}(3x + c) = 3$$

Example 23.5

Use Table 23.1 to find $\int x^7 \, dx$.

Solution   From Table 23.1, we note that

$$\int x^n \, dx = \frac{x^{n+1}}{n+1} + c$$

With $n = 7$, we find

$$\int x^7 \, dx = \frac{x^8}{8} + c$$

Example 23.6

Find $\int \frac{1}{x} \, dx$.

Solution   This integrand occurs in Table 23.1. So, directly from the table, we find

$$\int \frac{1}{x} \, dx = \ln |x| + c.$$

In more advanced work it is sometimes helpful to express this solution in an alternative form. Because the logarithm of a constant is still a constant, it is permissible to write the constant of integration as $\ln |A|$, where A is a constant. The solution would then take the form

$$\int \frac{1}{x} \, dx = \ln |x| + \ln |A|$$

Using the first law of logarithms, this can be written in the alternative form

$$\ln |Ax|$$

We can still use Table 23.1 when variables other than x are involved.

Example 23.7

Find $\int \cos 5t \, dt$.

Solution   In this example the independent variable is t. From Table 23.1,

$$\int \cos kx \, dx = \frac{\sin kx}{k} + c$$

and so, in terms of t, this result becomes

$$\int \cos kt \, dt = \frac{\sin kt}{k} + c$$

Therefore, with $k = 5$, we have

$$\int \cos 5t \, dt = \frac{\sin 5t}{5} + c$$

Note that the answer can always be checked by differentiation.

Example 23.8

Find $\int 16 \, dt$.

Solution   We note that the independent variable is t. From Table 23.1, we have $\int k \, dx = kx + c$. In terms of t, this result becomes

$$\int k \, dt = kt + c$$

and so, with $k = 16$, we have

$$\int 16 \, dt = 16t + c$$

Example 23.9

Find (a) $\int e^t \, dt$ (b) $\int e^{2t} \, dt$ (c) $\int e^{-3t} \, dt$

Solution   In all cases the independent variable is t.
(a)   From Table 23.1, we note that $\int e^x \, dx = e^x + c$. Hence, with independent variable t, we have

$$\int e^t \, dt = e^t + c$$

It is useful to remember that the integral of the exponential function e^t is the same function together with a constant of integration.
(b)   From Table 23.1, we can write

$$\int e^{kt} \, dt = \frac{e^{kt}}{k} + c$$

Hence, with $k = 2$, we find

$$\int e^{2t} \, dt = \frac{e^{2t}}{2} + c$$

(c)   With $k = -3$, we find

$$\int e^{-3t}\,dt = \frac{e^{-3t}}{-3} + c$$

which we would normally write as $-\frac{1}{3}e^{-3t} + c$.

Example 23.10

If $\dfrac{dy}{dx} = x^5$, find y.

Solution   We are given the derivative dy/dx, and must find y. Thus we must reverse the process of differentiation by finding an indefinite integral. We know that

$$\text{if } \frac{dy}{dx} = x^5 \quad \text{then} \quad y = \int x^5\,dx$$

From Table 23.1, we have

$$\int x^n\,dx = \frac{x^{n+1}}{n+1} + c$$

and so, with $n = 5$, we have

$$y = \int x^5\,dx = \frac{x^6}{6} + c$$

Note from Example 23.10 the following relationship:

$$\text{if } \frac{dy}{dx} = f(x) \quad \text{then} \quad y = \int f(x)\,dx$$

Example 23.11

If $\dfrac{dy}{dx} = e^{-0.5x}$, find y.

Solution   From Table 23.1, we note

$$\int e^{kx}\,dx = \frac{e^{kx}}{k} + c$$

So, with $k = -0.5$, we have

$$y = \int e^{-0.5x} \, dx = \frac{e^{-0.5x}}{-0.5} + c$$

$$= -2e^{-0.5x} + c$$

In the previous chapters on differentiation we have seen that there are several important formulae connecting acceleration, velocity and displacement. These have the following integral equivalents.

The velocity v of an object is the derivative of its displacement s; that is,

$$v = \frac{ds}{dt}$$

The displacement is therefore given by

$$s = \int v \, dt$$

The acceleration a is the derivative of the velocity v; that is,

$$a = \frac{dv}{dt}$$

The velocity is therefore given by

$$v = \int a \, dt$$

Example 23.12 Velocity of an object moving in a straight line with constant acceleration

An object moving with velocity u at time $t = 0$ accelerates with constant acceleration a. The velocity of the object at a later time t is v.

(a) Use indefinite integration to find an expression for v.

(b) Given that when $t = 0$ the velocity is u, find the constant of integration.

Solution   (a)   The velocity is found by integrating the acceleration; that is,

$$v = \int a \, dt$$

We are told that in this example the acceleration a is constant. From Table 23.1, we know that $\int k \, dt = kt + c$, and so

$$v = \int a \, dt = at + c$$

where c is a constant of integration. Hence

$$v = at + c \tag{23.1}$$

(b)   We are told that initially, that is when $t = 0$, the velocity is u. Therefore, evaluating Equation (23.1) with $t = 0$ and $v = u$, we have

$$u = a(0) + c$$

from which we obtain $c = u$. Therefore

$$v = at + u$$

or, in its more usual form,

$$v = u + at$$

Self-assessment questions 23.2

1.   Explain what is meant by an indefinite integral.
2.   Why does an indefinite integral always contain an added constant term?
3.   What is meant by the word 'integrand'.

Exercises 23.2

1.   Use Table 23.1 to find the following indefinite integrals:

(a) $\int x^7 \, dx$ (b) $\int x^9 \, dx$ (c) $\int x^{\frac{1}{2}} \, dx$

(d) $\int x^{0.2} \, dx$ (e) $\int x^{-3} \, dx$ (f) $\int \frac{1}{x^4} \, dx$

(g) $\int \frac{1}{x^{5/2}} \, dx$ (h) $\int 4 \, dx$ (i) $\int 4 \, dt$

(j) $\int 5 \, dz$

2.   Find the following indefinite integrals:

(a) $\int \sqrt{x} \, dx$ (b) $\int \frac{1}{x^2} \, dx$ (c) $\int \frac{1}{\sqrt{x}} \, dx$

3.   Find the indefinite integral of each of the following:

(a) e^{4x} (b) e^{5x} (c) e^{-6x} (d) e^{-7x}

(e) $\dfrac{1}{e^{-x}}$ (f) $\dfrac{1}{e^x}$

4.   Find the following indefinite integrals:

(a) $\int \cos 3x \, dx$ (b) $\int \sin(-2x) \, dx$

(c) $\int e^{0.5x} \, dx$

5.   Find the following indefinite integrals:

(a) $\int t^2 \, dt$ (b) $\int \tan 3t \, dt$ (c) $\int \frac{1}{t} \, dt$

(d) $\int \sin 7z \, dz$ (e) $\int \cos \frac{t}{2} \, dt$

(f) $\int \sin \frac{3\theta}{2} \, d\theta$ (g) $\int \cos(-x) \, dx$

6.   Use Table 23.1 to find the indefinite integral of each of the following functions:
(a) $f(x) = x^4$ (b) $f(x) = x^{-4}$
(c) $f(x) = \cos x$ (d) $f(x) = \sqrt[3]{x}$
(e) $f(x) = 5$ (f) $f(x) = \frac{1}{2}$ (g) $f(x) = \pi$
(h) $f(x) = x^{\frac{3}{2}}$

23.3 Some rules for finding indefinite integrals

To enable us to find integrals of a wider range of functions than those listed in Table 23.1, we can make use of several rules.

23.3.1 The indefinite integral of $kf(x)$

When the integrand is multiplied by a constant, it is permissible to take the constant factor outside the integral:

$$\int kf(x) \, dx = k \int f(x) \, dx$$

Example 23.13
    Determine $\int 3 \cos x \, dx$.

Solution   Using the rule with $k = 3$,

$$\int 3 \cos x \, dx = 3 \int \cos x \, dx$$

The constant factor 3 has been brought out in front of the integral sign. From Table 23.1, we note that

$$\int \cos x \, dx = \sin x + c$$

Therefore

$$\int 3 \cos x \, dx = 3 \int \cos x \, dx = 3(\sin x + c)$$

Note that the answer can be written $3 \sin x + 3c$ or simply $3 \sin x + A$, where the new constant $A = 3c$.

Example 23.14

Find $\int 7x^3 \, dx$.

Solution

$$\int 7x^3 \, dx = 7 \int x^3 \, dx = 7\left(\frac{x^4}{4} + c\right)$$

$$= \frac{7x^4}{4} + K$$

where we have written the constant $7c$ as K.

Example 23.15

Find $\int \frac{t}{3} \, dt$.

Solution

$$\int \frac{t}{3} \, dt = \frac{1}{3} \int t \, dt = \frac{1}{3}\left(\frac{t^2}{2} + c\right)$$

$$= \frac{t^2}{6} + K$$

where we have written the constant $\frac{1}{3}c$ as K.

Example 23.16

Find $\int \frac{\cos 3x}{2} \, dx$.

Solution

We have

$$\int \frac{\cos 3x}{2} \, dx = \frac{1}{2} \int \cos 3x \, dx = \frac{1}{2}\left(\frac{\sin 3x}{3} + c\right)$$

$$= \frac{\sin 3x}{6} + K$$

where we have written the constant of integration $\frac{1}{2}c$ as K.

Example 23.17 Voltage across a capacitor

The current i through a capacitor of capacitance C varies with time, and is given by

$$i = C\frac{dv}{dt}$$

where v is the voltage across the capacitor and t is time. Derive an expression for v.

Solution   We are given

$$i = C\frac{dv}{dt}$$

So, rearranging, we have

$$\frac{dv}{dt} = \frac{i}{C}$$

Therefore

$$v = \int \frac{i}{C}\, dt$$

Now C is a constant and so

$$v = \frac{1}{C}\int i\, dt$$

Note that whereas the capacitance, C, is constant, the current, i, is not and so it cannot be taken through the integral sign. In order to perform the integration we need to have an expression for i as a function of t.

Example 23.18 Sinusoidal current through a capacitor

The current through a capacitor of capacitance C is given by $i = 5\cos \omega t$. Find an expression for the voltage across the capacitor.

Solution   The voltage is found from $v = \dfrac{1}{C}\displaystyle\int i\, dt$. Hence

$$v = \frac{1}{C}\int 5\cos \omega t\, dt = \frac{5}{C\omega}\sin \omega t + K$$

where K is the constant of integration.

23.3.2 The indefinite integral of $f(x) \pm g(x)$

When the integrand is the sum or difference of two functions, the indefinite integral is found by integrating each term separately; that is,

$$\int [f(x) + g(x)] \, dx = \int f(x) \, dx + \int g(x) \, dx$$

and

$$\int [f(x) - g(x)] \, dx = \int f(x) \, dx - \int g(x) \, dx$$

Example 23.19

Find $\int (x^3 + \sin x) \, dx$

Solution   We can integrate each term separately to give

$$\int (x^3 + \sin x) \, dx = \int x^3 \, dx + \int \sin x \, dx$$

$$= \frac{x^4}{4} + c_1 - \cos x + c_2$$

$$= \frac{x^4}{4} - \cos x + c$$

Note that the constants of integration arising from the two separate integrals have been combined into a single constant c.

Example 23.20

Find $\displaystyle\int \frac{x + \cos 2x}{3} \, dx.$

Solution   The integrand can be written as $\frac{1}{3}(x + \cos 2x)$, and so

$$\int \frac{x + \cos 2x}{3} \, dx = \frac{1}{3} \int (x + \cos 2x) \, dx$$

$$= \frac{1}{3} \left(\frac{x^2}{2} + \frac{\sin 2x}{2} + c \right)$$

$$= \frac{x^2}{6} + \frac{\sin 2x}{6} + K$$

where $K = \frac{1}{3}c$.

Example 23.21 Displacement of an object moving in a straight line with constant acceleration

When an object having initial velocity u accelerates with constant acceleration a, the velocity of the object at time t is given by $v = u + at$. This was shown in Example 23.12. The displacement of the object at time t is s. The velocity is the derivative of displacement; that is,

$$v = \frac{ds}{dt}$$

(a)   Derive an expression for s.
(b)   Given that when $t = 0$ the displacement $s = 0$, find the value of the constant of integration.

Solution   (a)   Knowing that $v = u + at$, and that $v = \frac{ds}{dt}$, we can write

$$\frac{ds}{dt} = u + at$$

so that

$$s = \int (u + at) \, dt$$

$$= ut + \frac{1}{2}at^2 + c$$

where c is the constant of integration.

(b)   We are given that when $t = 0$, $s = 0$, and substituting these values gives

$$0 = u(0) + \frac{1}{2}a(0)^2 + c$$

so that $c = 0$. Therefore,

$$s = ut + \frac{1}{2}at^2$$

Example 23.22

An object moves with variable acceleration a given by the formula $a = 3t^2$.
(a)   Find an expression for the velocity of the object.
(b)   Find an expression for the displacement of the object.

Solution   We are given that $a = 3t^2$.

(a) $\qquad v = \int a \, dt = \int 3t^2 \, dt = t^3 + C$

(b) $\qquad s = \displaystyle\int v \, dt = \int (t^3 + C) \, dt = \tfrac{1}{4} t^4 + Ct + D$

where C and D are constants of integration.

Self-assessment question 23.3

1. State the rules for integrating functions of the form (a) $f(x) + g(x)$ (b) $f(x) - g(x)$
    (c) $kf(x)$, where k is a constant

Exercises 23.3

1. Use Table 23.1 and the rules given in this section to find the following indefinite integrals:

    (a) $\displaystyle\int 3x^3 \, dx$ (b) $\displaystyle\int 4x^2 \, dx$ (c) $\displaystyle\int 5x^{\frac{1}{2}} \, dx$

    (d) $\displaystyle\int 7x^{-2} \, dx$

2. Find the following indefinite integrals:

    (a) $\displaystyle\int 2x^4 - 7x^2 + 3x - 2 \, dx$

    (b) $\displaystyle\int (x+3)^2 \, dx$ (c) $\displaystyle\int (3x-2)^2 \, dx$

3. Find the following indefinite integrals:

    (a) $\displaystyle\int 2x^2 + 3x \, dx$ (b) $\displaystyle\int 3x^7 - 2x^2 \, dx$

    (c) $\displaystyle\int \cos x + \sin x \, dx$ (d) $\displaystyle\int 7 \cos 3x \, dx$

    (e) $\displaystyle\int 8 \sin 2x \, dx$ (f) $\displaystyle\int 16e^{-2x} \, dx$

    (g) $\displaystyle\int \cos 3x + \cos 2x \, dx$ (h) $\displaystyle\int 5e^{3x} \, dx$

4. An object rotates with a constant acceleration α. The initial angular velocity of the object is ω_0 and the angular velocity of the object at time t is ω. The angular displacement of the object at time t is θ. Derive expressions for ω and θ.

5. Find

    (a) $\displaystyle\int \left(\frac{3}{x} - 2x \right) dx$ (b) $\displaystyle\int \left(\frac{x}{3} - \frac{x}{2} \right) dx$

    (c) $\displaystyle\int \frac{3 - 2x}{x} \, dx$ (d) $\displaystyle\int (3 - 2x)^2 \, dx$

6. Find (a) $\displaystyle\int 3 \sin 4x \, dx$ (b) $\displaystyle\int 4 \sin 3x \, dx$

    (c) $\displaystyle\int \tfrac{1}{3} \sin 4x \, dx$ (d) $\displaystyle\int \frac{\sin 5x}{4} \, dx$

Review exercises 23

1   Use Table 23.1 to find the indefinite integral of each of the following:
(a) x^8 (b) $x^{\frac{1}{8}}$ (c) x^{-8} (d) $x^{0.25}$
(e) $\sin 3x$ (f) $\sin \dfrac{x}{3}$ (g) $\cos 0.25x$

2   Find the indefinite integral of each of the following:
(a) $x^4 + x^{-4}$ (b) $x^2 + x^{-2}$ (c) $x + x^{-1}$
(d) $x^{\frac{1}{4}} + \dfrac{1}{x}$

3   Find the following indefinite integrals:

(a) $\int x^9 \, dx$ (b) $\int t^3 + t^2 \, dt$

(c) $\int x^2 + \cos 3x \, dx$ (d) $\int x^7 - 7e^x \, dx$

4   Find

(a) $\int e^{3x} + e^{-3x} \, dx$ (b) $\int 7 \, dx$

(c) $\int -7 \, dx$ (d) $\int e^1 \, dx$

5   Find

(a) $\int \frac{1}{3} t \, dt$ (b) $\int \frac{t}{3} \, dt$ (c) $\int \frac{1}{3} \, dt$

(d) $\int \frac{1}{3\pi} \, dt$ (e) $\int \frac{\pi}{3} \, dt$

6   If $\frac{dy}{dx} = 3 + x$, and $y = 5$ when $x = 0$, find an expression for y.

7   If $\frac{dy}{dt} = e^t$, and $y = 0$ when $t = 0$, find an expression for y.

8   Find the indefinite integral of each of the following:

(a) x^{-3} (b) $\frac{7}{x^2}$ (c) $4e^{-3x}$

(d) $11x^{-1}$

9   Find the following indefinite integrals:

(a) $\int 3t + 7 \, dt$ (b) $\int 4t^2 + 3t - 2 \, dt$

(c) $\int 2t^{-1} \, dt$ (d) $\int \sin 3t + \cos 3t \, dt$

24

Areas under Curves and Definite Integrals

KEY POINTS

This chapter

- explains how areas under curves can be estimated using rectangular approximations

- explains how exact areas can be found by taking the 'limit of a sum'

- explains what is meant by a definite integral and shows how its value is calculated

- explains how a definite integral can be used to find the area under a curve

CONTENTS

24.1 Introduction

When a graph is used to represent a physical quantity, it is often useful to calculate the area under the graph, because this can represent a related physical quantity. In this chapter we consider the rectangle rule, which is one way in which such an area can be approximated. We then show how an improvement can be made to the rectangle rule that yields an exact value for the area under the curve. This uses a technique called definite integration. We shall see that not only can integration be used to reverse the process of differentiation, it can also be applied to finding the area underneath a curve.

Example 24.1 *Energy used by an electric motor*

 Figure 24.1 shows a graph of power supplied to an electric motor against time. It can be shown that the area under the graph between any two instants in time represents the energy used by the motor during that time interval. For example, the area under the graph between $t=1$ second and $t=3$ seconds is equal to the energy used in this 2 seconds time interval.

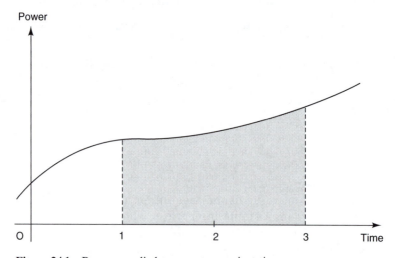

Figure 24.1   Power supplied to a motor against time.

Example 24.2 *Distance travelled by a car*

 Figure 24.2 shows a graph of the velocity of a car against time. It can be shown that the area under the graph between any two instants in time represents the distance travelled by the car in that time interval. For example, the distance travelled between $t=t_1$ and $t=t_2$ is equal to the area under the curve between these values of t.

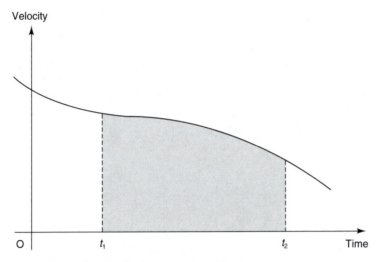

Figure 24.2 Graph of velocity against time.

24.2 The rectangle rule

Consider the graph of $y(x)$ shown in Figure 24.3. Suppose we wish to find the area under the curve and above the x axis between $x=a$ and $x=b$. There are many ways in which this area can be estimated. One particularly simple way

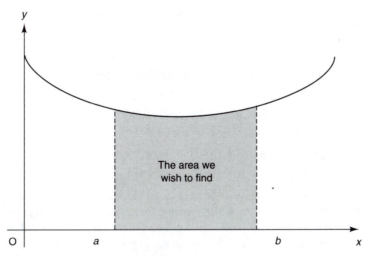

Figure 24.3 The area under $y(x)$ from $x=a$ to $x=b$.

is to divide the area into strips, each of the same width. Suppose we assume that each strip is rectangular in shape and that its height is equal to the value of y on its right-hand side. These rectangular strips are shown in Figure 24.4 for the case of four strips. Knowing the equation of the curve, $y(x)$, it is straightforward to calculate the height of each rectangle and hence its area. The areas of all such rectangles are then added, to estimate the total area required. This method of estimating the area is known as the **rectangle rule**. Let us consider a specific example.

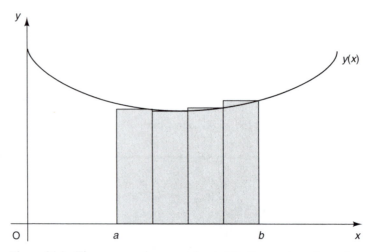

Figure 24.4   The area can be approximated by four rectangular strips.

Example 24.3

Estimate the area under the curve $y = 3x^2$ between $x = 1$ and $x = 2$ by splitting the area into four rectangular strips.

Solution   The graph of $y = 3x^2$ is shown in Figure 24.5, with the four strips indicated. Each rectangle has width 0.25, and its height is found from the equation $y = 3x^2$. For example,

rectangle 1 has height $3(1.25)^2 = 4.6875$
its area is then $4.6875 \times 0.25 = 1.1719$

rectangle 2 has height $3(1.50)^2 = 6.7500$
its area is then $6.7500 \times 0.25 = 1.6875$

Corresponding values have been calculated for all four rectangles, and are shown in Table 24.1. The sum of the rectangular areas is 8.1563. Of course, the sum of the rectangular areas is not the area under the curve. It is evident from Figure 24.5 that the total area of the rectangles is greater than the area under the curve. However, we can regard 8.1563 as an approximation to the required area.

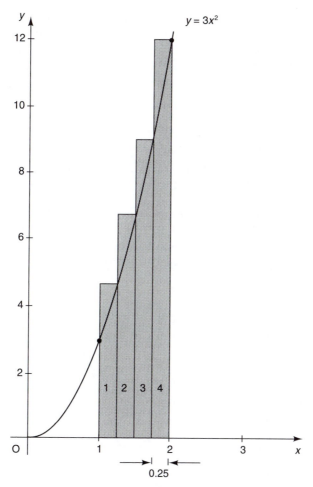

Figure 24.5 Graph of $y = 3x^2$.

Table 24.1

	x	height $= 3x^2$	area $= 0.25 \times$ height
Rectangle 1	1.25	4.6875	1.1719
Rectangle 2	1.50	6.7500	1.6875
Rectangle 3	1.75	9.1875	2.2969
Rectangle 4	2.00	12.0000	3.0000
Sum			8.1563

Our approximation can be improved by selecting thinner rectangles, but the price we pay is having to calculate a larger number of areas. By choosing 8 rectangles each of width 0.125, an improved estimate of the area under the curve can be shown to be 7.5703. The exact answer is in fact 7. We see that increasing the number of rectangles improves the accuracy of the estimate.

Example 24.4

(a)   Find a formula for estimating the area under the curve $y=x^2$ for values of x between 0 and 1 by constructing n rectangles each of width h.

(b)   Evaluate this estimated area (i) when $n=10$ and (ii) when $n=20$.

Solution   (a)   The graph of $y=x^2$ between $x=0$ and $x=1$ is shown in Figure 24.6. The interval from $x=0$ to $x=1$ is divided into n equal subintervals each of width $h=1/n$. A rectangle of width h is drawn on each subinterval. Some typical rectangles are shown in Figure 24.6. Using the function $y=x^2$,

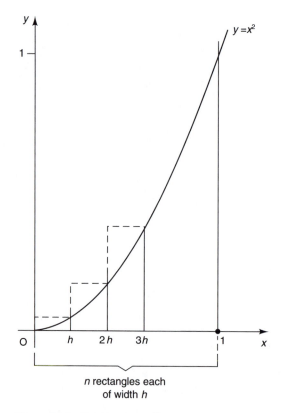

Figure 24.6   Graph of $y=x^2$.

we can calculate the height of each rectangle: the height of the 1st rectangle is h^2, so its area is

$$h^2 \times h = h^3$$

The height of the 2nd rectangle is $(2h)^2$, so its area is

$$(2h)^2 \times h = 2^2 h^3$$

The height of the 3rd rectangle is $(3h)^2$, so its area is

$$(3h)^2 \times h = 3^2 h^3$$

and so on.

There are n such rectangles altogether. By adding their areas, we obtain an expression for the total area of the rectangles. That is,

$$\text{total area of rectangles} = h^3 + 2^2 h^3 + 3^2 h^3 + \dots + n^2 h^3$$

$$= h^3 (1^2 + 2^2 + 3^2 + \dots + n^2)$$

The expression in brackets on the right-hand side is the sum of the squares of the first n whole numbers. It was stated in Section 20.7.2 that this sum equals $\frac{1}{6} n(n+1)(2n+1)$. We can therefore write the total area of the rectangles as

$$\text{total area of rectangles} = h^3 (1^2 + 2^2 + 3^2 + \dots + n^2)$$

$$= h^3 \frac{n}{6} (n+1)(2n+1)$$

Using the fact that $h = \dfrac{1}{n}$, we can write this as

$$\text{total area of rectangles} = \frac{1}{6n^2} (n+1)(2n+1)$$

The area under $y = x^2$ between $x = 0$ and $x = 1$ is approximately

$$\frac{1}{6n^2} (n+1)(2n+1)$$

where n is the number of rectangles used in the approximation.

(b)   (i)   When $n = 10$, the expression for the total area of 10 rectangles is

$$\frac{1}{6(10^2)} (10+1)(20+1) = 0.385$$

(ii)   When $n = 20$, the expression for the total area of 20 rectangles is

$$\frac{1}{6(20^2)} (20+1)(40+1) = 0.359$$

We have two estimates for the area under $y=x^2$: one is 0.385 and the other is 0.359. The more rectangles we use, the greater will be the accuracy with which we can estimate the area under the curve.

We shall now generalize the previous development to find an estimate of the area under an arbitrary curve $y(x)$ between $x=a$ and $x=b$. Consider the curve shown in Figure 24.7. Imagine that the area under the curve has been divided into n rectangles each of width h. The distance from $x=a$ to $x=b$ is $b-a$, and so each rectangle has a width of $(b-a)/n$.

The height of the 1st rectangle is $y(a+h)$, so its area is

$$y(a+h) \times h$$

The height of the 2nd rectangle is $y(a+2h)$, so its area is

$$y(a+2h) \times h$$

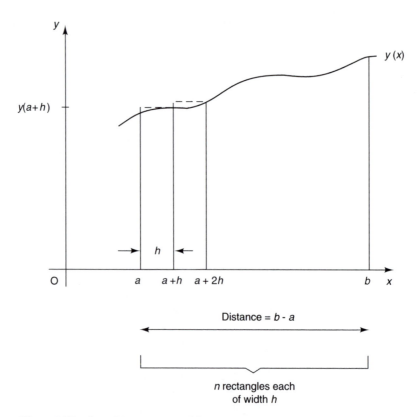

Figure 24.7 An arbitrary curve $y(x)$.

the height of the 3rd rectangle is $y(a+3h)$, so its area is

$$y(a+3h) \times h$$

and so on.

There are n such rectangles altogether. By adding their areas, we obtain an expression for the total area. We can write the rectangle rule as

$$\text{total area} = y(a+h) \times h + y(a+2h) \times h + \ldots + y(a+nh) \times h$$

Using sigma notation, we can write this more compactly as follows:

---

rectangle rule:

$$\text{total area of rectangles} = \sum_{k=1}^{n} y(a+kh) \times h$$

Self-assessment question 24.2

1. Will the rectangle rule overestimate or underestimate a required area? Give reasons for your answer.

Exercises 24.2

1. Use the rectangle rule to estimate the area under $y=x^3$ between $x=0$ and $x=2$ using (a) 4 rectangles and (b) 8 rectangles.

2. The rectangle rule built rectangular blocks to the height of the right-most point on the curve within each strip. Repeat the calculation in Example 24.3, but this time construct rectangles to the height of the left-most point on the curve within each strip. Does this overestimate or underestimate the area?

3. Plot a graph of the function $y=\dfrac{1}{x}$ for values of x between 1 and 3. Use the rectangle rule to estimate the area bounded by the curve and the x axis between $x=1$ and $x=3$, using (a) 2 strips and (b) 4 strips.

24.3 The exact value of the area under a curve

In order to use the rectangle rule to obtain very accurate estimates of the area under a curve, it is necessary to choose a large number of rectangles each of small width. The approximate area given by the rectangle rule will become exact if we let the width approach zero, and the number of rectangles become infinite.

In Example 24.4 we estimated the area under the curve $y=x^2$ between $x=0$ and $x=1$ by constructing n rectangular strips, and showed that the total area of the strips is given by

$$\text{total area of rectangles} = \frac{1}{6n^2}(n+1)(2n+1)$$

Note that we can rewrite the right-hand side of this expression to give

$$\text{total area of rectangles} = \frac{1}{6n^2}(2n^2+3n+1)$$

$$= \frac{2n^2}{6n^2} + \frac{3n}{6n^2} + \frac{1}{6n^2}$$

$$= \frac{1}{3} + \frac{1}{2n} + \frac{1}{6n^2}$$

Suppose we increase the number of rectangles by letting $n \to \infty$. We must observe the behaviour of

$$\frac{1}{3} + \frac{1}{2n} + \frac{1}{6n^2}$$

as n becomes larger and larger. We find

$$\lim_{n \to \infty} \left(\frac{1}{3} + \frac{1}{2n} + \frac{1}{6n^2} \right) = \frac{1}{3} + 0 + 0$$

That is, the exact value of the area under the curve is $\frac{1}{3}$. This process is known as **taking the limit of a sum**.

To generalize this to find the area under an arbitrary curve $y(x)$ between a and b, recall the rectangle rule from Section 24.2:

$$\text{total area of } n \text{ rectangles} = \sum_{k=1}^{n} y(a+kh) \times h$$

where h is the width of each rectangle. To obtain the area under the curve, we now let h become smaller and smaller and let the number of rectangles tend to infinity. In this context it is usual to let the width of each rectangle be δx. So, writing δx for h, we have the following formula:

---

the area under the curve between $x=a$ and $x=b$ is equal to

$$\lim_{n \to \infty} \sum_{k=1}^{n} y(a+k\,\delta x)\,\delta x$$

where

$$\delta x = \frac{b-a}{n}$$

The area under the curve has been expressed as the **limit of a sum**. In theory, we could evaluate this limit for almost any function $y(x)$ in order to find the area under its graph. Fortunately, such a tedious and time-consuming process is not necessary, as we shall see in the following section.

Exercises 24.3

1. Apply the method of the limit of a sum to find the area under the graph of the function $y(x)=7x+2$ for values of x between 2 and 4.

2. Apply the method of the limit of a sum to find the area under the graph of the function $y(x)=2x^2$ for values of x between 1 and 3.

3. Apply the method of the limit of a sum to find the area under the graph of the function $y(x)=4x^2$ for values of x between 3 and 4.

24.4 Introducing definite integrals

We have seen that to find the area under a curve, the area can be approximated by a large number of rectangles.

Consider the graph of $y(x)$ shown in Figure 24.8(a). Suppose that the graph of $y(x)$ lies entirely above the x axis, and that we are interested in the area A between the curve and the x axis up to the point with x coordinate x.

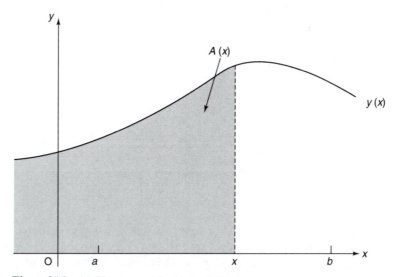

Figure 24.8 (a) The area under the graph up to x.

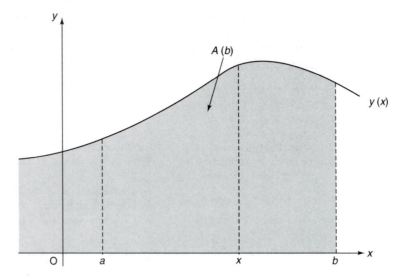

Figure 24.8   (b) The area under the graph up to b.

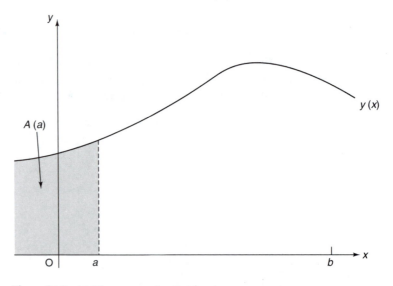

Figure 24.8   (c) The area under the graph up to *a*.

This area depends upon the value of *x*. If *x* is large, the area will be greater than if *x* is small. To show that the area depends upon *x*, we write $A(x)$. Thus the area up to the point where $x=b$ is equal to $A(b)$. Similarly, the area up to the point where $x=a$ is $A(a)$. These areas are shown in Figure 24.8(b) and (c).

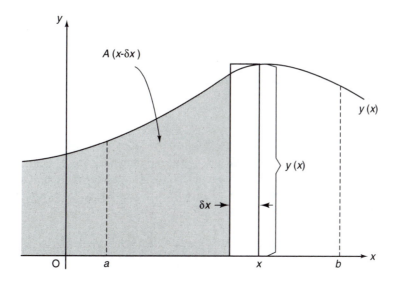

Figure 24.9 The area of the slice is $\delta A = A(x) - A(x - \delta x)$.

Now consider Figure 24.9. The area of the thin slice between $x - \delta x$ and x is then

$$A(x) - A(x - \delta x)$$

Let us call this area δA; that is, $\delta A = A(x) - A(x - \delta x)$. This area is approximately equal to the area of a rectangle of height $y(x)$ and width δx; that is,

$$\delta A \approx y(x)\,\delta x$$

so that

$$\frac{\delta A}{\delta x} \approx y(x)$$

where the symbol $\approx$ means 'is approximately equal to'. As we let the rectangle become thinner, δx tends to zero.

Recall from Section 21.5 that as $\delta x \to 0$, $\delta y/\delta x$ tends to the derivative dy/dx; that is,

$$\frac{\delta y}{\delta x} \to \frac{dy}{dx} \quad \text{as } \delta x \to 0$$

It follows that

$$\frac{\delta A}{\delta x} \to \frac{dA}{dx} \quad \text{as } \delta x \to 0$$

and so we have

$$\frac{\mathrm{d}A}{\mathrm{d}x} = y(x)$$

This result is important. It means that the derivative of the area under the curve is equal to $y(x)$. Equivalently,

$$A(x) = \int y(x)\,\mathrm{d}x$$

The area under the curve up to the point where $x=b$, namely $A(b)$, is this integral evaluated at $x=b$. Similarly, the area under the curve up to the point where $x=a$, namely $A(a)$, is this integral evaluated at $x=a$. Thus the area between $x=a$ and $x=b$ is

$$A(b) - A(a) = \int y(x)\,\mathrm{d}x \qquad \text{evaluated when } x=b$$

$$- \int y(x)\,\mathrm{d}x \quad \text{evaluated when } x=a$$

Consider the following example.

Example 24.5

Find the area under the curve $y = 3x^2$ between $x=1$ and $x=2$.

Solution

$$\text{area required} = \int 3x^2\,\mathrm{d}x \qquad \text{evaluated when } x=2$$

$$- \int 3x^2\,\mathrm{d}x \quad \text{evaluated when } x=1$$

$$= (x^3 + c) \qquad \text{evaluated when } x=2$$

$$- (x^3 + c) \qquad \text{evaluated when } x=1$$

$$= (8+c) - (1+c)$$

$$= 7$$

Thus the area under the curve $y = 3x^2$ between $x=1$ and $x=2$ is equal to 7.

Note from the previous example that the constant of integration cancels out. This will always happen, and so the constant can be neglected. To avoid the lengthy working in the previous example, a shorthand is now introduced.

To denote the limits of the area being considered, we place values on the integral sign.

The area under $y(x)$ between $x=a$ and $x=b$ is denoted by

$$\int_{x=a}^{x=b} y(x)\,dx, \quad \text{or simply} \quad \int_a^b y(x)\,dx$$

The quantity $\int_a^b y(x)\,dx$ is called the **definite integral** of $y(x)$ from a to b. The numbers a and b are the **lower** and **upper limits** of the integral. The calculation of Example 24.5 is written concisely as

$$\int_1^2 3x^2\,dx = [x^3]_1^2$$

$$= (2^3) - (1^3)$$

$$= 7$$

When the integration has been performed, the result is enclosed in square brackets and the limits of integration are written on the right-hand bracket. The integral is evaluated at the upper limit, then the lower limit, and the difference between the two results is calculated.

24.5 Evaluating definite integrals

In addition to finding areas under curves, definite integrals arise in many other applications. This section provides a number of examples in order to develop skills in evaluating them.

Example 24.6

Find $\int_1^3 x^2\,dx$.

Solution

$$\int_1^3 x^2\,dx = \left[\frac{x^3}{3}\right]_1^3$$

$$= \left(\frac{3^3}{3}\right) - \left(\frac{1^3}{3}\right)$$

$$= \left(\frac{27}{3}\right) - \left(\frac{1}{3}\right)$$

$$= \frac{26}{3}$$

Example 24.7

Find $\displaystyle\int_0^3 (x+7x^2)\,dx.$

Solution

$$\int_0^3 (x+7x^2)\,dx = \left[\frac{x^2}{2}+7\frac{x^3}{3}\right]_0^3$$

$$= (\tfrac{9}{2}+63)-(0)$$

$$= 67.5$$

Example 24.8

Evaluate $\displaystyle\int_0^{\pi/2} \cos x\,dx.$

Solution   We have

$$\int_0^{\pi/2} \cos x\,dx = [\sin x]_0^{\pi/2} = \sin(\pi/2)-\sin 0 = 1-0 = 1$$

Note that the angle x must be measured in radians.

Example 24.9

Find $\displaystyle\int_1^2 (2x^2+3x+7)\,dx.$

Solution   We have

$$\int_1^2 (2x^2+3x+7)\,dx = \left[\frac{2x^3}{3}+\frac{3x^2}{2}+7x\right]_1^2$$

$$= \left(\frac{2(2)^3}{3}+\frac{3(2)^2}{2}+7(2)\right)-\left(\frac{2}{3}+\frac{3}{2}+7\right)$$

$$= \left(\frac{76}{3}-\frac{55}{6}\right)$$

$$= \frac{97}{6} = 16.1667$$

Example 24.10 Centre of mass of a cone

The centre of mass of a body is the point at which the mass of the body can be considered to be located. Consider a circular cone of perpendicular height 1 m. The distance s of the centre of mass from the circular base is found by

evaluating the integral

$$s = 3 \int_0^1 x(1-x)^2 \, dx$$

Calculate the position of the centre of mass.

Solution   Expanding the brackets in the integrand, we can write

$$s = 3 \int_0^1 x(1-2x+x^2) \, dx = 3 \int_0^1 (x-2x^2+x^3) \, dx$$

$$= 3 \left[\frac{x^2}{2} - \frac{2x^3}{3} + \frac{x^4}{4} \right]_0^1$$

$$= 3 \left(\frac{1}{2} - \frac{2}{3} + \frac{1}{4} \right) - 3(0)$$

$$= 3 \left(\frac{6-8+3}{12} \right) - 0$$

$$= 3 \left(\frac{1}{12} \right)$$

$$= \tfrac{1}{4}$$

The centre of mass of the cone lies $\tfrac{1}{4}$ m from the base.

Example 24.11 Moment of inertia of a circular disc

The moment of inertia of a body is a measure of its resistance to rotation. The moment of inertia J of a circular disc of mass M and radius a about an axis perpendicular to the disc and passing through its centre of mass is given by

$$J = \frac{2M}{a^2} \int_0^a r^3 \, dr$$

Find an expression for J.

Solution

$$J = \frac{2M}{a^2} \int_0^a r^3 \, dr = \frac{2M}{a^2} \left[\frac{r^4}{4} \right]_0^a$$

$$= \frac{2M}{a^2} \left(\frac{a^4}{4} - 0 \right)$$

$$= \tfrac{1}{2} M a^2$$

Thus the moment of inertia of the disc is given by the formula $J = \tfrac{1}{2} M a^2$.

Exercises 24.5

1. Evaluate the following definite integrals:

(a) $\displaystyle\int_0^1 x^2\,dx$ (b) $\displaystyle\int_2^3 (2x-7)\,dx$

(c) $\displaystyle\int_0^2 \frac{x^3}{3}\,dx$ (d) $\displaystyle\int_2^3 \frac{1}{x^2}\,dx$

2. Find $\displaystyle\int_0^1 (x-1)(x-2)\,dx$.

3. A function $F(x)$ is such that

$$\frac{dF}{dx}=3x^3-7x^2 \quad \text{and} \quad F(0)=1$$

Find $F(x)$.

4. Using Table 23.1, evaluate the following definite integrals:

(a) $\displaystyle\int_3^5 x^4\,dx$ (b) $\displaystyle\int_0^\pi \sin x\,dx$

(c) $\displaystyle\int_1^2 \frac{1}{x}\,dx$ (d) $\displaystyle\int_0^3 \sqrt{x}\,dx$

(e) $\displaystyle\int_{-1}^1 x^7\,dx$

5. The distance of the centre of mass of a solid hemisphere of radius 2 metres from the centre of its plane base is found by evaluating

the integral

$$\tfrac{3}{16}\int_0^2 x(4-x^2)\,dx$$

Show that this distance is equal to $\tfrac{3}{4}$ m.

6. The moment of inertia about a diameter, of a sphere of radius 1 m and mass 1 kg is found by evaluating the integral

$$\tfrac{3}{8}\int_{-1}^1 (1-x^2)^2\,dx$$

Show that the moment of inertia of the sphere is $\tfrac{2}{5}$ kg m^2.

7. Evaluate the following definite integrals:

(a) $\displaystyle\int_1^2 e^x\,dx$ (b) $\displaystyle\int_{-1}^1 e^{-x}\,dx$

(c) $\displaystyle\int_{-1}^2 e^{2x}\,dx$

8. Evaluate

(a) $\displaystyle\int_0^\pi \cos x\,dx$ (b) $\displaystyle\int_0^{\pi/2} 2\sin 3x\,dx$

(c) $\displaystyle\int_{-\pi}^\pi 2\cos 4x\,dx$ (d) $\displaystyle\int_{\pi/2}^{3\pi/2} \tfrac{1}{2}\cos 5x\,dx$

(e) $\displaystyle\int_0^{\pi/4} (2\sin 4x + \cos 4x)\,dx$

24.6 Areas under curves

In this section we develop the application of definite integration to finding areas under curves, and show how the technique can still be used when part or all of the area lies below the x axis.

Example 24.12

A graph of the function $y=x^3-3x^2+4$ is shown in Figure 24.10. Find the area under the curve between $x=-1$ and $x=2$.

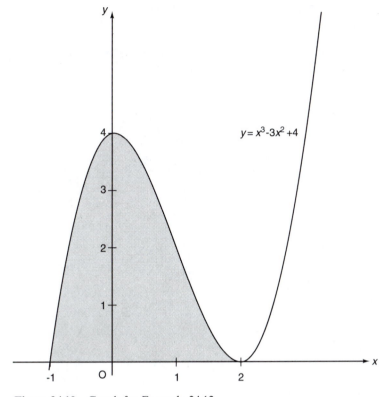

Figure 24.10 Graph for Example 24.12.

Solution　The required area is shaded in Figure 24.10, and is given by

$$\int_{-1}^{2} (x^3 - 3x^2 + 4)\, dx = \left[\frac{x^4}{4} - x^3 + 4x \right]_{-1}^{2}$$

$$= \left(\frac{2^4}{4} - 2^3 + 4(2) \right) - \left(\frac{(-1)^4}{4} - (-1)^3 + 4(-1) \right)$$

$$= (4) - (-2.75)$$

$$= 6.75$$

Example 24.13 **Liquid flowing into a chemical storage tank**

Suppose that liquid flows into a tank at a variable rate q. It can be shown that the area under the graph of q against t represents the volume of liquid in the tank. Consider a chemical storage tank with liquid entering with a rate

$$q = 3 - \frac{t}{20} + \frac{t^2}{400}$$

If at time $t=0$ the tank is empty, calculate the volume of liquid in the tank at time $t=20$.

Solution The volume of liquid in the tank is given by the area under the graph of q against t between $t=0$ and $t=20$. That is,

$$\int_0^{20} q\,dt$$

and so

$$\text{at } t=20 \text{ the volume in the tank} = \int_0^{20}\left(3-\frac{t}{20}+\frac{t^2}{400}\right)dt$$

$$=\left[3t-\frac{t^2}{40}+\frac{t^3}{1200}\right]_0^{20}$$

$$=\left(3(20)-\frac{20^2}{40}+\frac{20^3}{1200}\right)-(0)$$

$$=60-\frac{400}{40}+\frac{8000}{1200}$$

$$=60-10+\tfrac{20}{3}$$

$$=56.67 \text{ m}^3$$

Example 24.14
A graph of the function $y=x^2-4$ is shown in Figure 24.11. Evaluate the integral $\int_0^1 (x^2-4)\,dx$ and comment upon the result.

Solution Evaluating the given integral, we find

$$\int_0^1 (x^2-4)\,dx = \left[\frac{x^3}{3}-4x\right]_0^1$$

$$=\tfrac{1}{3}-4$$

$$=-3.6667$$

The integral is negative, reflecting the fact that the area bounded by the curve and x axis between $x=0$ and $x=1$ lies below the x axis. The actual area is 3.6667.

The previous example illustrates a very important point. If the area required lies below the x axis, the corresponding definite integral always yields a negative result. Sometimes parts of the required area lie above the x axis and

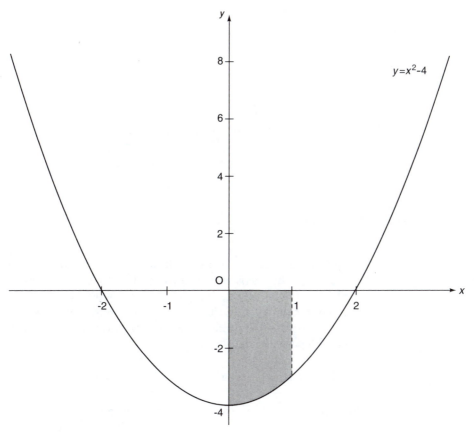

Figure 24.11 Graph for Example 24.14.

other parts lie below. In such cases, separate calculations are needed to find each area.

Example 24.15

(a) Evaluate $\displaystyle\int_{0}^{2\pi} \sin x \, dx$.

(b) Find the area enclosed by $y = \sin x$ and the x axis from $x=0$ to $x=2\pi$.

Solution (a) Performing the integration, we find

$$\int_{0}^{2\pi} \sin x \, dx = [-\cos x]_{0}^{2\pi}$$

$$= (-\cos 2\pi) - (-\cos 0)$$

$$= (-1) - (-1)$$

$$= 0$$

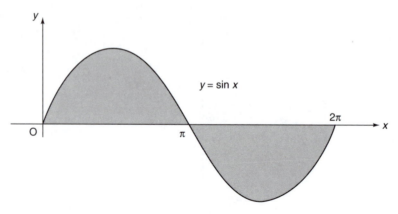

Figure 24.12   Areas are both above and below the *x* axis.

(b)   A graph of the integrand, sin *x*, is shown in Figure 24.12. We see that the area above the *x* axis and the area below are identical. However, the contribution to the definite integral from below the *x* axis is negative, and equal in magnitude to the positive contribution from above. Hence the integral equals 0. In order to find the shaded area, we must find the two areas separately and sum them.

$$\int_0^{\pi} \sin x \, dx = [-\cos x]_0^{\pi} = (-\cos \pi) - (-\cos 0) = 1 + 1 = 2$$

so the area above the *x* axis is 2 square units. Similarly,

$$\int_{\pi}^{2\pi} \sin x \, dx = [-\cos x]_{\pi}^{2\pi} = (-\cos 2\pi) - (-\cos \pi) = -1 - 1 = -2$$

The integral is -2, indicating an area of 2 square units below the *x* axis. Therefore the total area between 0 and 2π is 4 square units.

The previous example shows that to find an area when parts are above and parts below the axis, it is necessary to sketch a graph and identify these parts. Then integration is performed over each part separately.

Self-assessment questions 24.6

1.   If the area under a curve is evaluated by means of a definite integral and this yields a negative answer, does this mean that all the curve lies below the *x* axis?

2. If you are required to find the area under a curve using definite integrals, what important task should be carried out before any integration is performed?

Exercises 24.6

1. Find the area of the portion of the curve $y = x^2 - 5x + 6$ that lies below the x axis.

2. A chemical storage tank is initially empty at time $t = 0$. A chemical is poured into the tank at a rate

$$q = 6 - \frac{t}{10} + \frac{t^2}{100}$$

Calculate the volume of chemical in the storage tank after 4 seconds.

3. An acid storage tank has $60\,\text{m}^3$ of acid stored in it at time $t = 0$. Further acid is added to it at a flow rate given by $q = 6 + \frac{t^2}{4}$. Calculate the volume of acid in the tank at $t = 10$.

4. Find the area enclosed by $y = 4 - x^2$ and the x axis from
   (a) $x = 0$ to $x = 2$ (b) $x = -2$ to $x = 1$
   (c) $x = 1$ to $x = 3$

Review exercises 24

1 Use the rectangle rule with 10 strips to approximate the area under $y = x^4$ between 0 and 1.

2 Estimate the value of the integral $\int_0^1 x^3\, dx$ using the method of the limit of a sum with n rectangular strips. By letting $n \to \infty$, find the exact value of the integral. (Hint: You may use the result $\Sigma_{k=1}^n k^3 = [\frac{1}{2}n(n+1)]^2$.)

3 Find the areas under the following curves:
(a) $y = 3x + 2$ from $x = 1$ to $x = 5$
(b) $y = x + x^2$ from $x = 0$ to $x = 2$
(c) $y = 3 + \dfrac{1}{x}$ from $x = 1$ to $x = 100$
(d) $y = (x + 1)^2$ from $x = 0$ to $x = 2$
(e) $y = \dfrac{1}{x^2}$ from $x = -2$ to $x = -1$
(f) $y = \sin 3x$ from $x = 0$ to $x = 0.5$
(g) $y = e^x$ from $x = 0$ to $x = 3$
(h) $y = e^{-x}$ from $x = -1$ to $x = 1$
(i) $y = 3e^{2x}$ from $x = -2$ to $x = 0$

4 Find the area under $y = x^2 - 2x$ between
(a) $x = 2$ and $x = 3$ (b) $x = 1$ and $x = 3$
(c) $x = 0$ and $x = 2$ (d) $x = 0$ and $x = 3$

5 If $\int_0^k (2 - x)\, dx = 0$, find the value of k.

6 Evaluate
(a) $\displaystyle\int_0^1 2\sin x\, dx$ (b) $\displaystyle\int_0^1 \sin 2x\, dx$

(c) $\displaystyle\int_0^2 \cos 3x\, dx$ (d) $\displaystyle\int_0^2 3\cos x\, dx$

(e) $\displaystyle\int_0^1 e^{2x}\, dx$ (f) $\displaystyle\int_0^1 2e^x\, dx$

(g) $\displaystyle\int_1^2 (x^2 + 1)\, dx$ (h) $\displaystyle\int_1^2 (x + 1)^2\, dx$

(i) $\displaystyle\int_1^2 \frac{x}{2}\, dx$ (j) $\displaystyle\int_1^2 \frac{2}{x}\, dx$

7 Find the area enclosed by $y = \sin 2t$ and the t axis for (a) $t = 0$ to $t = \frac{1}{2}\pi$ and (b) $t = 1$ to $t = 3$.

Techniques of Integration

CONTENTS

25.1 Introduction

A number of techniques for integrating more complicated functions are introduced in this chapter. These techniques are useful in solving a variety of engineering problems, for example finding the root mean square value of a signal.

25.2 Integration by parts

It is sometimes necessary to integrate a function that is the product of two other functions. For example, we may wish to find

$$\int xe^x \, dx, \quad \int x^2 \sin 2x \, dx, \quad \int x \ln x \, dx$$

In each case notice that the integrand is a product. It is often possible to find such integrals using a method known as **integration by parts**. The formula for this method is as follows:

$$\text{integration by parts:} \quad \int u \frac{dv}{dx} \, dx = uv - \int v \frac{du}{dx} \, dx$$

To apply this formula, we must let one of the functions in the product equal u. It must be possible to differentiate this function. We let the other term equal dv/dx. It must be possible to find the integral of this function. The formula for integration by parts replaces the given integral by another integral which is frequently easier to evaluate. Consider the following example.

Example 25.1

Find $\int 3xe^{2x} \, dx$.

Solution

We must first match the given example with the left-hand side of the formula for integration by parts; that is, we must let

$$u \frac{dv}{dx} = 3xe^{2x}$$

Suppose we choose $u = 3x$; since $u = 3x$, we must then choose $dv/dx = e^{2x}$. Looking at the right-hand side of the formula we see that it is necessary to calculate du/dx and v. Clearly,

$$\text{if} \quad u = 3x \quad \text{then} \quad \frac{du}{dx} = 3$$

We find v by integration:

$$\text{if} \quad \frac{dv}{dx} = e^{2x} \quad \text{then} \quad v = \int e^{2x}\,dx = \frac{e^{2x}}{2}$$

We can now apply the formula:

$$\int 3xe^{2x}\,dx = (3x)\frac{e^{2x}}{2} - \int \frac{e^{2x}}{2}(3)\,dx$$

$$= \frac{3xe^{2x}}{2} - \frac{3}{2}\int e^{2x}\,dx$$

The final integral is easily found to be $\frac{1}{2}e^{2x}$, and so

$$\int 3xe^{2x}\,dx = \tfrac{3}{2}xe^{2x} - \tfrac{3}{4}e^{2x} + c$$

Note that there is no need to include a constant of integration at the stage when v is found. The introduction of c at the end of the calculation is sufficient.

Example 25.2

Find $\int x\cos x\,dx$.

Solution   To match this example with the formula, we must let

$$u\frac{dv}{dx} = x\cos x$$

Deciding which function should be u and which should be dv/dx is largely a matter of experience. Suppose we let

$$u = x, \quad \text{and so} \quad \frac{du}{dx} = 1$$

Letting

$$\frac{dv}{dx} = \cos x, \quad \text{we find} \quad v = \int \cos x\,dx = \sin x$$

Then, applying the formula for integration by parts, we find

$$\int x\cos x\,dx = x\sin x - \int (\sin x)(1)\,dx$$

$$= x\sin x - \int \sin x\,dx$$

$$= x\sin x + \cos x + c$$

Example 25.3

Find $\int 4x \ln x \, dx$.

Solution   To match the given example with the formula, we must let

$$u\frac{dv}{dx} = 4x \ln x$$

If we choose u to be $4x$ then we must choose dv/dx to equal $\ln x$. Unfortunately, the integral of $\ln x$ does not appear in Table 23.1, and so in this case we shall let $u = \ln x$ and $dv/dx = 4x$:

$$\text{if} \quad u = \ln x \quad \text{then} \quad \frac{du}{dx} = \frac{1}{x}$$

$$\text{if} \quad \frac{dv}{dx} = 4x \quad \text{then} \quad v = \int 4x \, dx = 2x^2$$

We can now apply the formula:

$$\int 4x \ln x \, dx = (\ln x)2x^2 - \int (2x^2)\left(\frac{1}{x}\right) dx$$

$$= 2x^2 \ln x - \int 2x \, dx$$

$$= 2x^2 \ln x - x^2 + c$$

Sometimes it is necessary to apply the formula more than once.

Example 25.4

Find $\int x^2 \sin x \, dx$.

Solution   Suppose we let $u = x^2$, so that $du/dx = 2x$. Letting $dv/dx = \sin x$, we find that $v = -\cos x$. Then, applying the formula for integration by parts, we find

$$\int x^2 \sin x \, dx = x^2(-\cos x) - \int (-\cos x)2x \, dx$$

$$= -x^2 \cos x + 2 \int x \cos x \, dx$$

The last integral has already been found using integration by parts in

Example 25.2. So, finally,

$$\int x^2 \sin x \, dx = -x^2 \cos x + 2x \sin x + 2 \cos x + c$$

When definite integrals need to be evaluated, the formula for integration by parts is modified as follows:

integration by parts: $\quad \int_a^b u \frac{dv}{dx} dx = [uv]_a^b - \int_a^b v \frac{du}{dx} dx$

Example 25.5

Evaluate $\displaystyle\int_0^{\pi/2} x \sin 2x \, dx$.

Solution   If we let $u = x$ then $du/dx = 1$. We let

$$\frac{dv}{dx} = \sin 2x, \quad \text{so that} \quad v = \int \sin 2x \, dx = -\frac{\cos 2x}{2}$$

Then, applying the formula, we find

$$\int_0^{\pi/2} x \sin 2x \, dx = \left[x \left(-\frac{\cos 2x}{2} \right) \right]_0^{\pi/2} - \int_0^{\pi/2} \left(-\frac{\cos 2x}{2} \right)(1) \, dx$$

$$= \left(\frac{\pi}{2} \left(-\frac{\cos \pi}{2} \right) - 0 \right) + \int_0^{\pi/2} \frac{\cos 2x}{2} \, dx$$

$$= \frac{\pi}{4} + \left[\frac{\sin 2x}{4} \right]_0^{\pi/2}$$

$$= \frac{\pi}{4} + \frac{\sin \pi}{4} - \frac{\sin 0}{4}$$

$$= \frac{\pi}{4}$$

Example 25.6

Find $\displaystyle\int_2^3 5x e^{-x} \, dx$

Solution   Let $u = 5x$, so that $du/dx = 5$. Let $dv/dx = e^{-x}$, so that $v = -e^{-x}$. Then, applying the formula, we find

$$\int_2^3 5xe^{-x} \, dx = [(5x)(-e^{-x})]_2^3 - \int_2^3 (-e^{-x})5 \, dx$$

$$= [-5xe^{-x}]_2^3 + 5\int_2^3 e^{-x} \, dx$$

$$= -15e^{-3} + 10e^{-2} + 5[-e^{-x}]_2^3$$

$$= -15e^{-3} + 10e^{-2} - 5e^{-3} + 5e^{-2}$$

$$= -20e^{-3} + 15e^{-2}$$

$$= 1.034$$

Self-assessment question 25.2

1.   State the formula for integration by parts used to evaluate indefinite integrals. How is the formula modified when evaluating definite integrals?

Exercises 25.2

1.   Use integration by parts to find the following integrals:

(a) $\int 4xe^x \, dx$ (b) $\int xe^{3x} \, dx$

(c) $\int 5xe^{2x} \, dx$ (d) $\int x \sin 3x \, dx$

(e) $\int 4x \ln 5x \, dx$ (f) $\int x \sec^2 x \, dx$

2.   By writing $\ln x$ as $1 \times \ln x$, use integration by parts to find $\int \ln x \, dx$.

3.   Use integration by parts to evaluate the following definite integrals:

(a) $\int_0^1 xe^x \, dx$ (b) $\int_1^2 \ln x \, dx$

(c) $\int_0^\pi x \cos 2x \, dx$ (d) $\int_0^{2\pi} x^2 \sin x \, dx$

4.   Find the area under the curve $y = xe^x$ between $x = 1$ and $x = 3$.

25.3 Integration by substitution

Suppose we wish to find the integral $\int (2x+5)^6 \, dx$. The technique known as **integration by substitution** involves changing the variable so that $2x+5$ is

George Green: mathematician and miller

George Green was one of the most important applied mathematicians of the 19th century. When he was 8 years old, he was sent to the school of a science teacher, Robert Goodacre, but left after only four terms to work in a bakery. Thereafter, he received no more formal education until the age of 40. When he was 14, his father built a windmill at Sneinton, just outside Nottingham, and George began working in the mill. In his spare time he developed an interest in mathematics and physics, and, astoundingly, in 1828 at the age of 35 he published his first scientific paper 'An essay on the application of mathematical analysis to the theories of electricity and magnetism'. He had invented revolutionary mathematical techniques to solve certain of the problems arising in these fields. Unfortunately, at the time, there was little interest in his work. Nevertheless, he went on to produce more papers in scientific journals, including work on optics and elasticity. At the age of 40, he entered Cambridge University to read for a mathematics degree. He was awarded his degree in 1837, and then held a fellowship at the university for two more years. Unfortunately, he then became ill and returned to Nottingham, where he died in 1841.

His new mathematical techniques were devised to solve a range of physical problems, particularly in electrostatics and elasticity. One particularly important piece of work is now known as Green's theorem. This relates properties of functions at the surface of a closed volume to other properties inside that volume. Figure 1 shows his original statement of the theorem taken from his essay in 1828.

(3.)   Before proceeding to make known some relations which exist between the density of the electric fluid at the surfaces of bodies, and the corresponding values of the potential functions within and without those surfaces, the electric fluid being confined to them alone, we shall in the first place, lay down a general theorem which will afterwards be very useful to us.   This theorem may be thus enunciated:

Let U and V be two continuous functions of the rectangular co-ordinates x, y, z, whose differential co-efficients do not become infinite at any point within a solid body of any form whatever; then will

$$\int dx\,dy\,dz\, U \delta V + \int d\sigma\, U \left(\frac{dV}{dw}\right) = \int dx\,dy\,dz\, V \delta U + \int d\sigma\, V \left(\frac{dU}{dw}\right);$$

the triple integrals extending over the whole interior of the body, and those relative to $d\sigma$, over its surface, of which $d\sigma$ represents an element: dw being an infinitely small line perpendicular to the surface, and measured from this surface towards the interior of the body.

To prove this let us consider the triple integral

$$\int dx\,dy\,dz \left\{ \left(\frac{dV}{dx}\right)\left(\frac{dU}{dx}\right) + \left(\frac{dV}{dy}\right)\left(\frac{dU}{dy}\right) + \left(\frac{dV}{dz}\right)\left(\frac{dU}{dz}\right) \right\}.$$

The method of integration by parts, reduces this to

$$\int dy\,dz\, V'' \frac{dU''}{dx} - \int dy\,dz\, V' \frac{dU'}{dx} + \int dx\,dz\, V'' \frac{dU''}{dy} - \int dx\,dz\, V' \frac{dU'}{dy}$$

$$+ \int dx\,dy\, V'' \frac{dU''}{dz} - \int dx\,dy\, V' \frac{dU'}{dz} - \int dx\,dy\,dz\, V \left\{ \frac{d^2 U}{dx^2} + \frac{d^2 U}{dy^2} + \frac{d^2 U}{dz^2} \right\};$$

the accents over the quantities indicating, as usual, the values of those quantities at the limits of the integral, which in the present case are on the surface of the body, over whose interior the triple integrals are supposed to extend.

Figure 1 *Green's original statement of the theorem (1828).*

While detailed analysis of this theorem is material for a more advanced course, study of this extract will reveal several areas of mathematics that have been developed in earlier chapters of this book. You will see that the theorem refers to two continuous functions U and V that are defined in terms of cartesian coordinates x, y and z. The triple integrals referred to are obtained by generalization of single integrals, with which you are now familiar. Integration by parts plays an important role in proving Green's theorem. The proof requires detailed knowledge of calculus, together with an understanding of vector quantities, which are introduced in Chapter 27.

Figure 2 *The windmill at Sneinton which now houses the science museum.*

Today, the windmill has been restored and houses a science museum in Green's memory (Figure 2).

In July 1993, about the time of the bicentenary of Green's birth, a memorial slab was unveiled in Westminster Abbey to commemorate his life and work.

A handsome hardback facsimile edition of Green's essay may be purchased from Professor Challis, Physics Department, Nottingham University, Nottingham NG7 2RD, price £12, including postage (cheque to George Green Memorial Fund). The edition is one of a limited edition of 1000 prepared in 1958 for Professor Ekelof, Chalmers Institute, Gothenburg, who generously donated copies to the Fund.

Green's Mill and Science Centre is open, without charge, from Tuesday to Sunday. Further details can be obtained from Mr D. Plowman, Keeper, Green's Mill and Science Centre, Belvoir Hill, Sneinton, Nottingham NG2 4LF.

replaced by a new variable, z say. That is, we let

$$z = 2x + 5, \quad \text{from which we see that} \quad \frac{dz}{dx} = 2$$

This substitution enables us to change the variable of integration to obtain a much simpler integral. The integrand $(2x+5)^6$ is replaced by z^6, and we must also replace the term dx. Since $dz/dx = 2$, we can write $dx = dz/2$, and so

$$\int (2x+5)^6 \, dx = \int z^6 \frac{dz}{2}$$

$$= \frac{1}{2} \int z^6 \, dz$$

$$= \frac{1}{2} \frac{z^7}{7} + c$$

$$= \frac{z^7}{14} + c$$

Finally, to revert to the original variable, x, we let $z = 2x + 5$. We find

$$\int (2x+5)^6 \, dx = \frac{(2x+5)^7}{14} + c$$

Example 25.7

By letting $z = 4 - 5x$, find

$$\int \frac{1}{(4-5x)^2} \, dx$$

Solution   By letting $z = 4 - 5x$, the integrand becomes $1/z^2$. We must also replace the term dx. Noting that $dz/dx = -5$, we can write $dx = -\frac{1}{5} dz$. The integral then becomes

$$\int \frac{1}{(4-5x)^2} \, dx = -\frac{1}{5} \int \frac{1}{z^2} \, dz$$

$$= -\frac{1}{5} \left(-\frac{1}{z} \right) + c$$

$$= \frac{1}{5z} + c$$

Finally, we let $z = 4 - 5x$ to return to the original variable. This gives

$$\int \frac{1}{(4-5x)^2} \, dx = \frac{1}{5(4-5x)} + c$$

Example 25.8

Find $\int \cos x \sin^2 x \, dx$ by letting $z = \sin x$.

Solution Writing $z = \sin x$, we see that $\sin^2 x = z^2$. Also, $dz/dx = \cos x$; that is, $dx = \dfrac{dz}{\cos x}$.

Then

$$\int \cos x \sin^2 x \, dx = \int \cos x \, z^2 \frac{dz}{\cos x}$$

$$= \int z^2 \, dz$$

$$= \frac{z^3}{3} + c$$

Finally, reverting to the original variable x, we have

$$\int \cos x \sin^2 x \, dx = \frac{1}{3} \sin^3 x + c$$

Example 25.9

Find

$$\int \frac{2x+2}{x^2+2x+7} \, dx$$

by letting $u = x^2 + 2x + 7$.

Solution If $u = x^2 + 2x + 7$ then $du/dx = 2x + 2$ so that $(2x+2)\,dx = du$. Then

$$\int \frac{2x+2}{x^2+2x+7} \, dx = \int \frac{du}{u}$$

$$= \ln|u| + c \qquad \text{(from Table 23.1)}$$

$$= \ln|x^2+2x+7| + c$$

Note that in Example 25.9 the numerator of the integrand is the derivative of the denominator, and that the required integral is equal to the natural logarithm of the modulus of the denominator. This result is true more generally:

$$\int \frac{f'(x)}{f(x)} \, dx = \ln|f(x)| + c$$

Example 25.10

$$\text{Find } \int \frac{1}{x+2} \, dx.$$

Solution   Consider the integrand $1/(x+2)$. Note that the numerator is the derivative of the denominator. Therefore we can immediately state

$$\int \frac{1}{x+2} \, dx = \ln |x+2| + c$$

Example 25.11

$$\text{Find } \int \frac{\cos x}{\sin x} \, dx.$$

Solution   Note that the numerator is the derivative of the denominator. Therefore

$$\int \frac{\cos x}{\sin x} \, dx = \ln |\sin x| + c$$

Example 25.12

$$\text{Find } \int \frac{1}{2x-1} \, dx.$$

Solution   Letting $u = 2x - 1$, we find that $du/dx = 2$, so that $dx = du/2$. Then

$$\int \frac{1}{2x-1} \, dx = \int \frac{1}{u} \frac{du}{2}$$

$$= \frac{1}{2} \int \frac{du}{u}$$

$$= \tfrac{1}{2} \ln |u| + c$$

$$= \tfrac{1}{2} \ln |2x-1| + c$$

An alternative way of finding this integral is to notice that the derivative of the denominator of the integrand is 2. We can force the numerator to be 2 by writing

$$\int \frac{1}{2x-1} \, dx = \frac{1}{2} \int \frac{2}{2x-1} \, dx$$

and then using the result

$$\int \frac{f'(x)}{f(x)}\, dx = \ln |f(x)| + c$$

we find

$$\int \frac{1}{2x-1}\, dx = \frac{1}{2}\int \frac{2}{2x-1}\, dx = \tfrac{1}{2}\ln |2x-1| + c$$

as before.

When using substitution to find definite integrals, care must be taken with the limits of integration.

Example 25.13

By letting $z = 3x + 4$, find $\int_0^1 (3x+4)^2\, dx$.

Solution

Note that the limits on x are 0 and 1. If we make the substitution $z = 3x + 4$ then $dz/dx = 3$ and so $dx = dz/3$. When writing the integral in terms of z, the limits must refer to z and not x. Now,

$$\text{when } x = 0, \qquad z = 3(0) + 4 = 4$$

and

$$\text{when } x = 1, \qquad z = 3(1) + 4 = 7$$

Hence the integral may be written in terms of z as follows:

$$\int_0^1 (3x+4)^2\, dx = \int_4^7 z^2 \frac{dz}{3}$$

Note that the limits 0 and 1 refer to x, and the limits 4 and 7 refer to z. Now

$$\int_4^7 z^2 \frac{dz}{3} = \frac{1}{3}\int_4^7 z^2\, dz$$

$$= \frac{1}{3}\left[\frac{z^3}{3}\right]_4^7$$

$$= \left[\frac{z^3}{9}\right]_4^7$$

$$= \left(\frac{7^3}{9}\right) - \left(\frac{4^3}{9}\right)$$

$$= 31$$

Exercises 25.3

1. By making the given substitution, find the following integrals:
   (a) $\int (7x+4)^2 \, dx$, let $z=7x+4$
   (b) $\int \cos(2x+1)\, dx$, let $z=2x+1$
   (c) $\int (14x+3)(7x^2+3x-2)\, dx$,
   let $z=7x^2+3x-2$
   (d) $\int e^{3x-9}\, dx$, let $z=3x-9$

2. Find the following integrals:
   (a) $\int \sqrt{x+1}\, dx$ (b) $\int \sqrt{2x+1}\, dx$
   (c) $\int \sqrt{ax+b}\, dx$ (d) $\int \dfrac{1}{x+1}\, dx$
   (e) $\int \dfrac{1}{(x+1)^2}\, dx$
   (f) $\int \dfrac{1}{(x+1)^n}\, dx$, where $n\neq 1$
   (g) $\int \dfrac{1}{ax+b}\, dx$
   (h) $\int \dfrac{1}{(ax+b)^n}\, dx$, where $n\neq 1$

3. Find the following integrals:
   (a) $\int \dfrac{1}{2x+5}\, dx$ (b) $\int \dfrac{2}{2x-1}\, dx$

   (c) $\int \dfrac{2}{2x+7}\, dx$ (d) $\int \dfrac{1}{x-1}\, dx$
   (e) $\int \dfrac{1}{2x+1}\, dx$ (f) $\int \dfrac{dx}{x\ln x}$

4. Find $\int \cot x\, dx$.

5. Evaluate the following integrals:
   (a) $\int_{-1}^{1} (9x-8)^2\, dx$
   (b) $\int_{0}^{\pi} \cos(2x-3)\, dx$ (c) $\int_{1}^{2} \dfrac{1}{3x+2}\, dx$
   (d) $\int_{0}^{4} \dfrac{1}{4x+7}\, dx$

6. Find
   (a) $\int \sin(ax+b)\, dx$ (b) $\int \cos(ax+b)\, dx$
   (c) $\int e^{ax+b}\, dx$

7. Find $\int 7xe^{x^2}\, dx$.

8. Find $\int \dfrac{3e^{3x}}{10+e^{3x}}\, dx$.

25.4 Integration by partial fractions

In order to integrate a rational function, it is often possible to express the function in partial fractions and then integrate each fraction separately. It is essential that you have mastered the techniques of finding partial fractions dealt with in Chapter 11 before continuing. Consider the following example.

Example 25.14

Find $\int \dfrac{3x+2}{(x-1)(x+2)}\, dx$.

Solution The integrand is first expressed in partial fractions. We have

$$\frac{3x+2}{(x-1)(x+2)} = \frac{A}{x-1} + \frac{B}{x+2} = \frac{A(x+2)+B(x-1)}{(x-1)(x+2)}$$

so that

$$3x+2 = A(x+2)+B(x-1)$$

Setting $x=1$, we find $5=3A$, so that $A=\frac{5}{3}$. Setting $x=-2$, we find $-4=-3B$, so that $B=\frac{4}{3}$. The problem becomes that of finding the integral

$$\int \left(\frac{5}{3(x-1)} + \frac{4}{3(x+2)} \right) dx$$

Both terms can be integrated using the method of Example 25.10. We have

$$\int \left(\frac{5}{3(x-1)} + \frac{4}{3(x+2)} \right) dx = \int \frac{5}{3(x-1)} dx + \int \frac{4}{3(x+2)} dx$$

$$= \frac{5}{3} \int \frac{1}{x-1} dx + \frac{4}{3} \int \frac{1}{x+2} dx$$

$$= \tfrac{5}{3} \ln|x-1| + \tfrac{4}{3} \ln|x+2| + c$$

Example 25.15

Find $\displaystyle\int_0^1 \frac{x}{(x+1)^2} dx$.

Solution The integrand is first expressed in partial fractions. Noting the repeated linear factor in the denominator, we have

$$\frac{x}{(x+1)^2} = \frac{A}{x+1} + \frac{B}{(x+1)^2}$$

so that

$$x = A(x+1)+B$$

Setting $x=-1$, we find $B=-1$. Equating coefficients of x, we find $A=1$. Hence the integral becomes

$$\int_0^1 \left(\frac{1}{x+1} - \frac{1}{(x+1)^2} \right) dx$$

Integrating, we find

$$\int_0^1 \left(\frac{1}{x+1} - \frac{1}{(x+1)^2} \right) dx = \left[\ln |x+1| + \frac{1}{x+1} \right]_0^1$$

$$= (\ln 2 + \tfrac{1}{2}) - (\ln 1 + 1)$$

$$= \ln 2 + \tfrac{1}{2} - 1 \qquad \text{(since } \ln 1 = 0\text{)}$$

$$= \ln 2 - \tfrac{1}{2}$$

$$= 0.193$$

Exercises 25.4

1. Use partial fractions to find the following integrals:

   (a) $\displaystyle\int \frac{4}{(x+1)(x-2)}\, dx$

   (b) $\displaystyle\int \frac{5x+2}{(x-1)(x+2)}\, dx$

   (c) $\displaystyle\int \frac{2x-3}{x^2+5x+6}\, dx$

2. Express $\dfrac{x^2}{x^2+2x+1}$ in partial fractions.

   Hence find $\displaystyle\int \frac{x^2}{x^2+2x+1}\, dx$.

3. Find $\displaystyle\int \frac{2x+7}{(x+3)^2}\, dx$.

4. Evaluate

   (a) $\displaystyle\int_1^3 \frac{2x}{(x+1)(x+3)}\, dx$

   (b) $\displaystyle\int_0^2 \frac{6}{x^2+7x+12}\, dx$

   (c) $\displaystyle\int_{-1}^0 \frac{4x}{x^2+4x+4}\, dx$

25.5 Integration of trigonometric functions

In engineering applications it is sometimes necessary to integrate expressions involving the trigonometric functions. The trigonometric identities may be used to rewrite an integrand in a way that allows easy integration. It is often not clear which identities are useful, and each case needs to be considered individually. Consider the following examples.

Example 25.16

Find $\int \sin^2 x\, dx$.

Solution   The integrand $\sin^2 x$ does not appear in the table of integrals, Table 23.1. Integrals of powers of the trigonometric functions are not given. However, Table 23.1 does give the integrals of $\sin kx$ and $\cos kx$, and so we attempt to rewrite the integrand using an appropriate identity. Using the trigonometric identity

$$\sin^2 \theta = \tfrac{1}{2}(1 - \cos 2\theta)$$

we can write

$$\int \sin^2 x \, dx = \int \tfrac{1}{2}(1 - \cos 2x) \, dx$$

$$= \frac{1}{2} \int (1 - \cos 2x) \, dx$$

$$= \frac{1}{2}\left(x - \frac{\sin 2x}{2} + c\right)$$

$$= \tfrac{1}{2}x - \tfrac{1}{4}\sin 2x + K$$

where $K = \tfrac{1}{2}c$.

Example 25.17

Find $\int \sin x \cos x \, dx$.

Solution   Using the trigonometric identity

$$\sin 2x = 2 \sin x \cos x$$

we can write the integrand as $\tfrac{1}{2}\sin 2x$. Then

$$\int \tfrac{1}{2}\sin 2x \, dx = -\tfrac{1}{4}\cos 2x + c$$

Exercises 25.5

1. Find (a) $\int \cos^2 x \, dx$ (b) $\int_0^{\pi/2} \cos^2 x \, dx$

   (Hint: Use $\cos^2 A = \tfrac{1}{2}(1 + \cos 2A)$.)

2. Find (a) $\int_0^{\pi} 3 \sin^2 x \, dx$

   (b) $\int_0^{\pi/4} 3 \sin^2 x \, dx$

3. Find $\int (\cos^2 x + \sin^2 x) \, dx$.

4. Find $\int \sin 3x \cos 2x \, dx$. (Hint: Consider the identity $2 \sin A \cos B = \sin (A + B) + \sin (A - B)$.)

5. Find $\int \sin 2x \cos 2x \, dx$: (a) by using a trigonometric identity and (b) by using the substitution $z = \cos 2x$.

6. Use a trigonometric identity to find $\int (1 + \tan^2 \theta) \, d\theta$.

25.6 The average value of a function

Suppose the function $f(t)$ is defined for values of t between a and b as shown in Figure 25.1. The area under the curve and above the t axis is $\int_a^b f(t)\,dt$.

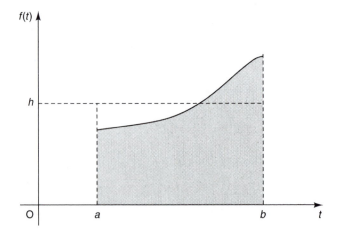

Figure 25.1   The area of the rectangle equals the area under the curve.

Suppose we now construct a rectangle of height h. The base of the rectangle is that part of the t axis between a and b. The area of this rectangle is $(b-a)\times h$. It is possible to choose the value of h so that the area of the rectangle and the area under the curve are the same; that is,

area of rectangle = area under curve

$$(b-a)h = \int_a^b f(t)\,dt$$

so that

$$h = \frac{1}{b-a}\int_a^b f(t)\,dt$$

This value of h is called the **average value** of the function $f(t)$ across the interval from a to b.

$$\text{average value of } f(t) \text{ over } [a, b] = h = \frac{1}{b-a}\int_a^b f(t)\,dt$$

Example 25.18

Find the average value of the function $f(t) = 3t^2$ across the intervals (a) from 1 to 2, and (b) from 0 to 1.5.

Solution   (a)   Using the formula for the average value, we find

$$h = \frac{1}{2-1} \int_1^2 3t^2 \, dt = [t^3]_1^2 = 8 - 1 = 7$$

This means that a rectangle of height 7 with the interval [1, 2] as its base will have the same area as that under the curve $y = 3t^2$ between $t = 1$ and $t = 2$.

(b)   Using the formula for the average value, we find

$$h = \frac{1}{1.5 - 0} \int_0^{1.5} 3t^2 \, dt = \frac{1}{1.5} [t^3]_0^{1.5} = 1.5^2 = 2.25$$

This means that a rectangle of height 2.25 with the interval [0, 1.5] as its base will have the same area as that under the curve $y = 3t^2$ between $t = 0$ and $t = 1.5$.

Note from (a) and (b) that the average value of a function depends upon the interval chosen.

Example 25.19 Average value of a signal

Electronics engineers are often interested in the average value of a signal. The signal shown in Figure 25.2 is known as a **rectified sine wave**. Calculate the average value of the rectified sine wave signal over one period.

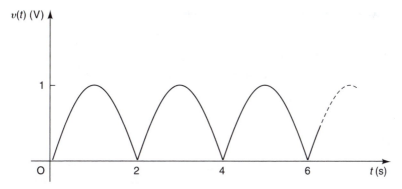

Figure 25.2   Rectified sine wave signal.

Solution   We must find the average value over one period. For convenience, we shall choose the interval $t = 0$ to $t = 2$. We need to take care when formulating a mathematical expression for v over this interval. We note that although the rectified sine wave repeats with a period of 2 seconds, the waveform in the interval [0, 2] can be thought of as part of a full sine wave with period 4 seconds. Recall from Chapter 16 that the angular frequency of such a sine wave is $\frac{1}{2}\pi$ radians per second, and so its equation is $v = \sin \frac{1}{2}\pi t$. Then

$$\text{average value of } v(t) = \frac{1}{b-a} \int_a^b v(t)\, dt$$

$$= \frac{1}{2-0} \int_0^2 \sin \tfrac{1}{2}\pi t\, dt$$

$$= \frac{1}{2} \int_0^2 \sin \tfrac{1}{2}\pi t\, dt$$

$$= \frac{1}{2} \left[\frac{-\cos \frac{1}{2}\pi t}{\frac{1}{2}\pi} \right]_0^2$$

$$= \frac{2}{2\pi}(-\cos \pi + \cos 0)$$

$$= \frac{1}{\pi}(1+1)$$

$$= \frac{2}{\pi}$$

$$= 0.637 \text{ V}$$

This value is also the average value of the signal over many periods.

Self-assessment questions 25.6

1. Explain the graphical interpretation of the 'average value' of a function across an interval.

2. State the formula for this average value.

Exercises 25.6

1. Find the average values of the following functions across the intervals stated:

   (a) $f(t) = 4t^3$, $t = 1$ to $t = 2$
   (b) $f(t) = \sin t$, $t = 0$ to $t = \pi$
   (c) $f(t) = t^2 + 7$, $t = 0$ to $t = 10$

2. Find the average value of the function

   $f(x) = x^2$ across the intervals (a) [0, 3]
   (b) [2, 4]

3. Find the average values of the signals shown in Figure 25.3 across the intervals shown.

4. Calculate the average value of $f(t) = e^{2t}$ across [1, 4].

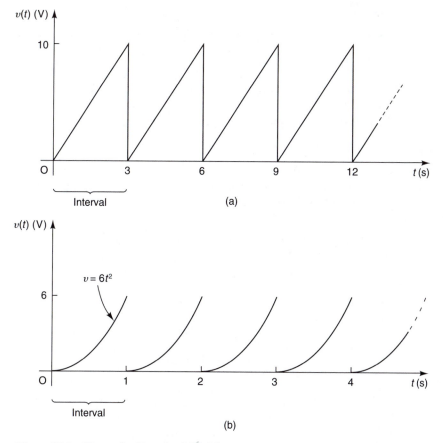

Figure 25.3 Figure for Exercise 25.6 Q3.

25.7 The root mean square (r.m.s.) value of a function

The **mean square value** of a function $f(t)$ across the interval from a to b is the average value of the square of the function across that interval; that is,

$$\text{mean square value across } [a,\, b] = \frac{1}{b-a} \int_a^b [f(t)]^2 \, dt$$

If we take the square root of this quantity, we obtain the **root mean square**

value; that is,

$$\text{root mean square value across } [a,\, b] = \sqrt{\dfrac{1}{b-a} \int_a^b [f(t)]^2 \, dt}$$

The phrase 'root mean square' is usually abbreviated to r.m.s.

Example 25.20

Find (a) the mean square value and (b) the r.m.s. value of the function $f(t)=t^3$ across the interval from 0 to 2.

Solution   (a)   Using the formula for mean square value with $f(t)=t^3$, we find

$$\text{mean square value} = \frac{1}{b-a} \int_a^b [f(t)]^2 \, dt$$

$$= \frac{1}{2-0} \int_0^2 (t^3)^2 \, dt$$

$$= \frac{1}{2} \int_0^2 t^6 \, dt$$

$$= \frac{1}{2} \left[\frac{t^7}{7} \right]_0^2$$

$$= \frac{1}{2} \left(\frac{2^7}{7} - 0 \right)$$

$$= \frac{18.286}{2}$$

$$= 9.143$$

   (b)   The r.m.s. value is the square root of the mean square value; that is, $\sqrt{9.143} = 3.024$.

Example 25.21

Consider Figure 25.4, which shows a voltage supply that varies with time, $v(t)$, connected across a load resistor R. Clearly, the instantaneous power dissipated in the resistor changes as the voltage of the supply changes. It is useful, however, to know the average power dissipated in the resistor. For example, an electric fire is supplied by an AC voltage, and yet it is useful to know the average power

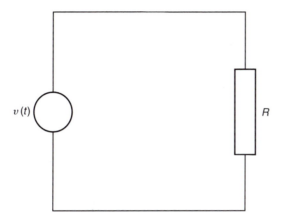

Figure 25.4   A voltage supply $v(t)$ connected to a load resistor R.

given out by the fire, since this enables a person buying the fire to know whether it will be sufficient to warm the room in which it will be used. Once the r.m.s. value of a waveform has been calculated, it is a simple matter to calculate the power dissipated in a load. This is given by

$$P = \frac{(v_{rms})^2}{R}$$

where v_{rms} is the root mean square voltage. For this reason, the r.m.s. value of a waveform can be thought of as the equivalent DC value for the purposes of calculating power. In the following example we shall calculate the r.m.s. value of a sinusoidal waveform. This is the most common type of signal found in engineering. For example, the electricity supply consists of a sinusoidal waveform.

Example 25.22 The r.m.s. value of a sinusoidal waveform

Calculate the r.m.s. value of the waveform $v(t) = V \sin \omega t$ over a single cycle, where V is the amplitude of the waveform and ω is the angular frequency.

Solution   We use an interval of $[0, 2\pi/\omega]$ because this corresponds to the first complete cycle:

$$\text{r.m.s. value of } v(t) = \sqrt{\frac{1}{2\pi/\omega} \int_0^{2\pi/\omega} V^2 \sin^2 \omega t \, dt}$$

$$= \sqrt{\frac{V^2 \omega}{2\pi} \int_0^{2\pi/\omega} \sin^2 \omega t \, dt}$$

Using the identity

$$\sin^2 \theta = \tfrac{1}{2}(1 - \cos 2\theta)$$

we can write

$$\sin^2 \omega t = \tfrac{1}{2}(1 - \cos 2\omega t)$$

Thus

$$\text{r.m.s. value of } v(t) = \sqrt{\frac{V^2\omega}{4\pi} \int_0^{2\pi/\omega} (1 - \cos 2\omega t) \, dt}$$

$$= \sqrt{\frac{V^2\omega}{4\pi} \left[t - \frac{\sin 2\omega t}{2\omega} \right]_0^{2\pi/\omega}}$$

$$= \sqrt{\frac{V^2\omega}{4\pi} \left(\frac{2\pi}{\omega} - \frac{\sin 2(2\pi)}{2\omega} \right)}$$

$$= \sqrt{\frac{V^2\omega}{4\pi} \frac{2\pi}{\omega}} \qquad (\text{since } \sin 4\pi = 0)$$

$$= \sqrt{\frac{V^2}{2}}$$

$$= \frac{V}{\sqrt{2}}$$

$$= 0.707V$$

The result obtained in the previous example is true more generally. The r.m.s. value of any sinusoidal waveform is $0.707 \times$ amplitude of the waveform.

Self-assessment question 25.7

1. Explain how the mean square value and r.m.s. value of a function are found.

Exercises 25.7

1. Calculate the r.m.s. value of $f(t) = t^2$ across the interval $[0, 2]$.

2. Show that the r.m.s. value of $f(t) = \cos \omega t$ across the interval $[0, 2\pi/\omega]$ is 0.707.

3. Calculate the r.m.s. value of $f(t) = \sqrt{t + 1}$ across the interval $[0, 2]$.

4. Calculate the r.m.s. value of the function $f(t) = \cos 3t$ across the interval $[0, 2\pi]$.

5. Calculate the root mean square value of the function $f(t) = e^t$ across the interval $[0, 1]$.

6. Calculate the r.m.s. value of each of the waveforms shown in Figure 25.5 across the intervals indicated. Graph 25.5(a) shows a rectified sine wave.

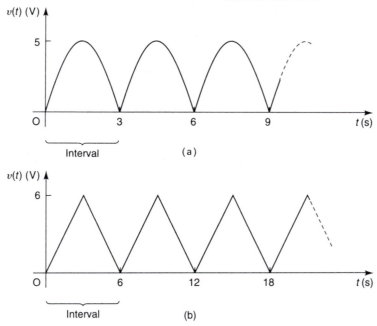

Figure 25.5 Waveforms for Exercise 25.7 Q6.

25.8 Using computer algebra packages

Computer algebra packages are able to carry out a wide range of calculations involving differentiation and integration. These range from finding limits of functions, and obtaining indefinite and definite integrals, through to much more advanced techniques. It will be useful for your future studies if you can familiarize yourself with such a package.

Computer and calculator exercises 25.8

1. Find $\int x^3 e^x \, dx$.

2. Find $\int x^4 e^{4x} \, dx$.

3. Find $\displaystyle\int_0^2 \ln(x^2 + 3) \, dx$.

4. Find $\displaystyle\int_0^{2\pi} \sin t^2 \, dt$.

Review exercises 25

1   Evaluate the following:

(a) $\displaystyle\int_1^2 \frac{2x+1}{x+1}\,dx$ (b) $\displaystyle\int_0^2 \frac{x^2+x}{x+3}\,dx$

(c) $\displaystyle\int_0^1 \frac{2x^2+3}{x^2+1}\,dx$ (d) $\displaystyle\int_3^4 \frac{x^2+x+1}{2x+1}\,dx$

2   Evaluate

(a) $\displaystyle\int_0^\pi x\cos 2x\,dx$ (b) $\displaystyle\int_0^{\pi/2} x\sin 3x\,dx$

(c) $\displaystyle\int_1^3 xe^{4x}\,dx$ (d) $\displaystyle\int_1^3 x\ln x\,dx$

(e) $\displaystyle\int_1^3 x^2\ln x\,dx$

3   Evaluate

(a) $\displaystyle\int_0^1 (2x+1)^9\,dx$ (b) $\displaystyle\int_1^5 (3-t)^7\,dt$

(c) $\displaystyle\int_0^1 \sin x\,(\cos x)^4\,dx$ (d) $\displaystyle\int_0^1 x^2 e^{x^3}\,dx$

4   Find the average value of each of the following functions across the intervals stated:

(a) $f(t)=(t-1)(t-2)$ across $[0,\,2]$

(b) $f(x)=\tan x$ across $[0,\,\tfrac{1}{4}\pi]$.

5   Calculate the r.m.s. value of $y=t+3$ across the interval from $t=2$ to $t=6$.

26

Matrices and Determinants

KEY POINTS

This chapter

- explains what is meant by a matrix

- explains how a matrix is multiplied by a number

- shows how to add and subtract matrices when this is possible

- shows how to multiply two matrices when this is possible

- explains the terms 'premultiplication' and 'postmultiplication'

- explains what is meant by the determinant of a matrix and how this is calculated

- explains what is meant by the inverse of a 2×2 matrix and how this is calculated

- shows how matrices can be used to solve simultaneous equations

CONTENTS

26.1 Introduction

In many engineering calculations it is necessary to handle large amounts of data. Matrices can be used to store this data in a form that makes these calculations straightforward. For example, when analysing an electrical circuit, it is usual to write down expressions relating voltage sources and currents in different parts of the circuit. These expressions can be written concisely using matrices. Knowing the voltages, matrix calculations can then be performed in order to calculate the various currents.

26.2 Matrices

A **matrix** is a rectangular pattern of numbers usually enclosed in brackets. The plural of matrix is **matrices**. For example,

$$\begin{pmatrix} 3 & 2 \\ 1 & 8 \end{pmatrix}, \quad \begin{pmatrix} 1 & 2 \\ 4 & -3 \\ -2 & 5 \end{pmatrix}, \quad (1 \quad 3 \quad -9)$$

are all rectangular patterns of numbers, and are therefore matrices.

In order to refer to a particular matrix, we often label it with a capital letter. For example, we might write

$$A = \begin{pmatrix} 3 & 2 \\ 1 & 8 \end{pmatrix}, \quad B = \begin{pmatrix} 1 & 2 \\ 4 & -3 \\ -2 & 5 \end{pmatrix}, \quad C = (1 \quad 3 \quad -9)$$

To describe the **size** of a matrix, we state its number of rows and columns, in that order. The matrix B has three rows and two columns, and is said to be a 'three-by-two' matrix. We write B is a 3×2 matrix. The matrix C has one row and three columns, and is therefore a 1×3 matrix. A matrix that has just one row and one column is a single number. Each number in a matrix is known as an **element** of the matrix.

Generally, if a matrix A has m rows and n columns, we can write

$$A = \begin{pmatrix} a_{11} & a_{12} & \cdots & a_{1n} \\ a_{21} & a_{22} & \cdots & a_{2n} \\ \vdots & \vdots & \ddots & \vdots \\ a_{m1} & a_{m2} & \cdots & a_{mn} \end{pmatrix}$$

where the symbol a_{ij} represents the element in the ith row and jth column.

Example 26.1

The matrix A is given by

$$A = \begin{pmatrix} 1 & -4 & 3 & -19 \\ 0 & 4 & -0.5 & 2 \\ 0 & 1 & 1.5 & -13 \\ 9 & -5 & 7 & 12 \end{pmatrix}$$

State the elements a_{21}, a_{43} and a_{34}.

Solution

The element a_{21} is that in the second row and first column, namely 0. The element a_{43} is that in the fourth row and third column, namely 7. Finally, the element a_{34} is that in the third row and fourth column, namely -13.

26.2.1 Some special types of matrix

There are some types of matrix that are particularly important and so have been given special names.

A **square** matrix has the same number of rows as columns. For example,

$$\begin{pmatrix} 1 & 4 & -2 \\ 12 & 3 & 1 \\ 3 & 7 & 2 \end{pmatrix} \quad \text{is a } 3 \times 3 \text{ square matrix}$$

A **diagonal** matrix is a square matrix with zeros everywhere except on the diagonal running from the top left to the bottom right. This diagonal is known as the **leading diagonal**. For example,

$$\begin{pmatrix} 1 & 0 & 0 \\ 0 & 3 & 0 \\ 0 & 0 & 2 \end{pmatrix} \quad \text{is a } 3 \times 3 \text{ diagonal matrix}$$

An **identity** or **unit** matrix is a diagonal matrix with all its diagonal elements equal to 1. For example

$$\begin{pmatrix} 1 & 0 & 0 \\ 0 & 1 & 0 \\ 0 & 0 & 1 \end{pmatrix} \quad \text{and} \quad \begin{pmatrix} 1 & 0 \\ 0 & 1 \end{pmatrix} \quad \text{are both identity matrices}$$

Self-assessment questions 26.2

1. When quoting the size of a matrix, which comes first – the number of rows or the number of columns?

2. What is meant by a square matrix?

3. Explain what is meant by a diagonal matrix. Give an example of a matrix that is diagonal and of one that is not.

4. What is meant by an identity matrix? Must an identity matrix be square?

5. Describe the position of element a_{32} in a matrix.

Exercises 26.2

1. Describe the size of each of the following matrices:

(a) $(1 \quad 7 \quad -2 \quad 4)$ (b) $\begin{pmatrix} 1 & 7 & 8 \\ 2 & 6 & -3 \end{pmatrix}$

(c) $\begin{pmatrix} 1 \\ 4 \\ -2 \\ -5 \\ 1 \end{pmatrix}$ (d) $\begin{pmatrix} 5 & 6 & 3 \\ 2 & -7 & 2 \\ 4 & 4 & 5 \\ -1 & 0 & 0 \end{pmatrix}$

(e) $\begin{pmatrix} a & b \\ c & d \\ e & f \end{pmatrix}$ (f) $\begin{pmatrix} 1 & 5 & 6 & 7 \\ -1 & 2 & 0 & 0 \\ 0 & 1 & 2 & -1 \end{pmatrix}$

2. How many elements does a 3×5 matrix have? How many elements does an $m \times n$ matrix have?

3. Write down the 4×4 identity matrix.

4. If $M = \begin{pmatrix} 0 & 3 & -2 & 1 \\ 4 & 6 & 1 & 5 \\ -1 & 7 & 1.5 & 0.5 \end{pmatrix}$ state the values of

(a) m_{11} (b) m_{22} (c) m_{21} (d) m_{12}
(e) m_{34} (f) m_{31}

26.3 Addition and subtraction of matrices

In order to add or subtract two matrices, they must have exactly the same size. When this is so, addition is performed by simply adding corresponding elements together. Subtraction is performed by subtracting corresponding elements. Consider the following examples.

Example 26.2

If $A = \begin{pmatrix} 2 & 7 \\ 3 & -1 \end{pmatrix}$, $B = \begin{pmatrix} 3 & 6 \\ 2 & 4 \end{pmatrix}$ and $C = \begin{pmatrix} 4 & 2 \\ -1 & 3 \end{pmatrix}$, find (a) $A + B$ (b) $A + B + C$

Solution (a) The matrices A and B have the same size, and so they can be added together:

$$A+B=\begin{pmatrix} 2 & 7 \\ 3 & -1 \end{pmatrix}+\begin{pmatrix} 3 & 6 \\ 2 & 4 \end{pmatrix}=\begin{pmatrix} 2+3 & 7+6 \\ 3+2 & -1+4 \end{pmatrix}=\begin{pmatrix} 5 & 13 \\ 5 & 3 \end{pmatrix}$$

(b) A, B and C all have the same size, and so can be added. We find

$$A+B+C=\begin{pmatrix} 2 & 7 \\ 3 & -1 \end{pmatrix}+\begin{pmatrix} 3 & 6 \\ 2 & 4 \end{pmatrix}+\begin{pmatrix} 4 & 2 \\ -1 & 3 \end{pmatrix}=\begin{pmatrix} 9 & 15 \\ 4 & 6 \end{pmatrix}$$

Example 26.3

If $C=(1\quad 2\quad 5\quad -4)$ and $D=(-2\quad 2\quad 0.5\quad -3)$, find $C-D$.

Solution The matrices C and D have the same size, and so they can be subtracted:

$$C-D=(1\quad 2\quad 5\quad -4)-(-2\quad 2\quad 0.5\quad -3)=(3\quad 0\quad 4.5\quad -1)$$

Example 26.4

If $A=\begin{pmatrix} 3 & 2 \\ -1 & 7 \end{pmatrix}$ and $B=\begin{pmatrix} 3 \\ 2 \end{pmatrix}$, find, if possible, their sum $A+B$.

Solution A and B do not have the same size, and so cannot be added together. We say that their sum does not exist.

Self-assessment questions 26.3

1. State the conditions under which it is possible to add or subtract two matrices.

2. Explain why it is not possible to add a 5×2 matrix to a 2×5 matrix.

Exercises 26.3

1. If $A=\begin{pmatrix} 9 & 6 & 5 \\ 18 & -2 & 5 \end{pmatrix}$, $B=\begin{pmatrix} 1 & 2 & 8 \\ 3 & -2 & 5 \end{pmatrix}$ and $C=\begin{pmatrix} 3 & 8 & -9 \\ 0 & 7 & 2 \end{pmatrix}$, find

   (a) $A+B$ (b) $A-B$ (c) $B+C$
   (d) $B-C$ (e) $A+C$ (f) $C-A$

   (g) $A+B+C$ (h) $A+B-C$
   (i) $A-B-C$

2. If $C=\begin{pmatrix} 7 & 8 \\ -2 & 7 \end{pmatrix}$ and $D=\begin{pmatrix} 9 & 11 \\ -9 & -9 \end{pmatrix}$, find

   (a) $C+D$ (b) $D+C$ (c) $C-D$
   (d) $D-C$

26.4 Multiplication of a matrix by a number

A matrix of any size can be multiplied by a single number. To do this, each element in the matrix is multiplied by this number. For example, if

$$A = \begin{pmatrix} 2 & -1 \\ 4 & 2 \\ 9 & -7 \end{pmatrix}$$

then

$$5A = \begin{pmatrix} 5 \times 2 & 5 \times -1 \\ 5 \times 4 & 5 \times 2 \\ 5 \times 9 & 5 \times -7 \end{pmatrix} = \begin{pmatrix} 10 & -5 \\ 20 & 10 \\ 45 & -35 \end{pmatrix}$$

Example 26.5

If $M = \begin{pmatrix} 2 & 0.5 & 6 \\ -3 & 0 & 1 \end{pmatrix}$ find (a) $7M$ (b) $-\frac{1}{2}M$

Solution (a) $\qquad 7M = 7\begin{pmatrix} 2 & 0.5 & 6 \\ -3 & 0 & 1 \end{pmatrix} = \begin{pmatrix} 14 & 3.5 & 42 \\ -21 & 0 & 7 \end{pmatrix}$

(b) $\qquad -\frac{1}{2}M = -\frac{1}{2}\begin{pmatrix} 2 & 0{:}5 & 6 \\ -3 & 0 & 1 \end{pmatrix} = \begin{pmatrix} -1 & -0.25 & -3 \\ 1.5 & 0 & -0.5 \end{pmatrix}$

Example 26.6

If $A = \begin{pmatrix} 1 \\ 2 \\ 9 \end{pmatrix}$ and $B = \begin{pmatrix} -1 \\ 3 \\ 5 \end{pmatrix}$, find (a) $7A + 2B$ (b) $2A - 3B$

Solution (a) $\quad 7A + 2B = 7\begin{pmatrix} 1 \\ 2 \\ 9 \end{pmatrix} + 2\begin{pmatrix} -1 \\ 3 \\ 5 \end{pmatrix} = \begin{pmatrix} 7 \\ 14 \\ 63 \end{pmatrix} + \begin{pmatrix} -2 \\ 6 \\ 10 \end{pmatrix} = \begin{pmatrix} 5 \\ 20 \\ 73 \end{pmatrix}$

(b) $\quad 2A - 3B = 2\begin{pmatrix} 1 \\ 2 \\ 9 \end{pmatrix} - 3\begin{pmatrix} -1 \\ 3 \\ 5 \end{pmatrix} = \begin{pmatrix} 2 \\ 4 \\ 18 \end{pmatrix} - \begin{pmatrix} -3 \\ 9 \\ 15 \end{pmatrix} = \begin{pmatrix} 5 \\ -5 \\ 3 \end{pmatrix}$

Self-assessment questions 26.4

1. Explain the procedure for multiplying a matrix by a number.

2.   If M is a matrix, explain what is meant by $\frac{1}{2}M$.

Exercises 26.4

1.   If $M = \begin{pmatrix} 4 & -3 \\ 5 & 5 \end{pmatrix}$ and $N = \begin{pmatrix} \alpha & \beta \\ \gamma & \delta \end{pmatrix}$, find

(a) $5N$ (b) $7M$ (c) $7M - 5N$
(d) $5N + 7M$

2.   If I is the 2×2 identity matrix, find
(a) $7I$ (b) $-3I$ (c) $0.5I$
(d) λI, where λ is a number

3.   For the matrices A, B and C of Exercise 26.3

Q1 find
(a) $5B$ (b) $4C$ (c) $5B - 4C$ (d) $0.5A$

4.   If $A = \begin{pmatrix} 1 & 3 \\ 2 & -1 \end{pmatrix}$, $B = \begin{pmatrix} 4 & 2 & 0 \\ 1 & 3 & 6 \end{pmatrix}$, $C = \begin{pmatrix} 3 & 1 \\ 4 & 0 \end{pmatrix}$

and $D = \begin{pmatrix} 1 & 0 & 1 \\ 2 & 1 & 1 \end{pmatrix}$, find, if possible,

(a) $A + C$ (b) $A + B$ (c) $3C - 2A$
(d) $4D - 5B$ (e) $A + 2B + 3C$
(f) $\frac{1}{2}(D - 2A)$

Computer and calculator exercise 26.4

1.   Your graphics calculator may be able to manipulate matrices. If so, find out how to enter a matrix and then, using the matrices of Exercise 26.3 Q1, find
(a) $4A$ (b) $3A - 7C$ (c) $-15A + 27B$

26.5 Multiplying two matrices together

Suppose A and B are two matrices and we wish to form the product AB. This product can only be found when the number of columns in the matrix on the left is the same as the number of rows in the matrix on the right. In the product AB, matrix A is on the left and matrix B is on the right. So, for example, if A is a 4×3 matrix and B is a 3×2 matrix then the number of columns in A is the same as the number of rows in B, and so the product AB can be found. The result is another matrix of size 4×2. In general, when we form the product AB, the resulting matrix has the same number of rows as A and the same number of columns as B. Mathematically, we can state this as follows.

> If A is a $p \times q$ matrix and B is an $r \times s$ matrix, we can only form the product AB if $q = r$. If q does equal r, the matrix AB has size $p \times s$.

Example 26.7

Given $A = (4 \quad 2)$ and $B = \begin{pmatrix} 3 & 7 & 6 \\ 5 & 2 & -1 \end{pmatrix}$ can the product AB be formed?

Solution A has size 1×2

B has size 2×3

Because the number of columns in A is the same as the number of rows in B, we can form the product AB. The resulting matrix will have size 1×3 because there is one row in A and three columns in B.

In the product AB we say that B has been **premultiplied** by A, or, alternatively, A has been **postmultiplied** by B.

The product is formed in a way that may seem rather strange at first but which turns out to be useful. Let us consider a very simple example first. Suppose we wish to find AB when $A = (4 \quad 2)$ and $B = \begin{pmatrix} 3 \\ 7 \end{pmatrix}$. A has size 1×2 and B has size 2×1, and so we can form the product AB. The result will be a 1×1 matrix, that is, a single number. We perform the calculation as follows:

$$AB = (4 \quad 2)\begin{pmatrix} 3 \\ 7 \end{pmatrix} = 4 \times 3 + 2 \times 7 = 12 + 14 = 26$$

Note that we have multiplied elements in the row of A with corresponding elements in the column of B, and added the results together.

Example 26.8

Find CD when $C = (1 \quad 3 \quad 5)$ and $D = \begin{pmatrix} 2 \\ 6 \\ 8 \end{pmatrix}$

Solution $CD = (1 \quad 3 \quad 5)\begin{pmatrix} 2 \\ 6 \\ 8 \end{pmatrix} = 1 \times 2 + 3 \times 6 + 5 \times 8 = 2 + 18 + 40 = 60$

Let us now extend this idea to general matrices A and B. Suppose we wish to find C, where $C = AB$. The element c_{11} is found by pairing each element in row 1 of A with the corresponding element in column 1 of B. The pairs are multiplied together and then the results are added to give c_{11}. Similarly, to find the element c_{12}, each element in row 1 of A is paired with the corresponding element in column 2 of B. Again, the paired elements are multiplied together and the results are added to form c_{12}. Other elements of C are found in a similar way. In general, the element c_{ij} is found by pairing elements in the ith row of A with those in the jth column of B. These are multiplied together and the results are added to give c_{ij}. Consider the following example.

Example 26.9

If $A = \begin{pmatrix} 1 & 2 \\ 4 & 3 \end{pmatrix}$ and $B = \begin{pmatrix} 5 \\ -3 \end{pmatrix}$, find, if possible, the matrix C where $C = AB$.

Solution We can form the product

$$C = AB = \begin{pmatrix} 1 & 2 \\ 4 & 3 \end{pmatrix}\begin{pmatrix} 5 \\ -3 \end{pmatrix}$$

$$\uparrow \qquad \uparrow$$
$$2 \times 2 \quad 2 \times 1$$

because the number of columns in A, namely 2, is the same as the number of rows in B. The size of the product is found by inspecting the number of rows in the first matrix, which is 2, and the number of columns in the second, which is 1. These numbers give the number of rows and columns respectively in C. Therefore C will be a 2×1 matrix. To find the element c_{11}, we pair the elements in the first row of A with those in the first column of B, multiply and then add these together. Thus $c_{11} = 1 \times 5 + 2 \times (-3) = 5 - 6 = -1$. Similarly, to find the element c_{21}, we pair the elements in the second row of A with those in the first column of B, multiply and then add these together. Thus $c_{21} = 4 \times 5 + 3 \times (-3) = 20 - 9 = 11$. The complete calculation is written as follows:

$$AB = \begin{pmatrix} 1 & 2 \\ 4 & 3 \end{pmatrix}\begin{pmatrix} 5 \\ -3 \end{pmatrix} = \begin{pmatrix} 1 \times 5 + 2 \times (-3) \\ 4 \times 5 + 3 \times (-3) \end{pmatrix}$$

$$= \begin{pmatrix} 5 - 6 \\ 20 - 9 \end{pmatrix}$$

$$= \begin{pmatrix} -1 \\ 11 \end{pmatrix}$$

Example 26.10

If $A = \begin{pmatrix} 1 & 3 \\ 4 & 7 \\ 2 & 5 \end{pmatrix}$ and $B = \begin{pmatrix} 4 & 7 \\ -3 & 6 \end{pmatrix}$, find, if possible, AB and BA.

Solution   We have

$$AB = \begin{pmatrix} 1 & 3 \\ 4 & 7 \\ 2 & 5 \end{pmatrix} \begin{pmatrix} 4 & 7 \\ -3 & 6 \end{pmatrix}$$

$$\underset{3 \times 2}{\uparrow} \qquad \underset{2 \times 2}{\uparrow}$$

We see that the number of columns in the first matrix is the same as the number of rows in the second, and so we can evaluate AB. The size of the product will be 3×2.

$$AB = \begin{pmatrix} 1 & 3 \\ 4 & 7 \\ 2 & 5 \end{pmatrix} \begin{pmatrix} 4 & 7 \\ -3 & 6 \end{pmatrix}$$

$$= \begin{pmatrix} 1 \times 4 + 3 \times (-3) & 1 \times 7 + 3 \times 6 \\ 4 \times 4 + 7 \times (-3) & 4 \times 7 + 7 \times 6 \\ 2 \times 4 + 5 \times (-3) & 2 \times 7 + 5 \times 6 \end{pmatrix}$$

$$= \begin{pmatrix} 4 - 9 & 7 + 18 \\ 16 - 21 & 28 + 42 \\ 8 - 15 & 14 + 30 \end{pmatrix}$$

$$= \begin{pmatrix} -5 & 25 \\ -5 & 70 \\ -7 & 44 \end{pmatrix}$$

On the other hand, if we write

$$BA = \begin{pmatrix} 4 & 7 \\ -3 & 6 \end{pmatrix} \begin{pmatrix} 1 & 3 \\ 4 & 7 \\ 2 & 5 \end{pmatrix}$$

$$\underset{2 \times 2}{\uparrow} \qquad \underset{3 \times 2}{\uparrow}$$

then this product does not exist, because the number of columns in the first matrix is not the same as the number of rows in the second. We can premultiply B by A but we cannot premultiply A by B. Alternatively, we can postmultiply A by B but we cannot postmultiply B by A. This distinction is important.

Example 26.11

If $A = \begin{pmatrix} 1 & 2 \\ 0 & 3 \end{pmatrix}$ and $B = \begin{pmatrix} -1 & 0 \\ 1 & 2 \end{pmatrix}$, find (a) AB (b) BA

Solution   (a) $AB = \begin{pmatrix} 1 & 2 \\ 0 & 3 \end{pmatrix}\begin{pmatrix} -1 & 0 \\ 1 & 2 \end{pmatrix} = \begin{pmatrix} 1 & 4 \\ 3 & 6 \end{pmatrix}$

(b) $BA = \begin{pmatrix} -1 & 0 \\ 1 & 2 \end{pmatrix}\begin{pmatrix} 1 & 2 \\ 0 & 3 \end{pmatrix} = \begin{pmatrix} -1 & -2 \\ 1 & 8 \end{pmatrix}$

Clearly, the product AB is not the same as BA.

The property exhibited in the previous example is very important and applies more generally:

In general, $AB \neq BA$.

Example 26.12

If $M = \begin{pmatrix} 8 & 5 \\ 4 & 2 \end{pmatrix}$, find MI and IM, where I is the 2×2 identity matrix.

Solution $MI = \begin{pmatrix} 8 & 5 \\ 4 & 2 \end{pmatrix}\begin{pmatrix} 1 & 0 \\ 0 & 1 \end{pmatrix} = \begin{pmatrix} 8 & 5 \\ 4 & 2 \end{pmatrix}$

$IM = \begin{pmatrix} 1 & 0 \\ 0 & 1 \end{pmatrix}\begin{pmatrix} 8 & 5 \\ 4 & 2 \end{pmatrix} = \begin{pmatrix} 8 & 5 \\ 4 & 2 \end{pmatrix}$

We see that multiplying the matrix M by the identity matrix leaves M unaltered.

The previous result is true more generally. Whenever a matrix is multiplied by an identity matrix, the result is unchanged, provided of course that it is possible to find the product.

Example 26.13

If $A = \begin{pmatrix} 1 & 5 \\ 8 & -3 \end{pmatrix}$ and $X = \begin{pmatrix} x \\ y \end{pmatrix}$, evaluate, if possible, AX.

Solution   We are required to find

$$AX = \begin{pmatrix} 1 & 5 \\ 8 & -3 \end{pmatrix} \begin{pmatrix} x \\ y \end{pmatrix}$$

$$\uparrow \qquad \uparrow$$
$$2 \times 2 \quad 2 \times 1$$

The first matrix has two columns and the second matrix has two rows. We can therefore find the product AX, the size of which will be 2×1:

$$AX = \begin{pmatrix} 1 & 5 \\ 8 & -3 \end{pmatrix} \begin{pmatrix} x \\ y \end{pmatrix} = \begin{pmatrix} x + 5y \\ 8x - 3y \end{pmatrix}$$

Self-assessment questions 26.5

1.   Describe carefully the circumstances under which two matrices can be multiplied together.

2.   Give an example of two matrices that cannot be multiplied together.

3.   Describe the distinction between premultiplication and postmultiplication.

4.   Given that it is possible to work out the product AB, is it true that it is also possible to work out the product BA?

5. A and B are two matrices. Both AB and BA can be found. What can you deduce about the sizes of A and B?

6.   When A is a matrix, we write A^2 for the product of A with itself. Under what circumstances does A^2 exist?

Exercises 26.5

1.   If $A = \begin{pmatrix} 1 & 3 \\ 3 & 4 \end{pmatrix}$ and $B = \begin{pmatrix} 3 & 7 \\ -6 & 0 \end{pmatrix}$, find, if possible, AB and BA, and comment upon the result.

2.   If $A = \begin{pmatrix} 4 & -2 \\ 1 & 5 \end{pmatrix}$, find A^2, that is, the product of A with itself.

3.   If $A = \begin{pmatrix} 5 & 4 \\ 6 & 5 \end{pmatrix}$ and $B = \begin{pmatrix} 5 & -4 \\ -6 & 5 \end{pmatrix}$, find AB and BA, and comment upon the result.

4.   If $M = \begin{pmatrix} 4 & 6 & -2 \\ 2 & 4 & 1 \end{pmatrix}$ and $N = \begin{pmatrix} 9 \\ 4 \\ -1 \end{pmatrix}$, find, if possible, MN and NM.

5.   If $P = \begin{pmatrix} 3 & 0 & 7 \\ 2 & 0 & 1 \\ -2 & 7 & 0 \end{pmatrix}$ and $Q = \begin{pmatrix} \alpha \\ \beta \\ \gamma \end{pmatrix}$, find, if possible, PQ and QP.

6.   Show that the product of $\begin{pmatrix} 2 & 3 \\ 5 & 1 \end{pmatrix}$ and

$-\frac{1}{13}\begin{pmatrix} 1 & -3 \\ -5 & 2 \end{pmatrix}$ is an identity matrix, regardless of the order in which the product is evaluated.

7. If $A = \begin{pmatrix} 3 & 0 & 0 \\ 0 & 4 & 0 \\ 0 & 0 & 5 \end{pmatrix}$ and $X = \begin{pmatrix} x \\ y \\ z \end{pmatrix}$, find AX.

Computer and calculator exercises 26.5

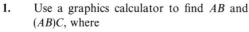

1. Use a graphics calculator to find AB and $(AB)C$, where

$$A = \begin{pmatrix} 3 & -2 & 7 \\ 2 & 1 & 0 \\ 4 & 3 & 1 \end{pmatrix}$$

$$B = \begin{pmatrix} 2 & -1 & -7 \\ 0 & 2 & -3 \\ 1 & 3 & -2 \end{pmatrix}$$

$$C = \begin{pmatrix} 9 & 0 & -7 \\ 2 & 8 & 0 \\ 1 & -2 & 2 \end{pmatrix}$$

Show that $(AB)C$ is the same as $A(BC)$.

2. If $A = \begin{pmatrix} 21 & 3 & 7 \\ -5 & 2 & 5 \\ 1 & 0 & 0 \end{pmatrix}$ and

$$B = \begin{pmatrix} 0 & 0 & 1 \\ 5 & -7 & -140 \\ -2 & 3 & 57 \end{pmatrix}, \text{find } AB \text{ and } BA,$$

and comment upon the result.

26.6 The determinant of a 2×2 matrix

Consider the square 2×2 matrix $A = \begin{pmatrix} a & b \\ c & d \end{pmatrix}$. An important quantity related to A is known as the **determinant** of A, and is defined to be the number

$$(a \times d) - (b \times c)$$

or simply $ad - bc$. The determinant of A is denoted by

$$\det A \quad \text{or} \quad |A| \quad \text{or} \quad \begin{vmatrix} a & b \\ c & d \end{vmatrix}$$

For example, the determinant of $A = \begin{pmatrix} 7 & 3 \\ 4 & 2 \end{pmatrix}$ is

$$(7 \times 2) - (3 \times 4) = 14 - 12 = 2$$

We write

$$\begin{vmatrix} 7 & 3 \\ 4 & 2 \end{vmatrix} = 2$$

Note that $\begin{vmatrix} a & b \\ c & d \end{vmatrix}$ is a determinant, but $\begin{pmatrix} a & b \\ c & d \end{pmatrix}$ is a matrix, and the two are not the same.

$$\det A = \begin{vmatrix} a & b \\ c & d \end{vmatrix} = ad - bc$$

It is possible to calculate the determinant of a square matrix of any size. For details of how to do this, you should refer to a more advanced text.

Example 26.14

If $A = \begin{pmatrix} 1 & 2 \\ 4 & 7 \end{pmatrix}$, $B = \begin{pmatrix} 1 & -1 \\ 2 & 4 \end{pmatrix}$, $C = \begin{pmatrix} 4 & 1 \\ 8 & 2 \end{pmatrix}$ and $I = \begin{pmatrix} 1 & 0 \\ 0 & 1 \end{pmatrix}$, find

(a) $|A|$ (b) $|B|$ (c) $\det C$ (d) $|I|$

Solution   (a)   Using the formula, we find

$$|A| = (1)(7) - (2)(4) = 7 - 8 = -1$$

Note that the determinant of A is a single number and not a matrix. We can also write this result as

$$\begin{vmatrix} 1 & 2 \\ 4 & 7 \end{vmatrix} = -1$$

(b) $|B| = (1)(4) - (-1)(2) = 4 - (-2) = 6$. In this example the value of the determinant of B is 6.

(c) $\det C = (4)(2) - (1)(8) = 8 - 8 = 0$

In this example the value of the determinant of C is zero.

(d) $|I| = (1)(1) - (0)(0) = 1 - 0 = 1$. The determinant of the identity matrix is 1.

If the determinant of a square matrix happens to be zero, we say it is a **singular** matrix. In the previous example C is a singular matrix.

Self-assessment questions 26.6

1. What is the distinction between a determinant and a matrix?

2. What notation is used to distinguish between a determinant and a matrix?

Exercises 26.6

1. Find

   (a) $\begin{vmatrix} 1 & 4 \\ -1 & 2 \end{vmatrix}$ (b) $\begin{vmatrix} 1 & 0 \\ 0 & 5 \end{vmatrix}$ (c) $\begin{vmatrix} 0 & 4 \\ 0 & 2 \end{vmatrix}$

   (d) $\begin{vmatrix} 11 & -2 \\ -1 & 1 \end{vmatrix}$ (e) $\begin{vmatrix} -5 & -2 \\ -4 & -1 \end{vmatrix}$

2. Find the determinant of each of the following matrices:

   (a) $\begin{pmatrix} 5 & -2 \\ 3 & 2 \end{pmatrix}$ (b) $\begin{pmatrix} 1 & -13 \\ -2 & 1 \end{pmatrix}$

   (c) $\begin{pmatrix} 0.5 & -2 \\ 1 & 0.5 \end{pmatrix}$ (d) $\begin{pmatrix} 15 & 9 \\ 2 & 16 \end{pmatrix}$

   (e) $\begin{pmatrix} 4 & -2 \\ 0 & -7 \end{pmatrix}$

3. For what value of t is the determinant of $\begin{pmatrix} t-1 & 2 \\ t-4 & 1 \end{pmatrix}$ equal to zero.

4. Write down a 2×2 matrix A none of whose elements are zero but is such that $|A|=0$.

5. If a 2×2 matrix has two rows the same, what can you deduce about its determinant? Give an example to illustrate your argument.

6. If A is the matrix $\begin{pmatrix} x-3 & 3 \\ 2 & x-4 \end{pmatrix}$, find values of x for which $|A|=0$.

7. Write down a 2×2 matrix that has a determinant equal to 3. Write down another that has a determinant equal to 4. Find the product of your two matrices, and show that its determinant is equal to 12.

Computer and calculator exercises 26.6

1. Find out if your graphics calculator can calculate the determinant of a 2×2 matrix. If so, use it to calculate the determinant of each of the matrices in Exercise 26.6 Q2.

2. The calculation of the determinant of a 3×3 matrix is beyond the scope of this book, but your graphics calculator or computer software can help.

   (a) Use it to find $|A|$ and $|B|$, where

   $$A = \begin{pmatrix} 1 & -1 & 2 \\ 0 & 8 & 2 \\ -1 & 0 & 3 \end{pmatrix} \quad \text{and}$$

   $$B = \begin{pmatrix} 4 & 0 & 2 \\ 1 & 6 & -2 \\ 2 & 2 & 3 \end{pmatrix}$$

   (b) Use your calculator to find AB.
   (c) Use your calculator to find $|AB|$, and verify that $|AB|=|A|\,|B|$.

3. Find $\begin{vmatrix} 1 & 3 & -8 \\ 18 & 1 & 0 \\ -1 & 3 & 4 \end{vmatrix}$

4. Find det A, where $A = \begin{pmatrix} 4 & 5 & 6 \\ 7 & 8 & 6 \\ -1 & 3 & 4 \end{pmatrix}$

26.7 The inverse of a 2×2 matrix

The **inverse** of a 2×2 matrix A is another matrix, which is given the symbol A^{-1} and has the property that $AA^{-1} = I$ and also $A^{-1}A = I$, where I is the 2×2 identity matrix. That is, multiplying a matrix by its inverse yields the identity matrix:

$$AA^{-1} = I \quad \text{and also} \quad A^{-1}A = I$$

For example, it can be shown that the inverse of matrix $A = \begin{pmatrix} 4 & 1 \\ 7 & 2 \end{pmatrix}$ is the matrix $A^{-1} = \begin{pmatrix} 2 & -1 \\ -7 & 4 \end{pmatrix}$. To verify that A^{-1} has the required properties to be the inverse of A, we need to multiply the two together. Thus

$$AA^{-1} = \begin{pmatrix} 4 & 1 \\ 7 & 2 \end{pmatrix}\begin{pmatrix} 2 & -1 \\ -7 & 4 \end{pmatrix} = \begin{pmatrix} 1 & 0 \\ 0 & 1 \end{pmatrix}$$

and also

$$A^{-1}A = \begin{pmatrix} 2 & -1 \\ -7 & 4 \end{pmatrix}\begin{pmatrix} 4 & 1 \\ 7 & 2 \end{pmatrix} = \begin{pmatrix} 1 & 0 \\ 0 & 1 \end{pmatrix}$$

as required.

There is a formula for calculating the inverse of a 2×2 matrix:

$$\text{if} \quad A = \begin{pmatrix} a & b \\ c & d \end{pmatrix} \quad \text{then its inverse is} \quad A^{-1} = \frac{1}{ad - bc}\begin{pmatrix} d & -b \\ -c & a \end{pmatrix}$$

Note that this formula contains the quantity $ad - bc$, which we recognize as the determinant of A. We could therefore write the formula as

$$A^{-1} = \frac{1}{|A|}\begin{pmatrix} d & -b \\ -c & a \end{pmatrix}$$

Example 26.15

Use the formula to find the inverse of each of the following matrices:

(a) $A = \begin{pmatrix} 4 & 2 \\ 3 & 4 \end{pmatrix}$ (b) $B = \begin{pmatrix} 2 & 0.5 \\ 1 & 0 \end{pmatrix}$

Solution (a) Here $a=4, b=2, c=3$ and $d=4$. Using the formula for the inverse, we find

$$A^{-1} = \frac{1}{(4)(4)-(2)(3)}\begin{pmatrix} 4 & -2 \\ -3 & 4 \end{pmatrix}$$

$$= \frac{1}{10}\begin{pmatrix} 4 & -2 \\ -3 & 4 \end{pmatrix}$$

$$= \begin{pmatrix} 0.4 & -0.2 \\ -0.3 & 0.4 \end{pmatrix}$$

(b) Here $a=2, b=0.5, c=1$ and $d=0$. Using the formula for the inverse, we find

$$B^{-1} = \frac{1}{(2)(0)-(0.5)(1)}\begin{pmatrix} 0 & -0.5 \\ -1 & 2 \end{pmatrix}$$

$$= \frac{1}{-0.5}\begin{pmatrix} 0 & -0.5 \\ -1 & 2 \end{pmatrix}$$

$$= -2\begin{pmatrix} 0 & -0.5 \\ -1 & 2 \end{pmatrix}$$

$$= \begin{pmatrix} 0 & 1 \\ 2 & -4 \end{pmatrix}$$

Note that the formula for the inverse of a matrix requires us to calculate $1/|A|$. If the determinant of A happens to be zero, that is, A is singular, we cannot evaluate $1/|A|$. In such a case the inverse of A does not exist.

Example 26.16

Suppose A is a 2×2 matrix such that $AX = B$, where X and B are 2×1 matrices. Find an expression for X in terms of A and B.

Solution To find an expression for X, use is made of the inverse of A. Let us premultiply both sides of the equation $AX = B$ by A^{-1}. Then

$$A^{-1}AX = A^{-1}B$$

But $A^{-1}A = I$, the 2×2 identity matrix. Therefore

$$IX = A^{-1}B$$

But $IX = X$, because multiplying by the identity matrix leaves X unaltered as was shown in Example 26.12, and so

$$X = A^{-1}B$$

Self-assessment questions 26.7

1. Explain what is meant by the inverse of a 2×2 matrix. Under what circumstances will such an inverse exist?

2. State the formula for finding the inverse of a 2×2 matrix.

3. Give an example of a matrix that does not have an inverse and one that does.

4. What is meant by the term 'singular matrix'?

Exercises 26.7

1. Verify that $B = \begin{pmatrix} 2 & 13 \\ 1 & 7 \end{pmatrix}$ is the inverse of $A = \begin{pmatrix} 7 & -13 \\ -1 & 2 \end{pmatrix}$ by directly evaluating AB and BA.

2. Find the inverse, if it exists, of each of the following matrices:

   (a) $\begin{pmatrix} 1 & 7 \\ 4 & 2 \end{pmatrix}$ (b) $\begin{pmatrix} 2 & -1 \\ 4 & -3 \end{pmatrix}$

   (c) $\begin{pmatrix} 4 & 7 \\ 14 & 1 \end{pmatrix}$ (d) $\begin{pmatrix} 0 & 0 \\ 4 & 2 \end{pmatrix}$

   (e) $\begin{pmatrix} 12 & 7 \\ -2 & 3 \end{pmatrix}$ (f) $\begin{pmatrix} 0 & 14 \\ 3 & 0 \end{pmatrix}$

3. Find the inverse of each of the matrices $\begin{pmatrix} 1 & 0 \\ 0 & 1 \end{pmatrix}$ and $\begin{pmatrix} 0 & 1 \\ 1 & 0 \end{pmatrix}$, if they exist.

4. Find the inverse of the matrix $\begin{pmatrix} \alpha & \beta \\ \gamma & \delta \end{pmatrix}$. Under what conditions would the inverse not exist?

Computer and calculator exercises 26.7

1. Use a graphics calculator to find, if it exists, the inverse of each of the matrices in Exercise 26.7 Q2.

2. The calculation of the inverse of a 3×3 matrix is beyond the scope of this book. Use a graphics calculator to determine the inverse, if it exists, of each of the following matrices:

   (a) $\begin{pmatrix} 1 & 0 & 0 \\ 0 & 2 & 0 \\ 0 & 0 & 3 \end{pmatrix}$ (b) $\begin{pmatrix} \frac{1}{3} & 0 & 0 \\ \frac{2}{3} & 1 & 0 \\ \frac{4}{3} & 1 & \frac{1}{3} \end{pmatrix}$

   (c) $\begin{pmatrix} 5 & 4 & 1 \\ -5 & 1 & 0 \\ 1 & 0 & 0 \end{pmatrix}$ (d) $\begin{pmatrix} 7 & 8 & 1 \\ 17 & 1 & 0 \\ 1 & 0 & 0 \end{pmatrix}$

   (e) $\begin{pmatrix} 0 & 0 & 1 \\ 5 & -7 & -1 \\ -2 & 3 & -2 \end{pmatrix}$

26.8 Solving simultaneous equations using matrices

We can use matrices to solve simultaneous equations. Suppose we wish to solve the equations

$$5x + 2y = 1$$
$$8x + 3y = 4$$

First note that

$$\begin{pmatrix} 5x + 2y \\ 8x + 3y \end{pmatrix} \quad \text{can be written as} \quad \begin{pmatrix} 5 & 2 \\ 8 & 3 \end{pmatrix}\begin{pmatrix} x \\ y \end{pmatrix}$$

so that the simultaneous equations can be expressed in the form

$$\begin{pmatrix} 5 & 2 \\ 8 & 3 \end{pmatrix}\begin{pmatrix} x \\ y \end{pmatrix} = \begin{pmatrix} 1 \\ 4 \end{pmatrix}$$

This is in the form

$$AX = B$$

where

$$A = \begin{pmatrix} 5 & 2 \\ 8 & 3 \end{pmatrix}, \quad X = \begin{pmatrix} x \\ y \end{pmatrix}, \quad B = \begin{pmatrix} 1 \\ 4 \end{pmatrix}$$

Here A is known as the **coefficient matrix** because it is the matrix of coefficients of x and y. We now make use of the result from Example 26.16. Recall that when

$$AX = B$$

we showed that

$$X = A^{-1}B$$

In other words, we can determine the solution of the simultaneous equations by premultiplying the matrix $\begin{pmatrix} 1 \\ 4 \end{pmatrix}$ by the inverse of A.

First, we calculate the inverse of A:

$$A^{-1} = \frac{1}{(5)(3) - (2)(8)}\begin{pmatrix} 3 & -2 \\ -8 & 5 \end{pmatrix}$$

$$= \frac{1}{-1}\begin{pmatrix} 3 & -2 \\ -8 & 5 \end{pmatrix}$$

$$= \begin{pmatrix} -3 & 2 \\ 8 & -5 \end{pmatrix}$$

Finally, using $X = A^{-1}B$, we have

$$X = \begin{pmatrix} x \\ y \end{pmatrix} = \begin{pmatrix} -3 & 2 \\ 8 & -5 \end{pmatrix}\begin{pmatrix} 1 \\ 4 \end{pmatrix} = \begin{pmatrix} 5 \\ -12 \end{pmatrix}$$

The solution of the simultaneous equations is therefore $x = 5$ and $y = -12$.

Example 26.17

Use matrices to solve the following pair of simultaneous equations:

$$2x + y = 9$$

$$x - 2y = -8$$

Solution   We first note that the equations can be expressed in matrix form as

$$\begin{pmatrix} 2 & 1 \\ 1 & -2 \end{pmatrix}\begin{pmatrix} x \\ y \end{pmatrix} = \begin{pmatrix} 9 \\ -8 \end{pmatrix}$$

which is in the form $AX = B$. The inverse of the coefficient matrix must now be found:

$$A^{-1} = \frac{1}{(2)(-2) - (1)(1)}\begin{pmatrix} -2 & -1 \\ -1 & 2 \end{pmatrix} = \frac{1}{-5}\begin{pmatrix} -2 & -1 \\ -1 & 2 \end{pmatrix} = \begin{pmatrix} \frac{2}{5} & \frac{1}{5} \\ \frac{1}{5} & -\frac{2}{5} \end{pmatrix}$$

Then we can find $\begin{pmatrix} x \\ y \end{pmatrix}$ from

$$\begin{pmatrix} x \\ y \end{pmatrix} = \begin{pmatrix} \frac{2}{5} & \frac{1}{5} \\ \frac{1}{5} & -\frac{2}{5} \end{pmatrix}\begin{pmatrix} 9 \\ -8 \end{pmatrix} = \begin{pmatrix} \frac{18}{5} - \frac{8}{5} \\ \frac{9}{5} + \frac{16}{5} \end{pmatrix} = \begin{pmatrix} \frac{10}{5} \\ \frac{25}{5} \end{pmatrix} = \begin{pmatrix} 2 \\ 5 \end{pmatrix}$$

We conclude that the solution is $x = 2$ and $y = 5$.

Example 26.18 *Analysis of an electrical network*

In the analysis of an electrical network it is often necessary to solve simultaneous equations. We shall consider a network consisting of resistors and voltage sources. The technique is similar for other types of network. Consider the network shown in Figure 26.1. Mesh currents have been drawn for both loops in the circuit. A **mesh** is defined as a loop that cannot contain a smaller closed current path. Each mesh current has been drawn in a clockwise direction, even though it may eventually turn out to be in the opposite direction when the calculations have been performed. The current in each branch of the circuit can be obtained by combining the mesh currents. Suppose we wish to calculate the mesh currents for this network.

The first stage in the calculation is to make use of Kirchhoff's voltage law for each of the meshes. This states that the sum of the voltages around any

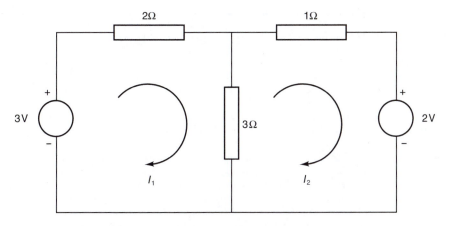

Figure 26.1 An electrical network with mesh currents shown.

closed loop in an electrical network is zero. Therefore the sum of the voltage rises must equal the sum of the voltage drops. When applying Kirchhoff's voltage law, it is important to use the correct sign for a voltage source, depending upon whether or not it is 'aiding' a mesh current. We use the convention that, when travelling in the direction of the mesh current, if a source goes from negative to positive then it is considered a positive source.

For mesh 1 we have

$$3 = 2I_1 + 3(I_1 - I_2)$$

For mesh 2 we have

$$-2 = 1I_2 + 3(I_2 - I_1)$$

Simplifying and rearranging these equations, we have

$$5I_1 - 3I_2 = 3$$

$$-3I_1 + 4I_2 = -2$$

In order to determine the mesh currents, we must solve these two simultaneous equations. In matrix form, we have

$$\begin{pmatrix} 5 & -3 \\ -3 & 4 \end{pmatrix}\begin{pmatrix} I_1 \\ I_2 \end{pmatrix} = \begin{pmatrix} 3 \\ -2 \end{pmatrix}$$

which is in the form $AX = B$ with $A = \begin{pmatrix} 5 & -3 \\ -3 & 4 \end{pmatrix}$, $X = \begin{pmatrix} I_1 \\ I_2 \end{pmatrix}$ and $B = \begin{pmatrix} 3 \\ -2 \end{pmatrix}$.

The inverse of A is first calculated:

$$A^{-1} = \frac{1}{11}\begin{pmatrix} 4 & 3 \\ 3 & 5 \end{pmatrix} = \begin{pmatrix} \frac{4}{11} & \frac{3}{11} \\ \frac{3}{11} & \frac{5}{11} \end{pmatrix}$$

The finite element method: an engineering revolution

The finite element method is a recent numerical technique now widely used by engineers and scientists. Originally developed for use in the aircraft industry, it has become a general-purpose engineering tool. Engineers are often interested in the stresses developed in a structure, for example a bridge, a submarine or an aircraft wing. If allowable stresses are exceeded, the structure may perform poorly or break completely. The finite element method was developed to calculate such stresses, and is now widely used to determine other variables, such as temperature, pressure and flow.

Analysing the stresses within a structure is indeed complex. The irregular shapes of structures, the diversity of materials with their different properties used to build structures and the nonuniformity of the loads applied all add to the difficulty of calculating stresses. Wind and temperature changes may also influence the stress within a structure.

When using the finite element method, the structure or region is divided into many small subregions called **elements**. These may be one-, two- or three-dimensional. Commonly, elements are triangles, parallelograms or straight lines. Irregular shapes can be approximated closely by using many thousands of elements.

Each element is defined by points on its boundary called **nodes**. A triangular element has three nodes corresponding to the vertices of the triangle, and a rectangular element has four nodes. In a finite element analysis the quantities of interest, for example stress, pressure and temperature, are calculated at the nodes. Finding the values of these unknown quantities at the nodal points results from the solution of sets of simultaneous equations. As there are often thousands of nodes, and hence thousands of equations to solve, the finite element method employs computer software.

Having determined the values at the nodal points, the values are then found throughout the element by a technique known as **interpolation**. This allows approximate function values to be found throughout an element from a finite number of known values on the boundary of the element. Knowing the values across each element allows the values to be determined throughout the entire region or structure.

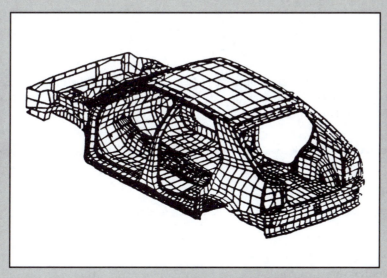

Figure 1 *Finite element mesh of a car body shell.*

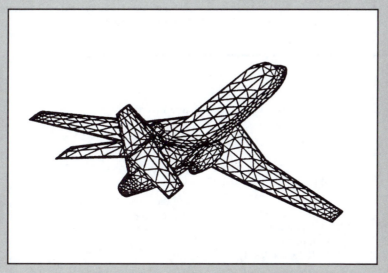

Figure 2 *Finite element mesh of an aeroplane.*

so that

$$\binom{I_1}{I_2} = \begin{pmatrix} \frac{4}{11} & \frac{3}{11} \\ \frac{3}{11} & \frac{5}{11} \end{pmatrix}\binom{3}{-2}$$

$$= \begin{pmatrix} \frac{6}{11} \\ -\frac{1}{11} \end{pmatrix}$$

We conclude that the mesh current I_1 is $\frac{6}{11}=0.545$ A, and the mesh current I_2 is $-\frac{1}{11}=-0.091$ A. The fact that I_2 is negative means that the direction is opposite to that indicated on the diagram.

Self-assessment question 26.8

1.   Explain how two simultaneous equations can be written in matrix form $AX = B$.

Exercises 26.8

1.   Use the matrix method to solve the following pairs of simultaneous equations:
   (a) $2x+y=5$, $x-3y=-1$
   (b) $7x-4y=11$, $x+y=0$
   (c) $x-4y=-13$, $3x+2y=17$
   (d) $x+y=1$, $7x+3y=7$
   (e) $2x+4y=32$, $x-y=4$
   (f) $3x-y=30$, $7x+2y=57$
   (g) $-3x+y=10$, $2x+2y=-12$

2.   Calculate the mesh currents of the circuits shown in Figure 26.2.

Review exercises 26

1   Find, if possible, PQ and QP when

$$P = \begin{pmatrix} 4 & 1 \\ 2 & 3 \\ 4 & 5 \end{pmatrix} \text{ and }$$

$$Q = \begin{pmatrix} 1 & 2 & 3 & 4 \\ -4 & -3 & -2 & -1 \end{pmatrix}.$$

2   If A is a 3×3 matrix and I is the 3×3 identity matrix then the inverse of A, denoted A^{-1}, if it exists, satisfies $AA^{-1}=A^{-1}A=I$. Verify that

$$\begin{pmatrix} 5 & 2 & 3 \\ 11 & 9 & 6 \\ 4 & 5 & 2 \end{pmatrix}$$

is the inverse of

$$\begin{pmatrix} -12 & 11 & -15 \\ 2 & -2 & 3 \\ 19 & -17 & 23 \end{pmatrix}$$

3   Find the inverse of the following matrices. If the inverse does not exist, state this:
   (a) $\begin{pmatrix} 1 & 4 \\ -1 & 3 \end{pmatrix}$ (b) $\begin{pmatrix} -1 & 2 \\ 3 & 4 \end{pmatrix}$
   (c) $\begin{pmatrix} 0 & 8 \\ 0 & -12 \end{pmatrix}$ (d) $\begin{pmatrix} 6 & 0.5 \\ 48 & 4 \end{pmatrix}$

4   Given $A = \begin{pmatrix} 4 & 2 \\ 1 & 3 \end{pmatrix}$, $B = \begin{pmatrix} 0 & 1 & 2 \\ 2 & -1 & 1 \end{pmatrix}$,

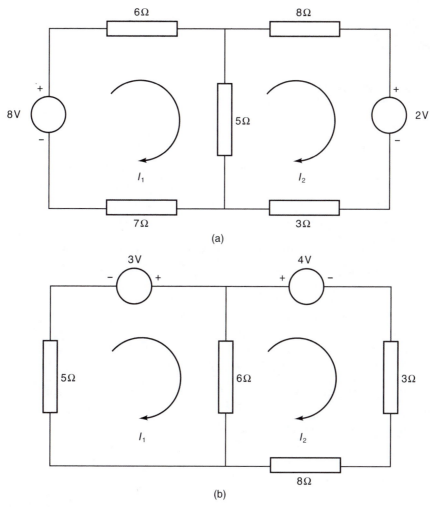

Figure 26.2 Electrical networks for Exercise 26.8 Q2.

$C = \begin{pmatrix} 3 \\ 2 \end{pmatrix}$ and $D = \begin{pmatrix} -1 \\ 1 \end{pmatrix}$, find, if possible, each of the following products:

(a) AB (b) CD (c) AC (d) AD
(e) CB (f) BA

5   Given $A = \begin{pmatrix} 3 & -6 \\ 2 & 6 \end{pmatrix}$ and $B = \begin{pmatrix} -1 & 2 \\ 4 & 0 \end{pmatrix}$,

calculate
(a) $3A$ (b) $5B$ (c) $A - B$
(d) $3A - 4B$ (e) $-7A + 2B$

6 A is an $m \times n$ matrix, B is a $p \times m$ matrix and C is an $n \times p$ matrix, where m, n and p are all different. State which of the following products can be formed:

(a) AB (b) BA (c) AC (d) CA
(e) BC (f) CB

7   Solve the following equations using the inverse matrix method:
(a) $2x - y = -5$, $4x + y = -1$
(b) $3x + 2y = 8$, $x - 3y = -12$
(c) $5x + 7y = -17$, $3x - 10y = 4$

27

Vectors

KEY POINTS

This chapter

- explains what is meant by a scalar and a vector

- shows how vectors can be represented pictorially

- explains how vectors can be added using the triangle law of addition

- shows how vectors can be subtracted

- shows how vectors can be multiplied by a scalar

- shows how vectors can be expressed in cartesian form

- shows how to calculate the magnitude of a vector

- shows how two vectors are multiplied using the scalar product

- explains what is meant by a transformation

CONTENTS

27.1 Introduction

In many engineering applications physical quantities have a direction as well as a magnitude. These quantities are known as vectors. A magnetic force, the velocity of an aeroplane and the momentum of a rocket are all examples of vector quantities.

27.2 Vectors and scalars

Some physical quantities are described by specifying a single number, known as the **magnitude** of the quantity, together with appropriate units. For example, we speak of a temperature of 28 °C, a time of 17 seconds and a speed of 15 m per second. Quantities that can be completely described in this way are called **scalars**. Certain other quantities are only fully described when, in addition to specifying a magnitude, a direction is given. Such quantities are called **vectors**. For example, suppose an object moves with a speed of 15 m per second in a direction that is due west. The quantity '15 m per second due west' is known as the **velocity** of the object, and this is a vector. A force of 12 N vertically downwards is another example of a vector quantity.

To obtain a pictorial representation of a vector, we draw a line. The length of the line represents the magnitude of the vector, and the orientation of the line shows the direction of the vector. We commonly refer to the magnitude as the 'length of a vector'. Consider Figure 27.1. The line joining the points A and B represents a vector whose magnitude is represented by the distance from A to B, and whose direction is that from A to B, as emphasized by an arrow on the line. Usually, the location of the line is of no consequence – only its length and direction are important. We denote the vector from A to B by $\overrightarrow{AB}$. We refer to the point B as the **head** of the vector $\overrightarrow{AB}$ and the point A as its **tail**. Sometimes it is useful to denote $\overrightarrow{AB}$ by a single letter, for example *a*, as shown in Figure 27.1. Note that in books a bold typeface is used to represent the vector *a*. However in written work it is common to underline vector quantities, and *a* would be written as $\underline{a}$ or $\underset{\sim}{a}$.

For example, a force of 3 N vertically downwards represented by a vector *F* is shown in Figure 27.2.

Given a vector *a*, its magnitude is represented by the symbol $|a|$ or simply *a*. If a vector is given in the form $\overrightarrow{AB}$, we write its magnitude as $|\overrightarrow{AB}|$. On a

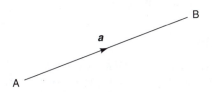

Figure 27.1   The vector *a*.

Figure 27.2 A force of 3 N vertically downwards.

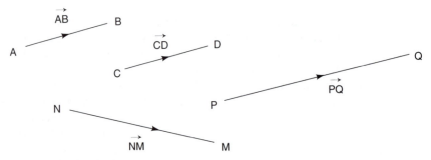

Figure 27.3 A selection of vectors.

diagram the magnitude of $\overrightarrow{AB}$ is the distance from A to B. Given a vector quantity such as 5 m per second in a north-westerly direction the magnitude of the vector is 5 m per second.

Two vectors that have the same magnitude and have the same direction are said to be **equal vectors**. In Figure 27.3 the vectors $\overrightarrow{AB}$ and $\overrightarrow{CD}$ have the same magnitude and the same direction, and so are equal. We write $\overrightarrow{AB} = \overrightarrow{CD}$. Clearly, the vector $\overrightarrow{NM}$ has a different direction and so is not equal to $\overrightarrow{AB}$ or $\overrightarrow{CD}$. The vector $\overrightarrow{PQ}$ has the same direction as $\overrightarrow{AB}$ but a different length. Therefore $\overrightarrow{AB}$ and $\overrightarrow{PQ}$ are not equal.

27.3 The arithmetic of vectors

In order to be able to use vectors, it is necessary to define ways in which they can be manipulated. In this section we look at how they can be added, subtracted and multiplied by a constant.

27.3.1 Addition of vectors

Consider the two vectors a and b shown in Figure 27.4(a). We add these together in the following way. Vector b is repositioned so that its tail coincides with the head of a, as shown in Figure 27.4(b). The sum $a+b$ is then defined to be the

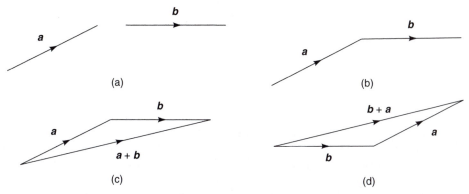

Figure 27.4 Vectors *a* and *b* can be added using the triangle law of addition.

vector from the tail of *a* to the head of *b*, as shown in Figure 27.4(c). This method of adding vectors is known as the **triangle law of addition**.

Suppose we now use the triangle law to find *b* + *a*. Vector *a* is repositioned so that its tail lies at the head of *b*. Then the sum *b* + *a* is the vector from the tail of *b* to the head of *a* as shown in Figure 27.4(d). Note from Figures 27.4(c) and (d) that the sum *a* + *b* is identical to the sum *b* + *a*, because both have the same magnitude and direction. This means that when adding two vectors, the order in which we place them does not matter. In some applications we refer to *a* + *b* as the **resultant** of *a* and *b*.

Example 27.1

Consider the rectangle ABCD shown in Figure 27.5. Simplify the following vector expressions:

(a) $\overrightarrow{DA} + \overrightarrow{AB}$ (b) $\overrightarrow{DB} + \overrightarrow{BC}$

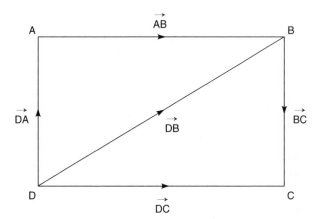

Figure 27.5 Rectangle ABCD.

Solution   (a)   The sum $\overrightarrow{DA} + \overrightarrow{AB}$ is found using the triangle law. We find

$$\overrightarrow{DA} + \overrightarrow{AB} = \overrightarrow{DB}$$

(b)   Using the triangle law, we find

$$\overrightarrow{DB} + \overrightarrow{BC} = \overrightarrow{DC}$$

Example 27.2 *Combining forces acting on an object*

Mechanical engineers often need to determine the effect of several forces acting on an object. For example, a space rocket travelling into space from Earth will be subject to a force due to the thrust of the engine as well as aerodynamic forces due to the rocket moving through the Earth's atmosphere. A force is a vector quantity.

Consider the case of two forces acting on an object as shown in Figure 27.6. A force F_1 acts vertically downwards and has a magnitude of 5 N. A force F_2 acts horizontally and has a magnitude of 8 N. The resultant force R is a single force that is equivalent to the sum of the forces F_1 and F_2. We can use the triangle law to calculate this force. Translating F_2 until its tail touches the head of F_1, we have

$$R = F_1 + F_2$$

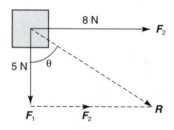

Figure 27.6   The resultant of F_1 and F_2 is R.

We can use Pythagoras' theorem to calculate the magnitude of R. We have

$$|R| = \sqrt{5^2 + 8^2} = \sqrt{25 + 64} = \sqrt{89} = 9.43 \text{ N}$$

The angle θ that the resultant force makes with the vertical is given by

$$\tan \theta = \frac{|F_2|}{|F_1|} = \frac{8}{5} = 1.6$$

Therefore

$$\theta = \tan^{-1} 1.6 = 58.0°$$

Example 27.3 The routeing of a robot vehicle

Robot vehicles are frequently used in factories to carry components to where they are required. The route a vehicle needs to take can be described by means of vectors. Consider Figure 27.7, which shows a simple factory floor layout. A robot vehicle carries components from the stores at the point A to various assembly areas at points B, C and D. The movement of the vehicle from point A to point B can be represented by a vector $\overrightarrow{AB}$, known as a **displacement vector**, whose magnitude is the distance between points A and B. Similarly, movement from point A to point D can be represented by the vector $\overrightarrow{AD}$. Since A, B, C and D are fixed points, these displacement vectors are fixed too. It is possible to combine the displacement vectors together. For example, the route from the stores to assembly area 2, denoted by the vector $\overrightarrow{AC}$, can be written as the sum of the route from the stores to assembly area 1, denoted by $\overrightarrow{AB}$, and the route from assembly area 1 to assembly area 2, denoted by $\overrightarrow{BC}$. In symbols, we can write

$$\overrightarrow{AC} = \overrightarrow{AB} + \overrightarrow{BC}$$

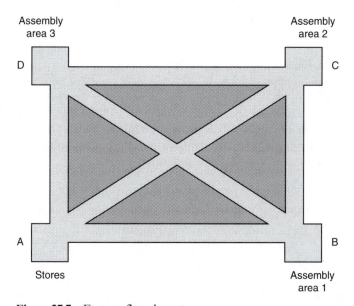

Figure 27.7 Factory floor layout.

27.3.2 Multiplication of a vector by a number

Given a vector a, then $2a$ is a vector having the same direction as a but twice the magnitude. Similarly, $\frac{1}{2}a$ is a vector having the same direction as a but half

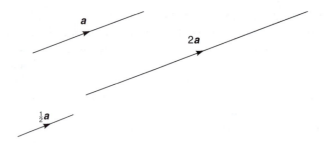

Figure 27.8   Multiplying a vector by a number.

the magnitude. These vectors are shown in Figure 27.8. Note that we can write

$$\tfrac{1}{2}a = \frac{a}{2}$$

and so we can also divide a vector by a number.

   In general, if k is a positive number, the vector ka has the same direction as the vector a, and k times the magnitude of a.

Example 27.4 *Magnetic vector quantities*

The magnetic field intensity H is a vector quantity with units of amperes per metre (A m^{-1}). The magnetic flux density B is another vector quantity, with units of webers per square metre (Wb m^{-2}). (Note that the unit Wb m^{-2} is also known as the tesla, T.) If we know H, the vector B can be calculated by multiplying H by a scalar quantity called the permeability, which has the symbol μ (and units of Wb A^{-1} m^{-1}); that is,

$$B = \mu H$$

This is an example of multiplication of a vector by a scalar.

If we multiply a vector a by -1, the result is a vector that points in the opposite direction to a, but has the same magnitude. Figure 27.9 shows the

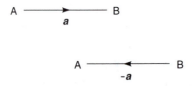

Figure 27.9   The vectors a and $-a$ have the same magnitude but opposite directions.

vectors a and $-a$. Note that whereas $a=\overrightarrow{AB}$, $-a=\overrightarrow{BA}$. Also note that a and $-a$ have the same magnitude. Similarly, $-3a$ has magnitude 3 times that of a and a direction opposite to that of a.

27.3.3 Subtraction of vectors

Suppose we wish to find $a-b$. We do this by writing

$$a-b=a+(-b)$$

We then use the triangle law of addition, adding $-b$ to a.

Example 27.5

The vectors a and b are shown in Figure 27.10. Show $a-b$ and $b-a$ on a diagram.

Figure 27.10   The vectors a and b.

Solution   We form the vector $-b$, which has the same length as b but the opposite direction. This is then added to a to form $a-b$, as shown in Figure 27.11(a). To form $b-a$, we add $-a$ to b. This is shown in Figure 27.11(b). Note from the figure that $b-a=-(a-b)$.

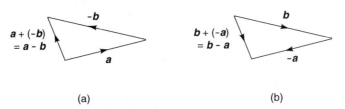

(a)   (b)

Figure 27.11   The vectors $a-b$ (a) and $b-a$ (b).

Example 27.6

Consider the square ABCD shown in Figure 27.12, where $\overrightarrow{AB}=p$ and $\overrightarrow{AD}=q$. Express (a) $\overrightarrow{AC}$ and (b) $\overrightarrow{DB}$ in terms of p and q.

Solution   (a)   Because ABCD is a square, $\overrightarrow{BC}$ is equal to q. Then, using the triangle

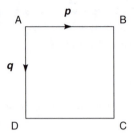

Figure 27.12 The square ABCD.

law of addition,

$$\overrightarrow{AC} = \overrightarrow{AB} + \overrightarrow{BC}$$

$$= p + q$$

(b)   The vector $\overrightarrow{DA}$ equals $-q$. Therefore, from the triangle law of addition, we have

$$\overrightarrow{DB} = \overrightarrow{DA} + \overrightarrow{AB}$$

$$= -q + p$$

$$= p - q$$

27.3.4 Unit vectors

Any vector that has a magnitude of 1 is called a **unit vector**. To distinguish unit vectors from other vectors it is common to use the 'hat' symbol ˆ. Figure 27.13

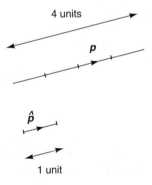

Figure 27.13 The unit vector $\hat{p}$.

shows a vector **p** that has length 4. A unit vector having the same direction as **p** is also shown. This is denoted by **p̂**. Because **p̂** has the same direction as **p** but only one-quarter of the magnitude, it follows that

$$\hat{p} = \tfrac{1}{4}p = \frac{p}{4}$$

In general, given any vector **a**, a unit vector having the same direction as **a** is found by dividing **a** by its magnitude. We have the following formula:

$$\hat{a} = \frac{a}{|a|}$$

Self-assessment questions 27.3

1. Describe the triangle law of addition of vectors.

2. Explain what is meant by a unit vector.

3. Given a vector p, explain how you would calculate $\hat{p}$.

Exercises 27.3

1. Draw a vector $a = \overrightarrow{AB}$ of length 2 cm. On the same figure show $3a$, $-a$, $\hat{a}$ and $\tfrac{1}{2}a$.

2. Draw a vector $p = \overrightarrow{CD}$ of length 3 cm. On the same figure show $-p$ and $-2p$.

3. M is the midpoint of PQ, and $\overrightarrow{PM} = a$. Express the following in terms of a:
   (a) $\overrightarrow{PQ}$ (b) $\overrightarrow{QM}$

4. PQRS is a four-sided figure with $\overrightarrow{PQ} = a$ and $\overrightarrow{QR} = b$. The midpoint of RS is labelled M,

and $\overrightarrow{MR} = c$.
   (a) Draw a diagram to depict this situation.
   (b) Express $\overrightarrow{SR}$, $\overrightarrow{PR}$, and $\overrightarrow{QM}$ in terms of a, b and c.

5. ABC is a triangle. P is the midpoint of AB, and Q is the midpoint of AC.
   (a) Draw a diagram to show this situation.
   (b) Show that BC is parallel to PQ, and find a vector relationship between $\overrightarrow{BC}$ and $\overrightarrow{PQ}$.

27.4 Cartesian components

Consider the (x, y) plane shown in Figure 27.14. The point P has coordinates (2, 5), and the point C has coordinates (2, 0). Suppose we join the origin to P by a vector $\overrightarrow{OP}$. Using the triangle law of addition, we can also express $\overrightarrow{OP}$ as

$$\overrightarrow{OP} = \overrightarrow{OC} + \overrightarrow{CP}$$

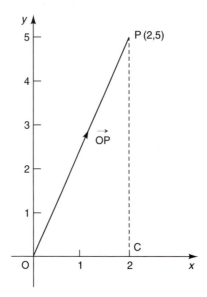

Figure 27.14 The vector $2i + 5j$.

We also note that $\overrightarrow{OC}$ has magnitude 2 and $\overrightarrow{CP}$ has magnitude 5. We can therefore think of the vector $\overrightarrow{OP}$ as being made up of 2 units in the x direction and 5 units in the y direction.

Suppose we let i represent a unit vector in the direction of the positive x axis and j a unit vector in the direction of the positive y axis. It is usual practice to omit the 'hat' here, even though i and j are unit vectors. Then we can write

$$\overrightarrow{OC} = 2i$$

and

$$\overrightarrow{CP} = 5j$$

The vector $\overrightarrow{OP}$ can then be written concisely as

$$\overrightarrow{OP} = 2i + 5j$$

We say that 2 and 5 are the **cartesian components** of the vector $\overrightarrow{OP}$, with 2 being the x component of $\overrightarrow{OP}$ and 5 the y component. Because the specific location of a vector is not important, the vector $2i + 5j$ represents any vector comprising 2 units in the x direction and 5 units in the y direction. Several such vectors are shown in Figure 27.15.

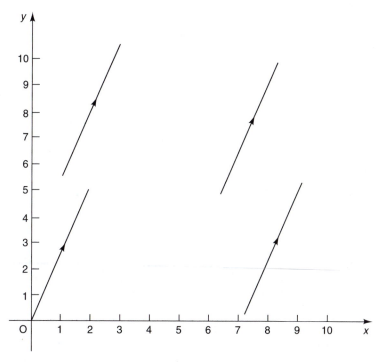

Figure 27.15 Several vectors equal to $2i + 5j$.

> The vector i is a unit vector in the x direction.
> The vector j is a unit vector in the y direction.

Example 27.7

Show on a diagram the vectors $a = 3i + 2j$ and $b = -i + 4j$. In both cases choose the origin as the tail of the vectors.

Solution Figure 27.16 shows the vectors a and b, both starting from the origin.

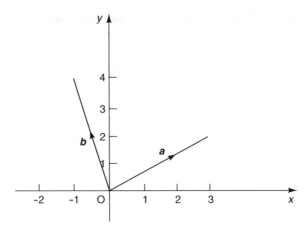

Figure 27.16   The vectors $a = 3i + 2j$ and $b = -i + 4j$.

Example 27.8

Show on a diagram the vectors $p = -3i + 4j$ and $q = 3i - 5j$. In both cases choose the point with coordinates $(2, -3)$ as the tail of the vector.

Solution   Figure 27.17 shows the vectors p and q, both starting from the point $(2, -3)$. Note that p comprises 3 units in the negative x direction and 4 units in the y direction. Similarly, q comprises 3 units in the x direction and 5 units in the negative y direction.

In general, the vector from the origin to the point P with coordinates (x, y), shown in Figure 27.18, can be written as $xi + yj$. When the tail of the vector is fixed at the origin, $xi + yj$ is said to be the **position vector** of the point P.

Example 27.9

State the position vector of the point with coordinates $(7, 9)$.

Solution   The position vector of the point with coordinates $(7, 9)$ is $7i + 9j$.

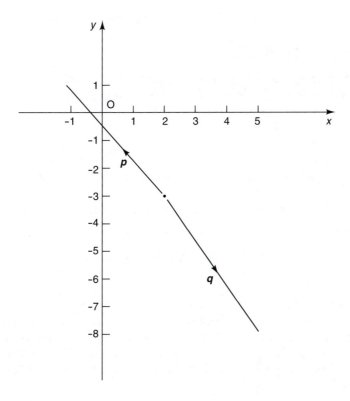

Figure 27.17 The vectors $p = -3i + 4j$ and $q = 3i - 5j$.

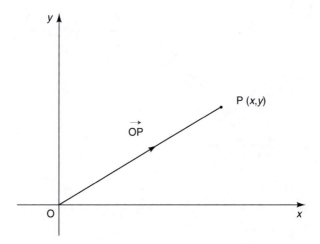

Figure 27.18 The vector $xi + yj$ is the position vector of P.

27.4.1 Addition and subtraction

Addition of vectors given in component form is particularly simple. We simply add the corresponding components. Similarly, to subtract two vectors, we subtract their components.

Example 27.10

Given $a = 3i + 4j$ and $b = 2i + 5j$, find (a) $a + b$ (b) $a - b$

Solution   (a)   We separately add the x components and the y components to obtain

$$a + b = 3i + 4j + 2i + 5j = 5i + 9j$$

(b) $$a - b = (3i + 4j) - (2i + 5j) = i - j$$

Example 27.11 Position of a point in a mechanism

Figure 27.19 shows a mechanism that consists of three pin-jointed links. The links are constrained to lie in the (x, y) plane. A mechanism constrained to move in a single plane is known as a **planar mechanism**. We can represent the length and orientation of each of the links by a vector. In vector form, these can be written as

$$a = a_1 i + a_2 j, \quad b = b_1 i + b_2 j, \quad c = c_1 i + c_2 j$$

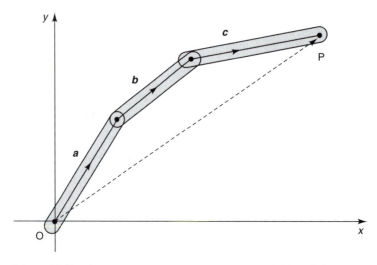

Figure 27.19   A mechanism consisting of three pin-jointed links.

where i and j are unit vectors based on a coordinate system with origin at the anchor point of the mechanism, O. The position of the point P of the mechanism relative to the point O can be represented by a vector p given by

$$p = a + b + c = (a_1 + b_1 + c_1)i + (a_2 + b_2 + c_2)j$$

The vector p is a position vector because the mechanism is fixed to the point O.

27.4.2 Multiplication of a vector by a number

To multiply a vector by a number, we simply multiply each component by that number. For example, if $p = 9i - 2j$ then $3p$ is given by

$$3p = 3(9i - 2j) = 27i - 6j$$

Example 27.12
      If $r = 3i - 2j$, find (a) $6r$ (b) $-2r$

Solution   (a) $6r = 6(3i - 2j) = 18i - 12j$
            (b) $-2r = -2(3i - 2j) = -6i + 4j$

27.4.3 Magnitude of a vector

The vector $r = xi + yj$ is shown in Figure 27.20. The magnitude of the vector r is the length of the line OP. We can use Pythagoras' theorem to find this length.

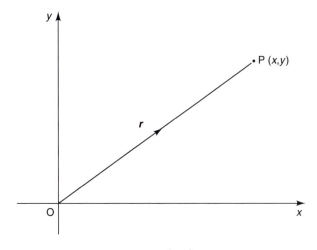

Figure 27.20 The vector $r = xi + yj$.

We have

$$\text{if} \quad \mathbf{r} = x\mathbf{i} + y\mathbf{j} \quad \text{then} \quad |\mathbf{r}| = \sqrt{x^2 + y^2}$$

Example 27.13

Find the magnitude of the vector $\mathbf{q} = -2\mathbf{i} + 7\mathbf{j}$.

Solution   The magnitude of $\mathbf{q}$ is given by

$$|\mathbf{q}| = \sqrt{(-2)^2 + 7^2} = \sqrt{4 + 49} = \sqrt{53} = 7.280$$

Example 27.14

If A is the point with coordinates (2, 3) and B the point with coordinates (5, 1),
(a) find an expression for the vector $\overrightarrow{AB}$,
(b) find $|\overrightarrow{AB}|$.

Solution   (a)   The points A and B are shown in Figure 27.21. From the figure, we note that to get from A to B requires a movement of 3 units in the x direction and 2 units in the negative y direction. Therefore

$$\overrightarrow{AB} = 3\mathbf{i} - 2\mathbf{j}$$

(b)   The magnitude of $\overrightarrow{AB}$ is given by

$$|\overrightarrow{AB}| = \sqrt{(3)^2 + (-2)^2} = \sqrt{9 + 4} = \sqrt{13} = 3.606$$

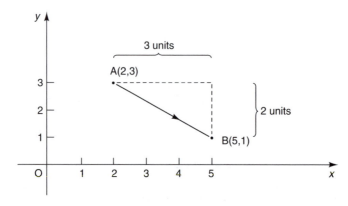

Figure 27.21 $\overrightarrow{AB} = (5-2)\mathbf{i} + (1-3)\mathbf{j} = 3\mathbf{i} - 2\mathbf{j}$.

Example 27.15 Distance of a point in a mechanism from a reference point
Recall Example 27.11. The position of the point P in the mechanism relative to a reference point is defined by means of the position vector p, where

$$p = (a_1 + b_1 + c_1)i + (a_2 + b_2 + c_2)j$$

The distance from the reference point to the point P is given by $|p|$; that is,

$$|p| = \sqrt{(a_1 + b_1 + c_1)^2 + (a_2 + b_2 + c_2)^2}$$

Self-assessment questions 27.4

1. Explain what is meant by the cartesian components of a vector.

2. Explain what is meant by the vectors i and j.

Exercises 27.4

1. A point P has coordinates (9, 2). Find the position vector of P and the distance of P from the origin.

2. If $a = 4i - 3j$, find $-4a$.

3. If $a = 7i + 2j$ and $b = 5i - 3j$, find
   (a) $a + b$ (b) $b + a$ (c) $a - b$
   (d) $b - a$
   Comment upon your results.

4. If $p = -2i + 6j$ and $q = 9i + 8j$, find
   (a) $9p - 6q$ (b) $2p + 7q$ (c) $-8p - 3q$

5. Find the magnitude of each of the following vectors:
   (a) $9j$ (b) $i - 9j$ (c) $-2i - 3j$
   (d) $0i + 0j$ (e) $-3i - 4j$ (f) $0.5i + 0.5j$

6. Find the magnitude of each of the following vectors:
   (a) $7i - 2j$ (b) $i - 2j$ (c) $-2j$
   (d) $5i + 3j$ (e) j (f) $0.5i + 7j$

27.5 The scalar product

There are several different ways in which the multiplication of one vector by another vector can be defined. We consider only one way, namely the 'scalar product'. The definition may seem strange at first, but it arises because it can be usefully applied in many physical situations.

Given two vectors a and b, as shown in Figure 27.22, their **scalar product**, denoted by $a \cdot b$, is defined as $|a| |b| \cos \theta$, where θ is the angle between a and b. That is,

scalar product: $a \cdot b = |a| |b| \cos \theta$

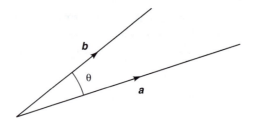

Figure 27.22   The scalar product of *a* and *b* is |*a*| |*b*| cos θ.

Clearly **b·a**=|*b*| |*a*| cos θ and so **a·b**=**b·a**. It is particularly important to use the dot in the formula. The dot is the symbol for the scalar product of *a* and *b*, and is the reason why this form of product is also known as a **dot product**. You must not use the symbol × here, because this is reserved for another form of product, namely the vector product, which is not discussed in this book.

   Note from the formula that to find the scalar product of *a* and *b*, we must determine the magnitude of each vector and the angle between them. The result of applying the formula is a scalar quantity, that is, just a single number.

Example 27.16
   Find **i·i**, where *i* is a unit vector in the direction of the positive *x* axis.

Solution   Because *i* is a unit vector, its magnitude is 1. Also, the angle between *i* and itself is zero. Therefore, using the formula for the scalar product, we find

$$\textbf{i·i}=(1)(1)\cos 0° = 1, \quad \text{since } \cos 0° = 1$$

So the scalar product of *i* with itself equals the number 1. It is easy to verify that **j·j** is also 1.

Example 27.17
   Find **i·j**, where *i* and *j* are the unit vectors in the directions of the *x* and *y* axes.

Solution   Because *i* and *j* are unit vectors, they both have a magnitude equal to 1. The angle between *i* and *j* is 90°. Therefore, using the formula for the scalar product, we find

$$\textbf{i·j}=(1)(1)\cos 90° = 0, \quad \text{since } \cos 90° = 0$$

That is **i·j**=0.

To summarize, we have the following important results that should be remembered:

$$i{\cdot}i=1, \quad j{\cdot}j=1, \quad i{\cdot}j=j{\cdot}i=0$$

If we are given the two vectors in component form, it is possible to find their scalar product without first finding the angle between the two vectors. Consider the following example.

Example 27.18

Find $a{\cdot}b$, where $a=2i+5j$ and $b=i-j$.

Solution

$$a{\cdot}b=(2i+5j){\cdot}(i-j)$$

$$=2i{\cdot}(i-j)+5j{\cdot}(i-j)$$

$$=2i{\cdot}i-2i{\cdot}j+5j{\cdot}i-5j{\cdot}j$$

Now, recalling that $i{\cdot}i=j{\cdot}j=1$ and $i{\cdot}j=j{\cdot}i=0$, this simplifies to

$$a{\cdot}b=2-0+0-5$$

$$=-3$$

That is, the scalar product of $2i+5j$ and $i-j$ is the number -3.

More generally, suppose that $a=a_1i+a_2j$ and $b=b_1i+b_2j$; then

$$a{\cdot}b=(a_1i+a_2j){\cdot}(b_1i+b_2j)$$

$$=a_1i{\cdot}(b_1i+b_2j)+a_2j{\cdot}(b_1i+b_2j)$$

$$=a_1b_1i{\cdot}i+a_1b_2i{\cdot}j+a_2b_1j{\cdot}i+a_2b_2j{\cdot}j$$

Again recalling that $i{\cdot}i=j{\cdot}j=1$ and $i{\cdot}j=j{\cdot}i=0$, this simplifies to

$$a{\cdot}b=a_1b_1+a_2b_2$$

That is,

$$\text{if} \quad a=a_1i+a_2j \text{ and } b=b_1i+b_2j \quad \text{then} \quad a{\cdot}b=a_1b_1+a_2b_2$$

Thus, to find the scalar product of two vectors, their i components are multiplied together, their j components are multiplied together, and the results are added.

Example 27.19
>   If $a=7i+2j$ and $b=4i+3j$, find $a\cdot b$.

Solution   Using the formula, we find
$$a\cdot b=(7i+2j)\cdot(4i+3j)=(7)(4)+(2)(3)=28+6=34$$

Example 27.20
>   If $a=-3i+2j$ and $b=4i-2j$, find $a\cdot b$.

Solution   Using the formula, we find
$$a\cdot b=(-3i+2j)\cdot(4i-2j)=(-3)(4)+(2)(-2)=-12-4=-16$$

Using the formula for the scalar product, we can determine the angle between two vectors.

Example 27.21
>   Find the angle between the vectors $a=3i+4j$ and $b=i+7j$.

Solution   We first find the scalar product $a\cdot b$. This is
$$a\cdot b=(3i+4j)\cdot(i+7j)=(3)(1)+(4)(7)=31$$

Then we find the magnitude of each vector:
$$|a|=\sqrt{3^2+4^2}=\sqrt{25}=5, \quad |b|=\sqrt{1^2+7^2}=\sqrt{50}$$

Then, using the formula for the scalar product, we have
$$a\cdot b=|a|\,|b|\cos\theta$$
$$31=5\sqrt{50}\cos\theta$$

so that
$$\cos\theta=\frac{31}{5\sqrt{50}}$$
$$=0.8768$$

Finally, using a calculator to take the inverse cosine yields $\theta=28.7°$. The angle between the two given vectors is therefore $28.7°$.

In general, the cosine of the angle between the two vectors a and b is given by

$$\cos\theta = \frac{a \cdot b}{|a|\,|b|}$$

Self-assessment question 27.5

1. Define the scalar product of two vectors a and b.

Exercises 27.5

1. If $a=4i+9j$, $b=5i+3j$ and $c=9i+11j$, find
    (a) $a \cdot b$ (b) $a \cdot c$ (c) $a \cdot a$ (d) $b \cdot b$
    (e) $c \cdot c$

2. Show that the vectors $2i+4j$ and $-i+\frac{1}{2}j$ are perpendicular.

3. Find the scalar product of the vectors $5i$ and j.

4. If $p=3i-j$ and $q=6i+4j$, find
    (a) $p \cdot q$ (b) $q \cdot p$ (c) $p \cdot p$ (d) $q \cdot q$

5. Points A, B and C have coordinates (3, 2), (4, −3), and (7, −5) respectively.
    (a)   Find $\overrightarrow{AB}$ and $\overrightarrow{AC}$.
    (b)   Find the scalar product of $\overrightarrow{AB}$ and $\overrightarrow{AC}$.
    (c)   Deduce the angle between $\overrightarrow{AB}$ and $\overrightarrow{AC}$.

27.6 Computer graphics

Points in the (x, y) plane can be described using vectors. It follows that if we imagine a computer screen to be the (x, y) plane then the location of any objects pictured on the screen can be described by vectors. For example, a rectangular shape can be described by four vectors, each giving the position of one of the corners. Sophisticated computer software, such as that found in computer games and flight simulators, is used to move graphic images around the screen. This movement is achieved by complex mathematical procedures involving matrices and vectors. We shall see how this can happen with some relatively simple examples. Before we do this, it is necessary to introduce **column vectors**. The vector $3i+2j$ is written in column vector form as $\begin{pmatrix} 3 \\ 2 \end{pmatrix}$. Similarly the vector

$xi + yj$ is written in column vector form as $\begin{pmatrix} x \\ y \end{pmatrix}$. Such column vectors can be thought of as 2×1 matrices.

Consider a point P with position vector $\begin{pmatrix} 3 \\ 2 \end{pmatrix}$, say, as shown in Figure 27.23.

Suppose the vector $\begin{pmatrix} 3 \\ 2 \end{pmatrix}$ is premultiplied by the matrix $M = \begin{pmatrix} 0 & 1 \\ 1 & 0 \end{pmatrix}$. We find

$$\begin{pmatrix} 0 & 1 \\ 1 & 0 \end{pmatrix}\begin{pmatrix} 3 \\ 2 \end{pmatrix} = \begin{pmatrix} 2 \\ 3 \end{pmatrix}$$

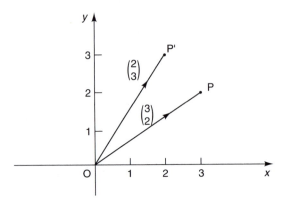

Figure 27.23   P′ is the image of P under transformation by $M = \begin{pmatrix} 0 & 1 \\ 1 & 0 \end{pmatrix}$.

The point P′ with position vector $\begin{pmatrix} 2 \\ 3 \end{pmatrix}$ is also shown in Figure 27.23. The effect on P of multiplying its position vector by M is to move it to P′. We say that P has undergone a **transformation**. We refer to P′ as the **image** of point P. This illustrates another application of matrices – they can be used to carry out transformations.

Example 27.22 Rotation

The points A, B and C have position vectors $\begin{pmatrix} 1 \\ 1 \end{pmatrix}$, $\begin{pmatrix} 1 \\ 2 \end{pmatrix}$ and $\begin{pmatrix} 2 \\ 1 \end{pmatrix}$ respectively. Each point is transformed by premultiplying its position vector by the matrix $M = \begin{pmatrix} 0 & -1 \\ 1 & 0 \end{pmatrix}$. Plot the triangle ABC and its image A′B′C′.

Solution   Premultiplying each position vector by M, we find

$$\begin{pmatrix} 0 & -1 \\ 1 & 0 \end{pmatrix}\begin{pmatrix} 1 \\ 1 \end{pmatrix} = \begin{pmatrix} -1 \\ 1 \end{pmatrix}, \quad \begin{pmatrix} 0 & -1 \\ 1 & 0 \end{pmatrix}\begin{pmatrix} 1 \\ 2 \end{pmatrix} = \begin{pmatrix} -2 \\ 1 \end{pmatrix},$$

$$\begin{pmatrix} 0 & -1 \\ 1 & 0 \end{pmatrix}\begin{pmatrix} 2 \\ 1 \end{pmatrix} = \begin{pmatrix} -1 \\ 2 \end{pmatrix}$$

The triangle ABC and its image are shown in Figure 27.24. Note in particular that the effect of multiplying by M has been to **rotate** the triangle anticlockwise through an angle of 90° about the origin.

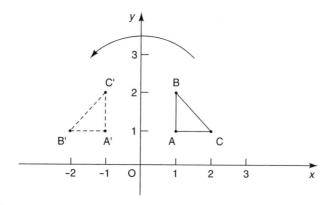

Figure 27.24   The triangle ABC is rotated by 90° about the origin.

Example 27.23 *Translation*

Figure 27.25 shows a rectangle with corners at the points with coordinates (1, 1), (1, 2), (3, 2) and (3, 1).
(a)   Express the position vector of each point as a column vector.
(b)   To each column vector in turn add the vector $\begin{pmatrix} 3 \\ 2 \end{pmatrix}$, and plot the resulting points.

Solution   (a)   The position vectors of A, B, C and D are respectively

$$a = \begin{pmatrix} 1 \\ 1 \end{pmatrix}, \quad b = \begin{pmatrix} 1 \\ 2 \end{pmatrix}, \quad c = \begin{pmatrix} 3 \\ 2 \end{pmatrix}, \quad d = \begin{pmatrix} 3 \\ 1 \end{pmatrix}$$

(b)   We now add the vector $\begin{pmatrix} 3 \\ 2 \end{pmatrix}$ to each in turn to give

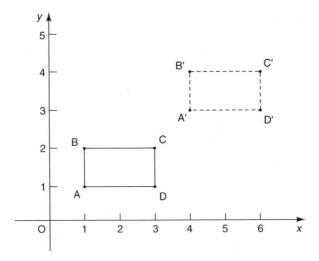

Figure 27.25   The rectangle ABCD has been translated.

$$a' = \begin{pmatrix} 4 \\ 3 \end{pmatrix}, \quad b' = \begin{pmatrix} 4 \\ 4 \end{pmatrix}, \quad c' = \begin{pmatrix} 6 \\ 4 \end{pmatrix}, \quad d' = \begin{pmatrix} 6 \\ 3 \end{pmatrix}$$

The resulting points are also shown in Figure 27.25. We see that the rectangular shape has been **translated**. This has involved a displacement in the x direction and a displacement in the y direction.

Example 27.24 Reflection

Consider the rectangle ABCD with vertices at (1, 1), (3, 1), (3, 2) and (1, 2). Transform each of the vertices by premultiplying its position vector by

(a) $\begin{pmatrix} 1 & 0 \\ 0 & -1 \end{pmatrix}$

(b) $\begin{pmatrix} -1 & 0 \\ 0 & 1 \end{pmatrix}$

Comment upon the resulting rectangles.

Solution   (a)   Performing the matrix multiplications, we have

$$\begin{pmatrix} 1 & 0 \\ 0 & -1 \end{pmatrix}\begin{pmatrix} 1 \\ 1 \end{pmatrix} = \begin{pmatrix} 1 \\ -1 \end{pmatrix}, \quad \begin{pmatrix} 1 & 0 \\ 0 & -1 \end{pmatrix}\begin{pmatrix} 3 \\ 1 \end{pmatrix} = \begin{pmatrix} 3 \\ -1 \end{pmatrix}$$

$$\begin{pmatrix} 1 & 0 \\ 0 & -1 \end{pmatrix}\begin{pmatrix} 3 \\ 2 \end{pmatrix} = \begin{pmatrix} 3 \\ -2 \end{pmatrix}, \quad \begin{pmatrix} 1 & 0 \\ 0 & -1 \end{pmatrix}\begin{pmatrix} 1 \\ 2 \end{pmatrix} = \begin{pmatrix} 1 \\ -2 \end{pmatrix}$$

The rectangle ABCD and its image A'B'C'D' are shown in Figure 27.26.

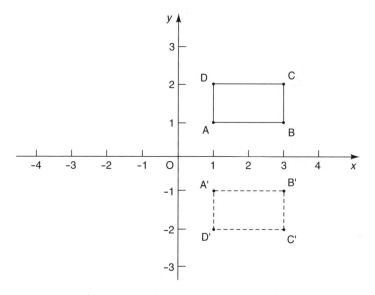

Figure 27.26   The rectangle ABCD has been reflected in the x axis.

Note from the figure that the effect of this matrix is to **reflect** the rectangle in the x axis.

(b)   Performing the matrix multiplications, we have

$$\begin{pmatrix} -1 & 0 \\ 0 & 1 \end{pmatrix}\begin{pmatrix} 1 \\ 1 \end{pmatrix} = \begin{pmatrix} -1 \\ 1 \end{pmatrix}, \quad \begin{pmatrix} -1 & 0 \\ 0 & 1 \end{pmatrix}\begin{pmatrix} 3 \\ 1 \end{pmatrix} = \begin{pmatrix} -3 \\ 1 \end{pmatrix}$$

$$\begin{pmatrix} -1 & 0 \\ 0 & 1 \end{pmatrix}\begin{pmatrix} 3 \\ 2 \end{pmatrix} = \begin{pmatrix} -3 \\ 2 \end{pmatrix}, \quad \begin{pmatrix} -1 & 0 \\ 0 & 1 \end{pmatrix}\begin{pmatrix} 1 \\ 2 \end{pmatrix} = \begin{pmatrix} -1 \\ 2 \end{pmatrix}$$

The rectangle ABCD and its image A′B′C′D′ are shown in Figure 27.27. Note from the figure that the effect of the matrix is to reflect the rectangle in the y axis.

Self-assessment question 27.6

1.   A transformation is such that each point and its image are identical. Give the transformation matrix.

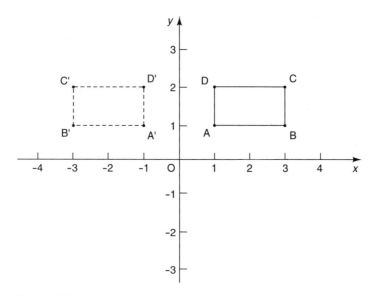

Figure 27.27   The rectangle has been reflected in the *y* axis.

Exercises 27.6

1.   The points A, B and C have coordinates (1, 1), (3, 1) and (3, 4) respectively. Find the coordinates of the image points when the triangle ABC is reflected (a) in the *x* axis (b) in the *y* axis

transformed by premultiplying again by *M*. Find the position vector of the final image point.

Show that the same result is achieved if $\begin{pmatrix} 5 \\ 9 \end{pmatrix}$ is premultiplied by M^2.

2.   A point P with position vector $\begin{pmatrix} 5 \\ 9 \end{pmatrix}$ is transformed by premultiplying its position vector by $M = \begin{pmatrix} 1 & 4 \\ 5 & 2 \end{pmatrix}$. The result is

3.   Points A and B have coordinates (3, −1) and (4, 2) respectively. Find the position of the image points A′ and B′ under the transformation $\begin{pmatrix} 1 & 2 \\ -1 & 1 \end{pmatrix}$.

Review exercises 27

1   Draw vectors $\overrightarrow{AB}$, $\overrightarrow{PQ}$ and $\overrightarrow{RS}$ such that $\overrightarrow{AB} = \frac{1}{2}\overrightarrow{PQ}$ and $\overrightarrow{RS} = 3\overrightarrow{AB}$.

2   Find $\overrightarrow{AB}$ when A has coordinates (4, 3) and B has coordinates (−1, −9).

3   If $a = 3i - j$ and $b = 7i + 4j$, find (a) $3a - 4b$ (b) $-3a + 2b$

4   Find (a) a vector that is parallel to the line $y = 7x + 1$ and (b) a unit vector that is parallel to this line.

5 Evaluate $a \cdot i$, where $a = 7i + 9j$. Hence find the angle that a makes with the x axis.

6 Which of the following vectors are parallel?

$$\begin{pmatrix} -4 \\ 2 \end{pmatrix}, \quad \begin{pmatrix} 4 \\ -2 \end{pmatrix}, \quad \begin{pmatrix} 6 \\ 9 \end{pmatrix}, \quad \begin{pmatrix} 4 \\ 11 \end{pmatrix}, \quad \begin{pmatrix} 18 \\ 27 \end{pmatrix}$$

7 If $p = 5i + 7j$ and $q = i - 9j$, find
(a) $|p|$ (b) $|q|$ (c) $\hat{p}$ (d) $\hat{q}$
(e) $p \cdot q$ (f) $|p - q|$

8 Find the angle between $p = 3i - 2j$ and $q = -i + j$.

9 A point P has position vector $\begin{pmatrix} -2 \\ 1 \end{pmatrix}$. It is transformed by premultiplication by $M = \begin{pmatrix} -1 & 0 \\ 1 & 2 \end{pmatrix}$. Find the position vector of the image of P.

10 If $a = 2i + j$, $b = 7i - 5j$ and $c = \frac{1}{2}i + j$, find
(a) $|a|$ (b) $2a + c$ (c) $|2a + c|$
(d) $a \cdot c$ (e) $b \cdot a$ (f) $4b$ (g) $c \cdot c$

An Introduction to
Complex Numbers

CONTENTS

28.1 Introduction

The numbers encountered so far within this book have all been real numbers. In order to solve certain types of mathematical problems, it is necessary to introduce further numbers with additional properties. These numbers are known as complex numbers. One of the most important applications of complex numbers is in the analysis of alternating current circuits. This chapter will introduce complex numbers and some of their properties.

28.2 The number j

Example 28.1

Evaluate (a) 3^2 (b) $(-3)^2$ (c) 4^2 (d) $(-4)^2$
Comment upon the sign of the answers.

Solution

(a) $3^2 = 9$ (b) $(-3)^2 = 9$ (c) $4^2 = 16$ (d) $(-4)^2 = 16$
We note that the result of squaring any nonzero number, whether positive or negative, is a positive number.

Now suppose we were asked to find $\sqrt{-9}$, that is, find a number that when squared gives -9. A limitation of the real number system is that it is impossible to find a real number that when squared is negative. As we have seen, squaring both positive and negative numbers yields positive answers. In some applications it is useful to overcome this limitation. To do this we introduce a new number, to which we give the symbol j, which has the property that

$$j^2 = -1, \quad \text{so that} \quad j = \sqrt{-1}$$

However, because there is no real number that when squared equals -1, the number j cannot be real. Instead, we call it an **imaginary** number. This concept seems strange when first met, but imaginary numbers can be very useful in engineering applications. Note that mathematicians and physicists conventionally use i rather than j for $\sqrt{-1}$.

Example 28.2

If $j^2 = -1$, find an expression for j^3.

Solution $j^3 = j^2 \times j = (-1) \times j = -j$

We are now in a position to be able to write down an expression for the square root of any negative number. For example,

$$\sqrt{-9} = \sqrt{(9)(-1)} = \sqrt{9}\sqrt{-1} = 3j$$

Once again, note that the number 3j is not a real number – it is imaginary.

Example 28.3

Write down an expression for (a) $\sqrt{-16}$ (b) $\sqrt{-13}$

Solution   (a) $\sqrt{-16} = \sqrt{(16)(-1)} = \sqrt{16}\sqrt{-1} = 4j$

(b) $\sqrt{-13} = \sqrt{(13)(-1)} = \sqrt{13}\sqrt{-1} = \sqrt{13}j = 3.61j$

Self-assessment questions 28.2

1.   Why is it helpful to use the number j that has the property that $j^2 = -1$?
2.   What is the distinction between $(-j)^2$ and $-j^2$?

Exercises 28.2

1.   Simplify
      (a) $3j^2$ (b) $(-2j)^2$ (c) $2j^2$ (d) $4j^3$
      (e) j^4 (f) $-3j^4$ (g) $-2j^2$

2.   Simplify the following:
      (a) $\sqrt{81}$ (b) $\sqrt{-81}$ (c) $\sqrt{-144}$
      (d) $\sqrt{0.25}$ (e) $\sqrt{-0.25}$ (f) $\sqrt{-\frac{1}{16}}$

28.3 The complex number $a + bj$

Recall the formula for solving the quadratic equation $ax^2 + bx + c = 0$, namely

$$x = \frac{-b \pm \sqrt{b^2 - 4ac}}{2a}$$

Now consider the problem of solving the equation $x^2 - 6x + 10 = 0$. Using the formula with $a = 1$, $b = -6$ and $c = 10$, we find

$$x = \frac{6 \pm \sqrt{(-6)^2 - 4(1)(10)}}{2}$$

$$= \frac{6 \pm \sqrt{36 - 40}}{2}$$

$$= \frac{6 \pm \sqrt{-4}}{2}$$

At this point, in order to continue, it is necessary to find the square root of -4. Using the symbol j, we can write $\sqrt{-4}$ as 2j. Then

$$x = \frac{6 \pm 2j}{2} = 3 \pm j$$

We can now write down the two solutions of the quadratic equation: $x = 3 + j$ and $x = 3 - j$. The numbers $3 + j$ and $3 - j$ are called **complex numbers**. Each complex number consists of two parts: a **real part** and an **imaginary part**. Generally, a complex number can be written as

$$a + bj$$

where a is its real part and b is its imaginary part. It is common to use the symbol z to stand for a complex number, in which case we write

$$z = a + bj$$

This is known as the **cartesian form** of the complex number.

---

cartesian form:

$$z = a + bj, \quad \text{where } j^2 = -1$$

$$\text{real part} = a, \quad \text{imaginary part} = b$$

Using complex numbers, it is always possible to find the solutions of the quadratic equation $ax^2 + bx + c = 0$. The solutions will be complex whenever $b^2 - 4ac$ is negative.

Example 28.4

State the real and imaginary parts of the following complex numbers:
(a) $z = 8 - 9j$ (b) $z = 8$ (c) $z = -\frac{1}{2}j$

Solution   (a)   If $z=8-9j$, the real part of z is 8 and the imaginary part is -9.
           (b)   If $z=8$, its real part is 8 and its imaginary part is zero.
           (c)   If $z=-\frac{1}{2}j$, its real part is zero and its imaginary part is $-\frac{1}{2}$.

Example 28.5

Solve the quadratic equation $4x^2-4x+17=0$.

Solution   Using the formula for solving a quadratic equation, we obtain

$$x=\frac{-(-4)\pm\sqrt{(-4)^2-(4)(4)(17)}}{(2)(4)}$$

$$=\frac{4\pm\sqrt{16-272}}{8}$$

$$=\frac{4\pm\sqrt{-256}}{8}$$

$$=\frac{4\pm16j}{8}$$

$$=\frac{1\pm4j}{2}$$

$$=\tfrac{1}{2}\pm2j$$

Hence the two solutions are $x=\frac{1}{2}+2j$ and $x=\frac{1}{2}-2j$.

Self-assessment questions 28.3

1.   Explain what is meant by a complex number.

2.   In what standard form is a complex number written?

3.   Under what conditions will the quadratic equation $ax^2+bx+c=0$ have complex solutions?

Exercises 28.3

1.   Identity the real and imaginary parts of each of the following complex numbers:
    (a) $3+7j$ (b) $5-2j$ (c) 7 (d) $5j$
    (e) $\alpha+j\beta$ (f) $2x+3jy$ (g) -1
    (h) $\cos\theta+j\sin\theta$

2.   Use the formula to solve the following quadratic equations:
    (a) $x^2+3x-4=0$ (b) $2x^2-3x-6=0$
    (c) $x^2+16=0$ (d) $x^2+4=0$
    (e) $x^2-4=0$ (f) $x^2-2x+5=0$
    (g) $x^2-14x+53=0$ (h) $2x^2-2x+1=0$
    (i) $x^2-6x+90=0$

28.4 Operations with complex numbers

Just as real numbers can be added, subtracted, multiplied and divided, so too can complex numbers. Addition or subtraction is performed in an obvious way by adding or subtracting their real parts and their imaginary parts.

Example 28.6

Add the complex numbers $3 + 2j$ and $8 + 4j$.

Solution

$$(3 + 2j) + (8 + 4j) = (3 + 8) + (2j + 4j)$$
$$= 11 + 6j$$

Example 28.7

Add the complex numbers $7 + 2j$ and $3j$.

Solution   The real part of the complex number $3j$ is zero. We find

$$(7 + 2j) + (3j) = 7 + 5j$$

Example 28.8

Suppose we have two complex numbers z_1 and z_2, where $z_1 = 6 - 3j$ and $z_2 = 9 + 7j$. Find
(a) $z_1 + z_2$ (b) $z_1 - z_2$

Solution   (a) $z_1 + z_2 = 6 - 3j + 9 + 7j = 15 + 4j$
(b) $z_1 - z_2 = (6 - 3j) - (9 + 7j) = -3 - 10j$

To multiply two complex numbers, we make use of the fact that $j^2 = -1$. Consider the following example.

Example 28.9

Multiply together $6 + 3j$ and $5 + 2j$.

Solution   We wish to find $(6 + 3j)(5 + 2j)$, so we start by removing the brackets in the usual way:

$$(6 + 3j)(5 + 2j) = 30 + 15j + 12j + 6j^2 = 30 + 27j - 6 = 24 + 27j$$

Note that to simplify the answer we have made use of the fact that $j^2 = -1$.

Example 28.10

Multiply (a) $2-4j$ and $5+j$ (b) $1+j$ and j

Solution (a) We have

$$(2-4j)(5+j)=10-20j+2j-4j^2=10-18j+4=14-18j$$

(b) Similarly,

$$(1+j)j=j+j^2=j-1=-1+j$$

If the complex number z is given by $z=a+bj$ then a related complex number is called the **complex conjugate** of z, and is defined to be

$$\bar{z}=a-bj$$

So the complex conjugate of $a+bj$ is found by changing the sign of the imaginary part. Note that we use the symbol $\bar{z}$, pronounced 'z bar', to stand for the complex conjugate. (Sometimes the symbol z^* is used instead of $\bar{z}$.)

> If $z=a+bj$, its complex conjugate is $\bar{z}$, defined as $\bar{z}=a-bj$.

Example 28.11

Find the conjugate of $z=4+3j$.

Solution To find the conjugate, we change the sign of the imaginary part. So if $z=4+3j$ then $\bar{z}=4-3j$.

Example 28.12

State the complex conjugates of (a) 1 (b) j (c) $1+j$ (d) $1-j$

Solution To find the complex conjugate, we change the sign of the imaginary part:
(a) The conjugate of 1 is simply 1, because this complex number has an imaginary part of 0.
(b) The conjugate of j is $-j$.
(c) The conjugate of $1+j$ is $1-j$.
(d) The conjugate of $1-j$ is $1+j$.

Example 28.13

If $z = 2 - 2j$, find $z\bar{z}$.

Solution

$$z\bar{z} = (2 - 2j)(2 + 2j) = 4 - 4j + 4j - 4j^2 = 4 + 4 = 8$$

Example 28.14

If $z = a + bj$ find (a) $\bar{z}$ (b) $z\bar{z}$

Solution   (a)   If $z = a + bj$ then $\bar{z} = a - bj$.
(b) $z\bar{z} = (a + bj)(a - bj) = a^2 + bja - abj - b^2j^2 = a^2 + b^2$.

We note from the last example that when we multiply a complex number by its conjugate, the result is a real number. In order to divide two complex numbers, we make use of the complex conjugate. Consider the following example.

Example 28.15

Find z_1/z_2, where $z_1 = 8 + 7j$ and $z_2 = 3 + j$.

Solution   We are required to find $(8 + 7j)/(3 + j)$. In order to do this, we multiply both the numerator and the denominator by the conjugate of the denominator, namely $3 - j$. Thus

$$\frac{8 + 7j}{3 + j} = \frac{8 + 7j}{3 + j} \times \frac{3 - j}{3 - j}$$

Note that multiplying both numerator and denominator by the same quantity does not change the value of the left-hand side. We then proceed to simplify the right-hand side:

$$\frac{8 + 7j}{3 + j} = \frac{8 + 7j}{3 + j} \times \frac{3 - j}{3 - j}$$

$$= \frac{24 + 21j - 8j - 7j^2}{9 + 3j - 3j - j^2}$$

$$= \frac{24 + 13j + 7}{9 + 1}$$

$$= \frac{31 + 13j}{10}$$

$$= \frac{31}{10} + \frac{13}{10}j$$

Example 28.16

If $z_1 = 8 - 2j$ and $z_2 = j$, find z_2/z_1.

Solution   We are required to find $j/(8-2j)$. To do this, we multiply both the numerator and the denominator by the conjugate of the denominator, namely $8 + 2j$. This gives

$$\frac{j}{8-2j} = \frac{j}{8-2j} \times \frac{8+2j}{8+2j}$$

$$= \frac{8j + 2j^2}{64 - 4j^2}$$

$$= \frac{8j - 2}{64 + 4}$$

$$= \frac{-2 + 8j}{68}$$

$$= -\frac{1}{34} + \frac{2}{17}j$$

Self-assessment questions 28.4

1.   How are complex numbers added and subtracted?

2.   What is meant by the complex conjugate of a complex number?

3.   How are two complex numbers divided?

Exercises 28.4

1.   If $z_1 = 4 + 7j$ and $z_2 = 2 + 3j$, find
     (a) $z_1 + z_2$ (b) $z_1 - z_2$ (c) $z_2 - z_1$
     (d) $z_1 z_2$ (e) z_1/z_2 (f) z_2/z_1 (g) $z_1 \bar{z}_1$
     (h) $(z_1)^2$

2.   If $z_1 = 8$ and $z_2 = 9j$, find
     (a) $z_1 + z_2$ (b) $z_1 z_2$ (c) z_1/z_2
     (d) $1/z_2$

3.   If $z = 5 - 3j$, find
     (a) $3z$ (b) jz (c) $\bar{z}$ (d) $j\bar{z}$ (e) z^2

4.   If $z_1 = 2 - j$ and $z_2 = 2 + j$ find $z_1 z_2$.

5.   Simplify (a) $\dfrac{1}{3j^3}$ (b) $\dfrac{1}{j^3 + 2j^4}$

6.   If $z_1 = x_1 + jy_1$ and $z_2 = x_2 + jy_2$, find
     (a) $z_1 + z_2$ (b) $z_1 - z_2$ (c) $z_2 - z_1$
     (d) $z_1 z_2$ (e) z_1/z_2 (f) z_2/z_1

7.   Simplify $j + 3j - 2j^2 + j^3$.

8.   In each case find the quadratic equation whose roots are
     (a) $x = 7$ and $x = -7$
     (b) $x = 7j$ and $x = -7j$

9.   Find the quadratic equation whose roots are $x = 3 - 2j$ and $x = 3 + 2j$.

28.5 The Argand diagram

A graphical interpretation of complex numbers can be found by thinking of the complex number $z = a + bj$ as the point with coordinates (a, b) as shown in Figure 28.1. Because the real part of z is plotted on the x axis, we often refer to this as the **real axis**. The imaginary part of z is plotted on the vertical axis, and so we refer to this as the **imaginary axis**. Such a diagram is called an **Argand diagram**.

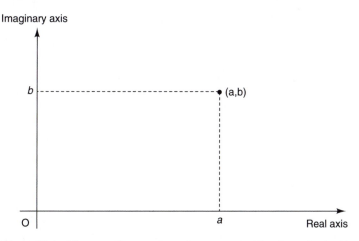

Figure 28.1   The complex number z is represented by the point (a, b).

Example 28.17

On an Argand diagram, represent the complex numbers $z_1 = 4 + j$, $z_2 = -3 + 2j$ and $z_3 = -5 - 5j$.

Solution   Figure 28.2 shows the three complex numbers.

Self-assessment questions 28.5

1. Explain what is meant by an Argand diagram.

2. How is the complex number $a + bj$ represented on an Argand diagram.

Exercise 28.5

1. Plot the following complex numbers on an Argand diagram:
   (a) $3 + 2j$ (b) $3 - 2j$ (c) $-5 - 8j$
   (d) $7j$ (e) 4 (f) j^2 (g) $\frac{1}{2} + \frac{1}{2}j$
   (h) $\dfrac{j}{2}$ (i) j^3

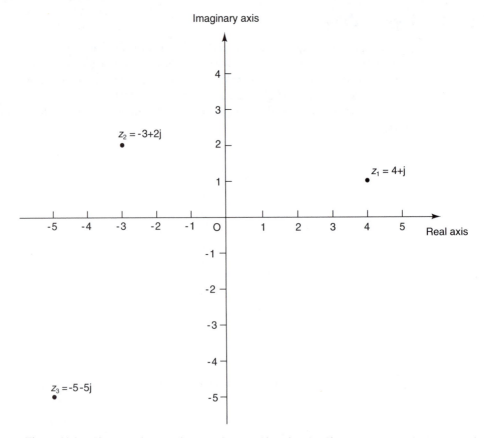

Figure 28.2 The complex numbers $4+j$, $-3+2j$ and $-5-5j$.

28.6 The polar form of a complex number

Various important quantities can be determined from an Argand diagram. The distance from the origin to the point representing the complex number is called the **modulus** of the complex number, and is given the symbol r. Alternatively, r is often written as $|z|$. The plural of modulus is **moduli**. This is illustrated in Figure 28.3.

The modulus of the complex number $z = a + bj$ can be found from the Argand diagram using Pythagoras' theorem. We see that

$$|z| = r = \sqrt{a^2 + b^2}$$

Note that the modulus of a complex number, being its distance from the origin, is never negative.

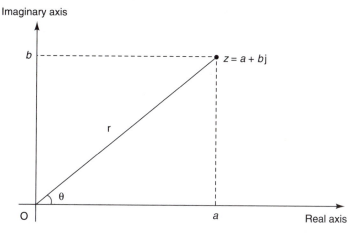

Figure 28.3 The modulus and argument of a complex number.

We can also calculate the angle between the positive x axis and a line joining the complex number to the origin. This angle is called the **argument** of the complex number. It is abbreviated to arg (z), and is given the symbol θ. We usually measure this angle so that $-\pi < \theta \leqslant \pi$. Angles measured anticlockwise are conventionally positive, while angles measured clockwise are negative. The argument can be found using trigonometry. If the argument of z is θ then

$$\tan \theta = \frac{b}{a}$$

Care must be taken when using this formula to determine the argument of a complex number. When evaluating the inverse tangent on your calculator, remember that it will only return values in the range $-\frac{1}{2}\pi < \theta < \frac{1}{2}\pi$. It is for this reason that a diagram is essential when attempting to find an argument. Consider the following example.

Example 28.18

(a) Plot the complex numbers $z_1 = 3 + 4j$ and $z_2 = -1 - j$ on an Argand diagram.
(b) Find their moduli.
(c) Find the argument of z_1.
(d) Find the argument of z_2.

Solution

(a) The Argand diagram showing z_1 and z_2 is shown in Figure 28.4.
(b) For each complex number we can use Pythagoras' theorem to determine its distance from the origin. For z_1 we find

$$|z_1| = \sqrt{3^2 + 4^2} = \sqrt{9 + 16} = \sqrt{25} = 5$$

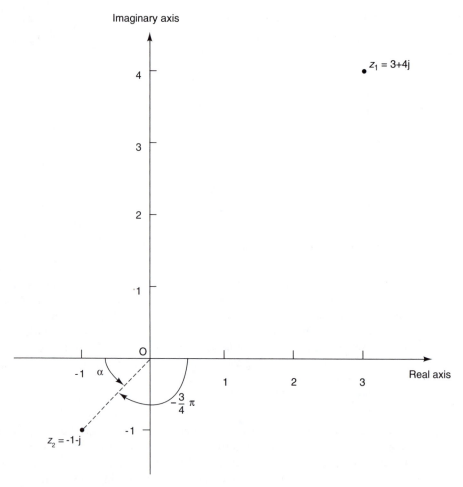

Figure 28.4 Argand diagram showing $3+4\mathrm{j}$ and $-1-\mathrm{j}$.

For z_2 we find

$$|z_2| = \sqrt{(-1)^2 + (-1)^2} = \sqrt{1+1} = \sqrt{2}$$

(c)   The argument of z_1 is the angle whose tangent is $\frac{4}{3}$; that is, $\tan^{-1}\frac{4}{3} = 0.927$ radians.

(d)   Care must be taken when finding the argument of z_2, because this complex number lies in the third quadrant. Referring to Figure 28.4, we see that

$$\alpha = \tan^{-1}\left(\frac{1}{1}\right) = \tfrac{1}{4}\pi \text{ radians}$$

Therefore the argument of z_2 is $-\tfrac{3}{4}\pi$ radians.

The position of a complex number is uniquely determined by giving its modulus and argument. This description of a complex number is known as its **polar form**. Engineers often find it easier to work with complex numbers in polar form, and in addition to writing $z = a + bj$ they use the notation $z = r \angle \theta$. Recall that this is the notation used for polar coordinates discussed in Chapter 18.

Polar form of a complex number:

$$z = r \angle \theta$$

where r is the modulus and θ is the argument.

Most calculators are able to convert complex numbers given in cartesian form into polar form, and vice versa. This is done using the facility to convert cartesian coordinates (x, y) into polar coordinates $r \angle \theta$. You should check that you can use your calculator to carry out these conversions.

Example 28.19

Express the complex numbers $z_1 = 5 + j$ and $z_2 = 5 - j$ in the form $r \angle \theta$.

Solution An Argand diagram showing $z_1 = 5 + j$ and $z_2 = 5 - j$ is given in Figure 28.5. From the figure, we find that $|z_1| = \sqrt{5^2 + 1^2} = \sqrt{26}$. Further, $\arg(z_1) = \tan^{-1}(\frac{1}{5}) = 0.197$ radian. By symmetry, $|z_2|$ is also $\sqrt{26}$ and $\arg(z_2) = -0.197$ radian. We write $z_1 = \sqrt{26} \angle 0.197$ and $z_2 = \sqrt{26} \angle -0.197$.

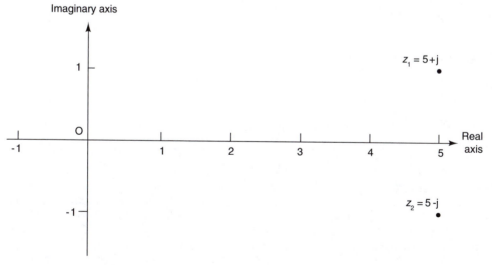

Figure 28.5 Argand diagram for Example 28.19.

Fractal geometry: the mathematics of nature

Over the last decade, fractal geometry has emerged as an important area of mathematical activity and discovery. Because of the startling and very graphic images associated with so-called Julia sets and Mandelbrot sets, fractal geometry has caught the attention and interest of the public at large – probably more so than any previous mathematical discoveries.

Traditional geometry, such as that of the line and the conics, has served us well in describing many artificial objects such as buildings, bridges and rockets. However its drawback is that it is unable to describe many naturally occurring objects such as clouds and trees. Fractal geometry can be thought of as providing a language with which we can describe, with the help of a computer, many of the forms found in nature.

An understanding of complex numbers is essential in order to appreciate how Julia and Mandelbrot sets are formed. We shall see that these are in fact Argand diagrams.

We start by considering what happens when we take a complex number z_0 and repeatedly square it. For example, repeatedly squaring $z_0 = 1 + 2j$, we obtain the sequence

$$z_1 = -3 + 4j, \quad z_2 = -7 - 24j, \quad z_3 = -527 + 336j, \quad \dots$$

Notice that the moduli of the complex numbers are getting larger and larger, and in fact continue to increase without bound. Similarly, repeatedly squaring $z_0 = 0.5 - 0.25j$ gives the sequence

$$z_1 = 0.1875 - 0.25j, \quad z_2 = -0.027 - 0.094j, \quad z_3 = -0.008 + 0.005j, \quad \dots$$

This time, the moduli of the complex numbers remain bounded. We can generate similar sequences starting with any other complex number. The process we have used for generating these sequences is called **iteration**, and we denote it mathematically by $z \to z^2$. We can plot the sequences on an Argand diagram as shown in Figure 1.

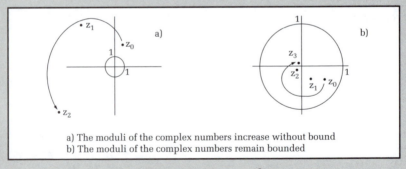

a) The moduli of the complex numbers increase without bound
b) The moduli of the complex numbers remain bounded

Figure 1 *Iterating $z \to z^2$.*

It appears that the complex plane is divided into two parts. One part, called the **escape set** is the set of complex numbers that move further and further away from the origin and escape to infinity. The other part, called the **prisoner set**, is the set of complex numbers that remain bounded forever. It can be shown that the boundary between these two sets is a circle of radius 1 centred at the origin, and this boundary is known as the **Julia set** of the iteration $z \to z^2$.

The Julia set becomes much more interesting if the more general iteration $z \rightarrow z^2 + c$ is used, where c is any complex constant. So, after squaring each complex number, we add on a constant as well. Similar behaviour is found, with some complex numbers escaping to infinity and others remaining bounded. Again, the Julia set marks the boundary. Two Julia sets obtained by choosing different values for c are shown in Figure 2.

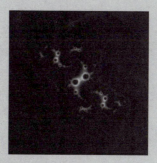

Figure 2 *Julia sets.*

The Mandelbrot set is another Argand diagram formed in a different way. Julia sets can come in one piece, and these are said to be **connected**. Others are totally disconnected. For each value of c in the complex plane we can ask 'is the Julia set connected?' If it is, we mark the complex number c by a black point on an Argand diagram. If not, the point is left white. Doing this for all points produces a **Mandelbrot set** as shown in Figure 3. Zooming in to a Mandelbrot set reveals remarkable and intricate patterns with enormous detail. These patterns have recently been used in the design of posters and T-shirts and so on.

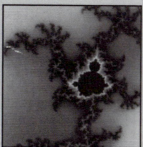

Figure 3 *A Mandelbrot set.*

Applications of fractal geometry are widespread, and include areas such as electrical circuit theory, fluid turbulence, meteorology and biology. This emerging science, often called **chaos**, is being used to see order where previously only random and chaotic behaviour had been observed.

For more information see H.-O. Peitgen, H. Jurgens and D. Saupe, *Chaos and Fractals. New Frontiers of Science* (Springer-Verlag, Berlin, 1992).

Self-assessment questions 28.6

1.　Explain what is meant by the modulus of a complex number.

2.　Can the modulus of a complex number have an imaginary part?

3.　Explain what is meant by the argument of a complex number.

4.　Explain what is meant by the polar form of a complex number.

5.　Explain what is meant by the notation $z = r \angle \theta$.

6.　Explain what is meant by writing $|z| = 5$ and $\arg(z) = \frac{1}{3}\pi$.

7.　Explain what is meant by writing $|z| = 5$ and $\arg(z) = -\frac{1}{3}\pi$.

Exercises 28.6

1.　Plot the complex numbers $z_1 = 1 + 2j$ and $z_2 = -1 - 2j$ on an Argand diagram. Find their moduli and arguments.

2.　Plot the following complex numbers on an Argand diagram and find the modulus and argument of each:
(a) $3j$　(b) 4　(c) $-j$　(d) -1
(e) $0.5j$　(f) j^2　(g) j^3　(h) $1/j$
(i) $1/j^2$

3.　A complex number z has a modulus r and an argument of θ, where $0 \leqslant \theta \leqslant \frac{1}{2}\pi$. Plot this complex number on an Argand diagram. Find its real part and its imaginary part.

4.　Draw an Argand diagram to depict the following complex numbers. Express each number in the form $a + bj$.
(a) $z = 3 \angle \frac{1}{2}\pi$　(b) $z = 5 \angle \frac{1}{6}\pi$
(c) $z = 2 \angle (-\pi)$　(d) $z = 5 \angle (-\frac{3}{4}\pi)$

Computer and calculator exercises 28.6

Many computer algebra packages allow the user to input and manipulate complex numbers.

1.　Use a package to simplify
(a) $(1+j)^4$　(b) $(1-j)^3(3-4j)^5$

(c) $\dfrac{(2+8j)^6}{(1-7j)^7}$

2.　Use the package to completely solve the following equations:
(a) $x^2 + x + 7 = 0$　(b) $x^3 + x^2 + x + 1 = 0$

Review exercises 28

1　Identify the real and imaginary parts of the following complex numbers:
(a) $19 - 3j$　(b) $j^2 - 2j^3$　(c) $6y + 2xj$
(d) $\cos 3\theta + j \sin 3\theta$　(e) $\cos \omega t + j \sin \omega t$
(f) $\cos \omega t - j \sin \omega t$

2　If $z_1 = 8 + 5j$, $z_2 = -9 + 2j$ and $z_3 = 14 - 2j$,

find
(a) $z_1 + z_2 + z_3$　(b) $z_1 - (z_2 + z_3)$
(c) $z_1 z_2 z_3$

3　Find
(a) $(3 - 2j)(1 + 6j)$　(b) $\dfrac{3 + 4j}{1 - 6j}$

(c) $j(1+j)(3-j)$ (d) $(8+j)(7-2j)(8-j)$

(e) $\dfrac{j}{j(1+j)}$

4 Solve the equation $x^2-4x+20=0$ and plot your solutions on an Argand diagram. Express these solutions in polar form.

5 Express the following complex numbers in polar form:

(a) $3+3j$ (b) $4-4j$ (c) $-j$ (d) 9
(e) $-1-j$ (f) -1 (g) $4+5j$ (h) -4
(i) 1

6 Plot the following complex numbers given in polar form on an Argand diagram. Express each of them in the form $a+bj$.

(a) $|z|=4$, $\arg z=\frac{1}{4}\pi$
(b) $|z|=6$, $\arg z=\frac{1}{2}\pi$
(c) $|z|=2$, $\arg z=-\frac{1}{6}\pi$

7 Express the following complex numbers in the form $r\angle\theta$:

(a) $\cos\frac{1}{3}\pi+j\sin\frac{1}{3}\pi$
(b) $3(\cos\frac{1}{3}\pi+j\sin\frac{1}{3}\pi)$

29

Statistics and Probability

KEY POINTS

This chapter

- explains the meaning of discrete and continuous variables

- explains what is meant by a frequency distribution

- shows how to plot a pie chart, a line chart and a histogram

- shows how to calculate the mean, the median and the mode

- shows how to calculate the variance and the standard deviation

- introduces basic ideas concerning probability

- explains the terms 'mutually exclusive', 'complementary' and 'independent' events

CONTENTS

29.1 Introduction

In order to make sensible decisions about an engineering project, it is usually necessary to draw information from a variety of sources and analyse it. Such information is called **data**.

Data often needs to be put into an easily understandable format. This is achieved by the use of charts and histograms.

When there is a large amount of data, the important characteristics of the data are often represented by a few quantities, called **parameters**, calculated from the data itself. The parameters commonly used to characterize a set of data are the mean, median, mode and standard deviation.

A knowledge of probability helps us to calculate the likelihood of an event occurring. For example, a quality control engineer needs to know the likelihood of a component meeting a specification. The later part of the chapter deals with the calculation of probability.

29.2 Discrete and continuous data

Suppose we measure the numbers of components produced by a machine in periods of 1 h. These could be, say, 17, 21, 15, 18, 23, ... The number produced can have any non-negative integer value. Variables such as these, which can assume only particular values, are called **discrete variables**. Examples of discrete variables are the mark obtained in an examination, shoe size and the amount of money in your pocket.

Now consider measuring the time for which a machine runs before breaking down. The length of time can have any non-negative value, and so we say that this variable is continuous. A **continuous variable** can have any value in a specified range. Examples of continuous variables include the volumes of tanks, the weights of components and the diameters of bearings. The variable may only be recorded to, say, the nearest integer or to one decimal place, but the variable is still continuous. The accuracy of the recording depends upon the recording device and the use to which the data will be put.

Example 29.1

State whether the following are discrete or continuous variables:
(a) monthly earnings of an employee;
(b) time taken to assemble a circuit board;
(c) the number of bricks in a building;
(d) the weight of steel in an ingot.

Solution   (a)   An amount of money can be recorded to 2 d.p., for example £926.14. However, an amount such as £926.143 does not exist, and so monthly earnings is a discrete variable.

(b)   The time taken to assemble a circuit board is a continuous variable. It may be recorded to say the nearest 10 seconds or the nearest minute.

(c)   The number of bricks in a building is a discrete variable. Although the number may be very large, it is still discrete.

(d)   The weight of steel in an ingot is a continuous variable. It may be recorded to the nearest kilogram.

Self-assessment question 29.2

1.   Explain the difference between a continuous variable and a discrete variable. Give two examples of each.

Exercise 29.2

1.   Determine whether the following variables are discrete or continuous:
(a) the area of a metal plate;
(b) the temperature at which an alloy melts;
(c) the number of gaskets used by a car plant in one year;
(d) the length of copper pipes;
(e) the cost of installing a computer system.

29.3 Frequency distributions

Suppose that in a survey 900 engineers were asked to classify themselves as one of the following: chemical engineer, electrical engineer, electronic engineer, mechanical engineer and civil engineer. The results are given in Table 29.1, from which we see that 146 engineers classify themselves as chemical engineers, and so on.

Table 29.1

Type of engineer	Frequency
Chemical engineer	146
Electrical engineer	301
Electronic engineer	236
Mechanical engineer	150
Civil engineer	67

The number of engineers in a particular category is usually called the **frequency**. A table such as Table 29.1, listing types of engineer and their frequency, is called a **frequency distribution**. As another example of a frequency distribution, consider Table 29.2, which shows the lifespan of 200 components, measured to the nearest week. In general, a frequency distribution lists values or measurements of some variable, for example 'type of engineer' and 'lifespan', and the frequency with which that value or measurement occurs.

Sometimes the measurements are grouped into **classes**. The number of measurements falling into a particular class is the **frequency** of that class. Suppose, for example, that the length of time a machine runs before breaking down is measured. The times are recorded to the nearest hour. This is repeated with several hundred machines. The results are given in Table 29.3. The classes are given in the left-hand column, that is, 120–124, 125–129 and so on. The **class width** is the difference between the lower values of two consecutive classes. In this example, the class width is $125 - 120 = 5$. All the classes of this frequency distribution are of equal width. This is commonly true, but is not a requirement of a frequency distribution. In this chapter all frequency distributions have a single class width. For grouped data, the **midpoint** of a class is often needed. In this example the class midpoints are 122, 127, 132, 137 and 142.

Table 29.2

Lifespan (weeks)	Frequency
6	13
7	41
8	93
9	42
10	11

Table 29.3

Running time (h)	Number of machines
120–124	76
125–129	83
130–134	42
135–139	117
140–144	23

Self-assessment questions 29.3

1. Explain what is meant by a frequency distribution.

2. Explain why it is sometimes beneficial to group data into classes, rather than simply list the actual measurements that have been made. Can you think of any disadvantages of using grouped data?

Exercise 29.3

1. Table 29.4 shows a frequency distribution of the running time of a number of machines before breakdown. The running times are measured to the nearest 100 hours.
   (a) State the class width.
   (b) List the midpoints of the classes.

Table 29.4

Running time (h)	Frequency
3000–3400	36
3500–3900	42
4000–4400	17
4500–4900	49
5000–5400	19

29.4 Representation of data

There are a variety of methods for representing frequency distributions in an easily understood form. We look at pie charts, line charts and histograms. In a **pie chart** data is represented by sectors of a circle. The angle of each sector is proportional to the frequency of the measurement or class. To find the angle of each sector we proceed as follows. The total of all the frequencies is found. This is represented by a full circle. Each frequency is expressed as a fraction of the total frequency. The angle representing a frequency is then found by multiplying this fraction by 360°.

Example 29.2

Express the frequency distribution given by Table 29.1 as a pie chart.

Solution The total of all the frequencies is 900. Consider the category of chemical engineers. There are 146 of these, and so the angle representing this category is $\frac{146}{900} \times 360 = 58.4°$. The angles for all categories are found in a similar way:

chemical engineer: $\frac{146}{900} \times 360 = 58.4°$
electrical engineer: $\frac{301}{900} \times 360 = 120.4°$
electronic engineer: $\frac{236}{900} \times 360 = 94.4°$
mechanical engineer: $\frac{150}{900} \times 360 = 60.0°$
civil engineer: $\frac{67}{900} \times 360 = 26.8°$

Finally, the pie chart is drawn as shown in Figure 29.1.

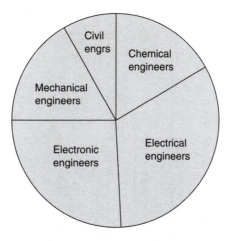

Figure 29.1   Pie chart showing a breakdown of types of engineer.

If the pie chart is drawn by hand a protractor is needed to obtain the correct angles. Alternatively, computer packages are now available to draw pie charts.

In a **line chart** data is represented by lines whose lengths are proportional to the frequency of the variable.

Example 29.3

Represent the data given in Table 29.1 by a line chart.

Solution   The line chart is shown in Figure 29.2.

A **histogram** is a graphical representation of a frequency distribution. Each class is represented by a rectangle. The width of the rectangle equals the class width, and when all class widths are equal the height of the rectangle represents the frequency. The centre of the base of the rectangle coincides with the midpoint of the class.

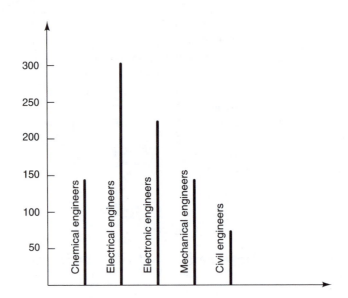

Figure 29.2   A line chart showing the breakdown of different types of engineer.

Example 29.4

Draw the histogram for the frequency distribution given in (a) Table 29.2 (b) Table 29.3

Solution   (a)   Figure 29.3 shows the histogram for Table 29.2. Note that each rectangle has a width of 1, and the height of each rectangle represents the frequency.

(b)   Figure 29.4 shows the histogram for Table 29.3. Note that the midpoints of the bases of the rectangles are located at the midpoints of the classes.

Example 29.5

The diameters of 180 pistons were measured to the nearest millimetre. Table 29.5 shows the results.

Table 29.5

Diameter of piston (mm)	Number of pistons
50–53	37
54–57	42
58–61	63
62–65	21
66–69	17

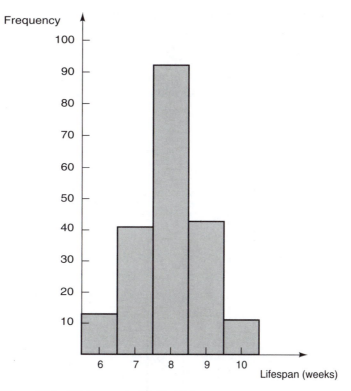

Figure 29.3   Histogram for Table 29.2.

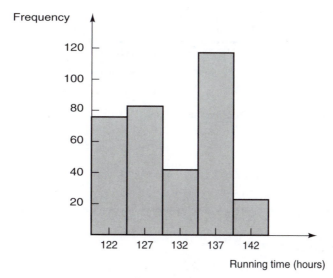

Figure 29.4   Histogram for Table 29.3.

(a) State the class width   (b) Represent the frequency distribution as a histogram.

Solution   (a)   The class width is $54 - 50 = 4$.
    (b)   Figure 29.5 shows the histogram. Note that the midpoints of the classes are 51.5, 55.5, 59.5, 63.5 and 67.5.

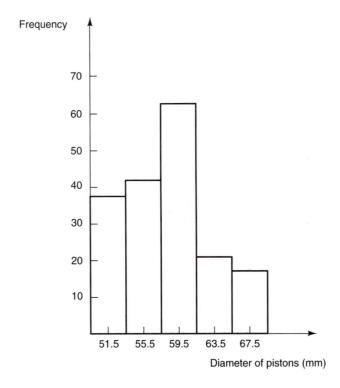

Figure 29.5   Histogram for Example 29.5.

Self-assessment question 29.4

1.   State the various ways in which a frequency distribution can be represented graphically.

Exercises 29.4

1. Express the following data as (a) a pie chart and (b) a line chart.

Components used in the construction of a machine	Number
Component A	7
Component B	4
Component C	11
Component D	1
Component E	5
Component F	3

2. Draw histograms for the following frequency distributions.

(a)

Lengths of metal rods (cm)	Number of rods
50–52	14
53–55	21
56–58	13
59–61	10
62–64	18
65–67	8

(b)

Diameters of ball bearings (mm)	Number of bearings
7.0–7.4	37
7.5–7.9	20
8.0–8.4	41
8.5–8.9	17
9.0–9.4	26

29.5 Mean, median and mode

A single value that is representative of a set of data is called an **average**. An average gives a measure of the centre of a set of values. There are several different averages, each with advantages and disadvantages. We look at three types: the mean, the median and the mode.

29.5.1 The mean

The **mean** is calculated by adding up all the data values and dividing this total by the number of items of data. Many scientific calculators have a facility for calculating the mean of a set of data values.

Example 29.6

Calculate the mean of 16, 21, 17, 20, 19.

Solution

sum of values $= 16 + 21 + 17 + 20 + 19 = 93$
number of values $= 5$
mean $= \frac{93}{5} = 18.6$

Note that the mean need not be one of the actual data values.
In general, if $x_1, x_2, x_3, \ldots, x_n$ are the values of n items of data then the mean is found using

$$\text{mean value} = \bar{x} = \frac{\sum_{i=1}^{n} x_i}{n}$$

The symbol $\bar{x}$, read as 'x bar', is used to denote the mean value.
The mean of a frequency distribution may be found as illustrated by the following examples.

Example 29.7

The diameters of some copper pipes are measured and recorded to the nearest tenth of a millimetre, as detailed in Table 29.6. Calculate the mean diameter of the copper pipes.

Table 29.6

Diameter of copper pipe (mm)	Frequency
35.6	2
35.7	4
35.8	1
35.9	6

Solution

There are 2 pipes of diameter 35.6 mm, 4 pipes of diameter 35.7 mm, 1 pipe of diameter 35.8 mm and 6 pipes of diameter 35.9 mm. The total diameter of all the pipes is then found from

$$\text{total diameter} = (35.6 \times 2) + (35.7 \times 4) + (35.8 \times 1) + (35.9 \times 6)$$

$$= 71.2 + 142.8 + 35.8 + 215.4$$

$$= 465.2$$

The total number of pipes $=2+4+1+6=13$. Hence the mean diameter is

$$\text{mean diameter} = \frac{465.2}{13} = 35.78 \text{ mm}$$

We introduce a general notation for calculating the mean of a frequency distribution. Let x_i, $i=1,\ldots,n$, be the distinct values taken by the variable and let f_i be the corresponding frequencies. In Example 29.7 we have $x_1=35.6$, $f_1=2$; $x_2=35.7$, $f_2=4$; $x_3=35.8$, $f_3=1$; $x_4=35.9$, $f_4=6$. Then the sum of all the data values is given by

$$\sum_{i=1}^{n} x_i f_i$$

The total number of data items is $\sum_{i=1}^{n} f_i$, and so

$$\text{mean} = \frac{\sum_{i=1}^{n} x_i f_i}{\sum_{i=1}^{n} f_i}$$

Example 29.8

The diameters of some pistons are measured, and the results are as recorded in Table 29.7. Calculate the mean diameter.

Table 29.7

Diameters of pistons (mm)	Number
30.0–30.4	6
30.5–30.9	3
31.0–31.4	4
31.5–31.9	2

Solution   Consider the first class: 30.0–30.4. We do not know the actual values in this class, so we assume that all values are equal to the midpoint value, namely 30.2. Hence we assume that there are 6 values of 30.2. Similarly, for the remaining classes we assume that all values occur at the midpoint of the class. These midpoint values are 30.7, 31.2 and 31.7. So we have

6 diameters of 30.2 mm
3 diameters of 30.7 mm

4 diameters of 31.2 mm

2 diameters of 31.7 mm

The total of these values is

$$\sum_{i=1}^{4} x_i f_i = (6 \times 30.2) + (3 \times 30.7) + (4 \times 31.2) + (2 \times 31.7) = 461.5$$

The number of values is $\sum_{i=1}^{4} f_i = 6 + 3 + 4 + 2 = 15$, and so

$$\text{mean diameter} = \frac{\sum_{i=1}^{4} x_i f_i}{\sum_{i=1}^{4} f_i} = \frac{461.5}{15} = 30.77$$

The mean diameter is 30.77 mm.

29.5.2 The median

The **median** is the middle value in the data set, when all the values have been arranged in ascending order. To calculate the median, the set of data values must be arranged in ascending order. If there is an odd number of values, the median is the middle one. If there is an even number of values, the median is the mean of the middle two.

Example 29.9

Calculate the medians of the following sets of data:

(a) 3, 1, -1, 6, 9, 7, 1

(b) -2, 6, 3, 1, 4, 0

Solution   (a)   First, we arrange the data in ascending order:

$$-1, 1, 1, 3, 6, 7, 9$$

There are 7 values; the fourth value is the middle one. Hence the median is 3.

(b)   The data is first arranged in ascending order:

$$-2, 0, 1, 3, 4, 6$$

There are 6 values. The median is the mean of the middle two. Hence

$$\text{median} = \frac{1+3}{2} = 2$$

29.5.3 The mode

The **mode** is the value with the greatest frequency. The mode may not be unique.

Example 29.10

State the modes of (a) 4, 5, 1, 4, 6, 2, 4, 3, 6   (b) 5, 3, 7, 0, 5, 6, 7

Solution   (a)   The value 4 occurs more than any other value, and so 4 is the mode.
            (b)   The highest frequency is 2. Both 5 and 7 have a frequency of 2, and so
                  this set of data has 2 modes: 5 and 7.

Self-assessment questions 29.5

1.   What assumption is made when calculating the mean of a grouped frequency distribution?

2.   The median of a set of data is always one of the data values. True or false?

3.   By considering the mean, median and mode of 1, 1, 1, 1, 1, 1, 10, state one advantage of the
     median and mode over the mean as a representative measure of a data set.

Exercises 29.5

1.   Calculate the mean of $-6, -2, -3, 0, 4$.

2.   Calculate the mean of the frequency
     distribution given in Table 29.8.

3.   Calculate the mean, median and mode of
     (a) 2, 1, 4, 6, 5, 3   (b) 4, 3, 2, 1
     (c) 3, 4, 4, 6, 6, 6, 9
     (d) $-2, -2, 1, -1, 3, 1, 2$

Table 29.8

Weight of a component (g)	Number
205–214	21
215–224	18
225–234	10
235–244	12
245–254	4

29.6 Variance and standard deviation

In Section 29.5 we looked at the mean, median and mode. These values give
an indication of the centre of a set of values. However, an average does not
convey how dispersed the values may be. To measure the spread or deviation
from the centre, we introduce the **variance** and **standard deviation**. The following
example illustrates the need for a measure of spread of a set of data.

Example 29.11

Calculate the means of (a) -200, 0, 209   (b) 2.7, 3.1, 3.2

Solution   (a) $$\text{mean} = \frac{-200 + 0 + 209}{3} = 3$$

(b) $$\text{mean} = \frac{2.7 + 3.1 + 3.2}{3} = 3$$

Both sets of data have a mean of 3. This one parameter does not fully describe the data; the first set clearly has a much greater spread than the second set. The standard deviation and variance address this shortcoming. These parameters give a measure of the spread of the data set. In Example 29.11 the spread of set (a) is greater than that of set (b). The following example illustrates how to quantify this spread by calculating the variance and standard deviation.

Example 29.12

Calculate the variance and standard deviation of (a) -200, 0, 209 (b) 2.7, 3.1, 3.2

Solution   (a)   First, the mean is found:

$$\text{mean} = \frac{-200 + 0 + 209}{3} = 3$$

The deviation from the mean is found for each value. The deviation of a value is found by subtracting the mean from it.

deviation of -200 from mean $= -200 - 3 = -203$
deviation of 0 from mean $= 0 - 3 = -3$
deviation of 209 from mean $= 209 - 3 = 206$

These deviations are then squared. The mean of the squared deviations is called the **variance**. This is now calculated:

sum of squared deviations $= (-203)^2 + (-3)^2 + (206)^2 = 83\,654$

Hence

$$\text{mean of the squared deviations} = \frac{83\,654}{3} = 27\,885$$

This is the variance. The **standard deviation** is the square root of the variance:

$$\text{standard deviation} = \sqrt{\text{variance}} = \sqrt{27\,885} = 167$$

(b)
$$\text{mean} = \frac{2.7 + 3.1 + 3.2}{3} = 3$$

The deviation from the mean is found for each value. This gives $2.7 - 3$, $3.1 - 3$, $3.2 - 3$; that is,

$$-0.3, \quad 0.1, \quad 0.2$$

The deviations are squared. The variance is the mean of the squared deviations:

$$\text{sum of squared deviations} = (-0.3)^2 + (0.1)^2 + (0.2)^2 = 0.14$$

$$\text{variance} = \frac{0.14}{3} = 0.047$$

The standard deviation is the square root of the variance:

$$\text{standard deviation} = \sqrt{\text{variance}} = \sqrt{0.047} = 0.22$$

As expected, the standard deviation of the first data set is significantly larger than that of the second.

Example 29.13 Manufacture of a miniature stainless-steel pump housing

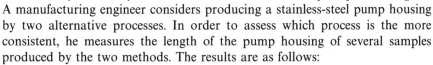

A manufacturing engineer considers producing a stainless-steel pump housing by two alternative processes. In order to assess which process is the more consistent, he measures the length of the pump housing of several samples produced by the two methods. The results are as follows:

Process A: length (mm)	Process B: length (mm)
25.03	25.01
24.98	24.99
25.01	25.01
25.04	25.01
24.96	24.99
25.02	24.98
24.98	25.02
25.02	25.02
25.03	24.99
24.99	24.98

Calculate the mean and standard deviation of the pump housing lengths for each of the two processes, and hence decide which process is the more consistent.

Solution   For process A we have

$$\text{mean} = \frac{25.03 + 24.98 + 25.01 + 25.04 + 24.96 + 25.02 + 24.98 + 25.02 + 25.03 + 24.99}{10}$$

$$= 25.006$$

The deviations from the mean are found, squared and added. The results are as follows:

Process A: length (mm)	Deviation from mean	Squared deviation
25.03	0.024	5.76×10^{-4}
24.98	-0.026	6.76×10^{-4}
25.01	0.004	1.6×10^{-5}
25.04	0.034	1.156×10^{-3}
24.96	-0.046	2.116×10^{-3}
25.02	0.014	1.96×10^{-4}
24.98	-0.026	6.76×10^{-4}
25.02	0.014	1.96×10^{-4}
25.03	0.024	5.76×10^{-4}
24.99	-0.016	2.56×10^{-4}

The sum of the squared deviations is found to be 6.44×10^{-3}. The variance is the mean of the squared deviations:

$$\text{variance} = \frac{6.44 \times 10^{-3}}{10} = 6.44 \times 10^{-4}$$

The standard deviation is the square root of the variance:

$$\text{standard deviation} = \sqrt{6.44 \times 10^{-4}} = 0.025\,38 = 2.538 \times 10^{-2}$$

The mean length is 25.006 mm and the standard deviation is 2.538×10^{-2} mm.
    We perform a similar calculation for process B. A summary of the results is given:

$$\text{mean} = 25.00$$

$$\text{sum of squared deviations} = 2.2 \times 10^{-3}$$

$$\text{variance} = 2.2 \times 10^{-4}$$

$$\text{standard deviation} = 0.014\,83 = 1.483 \times 10^{-2}$$

The mean is 25.00 mm and the standard deviation is 1.483×10^{-2} mm. We see that process B has a lower standard deviation, and therefore conclude that it is the more consistent production process.

We can state a general formula for variance and standard deviation using the steps of Example 29.12.

Suppose we have a set of n values: $x_1, x_2, \ldots, x_n$. We calculate the mean of the values: this is denoted by $\bar{x}$. The deviations from the mean are found: $x_1 - \bar{x}, x_2 - \bar{x}, \ldots, x_n - \bar{x}$. Each deviation is squared to give $(x_1 - \bar{x})^2, (x_2 - \bar{x})^2, \ldots$. The mean of the squared deviations is the **variance**, usually denoted by σ^2:

$$\text{variance} = \sigma^2 = \frac{\sum_{i=1}^{n} (x_i - \bar{x})^2}{n}$$

The **standard deviation** is the square root of the variance. The standard deviation is denoted by σ:

$$\text{standard deviation} = \sigma = \sqrt{\frac{\sum_{i=1}^{n} (x_i - \bar{x})^2}{n}}$$

Note that standard deviation has the same units as the original data.

Many scientific calculators have a facility for calculating standard deviation and variance. It is worth investigating how to use this facility on your calculator.

Self-assessment question 29.6

1. Explain why the mean of a data set does not adequately characterize the data set. How does the standard deviation rectify this?

Exercises 29.6

1. Calculate the standard deviations of the following data sets:
   (a) 0, 1, 2, 3, 4, 5
   (b) −2, 3, −1, 0, 7, 4
   (c) 150, 142, 210, 96, 110
   (d) 36, 40, 29, 51, 37, 40

2. (a) Calculate the mean of 1, 2, 3.
   (b) Calculate the mean of k, $2k$, $3k$.

(c) Calculate the standard deviation of 1, 2, 3.

(d) Calculate the standard deviation of k, $2k$, $3k$.

(e) A set of data $\{x_1, x_2, x_3, \ldots, x_n\}$ has mean $\bar{x}$ and standard deviation σ. What can you say about the mean and standard deviation of the data set $\{kx_1, kx_2, kx_3, \ldots, kx_n\}$, where k is a constant.

29.7 Basic concepts of probability

Suppose we toss a coin. The act of tossing a coin is an example of a trial. In general, any action or experiment in which various outcomes are possible is called a **trial**. When tossing a coin, there are two possible outcomes: 'throw a head' and 'throw a tail'. When throwing a die, there are six possible outcomes: 'throw a 1', 'throw a 2', and so on. For a given trial the set of all possible outcomes is called the **sample space**. Any subset of the sample space is called an **event**. For example, if the trial is throwing a die, the sample space is

sample space = {throw a 1, throw a 2, throw a 3, ..., throw a 6}

Examples of events are 'throw an even number', 'throw an odd number' and 'throw a number greater than 3'.

We need to measure how likely events are to happen. To quantify the notion of likelihood, we introduce the term **probability**. The probability of an event is a number between 0 and 1. A probability of 0 means that the event is impossible; a probability of 1 means that the event is a certainty. An event that is neither an impossibility nor a certainty has a probability between 0 and 1. An event that is as likely to happen as not has a probability of 0.5. Thus the probability of throwing a head is 0.5, which is the same as the probability of throwing a tail.

We have a notation for probability. If H is the event 'throw a head' then the probability of throwing a head is written as $P(H)$. Similarly, if T is the event 'throw a tail', the probability of throwing a tail is written as $P(T)$. We have seen that

$$P(H) = P(T) = 0.5$$

If E is any event, the probability of E, written as $P(E)$, is such that $0 \leqslant P(E) \leqslant 1$.

In any trial, the sum of the probabilities of the possible outcomes is always 1. If the trial is throwing a coin then the possible outcomes are H: 'throw a head', and T: 'throw a tail'. Thus

$$P(H) + P(T) = 1$$

If the trial is throwing a die, the possible outcomes are 'throw a 1', 'throw a 2', 'throw a 3', 'throw a 4', 'throw a 5' and 'throw a 6'. The sum of the six probabilities is 1; that is,

$$P(\text{throw a } 1) + P(\text{throw a } 2) + \ldots + P(\text{throw a } 6) = 1$$

Example 29.14

A die is thrown. What is the probability of throwing a 6?

Solution

The possible scores obtained on throwing a die are 1, 2, 3, 4, 5 and 6. Their probabilities must sum to 1. Also, the outcomes are all equally likely, and so the probability of each is $\frac{1}{6}$:

P(throwing a 6)$=\frac{1}{6}$

Example 29.15

What is the probability of the event E: drawing the ace of spades from a full pack of 52 playing cards?

Solution

There is only one ace of spades in a pack of cards. There are 52 possible outcomes when a card is drawn. All are equally likely. Hence

$P(E)=\frac{1}{52}$

Sometimes an event can happen in many ways, for example, throwing an even score with a die can be achieved by throwing a 2, 4 or 6. So 3 out of the 6 equally likely outcomes result in an even score. Hence

$$P(\text{even score})=\frac{3}{6}=\frac{1}{2}$$

Example 29.16

What is the probability of the event E: drawing a heart from a full pack of 52 playing cards?

Solution

There are 13 hearts in a pack of cards and each card has an equal probability of being drawn:

$$P(E)=\frac{13}{52}=\frac{1}{4}$$

Sometimes it may not be obvious what the probability of an event is. For example, we may want to know the probability that a machine will break down at some time during the next 24 h, or the probability that a wire can withstand a certain strain. In such cases we need to repeat the trial a very large number of times, say n. Suppose we count the number of times the event

E occurs, say *m*. Then, the probability of event *E* is given by

$$P(E) = \frac{\text{number of trials in which } E \text{ occurs}}{\text{total number of possible trials}} = \frac{m}{n}$$

If an event *E* is certain then it occurs with every trial; that is, there are *n* occurrences of *E* with *n* trials, so

$$P(E) = \frac{n}{n} = 1$$

as we previously stated. Similarly, if the event *E* is an impossibility, it will never occur, so in *n* trials there will be no occurrences and hence

$$P(E) = \frac{0}{n} = 0$$

as previously stated.

Example 29.17 Washing machine reliability

Out of 10 000 washing machines manufactured by a well-known company, 137 were reported as being faulty within one year of sale. Find the probability that a washing machine selected at random in a showroom will develop a fault within one year.

Solution

$$P(\text{the machine will develop a fault within one year}) = \frac{137}{10\,000}$$

$$= 0.0137$$

Self-assessment questions 29.7

1. Explain the meaning of the terms 'trial', 'sample space' and 'event'.

2. Give one example each of an impossible event and a certain event.

3. Which of the following cannot be the probability of an event occurring: (a) 0.99   (b) 1.3   (c) -0.5 (d) $\frac{2}{11}$

Exercises 29.7

1. A component is manufactured by one of two machines: 65% of components come from machine A, the remainder from machine B. Find the probability that a component chosen at random was made by machine B.

2. Components are manufactured by one of three machines: machine P makes 35%, machine Q makes 40% and machine R makes the remainder. Find the probability that a component chosen at random was manufactured by machine R.

3. Out of 1500 light bulbs sampled, 987 were manufactured by company M and the remainder by company N. What is the probability that a light bulb chosen at random was manufactured by company N?

4. Of resistors manufactured by a particular method, 7.5% are found to be unacceptable. Find the probability that a resistor manufactured by this method and chosen at random is acceptable.

5. A playing card is drawn at random from a pack of 52. Find the probability of each of the following events:
(a) E: the card is a jack of spades
(b) E: the card is a king

(c) E: the card is blank

6. A motor vehicle manufacturer wishes to give a warranty on all new vehicles sold. In a survey of vehicles sold in the previous two years, 1253 out of the 4897 sold developed a fault or faults within 12 months. Calculate the probability that a vehicle sold will develop one or more faults within 12 months.

7. A machine in a soap powder factory is used to deposit 1 kg of soap flakes in each box. Out of a sample of 580 boxes weighed, 13 are found to contain less than 1 kg of soap flakes. What is the probability that a box chosen at random contains at least the required amount of flakes?

29.8 Compound events and Venn diagrams

At this point it is useful to introduce set notation and Venn diagrams. Recall that Venn diagrams were introduced in Chapter 3. For every trial there are various possible outcomes, and the set of all possible outcomes is called the **sample space**. The universal set $\mathscr{E}$ represents the sample space. Recall that an event is a subset of the sample space, and so events are represented as subsets of the universal set.

Example 29.18

Describe the sample space and various events associated with tossing a coin.

Solution   When a coin is tossed, it may land with the head uppermost, or the tail uppermost. The possible outcomes can therefore be expressed as

H: the outcome is a head, T: the outcome is a tail

We can represent this on a Venn diagram using the set H to stand for the event of throwing a head and the set T to stand for the event of throwing a tail. The Venn diagram is shown in Figure 29.6. The sample space is the set of all possible outcomes, that is $\mathscr{E} = \{\text{head, tail}\}$.

Note that there are no set elements outside both H and T, and also that we can write $\mathscr{E} = H \cup T$.

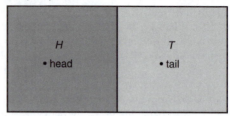

Sample space ε

Figure 29.6 The possible outcomes are represented as sets.

The set notation of union, $\cup$, and intersection, $\cap$, can be used in probability theory. If A and B are two events then the event $A \cup B$ is the event A occurs or B occurs or both occur. We can associate the symbol $\cup$ with the word 'or'. Similarly, the event $A \cap B$ is the event A occurs and also B occurs. We can associate the symbol $\cap$ with the word 'and'. Events such as $A \cap B$ and $A \cup B$ are known as **compound events**.

Example 29.19

From a pack of 52 playing cards, one card is selected at random. Define the events A and B as follows:

A: the card is a diamond
B: the card is a queen.

(a)   Find $P(A)$ and $P(B)$.
(b)   Explain the meaning of the event $A \cup B$ and find its probability.
(c)   Explain the meaning of the event $A \cap B$ and find its probability.

Solution   (a)   There are 13 diamonds in the pack and so $P(A) = \frac{13}{52} = \frac{1}{4}$. There are 4 queens in the pack, and so $P(B) = \frac{4}{52} = \frac{1}{13}$.

(b) $A \cup B$ is the event 'the card is a diamond or a queen'. There are 13 diamonds and 4 queens. One of these cards is the queen of diamonds, which is a queen and a diamond. Therefore there are 16 cards in $A \cup B$. We find $P(A \cup B) = \frac{16}{52} = 0.308$.

(c) $A \cap B$ is the event 'the card is a diamond and also a queen'. For this event to occur, we must select the queen of diamonds. Therefore $P(A \cap B) = \frac{1}{52}$.

Example 29.20

A die is thrown.
(a)   Find the probability of obtaining each of the following events,
(i) E_1: a score less than 2
(ii) E_2: a score more than 4

(b)   Show these events on a Venn diagram.

(c)   Explain the meaning of the event $E_1 \cup E_2$, and find its probability.

Solution   (a)   (i)   A score less than 2 occurs only if a 1 is scored. Therefore

$$P(E_1) = \tfrac{1}{6}$$

(ii)   A score more than 4 occurs only if a 5 or a 6 is scored. Therefore

$$P(E_2) = \frac{2}{6} = \frac{1}{3}$$

(b)   A Venn diagram is shown in Figure 29.7. Note that the events of throwing a 2, 3 or a 4 fall into neither set E_1 nor set E_2.

(c) $E_1 \cup E_2$ is the event 'a score of 1, 5 or 6 is obtained':

$$P(E_1 \cup E_2) = \frac{3}{6} = \frac{1}{2}$$

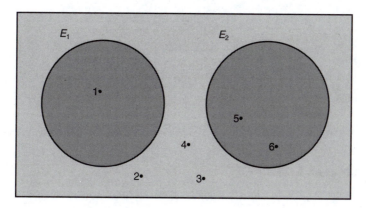

Figure 29.7   Venn diagram for Example 29.20.

Sometimes events overlap or occur at the same time, and in such cases their corresponding sets intersect.

Example 29.21

Consider again Example 29.20 with the additional event E_3: an odd score is obtained.

(a)   Draw a Venn diagram to represent this.

(b)   Describe the events $E_3 \cap E_2$ and $E_3 \cap E_1$.

(c)   Describe the event $E_3 \cup E_2$, and find the probability $P(E_3 \cup E_2)$.

Solution   (a)   The Venn diagram is shown in Figure 29.8.

(b)   The event of throwing a 5 lies in the intersection of E_3 and E_2, because 5 is both odd and more than 4. The event E_1 lies entirely within E_3, because a score less than 2 must be odd. Hence

$$E_3 \cap E_2 = \{5\}, \qquad E_3 \cap E_1 = \{1\}$$

(c)   The event $E_3 \cup E_2$ occurs when either E_2 or E_3 occurs. This occurs when either the score is odd or more than 4, or both. Specifically, $E_3 \cup E_2 = \{1, 3, 5, 6\}$. Therefore $P(E_3 \cup E_2) = \frac{4}{6} = \frac{2}{3}$.

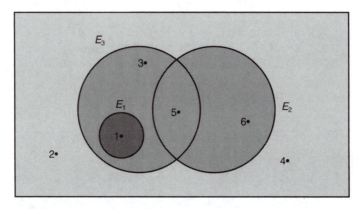

Figure 29.8   Venn diagram for Example 29.21.

Self-assessment question 29.8

1.   Explain how the set theory symbols $\cap$ and $\cup$ are used in probability theory.

Exercises 29.8

1.   A fair die is thrown. Calculate the probability of each of the following events.
   (a)   E: an odd score is obtained
   (b)   E: a score less than 5 is obtained
   (c)   E: a score more than 5 is obtained
   (d)   E: a score of 7 is obtained
   (e)   E: a score less than 10 is obtained

2.   A die is thrown. With

   E_1: an even score is obtained

   E_2: an odd score is obtained
   E_3: a score greater than 3 is obtained
   E_4: a score less than 5 is obtained
   calculate the probabilities of the following events:
   (a) $E_1 \cup E_2$ (b) $E_1 \cup E_3$
   (c) $E_1 \cup E_4$ (d) $E_1 \cap E_2$
   (e) $E_1 \cap E_3$ (f) $E_1 \cap E_4$
   (g) $E_2 \cup E_3$ (h) $E_2 \cap E_3$
   (i) $E_2 \cup E_4$ (j) $E_2 \cap E_4$
   (k) $E_3 \cup E_4$ (l) $E_3 \cap E_4$

29.9 Independent events

Two events are said to be **independent** if the occurrence of either one of them in no way affects the probability of the other occurring. For example, suppose a coin is tossed twice. The outcome of the first toss in no way affects the outcome of the second. The two events are independent. When events E_1 and E_2 are independent, the **multiplication law** of probability applies.

The multiplication law

If E_1 and E_2 are independent events then

$$P(E_1 \text{ and } E_2) = P(E_1 \cap E_2)$$

$$= P(E_1)P(E_2)$$

This law says that the probability of two independent events occurring is the product of the probabilities of the individual events.

Example 29.22

A coin is tossed twice. What is the probability of obtaining two heads?

Solution   Let E_1 be the event 'a head is obtained on the first toss'. Clearly, $P(E_1) = 0.5$. Let E_2 be the event 'a head is obtained on the second toss'. Clearly, $P(E_2)$ is also 0.5. The event that two heads are obtained is given by $E_1 \cap E_2$. Now events E_1 and E_2 are independent. The outcome of the first toss in no way can affect the outcome of the second. Therefore, using the multiplication law, we find

$$P(\text{getting 2 heads}) = P(E_1 \text{ and } E_2)$$

$$= P(E_1 \cap E_2)$$

$$= P(E_1)P(E_2)$$

$$= 0.5 \times 0.5 = 0.25$$

When several events are all independent of each other, the multiplication law is extended as follows:

$$P(E_1 \text{ and } E_2 \text{ and } E_3 \text{ and} \ldots) = P(E_1 \cap E_2 \cap E_3 \cap \ldots)$$

$$= P(E_1)P(E_2)P(E_3) \cdots$$

Example 29.23
>   Three dice are thrown. What is the probability of obtaining 3 sixes.

Solution   The scores obtained on each of the three dice are clearly independent of each other. Therefore

$$P(3 \text{ sixes}) = \tfrac{1}{6} \times \tfrac{1}{6} \times \tfrac{1}{6} = \tfrac{1}{216}$$

So there is 1 chance in 216 of obtaining 3 sixes.

Self-assessment questions 29.9

1.   Explain the meaning of the phrase 'Events E_1 and E_2 are independent events'.

2.   State the multiplication law and give the condition necessary before it can be applied.

Exercises 29.9

1.   A coin is tossed and a die is thrown. Find the probability of the event E: a tail and a 6 are obtained.

2.   A component is made by one of two machines: machine A manufactures 55% while machine B manufactures the remainder. A small percentage of components made by each machine is unacceptable; in both cases this percentage equals 2%. Find the probability that a component selected at random is made by machine A and is unacceptable.

3.   State, with reasons, which of the following pairs of events might be classed as independent:
  (a) A: a 6 is obtained on the first throw of a die, B: a 6 is obtained on the second throw;
  (b) A: low oil warning light comes on in a car, B: the car breaks down.

4.   In a consignment containing 50 000 washers 6% are either oversize or undersize.
  (a) A washer is picked at random. Find the probability that its size is acceptable.
  (b) Three washers are picked at random. Find the probability that they are all acceptable. Comment upon your answer.

29.10 Mutually exclusive events

>   Suppose two events are such that they cannot possibly both occur at the same time. Such events are said to be **mutually exclusive**. For example, a coin is tossed once. The events E_1: a head is thrown and E_2: a tail is thrown are mutually exclusive, because they cannot both occur together. If we have thrown a tail

then we cannot have thrown a head, and vice versa. The Venn diagram for two mutually exclusive events E_1 and E_2 is shown in Figure 29.9. Because the events are mutually exclusive the two sets are disjoint. Note that because they do not intersect, $E_1 \cap E_2 = \emptyset$, the empty set. For mutually exclusive events, the addition law can be applied.

The addition law

If events E_1 and E_2 are mutually exclusive then the probability of either E_1 or E_2 occurring is given by

$$P(E_1 \text{ or } E_2) = P(E_1 \cup E_2) = P(E_1) + P(E_2)$$

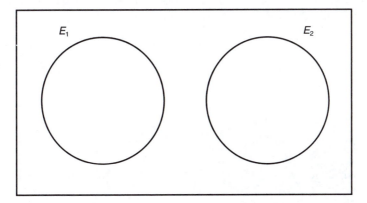

Figure 29.9 Venn diagram for two mutually exclusive events.

Example 29.24

A die is thrown. E_1 is the event 'a 1 is thrown'. E_2 is the event 'a 6 is thrown'. Find the probability of getting a 1 or a 6.

Solution   Clearly, the events E_1 and E_2 are mutually exclusive, because if a 6 is thrown then we cannot have thrown a 1, and vice versa. Furthermore, $P(E_1) = P(E_2) = \frac{1}{6}$. Therefore, using the addition law, we find

$$P(\text{getting a 1 or a 6}) = P(E_1 \cup E_2)$$
$$= P(E_1) + P(E_2)$$
$$= \tfrac{1}{6} + \tfrac{1}{6} = \tfrac{2}{6} = \tfrac{1}{3}$$

Example 29.25

A coin is tossed. H is the event of tossing a head and T is the event of tossing a tail. Find $P(H \cup T)$.

Solution $H \cup T$ is the event of tossing a head or tossing a tail, one of which must happen. Clearly, the probability of this event is therefore 1. Alternatively, we can argue as follows. H and T are mutually exclusive events with $P(H) = P(T) = 0.5$. Using the addition law, $P(H \cup T) = P(H) + P(T) = 0.5 + 0.5 = 1$ as before.

29.10.1 Complementary events

Suppose A is the event 'a 6 is scored when throwing a die'. The **complement** of A, which is written as $\bar{A}$ is the event 'a score other than 6 is thrown'. The events A and $\bar{A}$ are said to be **complementary**. They are mutually exclusive, and one of them must happen. This situation is illustrated in the Venn diagram shown in Figure 29.10. Clearly $P(A) = \frac{1}{6}$ and $P(\bar{A}) = \frac{5}{6}$. Note also that

$$P(A) + P(\bar{A}) = 1$$

The sum of the probabilities of all possible events is 1, corresponding to certainty.

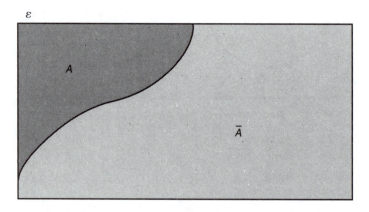

Figure 29.10   Venn diagram illustrating complementary events.

Example 29.26

The event A is 'the selected component is faulty'. Describe the complementary event $\bar{A}$.

Solution   The complement of A is the event $\bar{A}$: the selected component is not faulty. Clearly, a component is either faulty or it is not, and hence the two events are mutually exclusive. Furthermore, one of A and $\bar{A}$ must happen.

Self-assessment questions 29.10

1. Explain what is meant by saying events A and B are mutually exclusive.
2. Explain what is meant by saying A and B are complementary events.

Exercises 29.10

1. Illustrate on a Venn diagram the situation where E_1, E_2 and E_3 are all mutually exclusive.

2. A component is picked at random from a production line. The event E is

    E: the component is within the manufacturer's tolerance

    (a) State the complement of E.
    (b) If $P(\bar{E})=0.03$, state $P(E)$.

3. A component is classed as one of the following: (i) first-grade, (ii) acceptable, (iii) unacceptable. The events E_1, E_2 and E_3 are

    E_1: a component is first-grade
    E_2: a component is acceptable
    E_3: a component is unacceptable

    and $P(E_1)=0.1$ and $P(E_2)=0.85$.

    (a) State $\bar{E_1}$.
    (b) State $\bar{E_2}$.
    (c) State $\bar{E_3}$.
    (d) Calculate $P(E_3)$.
    (e) Calculate $P(E_1 \cup E_2)$.
    (f) Calculate $P(E_2 \cup E_3)$.

29.11 Non-mutually exclusive events

Two events are not mutually exclusive when it is possible for them both to occur simultaneously. For example, suppose that a die is thrown and events A and B are:

A: an even score is obtained
B: a score greater than 4 is obtained

The events A and B are not mutually exclusive, because if a 6 is thrown then both events occur simultaneously. When events are not mutually exclusive, the corresponding Venn diagram is as shown in Figure 29.11.

A and B will have a non-empty intersection, and the addition law must be modified as follows:

> **The addition law for non-mutually exclusive events**
> If events E_1 and E_2 are not mutually exclusive then the probability of either E_1 or E_2 occurring is given by
> $$P(E_1 \text{ or } E_2)=P(E_1 \cup E_2)=P(E_1)+P(E_2)-P(E_1 \cap E_2)$$

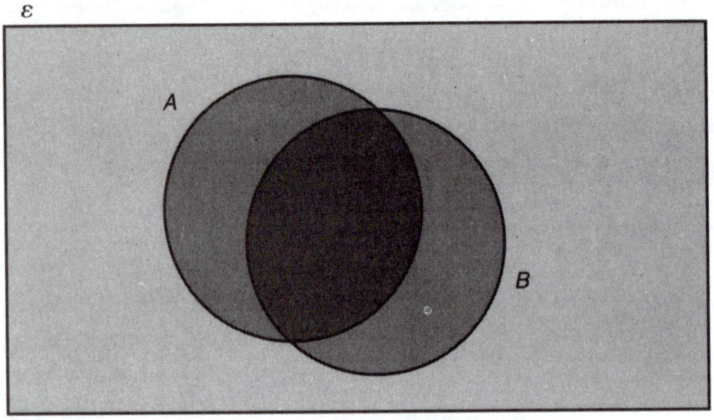

Figure 29.11   Non-mutually exclusive events can occur simultaneously, and so the sets intersect.

Example 29.27

A die is thrown. The events A and B are defined as

A: an even score is obtained
B: a score greater than 4 is obtained

Find the probability of obtaining either an even number or a score greater than 4.

Solution   We have $P(A)=\frac{3}{6}=\frac{1}{2}$ and $P(B)=\frac{2}{6}=\frac{1}{3}$. Also, $P(A\cap B)$ is the probability of obtaining an even score greater than 4. Only a score of 6 satisfies this requirement. Hence $P(A\cap B)=\frac{1}{6}$. Then, using the addition law, we find

$$P(A \text{ or } B)=P(A\cup B)=P(A)+P(B)-P(A\cap B)$$

$$=\tfrac{1}{2}+\tfrac{1}{3}-\tfrac{1}{6}=\tfrac{2}{3}$$

Example 29.28

A particular type of spring is manufactured by one of two machines, A or B: 60% of the springs are made by machine A and the remainder by machine B. Of the springs manufactured, 3% are defective in some way, regardless of which machine made them. Find

(a) the probability that a spring chosen at random is manufactured by machine A and is defective;
(b) the probability that a spring chosen at random is manufactured by machine A or is defective.

Solution   (a)   Let A be the event 'a spring chosen at random is manufactured by machine A'. Let B be the event 'a spring chosen at random is manufactured by machine B'. Let F be the event 'a spring chosen at random is defective'.

Then we are given that $P(A)=0.6$ and $P(B)=0.4$. Now 3% of springs are defective irrespective of which machine made them. Therefore $P(F)=0.03$. Because events A and F are independent, we can use the multiplication law for independent events; that is, the probability a spring is manufactured by machine A and is defective is

$$P(A \cap F) = P(A)P(F) = 0.6 \times 0.03 = 0.018$$

(b) Events A and F are not mutually exclusive, because a spring can be both defective and made by machine A. Using the addition law for non-mutually exclusive events, we find that the probability that a spring is manufactured by machine A or is defective is

$$P(A \cup F) = P(A) + P(F) - P(A \cap F) = 0.6 + 0.03 - 0.018 = 0.612$$

Self-assessment question 29.11

1. State the addition law appropriate to non-mutually exclusive events.

Exercises 29.11

1. A die is thrown. The following events are defined:

E_1: an even score is obtained
E_2: an odd score is obtained
E_3: a score greater than 4 is obtained
E_4: a score of 4 is obtained

(a) State which pairs of events are not mutually exclusive.

(b) Calculate the probabilities of the following events:

(i) $E_1 \cup E_2$ (ii) $E_1 \cup E_3$
(iii) $E_1 \cup E_4$ (iv) $E_2 \cup E_3$
(v) $E_2 \cup E_4$ (vi) $E_3 \cup E_4$

2. A component is manufactured by machines A and B: machine A makes 70% of the components and the remainder are made by machine B. Of those components made by machine A, 3% are defective, and of those made by machine B, 4% are defective. A component is picked at random. Calculate the following probabilities:
(a) the component is manufactured by machine A and is acceptable;
(b) the component is manufactured by machine A or is acceptable;
(c) the component is manufactured by machine B and is defective;
(d) the component is manufactured by machine B or is defective.

Review exercises 29

1 A box of resistors contains thirty $10\,\Omega$ resistors, fifty $5\,\Omega$ resistors, ten $3\,\Omega$ resistors and ninety $33\,\Omega$ resistors.
(a) A resistor is selected at random. Find the probability that its resistance is less than $5\,\Omega$.

(b) After the first resistor has been selected, it is found to have resistance $5\,\Omega$. It is not replaced. A second resistor is selected. Find the probability that this one has a resistance less than $5\,\Omega$.

2 State which of the following are discrete variables and which are continuous variables:
(a) the temperature of a chemical reaction;
(b) the volume of gas produced by a chemical reaction;
(c) the number of molecules of air in an air-tight room;
(d) the length of a pencil.

3 An experiment was repeated 100 times, and the temperature at which the reaction took place was noted:

Temperature at which reaction took place (°C)	Number of experiments
220	3
230	7
240	21
250	47
260	5
270	6
280	11

Represent the data as (a) a pie chart and (b) a line chart.

4 A mass is suspended from the centre of a bar. The mass required to bend the bar is noted, and the experiment is repeated 120 times. The results are recorded as follows:

Mass required to bend bar (kg)	Number of occurrences
70.1–70.2	36
70.3–70.4	17
70.5–70.6	21
70.7–70.8	30
70.9–80.0	16

Represent the data as a histogram.

5 Calculate the mean, median and mode of
(a) $-11, -3, 0, 1, 0, -2, -6$
(b) $0.36, 0.24, 0.36, 0.31, 0.25$

6 Calculate the standard deviation of the data given in Q5.

7 The event E is defined by

   E: the machine will break down within 3 months

(a) State $\bar{E}$.
(b) If $P(E)=0.2$, calculate $P(\bar{E})$.

8 Out of 20 000 gearboxes produced by a company, 217 were reported faulty within 12 months of being sold.
(a) Calculate the probability that a single gearbox picked at random will be faulty within 12 months of being sold.
(b) If two gearboxes are picked at random, calculate the probability that both will be faulty within 12 months of being sold.
(c) If two gearboxes are picked at random, calculate the probability that neither will develop any faults within 12 months of being sold.
(d) If two gearboxes are picked at random, calculate the probability that exactly one will develop faults within 12 months of being sold.

9 Components are manufactured by machines A, B and C: machine A makes 40% of the components, machine B makes 25% of the components and the remainder are made by machine C. Of those components made by machine A, 7% are faulty, of those made by machine B, 6% are faulty, and of those made by machine C, 2% are faulty. A component is picked at random. Calculate the probabilities of the following events:
(a) the component was manufactured by machine B;
(b) the component was not manufactured by machine A;
(c) the component was manufactured by machine A or is faulty;

(d) the component was manufactured by machine A and is not faulty;

(e) the component was manufactured either by machine A or by machine C;

(f) the component is not faulty and was made by machine C;

(g) the component is not faulty and was made either by machine A or by machine C.

10 The probability that a telephone line will develop a fault in a 12-month period is 0.02. Three lines are examined. E is the event 'none of the three lines will develop a fault within the next 12 months'.

(a) Calculate $P(E)$.

(b) State $\bar{E}$.

SI units and prefixes

SI units have been used throughout this book. The following is a list of these units together with their symbols:

Quantity	SI unit	Symbol	Quantity	SI unit	Symbol
Length	metre	m	Force	newton	N
Mass	kilogram	kg	Power	watt	W
Time	second	s	Electric charge	coulomb	C
Frequency	hertz	Hz	Potential difference	volt	V
Electric current	ampere	A	Resistance	ohm	Ω
Magnetic flux	weber	Wb	Magnetic flux density	tesla	T
Temperature	kelvin	K	Capacitance	farad	F
Energy	joule	J	Inductance	henry	H

Multiples of units are indicated by the following prefixes:

Multiple	Prefix	Symbol
10^{18}	exa	E
10^{15}	peta	P
10^{12}	tera	T
10^{9}	giga	G
10^{6}	mega	M
10^{3}	kilo	k
10^{2}	hecto	h
10^{1}	deca	da
10^{-1}	deci	d
10^{-2}	centi	c
10^{-3}	milli	m
10^{-6}	micro	μ
10^{-9}	nano	n
10^{-12}	pico	p
10^{-15}	femto	f
10^{-18}	atto	a

Unit Conversions

Length

1 millimetre (mm) = 0.0394 in
1 centimetre (cm) = 10 mm
1 metre (m)   = 100 cm
1 kilometre (km) = 1000 m

Surface or area

1 sq cm (cm^2)   = 100 mm^2
1 sq metre (m^2) = 10 000 cm^2

Capacity

1 cu cm (cm^3)   = 0.0610 cu in
1 cu decimetre (dm^3) = 1000 cm^3
1 cu metre (m^3)   = 1000 dm^3
1 litre (l)   = 1 dm^3

Weight

1 gram (g)   = 1000 mg
1 kilogram (kg) = 1000 g
1 tonne (t)   = 1000 kg

Solutions to Exercises

Self-assessment questions 1.2

7. False

Exercises 1.2

1. (a) 11 (b) -8 (c) -8 (d) -8 (e) 19
2. (a) -36 (b) -36 (c) 36 (d) -9 (e) -4 (f) 4
   (g) -24 (h) 2
3. 8400
4. 34
5. (a) 6, 14 (b) 4, 6 (c) 28, 52 (d) -4, 0 (e) -12, -8
   (f) 6 (g) 24
6. Negative
7. Positive
8. Negative
9. No
10. One or both must be zero
11. 8230, 7770
12. 352.93 mm, 352.47 mm
13. 2525 ± 75
14. 35.175 ± 0.152
15. (a) 16 (b) -16 (c) 8 (d) -20 (e) 24 (f) 20
    (g) $7! = 5040$

Self-assessment questions 1.3

2. True
3. (a) h.c.f. $= 1$ (b) l.c.m. $=$ product of the numbers
5. True
6. 1
7. Multiply them all together

Exercises 1.3

1. (a) 90 (b) 30 (c) 720 (d) 1200 (e) 600 (f) 1820
2. (a) 3 (b) 8 (c) 5 (d) 12 (e) 21 (f) 9
3. Only 13 is prime

Computer and calculator exercises 1.3

1. (a) $3 \cdot 3 \cdot 5 \cdot 7 \cdot 7 \cdot 11 \cdot 11$ (b) $3 \cdot 5 \cdot 7 \cdot 11 \cdot 13 \cdot 17 \cdot 19 \cdot 23 \cdot 29 \cdot 31 \cdot 37 \cdot 41$
   (c) 1 285 739 648 911
2. 197, 199, 211, 223

Self-assessment questions 1.4

3. False
4. True

Exercises 1.4

1. (a) $\frac{4}{9}$ (b) $\frac{4}{9}$ (c) -2 (d) $\frac{9}{8} = 1\frac{1}{8}$ (e) 1 (f) $\frac{17}{21}$
   (g) $-\frac{7}{5} = -1\frac{2}{5}$ (h) 3
2. (a) $\frac{5}{6}$ (b) $\frac{1}{6}$ (c) $\frac{17}{12}$ (d) $\frac{1}{6}$ (e) $\frac{113}{90}$ (f) $\frac{23}{70}$ (g) $\frac{6}{5}$
   (h) $\frac{27}{21}$ (i) $\frac{17}{8} = 2\frac{1}{8}$
3. (a) $\frac{3}{20}$ (b) $\frac{3}{2}$ (c) $\frac{9}{16}$ (d) $\frac{8}{3} = 2\frac{2}{3}$ (e) $\frac{3}{4}$ (f) $\frac{1}{3}$ (g) $\frac{27}{16}$
   (h) $\frac{81}{16} = 5\frac{1}{16}$
4. (a) 6 (b) 2 (c) $\frac{9}{8}$ (d) $\frac{3}{16}$ (e) $\frac{9}{2}$ (f) $\frac{9}{16}$ (g) $10\frac{1}{2}$
   (h) $2\frac{1}{3}$
5. (a) (b) proper (c) (d) (e) improper
6. (a) $2\frac{1}{2}$ (b) $2\frac{1}{3}$ (c) $-2\frac{3}{4}$ (d) $1\frac{1}{5}$ (e) $2\frac{2}{5}$ (f) $2\frac{4}{7}$
   (g) $5\frac{1}{3}$ (h) $9\frac{2}{9}$
7. (a) $\frac{9}{4}$ (b) $\frac{7}{2}$ (c) $\frac{17}{3}$ (d) $-\frac{17}{5}$ (e) $\frac{35}{3}$ (f) $\frac{74}{9}$
   (g) $\frac{67}{4}$ (h) $\frac{625}{7}$

Self-assessment questions 1.5

2. False. Division and multiplication have the same precedence
3. False

Exercises 1.5

1. (a) 4   (b) 4   (c) $10\frac{2}{5}$ (d) 4   (e) -3 (f) -10 (g) 65
   (h) $12\frac{2}{3}$ (i) $4\frac{1}{2}$ (j) $\frac{1}{2}$ (k) 2
2. (a) 36 900   (b) 900 000   (c) 490 000
3. (a) 3.5   (b) $\frac{108}{7} = 15.429$ (c) 12   (d) $\frac{1080}{11} = 98.182$ (e) 20
4. (a) $\frac{7}{36}$ (b) $-\frac{3}{8}$ (c) $2\frac{3}{28}$ (d) $1\frac{1}{14}$
5. (a) -2 (b) -4 (c) 2   (d) 4

Self-assessment questions 1.6

2. False

Exercises 1.6

1. (a) 1:2   (b) 3:20   (c) 1:2   (d) 6:4:1   (e) 2:3:10
   (f) 9:16
2. 21 type A and 56 type B
3. 840 kg m^{-3}
4. 0.8, 2.4, 3.2 m
5. 450 kg copper, 300 kg zinc
6. 361 kg copper, 19 kg aluminium
7. 1:1

Self-assessment questions 1.7

2. False
3. False
4. False

Exercises 1.7

1. (a) 2.7   (b) 76.5   (c) 82.5   (d) 375   (e) 60
2. max. 9.3279, min. 9.2721
3. (a) 20%   (b) 16%   (c) 80%   (d) 87.5%   (e) 120%
   (f) 27.27%   (f) 125%
4. 1.429% decrease
5. (a) 64.41   (b) 134.912   (c) 959.1   (d) 19.5
6. 2461 revs min^{-1}
7. max. 307.5 V, min. 292.5 V
8. 2.621 m
9. 4410 acceptable
10. max. 10 404 kg, min. 9996 kg
11. (a) 25.65 Ω, 28.35 Ω   (b) 423 Ω, 517 Ω   (c) 32.67 kΩ, 33.3 kΩ
    (d) 2.16 kΩ, 3.24 kΩ   (e) 666.4 kΩ, 693.6 kΩ
    (f) 2.178 Ω, 2.222 Ω

Review exercises 1

1. 56:24:16
2. (a) 7   (b) $3\frac{1}{3}$ (c) 18   (d) $3\frac{1}{2}$ (e) 25
3. (a) $\frac{2}{5}$ (b) $\frac{4}{5}$ (c) $\frac{4}{25}$ (d) $\frac{7}{6}$ (e) $\frac{2}{7}$
4. (a) $\frac{33}{26}$ (b) $\frac{4}{21}$ (c) $\frac{27}{24}$ (d) $\frac{1}{3}$ (e) $\frac{2}{3}$ (f) $\frac{53}{12}$ (g) $\frac{64}{35}$
5. 292.8 V
6. 11.85 V to 12.15 V
7. (a) 0.99 kΩ, 1.01 kΩ   (b) 31.35 Ω, 34.65 Ω
   (c) 26.73 Ω, 27.27 Ω   (d) 0.9 MΩ, 1.1 MΩ
   (e) 6.46 kΩ, 7.14 kΩ   (f) 98 Ω, 102 Ω
8. (a) 30.4   (b) -60.5 (c) 67.99   (d) 0.9   (e) 3.5
   (f) 0.9333
9. (a) 1   (b) 8   (c) 15
10. (a) 12   (b) 24   (c) 24   (d) 60
11. (a) 175 cm, 75 cm   (b) 125 cm, 125 cm
    (c) 62.5 cm, 125 cm, 62.5 cm   (d) 113.64 cm, 136.36 cm
12. (a) $\frac{13}{6}$ (b) $\frac{4}{9}$ (c) 1   (d) $-\frac{5}{6}$

13. (a) $\frac{4}{3}$ (b) $\frac{27}{4}$ (c) $\frac{8}{9}$ (d) $\frac{27}{64}$ (e) $\frac{3}{16}$ (f) $\frac{4}{3}$ (f) $2\frac{2}{3}$
(h) -1 (i) $2\frac{1}{2}$ (j) $\frac{3}{4}$ (k) $\frac{5}{2}$ (l) $\frac{125}{8}$

14. 2.96%

15. (a) $4\frac{4}{7}$ (b) $6\frac{1}{4}$ (c) $1\frac{1}{14}$ (d) $1\frac{1}{2}$

16. (a) 1 (b) $-1\frac{1}{3}$ (c) -1 (d) 11 (e) 17 (f) 0.75

Self-assessment questions 2.4

2. False

Exercises 2.4

1. (a) 216 (b) 16 (c) 16 (d) $\frac{3}{8}$ (e) -27 (f) 5184
(g) 12 (h) 100 (i) 1000 (j) 10000

2. (a) 11^3 (b) $5^2 6^2 7^2$ (c) 0.3^4 (d) $\frac{5^3}{6^3}$ (e) $(4)5^2 6^3$
(f) $\frac{3^2}{4^5}$ (g) $(\frac{1}{4})^3$ (h) 0.75^2

3. (a) $z^4 y^2$ (b) $a^3 b^2 x^4$ (c) $\frac{x^3}{y^4}$

7. (a) $16a^4$ (b) $9x^2$ (c) $\frac{x^2}{y^2}$ (d) $-64k^3$ (e) $16a^2 b^2$

Exercises 2.5

1. (a) 6^9 (b) 3^7 (c) 48^5 (d) 10^8 (e) 5^5 (f) 10^{23}
(g) $(-9)^{16}$ (h) $(-6)^7$

2. (a) $3^8 2^{12}$ (b) $4^4 3^6$ (c) $2^5 5^3$ (d) $3^7 4^8$

3. (a) x^7 (b) y^{10} (c) z^6 (d) t^{13} (e) a^4 (f) t^7 (g) b^{10}
(h) z^{14}

Exercises 2.6

1. (a) 100 (b) 81 (c) 0.5 (d) 1 (e) 124 (f) 361
(g) -27

2. (a) 36^4 (b) 17^5 (c) $7^3 \cdot 8$ (d) $9^2 5^6 7^3$ (e) $7^3 \cdot 12$
(f) $-5^2 \cdot 4$

3. (a) x^4 (b) y^4 (c) t^4 (d) z (e) v^7 (f) x^3

4. (a) 10 (b) 10^3 (c) x^3 (d) $\frac{x^7}{y^4}$ (e) $a^2 b^2$ (f) $9^9 \cdot 10$
(g) $x^3 y$ (h) abc

Self-assessment questions 2.7

1. False
2. False
3. False

Exercises 2.7

1. (a) $\frac{1}{6}$ (b) $\frac{1}{4}$ (c) $\frac{1}{81}$ (d) 64 (e) $\frac{1}{16}$ (f) $\frac{1}{125}$ (g) $-\frac{1}{27}$
(h) -64 (i) 0.1 (j) 0.01 (k) 0.001

2. (a) $\frac{1}{x^3}$ (b) $3x^5$ (c) t (d) $\frac{12}{ab^2}$ (e) $\frac{5^2}{x^3}$ (f) $\frac{y^2}{27x}$ (g) $\frac{1}{5^4}$

3. (a) 48 (b) $\frac{4}{9}$ (c) 1 (d) 4 (e) 25 (f) 1000 (g) 4
(h) $-\frac{1}{8}$

4. (a) t^{-3} (b) y^{-3} (c) $\frac{y}{2}$ (d) $-24t^{-6}$ (e) $\frac{1}{2t^5}$ (f) $\frac{4}{3}t^{-5}$
(g) $-\frac{t}{2}$

Exercises 2.8

1. (a) 5^{15} (b) 3^9 (c) 17^8 (d) y^{18} (e) y^3 (f) t^{-18}
(g) k^{12} (h) $(-1)^{12}=1$ (i) $(-1)^{12}=1$

2. (a) $\frac{1}{16}$ (b) $\frac{1}{4}$ (c) 81 (d) 36 (e) $\frac{25}{2}$ (f) $-\frac{1}{2}$ (g) $\frac{9}{4}$

3. (a) $4^5 6^9$ (b) $\dfrac{9a^2b^2}{c^6}$ (c) $\dfrac{b^2}{4^4a^6}$ (d) $8a^6b^3$ (e) $9x^2y^4z^6$
(f) $\dfrac{36}{a^2b^4}$ (g) $\dfrac{9}{x^4}$ (h) $\dfrac{8z^6}{27t^3}$ (i) $4x^2$ (j) $\dfrac{1}{4x^4}$ (k) $-\dfrac{x^6}{8}$

Self-assessment questions 2.9

1. $7^{0.3} = 7^{\frac{3}{10}} = \sqrt[10]{7^3} = \sqrt[10]{343}$
2. False
3. Yes. $2^{-3} = \frac{1}{8},\ (-2)^3 = -8$

Exercises 2.9

1. (a) 5.623   (b) 13.753   (c) 13.846   (d) 347.15   (e) $\frac{1}{3}$
(f) 0.0390   (g) 0.705
2. (a) $6^{\frac{3}{2}}$ (b) 5^2 (c) $10^{2.4}$ (d) $x^{\frac{2}{3}}$ (e) $2^{\frac{1}{3}}x^{\frac{2}{3}}$ (f) $a^{\frac{3}{2}}$
(g) $a^{\frac{1}{2}}b$
3. (a) $4^{-\frac{3}{2}}$ (b) $3^{\frac{1}{4}}$ (c) $7^{\frac{8}{3}}$ (d) $19^{\frac{1}{2}}$ (e) $a^{-3}b^{\frac{1}{2}}$ (f) k^4
4. (a) $5^{\frac{1}{6}}b^{\frac{1}{6}}$ (b) $27x^2$ (c) $3x^{\frac{3}{2}}$ (d) $3^{\frac{3}{2}}x^{\frac{3}{2}}$
5. (a) $x^{\frac{5}{6}}$ (b) $x^{\frac{1}{6}}$ (c) $x^{\frac{1}{6}}$ (d) $2x$ (e) $\pm 5y$ (f) $\frac{3}{t}$ (g) $\pm 2y$
(h) x^3 (i) a^4 (j) $a^{-\frac{3}{2}}$

Review exercises 2

1. $\delta p = 0.2$
2. $\delta x = -0.1$
3. $\delta x = x_2 - x_1$
4. (a) 10^{17} (b) 10^{11} (c) 6^{12} (d) 7^3 (e) $\frac{1}{49}$ (f) $5^{5.5}$
(g) $6^{3.5}$ (h) $3^{\frac{10}{3}}$
5. (a) 4096   (b) ± 4 (c) ± 4 (d) $\frac{1}{25}$ (e) $\pm\frac{1}{5}$ (f) $\pm\frac{1}{216}$
(g) 25   (h) $\frac{1}{5}$ (i) ± 1000 (j) 625   (k) $\frac{1}{25}$

6. (a) x^{10} (b) y^{15} (c) $3t^{-2}$ (d) $\dfrac{y^2}{5}$ (e) $4x$ (f) $\dfrac{1}{2x}$
(g) $\dfrac{1}{3a^2}$ (h) ab^3c^2 (i) k^2 (j) y^6 (k) $4t$ (l) $16t^2$

7. (a) c^{-6} (b) x^{13} (c) z^{-6} (d) 10^{17} (e) $\dfrac{1}{10^3}$ (f) y^6
(g) x^{-1} (h) $8ab^2c^3$

9. (a) 9   (b) $\frac{1}{9}$ (c) $\pm\frac{1}{4}$ (d) ± 12 (e) ± 8 (f) $\pm\frac{5}{8}$
(g) 4   (h) 8   (i) ± 4 (j) ± 8 (k) ± 27 (l) 100
(m) ± 0.1 (n) 100
10. (a) $9x^2$ (b) $-27x^3$ (c) $-27x^6$ (d) $3^{-2}x^{-2}$
(e) $-3^{-3}x^{-3}$ (f) $\dfrac{1}{9x^4}$

Self-assessment questions 3.2

1. (a) T   (b) F   (c) T   (d) F   (e) F   (f) T   (g) F
(h) F   (i) T   (j) F   (k) F   (l) T

Exercises 3.2

1. (a) $\{a,b,c,d,e,f,h\}$ (b) $\{f\}$ (c) $\{a,c,d,e,f\}$ (d) $\{e,f\}$
(e) $\{f\}$ (f) $\{a,c,d,e,f,h\}$
2. (a) $\{1,3,5,7,9,11\}$ (b) $\{20,22,24,26,28,30\}$
3. (a) $\{0,1,2,3,4,8\}$ (b) $\{1,3,5,7,9\}$ (c) $\{0,1,2,3,4,5,7,8,9\}$
(d) $\{1,3\}$ (e) $\{0,2,4,5,6,7,8,9\}$ (f) $\{1,3\}$
(g) $\{0,1,2,3,4,5,7,8,9\}$ (h) $\{1,3,5,6,7,9\}$ (i) $\{0,2,4,8\}$
4. Figure 3.2.4
5. (a) $\{3,4\}$ (b) $\{0,1,2,3,4,5,6,7,8\}$ (c) $\{3\}$
(d) $\{0,3,4,6,9,12,15,18,21,24\}$ (e) $\{3,4,6\}$

Self-assessment questions 3.3

1. (a) $\mathbb{N}$ (b) $\mathbb{Z}^+$ (c) $\mathbb{Z}$
2. T

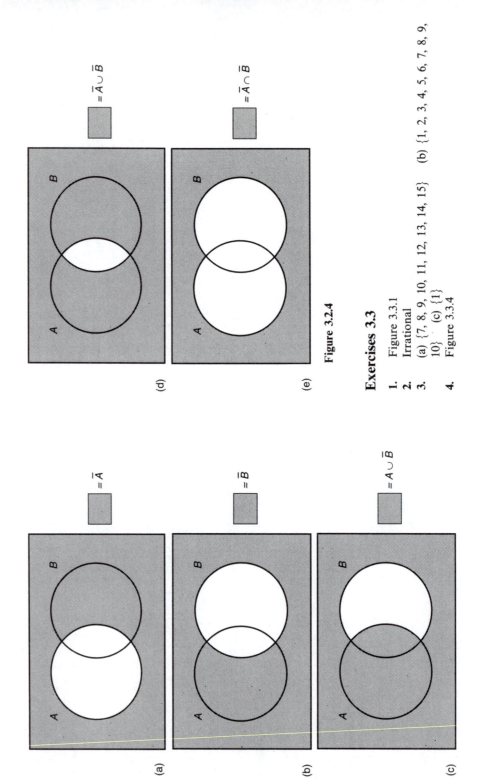

Figure 3.2.4

(a) $= \bar{A}$

(b) $= \bar{B}$

(c) $= A \cup \bar{B}$

(d) $= \bar{A} \cup \bar{B}$

(e) $= \bar{A} \cap \bar{B}$

Exercises 3.3

1. Figure 3.3.1
2. Irrational
3. (a) $\{7, 8, 9, 10, 11, 12, 13, 14, 15\}$ (b) $\{1, 2, 3, 4, 5, 6, 7, 8, 9, 10\}$ (c) $\{1\}$
4. Figure 3.3.4

5. (a) open (b) open (c) half-open/half-closed (d) closed
(e) open (f) open (g) half-open/half-closed

Self-assessment questions 3.4

1. F
2. F

Exercises 3.4

1. (a) 3.9×10^3 (b) 4.0010×10^4 (c) 1.06×10^{-3}
(d) -1.0×10^{-6} (e) 5×10^{-3} (f) 7.6×10^{-1} (g) 1.316
2. (a) 8×10^5 (b) 5.175×10^1 (c) 6.426×10^{-3} (d) 2×10^4
(e) 2.421 (f) 1.3434×10^2
3. (a) 15 000 (b) 18 000 (c) 22 000 (d) 47 000 (e) 100 000
(f) 120 000 (g) 150 000 (h) 180 000 (i) 220 000
(j) 330 000 (k) 470 000 (l) 1 000 000 (m) 1 200 000
(n) 1 500 000 (o) 2 200 000 (p) 3 300 000 (q) 4 700 000

Self-assessment questions 3.5

1. F
2. F

Exercises 3.5

1. (a) 73.09 (b) 73.1 (c) 73.1 (d) 73 (e) 70
2. (a) 0.102 55 (b) 0.1025 (c) 0.103 (d) 0.10 (e) 0.1
3. (a) 0.009 (b) 0.0091 (c) 0.009 07 (d) 0.009 075

Review exercises 3

1. (a) $\{17, 19, 21\}$ (b) $\{18, 20\}$ (c) $\varnothing$ (d) $\mathbb{Z}$ (e) $\mathbb{Z}$
2. Figure R3.2

(a)

$A = \{x : x \in \mathbb{R}, -1 \leqslant x \leqslant 7\}$

(b)

$C = \{x : x \in \mathbb{R}^+, x \leqslant 7\}$

Figure 3.3.1

(a)

(b)

(c)

(d)

Figure 3.3.4

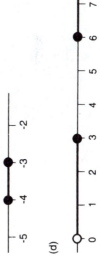

3. (a) $\{-5, -4, -3, -2, -1, 0, 1, 2, 3, 4, 5, 6, 7, 8, 9, 10\}$ (b) $\{-1\}$
(c) $\{-2, -1, 0, 1, 2, 5, 7\}$ (d) $\{-5, -4, -3, -2, 0, 3, 4, 6, 8, 9, 10\}$
(e) $\{-2, 0\}$ (f) $\{-5, -4, -2, 0, 3, 6, 8, 10\}$ (g) $\{-1\}$
(h) $\{-5, -4, -3, -2, 0, 1, 2, 3, 4, 5, 6, 7, 8, 9, 10\}$

4. (a) $\{0, 1, 2, 5, 7, 8, 9\}$ (b) $\{1, 2, 9\}$ (c) $\{3, 4, 6, 7, 8\}$
(d) $\{0, 3, 4, 5, 6\}$ (e) $\{3, 4, 6\}$ (f) $\{3, 4, 5, 6, 7, 8\}$ (g) $\{0, 3, 4, 5, 6, 7, 8\}$
(h) $\{0, 3, 4, 5, 6, 7, 8\}$
Note that $\overline{A \cup B} = \bar{A} \cap \bar{B}$ and $\overline{A \cap B} = \bar{A} \cup \bar{B}$. These are De Morgan's laws.

7. (a) $\{7, 8, 9, 10\}$ (b) $\{1, 2, 3, 4, 5, 6, 7, 8, 9, 10\}$
(c) $\{-2, -1, 0, 1, 2, 3, 4, 5, 6, 7, 8, 9, 10\}$

8. Figure R3.8

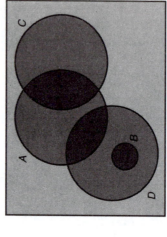

Figure R3.8

9. (a) -6.5 (b) -7 (c) -6.546 (d) -6.546 (e) -6.55
10. (a) 10.00 (b) 10.0 (c) 10 (d) 10
11. (a) 0.10 (b) 0.1 (c) 0.099 (d) 0.1
12. (a) 7.6×10^1 (b) 7.63×10^1 (c) 7.63×10^2 (d) 7.6×10^{-1}
(e) 7.63×10^{-1} (f) 3.960×10^3 (g) 1×10^{-6}
(h) 4.000001×10^6
13. (a) 3.6×10^5 (b) 9.6×10^4 (c) 1.10515×10^{10}
(d) 5.04×10^{-4} (e) 4.6875×10^{-1} (f) -1.4391×10^1
(g) 1.8934×10^{-8}

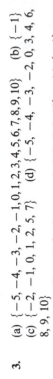

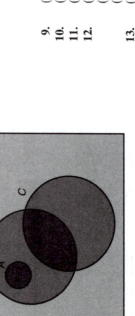

Figure R3.2

Self-assessment question 4.2

1. F

Exercises 4.2

1. (a) $8x$ (b) $4t$ (c) $3a+2b$ (d) $2ab$ (e) $\dfrac{7a}{6}$ (f) $\dfrac{2b}{3}$

   (g) $xy-xyz$ (h) $27xy$ (i) $3x^3$ (j) $5yx$ (k) $-13ab^2$

   (l) $a+\dfrac{5b}{2}$ (m) $-4a^2b^2c$ (n) 0

2. (a) x (b) $\dfrac{y}{6}$ (c) $\dfrac{2t}{3}$ (d) $\dfrac{ab}{6}$ (e) $\dfrac{2z}{7}$ (f) $\dfrac{7y}{2}$ (g) $\dfrac{13}{6}t$

Exercises 4.3

1. (a) $3x+6$ (b) $-3x-6$ (c) $3x+x^2$ (d) $-3x-x^2$
   (e) $-3x+6$ (f) x^2+2x (g) $6+5x+x^2$ (h) $6+x-x^2$
   (i) x^2-x-6 (j) $3+4x-4x^2$ (k) $12x^2-25x-7$
   (l) $-12x^2-41x-15$ (m) a^2-b^2 (n) x^2+8x+7
   (o) $x^2+8x+15$ (p) y^2+5y+6 (q) t^2+2t-3
   (r) z^3+2z^2+2z+4 (s) $6-v-v^2$ (t) $8x^2+10x+3$
   (u) $9y^2-1$ (v) $-2t^2+15t-7$

2. (a) x^2+3x+2 (b) $x^3+6x^2+11x+6$

3. (a) t^3+3t^2+2t (b) $2a^2-2a-12$ (c) $6t^2-3t-3$
   (d) t^3-2t^2-t+2 (e) $6x^3-17x^2-5x+6$ (f) x^3+7x^2+12x

4. $18*Y/4+6*Z/4-0.5$

5. (a) $5x+13$ (b) $10t-9$ (c) $11y+22$ (d) $9v-2$ (e) $x-\dfrac{7}{4}$
   (f) $\dfrac{31y}{6}-2$

6. (a) $2x+4$ (b) $t+4$ (c) $4x+5$ (d) $8y^2-17y+22$
   (e) $2a^2+2b^2$ (f) $4ab$

Self-assessment questions 4.4

3. For example, x^2+x+1 cannot be factorized.

Exercises 4.4

1. (a) $3(1+2x)$ (b) $3(2x-1)$ (c) $8(x+1)$ (d) $3(2+x+3y)$
   (e) $2(x+2t+3v)$ (f) $x(y+z)$ (g) $b(a-2c)$ (h) $5s(1-3t)$

2. (a) $(x+3)(x+2)$ (b) $(x-3)(x-2)$ (c) $(x+3)(x-2)$
   (d) $(x+2)(x-3)$ (e) $(x+7)(x+2)$ (f) $(x+7)(x-2)$
   (g) $(x-7)(x+2)$ (h) $(x-7)(x-2)$ (i) $(x+4)(x+4)=(x+4)^2$
   (j) $(x+5)(x-5)$ (k) $(x+5)(x-3)$ (l) $(x+5)(x+3)$
   (m) $(x-5)(x-3)$ (n) will not factorize (o) will not factorize

3. (a) $y(y^2+1)$ (b) $y^2(y+1)$ (c) $x^2(2x^2+1)$ (d) $v^2(v-2)$
   (e) $2v^2(3v-1)$ (f) $3a(a-2+4a^2)$ (g) $ab(c-a+2b)$
   (h) $5xyz(2y-4x+3z)$

4. (a) $2(x+5)(x-5)$ (b) $(3x-2)(x+4)$ (c) $(4x+3)(2x-3)$
   (d) $(x-4)(9x-1)$ (e) $(2x-3)(2x+1)$ (f) $(2x+1)(2x-1)$
   (g) $(x+6)(2x+3)$ (h) $(x-1)(3x+2)$

Exercises 4.5

1. (a) $\dfrac{3t+2}{t}$ (b) $\dfrac{x^2+1}{x}$ (c) $\dfrac{5z+3}{2(z+1)}$ (d) $\dfrac{a^3+2a-1}{a^2}$

2. (a) $\dfrac{3t^2+2}{t^3}$ (b) $\dfrac{7x}{6}$ (c) $\dfrac{3xy-2+y^2}{y^2}$ (d) $\dfrac{2x+3}{x^2+3x+2}$

   (e) $\dfrac{5x+3}{(x+1)(x+4)}$ (f) $\dfrac{x+10}{(x-2)^2}$ (g) $\dfrac{7x+41}{(x-3)(x-1)(x+7)}$

   (h) $-\dfrac{x(x+8)}{(x-1)^2(x+2)}$ (i) $-\dfrac{x}{(x+5)(x+1)(x+2)}$

3. (a) $\dfrac{3x}{2}$ (b) $3a^2b$ (c) $2x^2$ (d) $\dfrac{3y}{2x}$ (e) $3z$ (f) z^2

   (g) $-\dfrac{3x(y^2-1)}{y}$

4. (a) $\dfrac{x+2}{x+3}$ (b) $\dfrac{x-1}{x+1}$ (c) $\dfrac{x+1}{x+2}$ (d) $\dfrac{a-4}{a-3}$ (e) $\dfrac{2(x+2)}{x+1}$

5. (a) $\dfrac{2x}{x-1}$ (b) $\dfrac{2(x+2)}{(x+3)(x+1)}$ (c) $-\dfrac{2(x-1)}{x(x+2)}$ (d) $\dfrac{x}{x+y}$

   (e) $\dfrac{x^2+1}{x^2-1}$

Computer and calculator exercises 4.6

1. (a) $(x-4)^2(x-2)$ (b) $(x-2)(x+8)(2x-1)$
   (c) $(x-4)(x-3)(x+3)(x+4)$ (d) $(s-7)(s-2)(s+3)$

2. $x(x+1)(x+8)$

3. $\dfrac{x(3x^2+16x+8)}{(x+5)(x+4)(x+2)(x-13)}$

4. $(a-b)(a^2+ab+b^2)$

Review exercises 4

1. (a) $3(r+5)$ (b) $t^4(t-1)(t+1)$ (c) $(x-5)(x+3)$
   (d) $x(x-5)(x+3)$ (e) $y^3(y+1)(y+5)$ (f) $(x-7)(x+5)$
   (g) $(t-6)(t-3)$ (h) $(4y+1)(3y-4)$ (i) $\dfrac{1}{t}\left(\dfrac{1}{t}+1\right)$
   (j) $3(y-1)(y+4)$

2. (a) $10xy^2$ (b) $\dfrac{8}{t}$ (c) $\dfrac{1}{x+1}$ (d) $\dfrac{x+1}{x+2}$ (e) $\dfrac{3}{x+7}$ (f) $x+3$
   (g) $\dfrac{x-2}{x}$

3. (a) $3a+3b$ (b) $-12a+18b$ (c) t^2-5t-6 (d) $t+1$
   (e) $3t+2+\dfrac{6}{t}$ (f) $ac+bc+ad+bd$ (g) a^2-b^2
   (h) b^2-a^2 (i) $x^3+3x^2-4x-12$

4. (a) $\left(\dfrac{1}{t}-9\right)\left(\dfrac{1}{t}-4\right)$ (b) $\dfrac{(9t-1)(4t-1)}{t^2}$

5. (a) $\dfrac{3}{2x}$ (b) $\dfrac{3}{x}$ (c) $\dfrac{1+2x^2}{x}$ (d) $\dfrac{2+x^2}{2x}$ (e) $\dfrac{x^2-x-3}{(x+1)(x+2)}$
   (f) $-\dfrac{2}{y^2-1}$ (g) $\dfrac{7x-10}{(2x+1)(x-4)}$ (h) $\dfrac{4y^2-3x+y}{y^2}$
   (i) $-\dfrac{x}{2x+1}$

6. (a) $3ab(1+2ab)$ (b) $(a+x)(b+c)$ (c) $x^2(x-3)(x+2)$
   (d) $\left(\dfrac{1}{y}-1\right)\left(\dfrac{1}{y}+1\right)$ (e) $x^{\frac{1}{2}}(x+3)(x+1)$

Self-assessment question 5.2

1. Both are correct if appropriate units are quoted.

Exercises 5.2

1. (a) 18.85 m (b) 31.42 cm
2. (a) 30 V (b) 2.1 V
3. (a) $A=lw$ (b) 15 cm² (c) 1 m²
4. (a) 24 cm² (b) 2.25 m²
5. (a) 18 m (b) 32.5 m
6. (a) 4.91 s (b) 1.42 s
7. (a) 2 kg m² (b) 0.000 075 kg m² (c) 0.000 296 kg m²
8. (a) 7500 J (b) 592.064 J (c) 375.85 J
9. (a) 49 J (b) 1.536 J (c) 1875.6 J

Computer and calculator exercises 5.2

2. 797.6825 J

Self-assessment questions 5.3

3. For example, $y=x^3+x$.

Exercises 5.3

1. (a) $r=\dfrac{C}{2\pi}$ (b) $R=V/I$ (c) $h=2A/b$ (d) $b=\sqrt{\dfrac{3V}{h}}$
   (e) $x=9y^2$ (f) $l=\dfrac{gT^2}{4\pi^2}$ (g) $y=\dfrac{7-x}{x}$ (h) $x=\dfrac{7}{1+y}$
   (i) $b=\dfrac{Ra}{a-R}$ (j) $b=\dfrac{a}{c-a}$

2. (a) $t=\dfrac{c-b}{a}$ (b) $t=\sqrt{\dfrac{c-b}{a}}$ (c) $t=\dfrac{a}{c-b}$ (d) $t=\dfrac{c}{a+b}$
   (e) $t=\dfrac{c-a}{b}$ (f) $t=\dfrac{c-a}{b}$ (g) $t=\sqrt{\dfrac{c-a}{b}}$ (h) $t=\sqrt{\dfrac{b}{c-a}}$
   (i) $t=\sqrt[3]{\dfrac{c-b}{a}}$ (j) $t=\sqrt[n]{\dfrac{c-b}{a}}$ (k) $t=\sqrt{\dfrac{c-a}{b}}$

3. (a) $y=3/x$ (b) $y=\dfrac{3}{x}-1$ (c) $y=\sqrt{\dfrac{3-x^2}{x}}$ (d) $y=\dfrac{x}{1-x}$
   (e) $y=\dfrac{x-2}{1-x}$

4. (a) $P=V^2/R$ (b) $P=I^2R$

5. $v_2=\sqrt{2g\left(\dfrac{p_1}{\rho g}-\dfrac{p_2}{\rho g}+\dfrac{v_1^2}{2g}+h_1-h_2\right)}$

6. $T_2=T_1-\dfrac{Qd}{kA}$

7. $T=\sqrt[4]{\dfrac{R}{k}}$.

Computer and calculator exercises 5.3

1. (a) $I=\sqrt{\dfrac{P}{R}}$ (b) $R=P/I^2$

2. $u=\dfrac{s-0.5at^2}{t}$

Exercises 5.4

1. (a) $V=\dfrac{kx^2}{\sqrt{y}}$ (b) $k=3$ (c) 16

2. $k=10$ (a) 0.4 W   (b) 0.91 m

3. $\frac{4}{3}$

Review exercises 5

1. (a) 15.71   (b) 8.98   (c) 56.25   (d) 3.61   (e) 1.31

2. $h=\dfrac{1}{2g}\left[\left(\dfrac{A_1}{A_2}\right)^2-1\right]\left(\dfrac{q}{A_1}\right)^2$

3. $N=\sqrt{\dfrac{IL}{\mu A}}$

4. (a) $a=(H-c)/b$ (b) $h=\dfrac{S-2\pi r^2}{2\pi r}$ (c) $b=V^2/a$
   (d) $b=V^2-a$ (e) $a=\dfrac{m-bc}{n^2}$ (f) $n=\sqrt{\dfrac{m-bc}{a}}$
   (g) $a=\dfrac{b(1+P^2)}{P^2-1}$ (h) $a=\dfrac{b}{1-b}$ (i) $x=5y-6$

5. (a) $L=\dfrac{kab^2}{c}$ (b) $k=2, L=9$

Self-assessment questions 6.2

1. F

Exercises 6.2

1. Multiply the input by -12 and add 12 to the result.
   (a) 0   (b) -12 (c) 48   (d) 12   (e) 9

2. (a) $f(x)=x^2-5$ (b) $g(x)=-\dfrac{x^3}{7}$ (c) $v(x)=2x+x^2$

3. Multiply the input by ρg. Figure 6.2.3

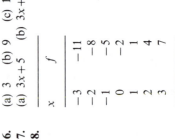

Multiply by
ρg

$h \longrightarrow \boxed{\quad} \longrightarrow p = \rho g h$

Figure 6.2.3

Self-assessment questions 6.4

4. The dependent variable is plotted on the vertical axis.

Exercises 6.4

1. Domain $[-3, 8]$, range $[-6, 5]$
   Figure 6.4.1
2. Domain $[-3, 8]$, range $[-1, 10]$
3. Domain of both g and h is $[-5, 4]$. Range of g is $[-8, 19]$; range of h is $[-13, 23]$.
4. Domain of both g and h is $[-4, 4]$. Range of g is $[1, 17]$; range of h is $[-1, 15]$. Figure 6.4.4
5. Domain $[0.5, 5]$, range $[0.04, 4]$

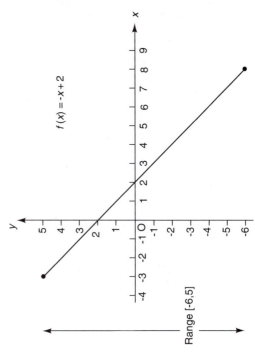

$f(x) = -x + 2$

Range $[-6,5]$

Figure 6.4.1

Exercises 6.3

1. (a) x; subtract 1 from the input and cube the result.
   (b) t; cube the input and subtract 1 from the result.
   (c) x; subtract 1 from the input and find the reciprocal of the result.
   (d) x; find the reciprocal of the input and subtract 1 from it.
   (e) x; divide the input by one less than the input.

2. (a) 40 (b) 1 (c) $20x^2 - 4x + 1$ (d) $5x^4 - 2x^2 + 1$
   (e) $20x^2 - 24x + 8$

3. (a) 6 (b) 71 (c) $\dfrac{1}{x^3} + 7$ (d) $x^6 + 7$ (e) $x^3 + 3x^2 + 3x + 8$

4. (a) $3t + 10$ (b) $3t - 2$ (c) $3t + 6$ (d) $3t + 2$
6. (a) 3 (b) 9 (c) 17 (d) 9
7. (a) $3x + 5$ (b) $3x + 8$ (c) $3x + 14$
8.

x	f
-3	-11
-2	-8
-1	-5
0	-2
1	1
2	4
3	7

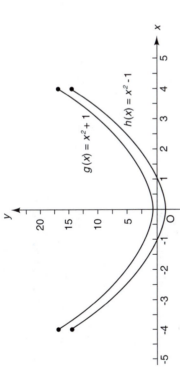

Figure 6.4.4

Graph showing $g(x) = x^2 + 1$ and $h(x) = x^2 - 1$

Review exercises 6

1. (a) 6   (b) 17   (c) $3x^2 + 2x + 1$ (d) $3\alpha^2 + 2\alpha + 1$

   (e) $3\alpha^2 + 8\alpha + 6$ (f) $\dfrac{3}{\alpha^2} + \dfrac{2}{\alpha} + 1$

2. (a) $\dfrac{3}{2-t}$ (b) $\dfrac{3}{2-\lambda}$ (c) $\dfrac{3}{2-2\lambda}$

3. Domain $[-3, 3]$, range $[-15, 3]$

4. Intersect at $(3, 2)$

5. Domain $\{x : x \in \mathbb{R}, x > 0\}$, i.e. $(0, \infty)$. Range $\{y : y \in \mathbb{R}, y > 0\}$, i.e. $(0, \infty)$

6. (a) 1   (b) $\frac{1}{2}$ (c) $\dfrac{1}{\alpha}$ (d) $\dfrac{1}{2\alpha}$ (e) $\dfrac{1}{\alpha^2}$ (f) $\dfrac{1}{\beta-1}$ (g) $\dfrac{1}{\beta+1}$

   (h) t

Self-assessment questions 7.1

4. 6

Exercise 7.1

1. (a) 2   (b) is not   (c) 1   (d) 0   (e) 0   (f) 2   (g) 4
    (h) is not   (i) 6   (j) is not

Self-assessment questions 7.2

2. All polynomials are rational functions.

3. A rational function is a polynomial if the denominator polynomial
    is a constant.

Self-assessment questions 7.3

1. No. When $x = 0$, $|x| = 0$

Exercises 7.3

1. Figure 7.3.1
2. Figure 7.3.2
3. $\frac{1}{2}$

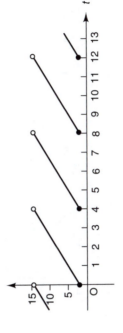

$f(t) = 3t + 2 \qquad 0 \leqslant t < 4$

Period 4

Figure 7.3.1

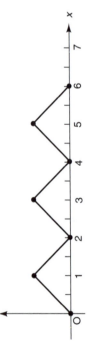

$$f(x) = \begin{cases} x & 0 \leqslant x < 1 \\ 2-x & 1 \leqslant x < 2 \end{cases} \quad \text{Period } T = 2$$

Figure 7.3.2

Computer and calculator exercises 7.3

1. (b) $-5, -2, 0, 2$ (c) 1 peak, 2 troughs
2. (b) $-2, 3.5, 3$
3. (b) When x is close to -3 and $+2$ values of y become infinitely large and the graph jumps at these points.
4. Intersect at $(5, -2)$
5. $(1.5, 4.25)$

Review exercises 7

1. (b), (c), (d) are polynomials

Exercises 8.2

1. (a) many-to-one   (b) one-to-one   (c) one-to-one
   (d) many-to-one
2. (a) one-to-one   (b) many-to-one   (c) many-to-one
   (d) one-to-one

Computer and calculator exercise 8.2

1. (a) one-to-one   (b) many-to-one   (c) one-to-one

Self-assessment questions 8.3

2. F
3. $f(f(t)) = t$

Exercises 8.3

1. Figure 8.3.1
2. (a) $x^2 + 3$ (b) $6x^2 + 18$ (c) $(6x+3)^2$ (d) $x+6$
3. $f(g(x)) = \dfrac{5x-7}{10x-13}$ $g(f(x)) = \dfrac{5x}{2x+1} - 7$
4. (a) no   (b) yes
5. $g(f(v)) = 160v$
6. (a) x (b) $x^{\frac{3}{2}}$ (c) $x^{\frac{1}{4}}$ (d) x (e) x^6 (f) x^4 (g) $x^{\frac{3}{2}}$
   (h) x^6 (i) x^9

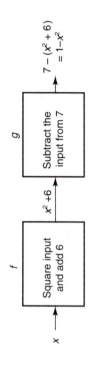

$$g(f(x)) = 1 - x^2$$

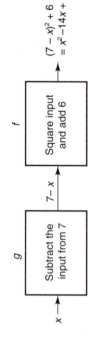

$$f(g(x)) = x^2 - 14x + 55$$

Figure 8.3.1

Self-assessment questions 8.4

1. Only if f is one-to-one
2. Yes, e.g. $f(t) = 2/t$

Exercises 8.4

1. (a) $f^{-1}(x) = \dfrac{x-2}{7}$ (b) $g^{-1}(x) = 3/x$ (c) $h^{-1}(x) = \dfrac{1+x}{x}$

2. (a) $f^{-1}(x) = \dfrac{x+7}{4}$ (b) has no inverse   (c) $g^{-1}(x) = \sqrt[3]{x}$

   (d) $f^{-1}(t) = \dfrac{1-4t}{2t}$ (e) and (f) have no inverses

3. (a) $f^{-1}(x) = \dfrac{3-x}{7}$ (b) $f^{-1}(x) = \dfrac{1}{x-5}$ (c) $f^{-1}(x) = x$

4. (a) $y^{-1}(x) = 2 - x$ (b) $y^{-1}(x) = 6 - 2x$ (c) $y^{-1}(x) = 3 - 2x$

   (d) $y^{-1}(x) = \frac{3}{2}(1-x)$ (e) $y^{-1}(x) = \dfrac{3-x}{x}$ (f) $y^{-1}(x) = \dfrac{3}{x-1}$

   (g) $y^{-1}(x) = \dfrac{3x}{1-x}$

Computer and calculator exercises 8.5

1. (a) $-3, -2, -1$ (b) $0, 1, 2$ (c) $-6, -5, -4$

Self-assessment questions 8.6

1. No
2. No

Exercises 8.6

1. Figure 8.6.1. Discontinuous at $x = 1$
2. Figure 8.6.2. Discontinuous at $x = 3$

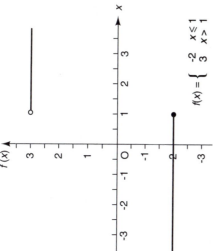

$$f(x) = \begin{cases} -2 & x \le 1 \\ 3 & x > 1 \end{cases}$$

Figure 8.6.1

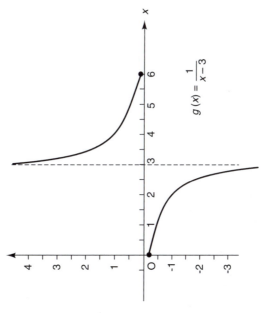

$$g(x) = \frac{1}{x-3}$$

Figure 8.6.2

3. Discontinuous at $t=0$

4. (a) 4 (b) 6 (c) no limit (d) 3 (e) 3 (f) 3

5. $k=3$

6. (a) Figure 8.6.6 (b) (i) 0.5 (ii) 0.5 (iii) 1 (iv) 0.5

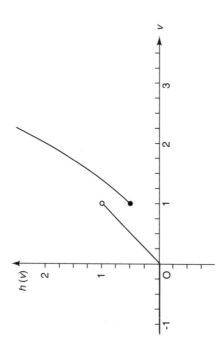

Figure 8.6.6

Review exercises 8

1. (a) many-to-one (b) many-to-one (c) one-to-one
(d) one-to-one (e) many-to-one

2. (a) $x^{-1}(t)=2/t$ (b) $y^{-1}(t)=3t-1$ (c) $z^{-1}(t)=1-t$

(d) $\dfrac{6}{t+1}$ (e) $\dfrac{2-t}{3}$ (f) $\dfrac{6}{2-t}$

3. Discontinuous at $t=1, t=3$

5. (a) $g^{-1}(x)=2(x-1)$ (b) $h^{-1}(x)=\dfrac{x+2}{4}$ (c) $q(x)=2x$

(d) $q^{-1}(x)=x/2$ (e) $h^{-1}(g^{-1}(x))=x/2$

---

6. (a) 3 (b) 5 (c) 3 (d) 3 (e) 3 (f) 5 (g) 5 (h) 3
(i) 4. Discontinuous at $x=4$

Exercises 9.3

1. (a) $x=\tfrac{1}{2}$ (b) $x=-2$ (c) $x=0$ (d) $x=\tfrac{1}{10}$ (e) $x=b/a$
(f) $x=\tfrac{7}{5}$ (g) $t=-\tfrac{17}{2}$

2. (a) $x=4$ (b) $x=7$ (c) $s=7$ (d) $t=29$ (e) $x=\tfrac{8}{3}$
(f) $x=\tfrac{8}{3}$ (g) $x=-\tfrac{2}{11}$

3. (a) $-\tfrac{7}{4}$ (b) $t=\tfrac{20}{3}$ (c) $t=\tfrac{3}{17}$ (d) $x=\tfrac{1}{2}$ (e) $x=\tfrac{15}{4}$
(f) $x=\tfrac{14}{3}$ (g) $x=\tfrac{9}{5}$ (h) $p=4$

4. $\delta T = 21.3\,^{\circ}\text{C}$

5. $\delta T = -133.3\,^{\circ}\text{C}$

6. (a) $t=-\tfrac{5}{4}$ (b) $z=-25$ (c) $x=-\tfrac{1}{2}$

7. (a) $x=-3$ (b) $x=10$ (c) $x=-\tfrac{47}{3}$ (d) $x=\tfrac{1}{23}$
(e) $x=-\tfrac{5}{2}$ (f) $x=-\tfrac{5}{4}$ (g) $x=-1$

Self-assessment questions 9.4

3. When $b^2-4ac=0$

Exercises 9.4

1. (a) 0, -16 (b) 0, -4 (c) 0, 2 (d) -1, 3 (e) -2, 3
(f) 8, -9 (g) -4, -5 (h) 7, -6 (i) -3, 7 (j) -4 twice
(k) 3, 6 (l) -1, -10 (m) 4, 8 (n) 3, -6 (o) $\tfrac{1}{3}$, $-\tfrac{2}{3}$
(p) $-\tfrac{4}{3}$, 3 (q) 11, -11 (r) $\tfrac{1}{3}$, -4 (s) $-\tfrac{3}{4}$, $\tfrac{3}{2}$
(t) $-\tfrac{1}{6}$, $-\tfrac{3}{2}$ (u) $-\tfrac{4}{3}$, $\tfrac{1}{2}$ (v) $-\tfrac{5}{4}$, $-\tfrac{3}{2}$

2. (a) -0.257, 2.591 (b) 1.702, -4.702 (c) -0.386, 3.886
(d) 6, -11 (e) -1.541, 4.541 (f) -0.209, -4.791
(g) 0.193, -5.193 (h) 0.209, 4.791 (i) -0.193, 5.193
(j) 0.314, 3.186 (k) -0.266, 3.766 (l) -0.314, -3.186
(m) 0.266, -3.766 (n) -0.258, -7.742 (o) 0.243, -8.243
(p) 0.258, 7.742 (q) -0.243, 8.243

3. $7\,\text{m} \times 10\,\text{m}$

4. $8.15\,\text{m}$

5. (a) 3, 3 (b) −0.618, 1.618
6. (a) −1.541, 4.541 (b) 2, 5 (c) 8, −16

Exercises 9.5

1. (a) $x = \frac{11}{7}, y = \frac{27}{7}$ (b) $x = \frac{8}{15}, y = \frac{8}{5}$ (c) $x = \frac{9}{13}, y = -\frac{37}{13}$
 (d) $x = 3, y = 21$ (e) $x = 4, y = 3$ (f) $x = 3, y = 1$
 (g) $x = -2, y = 3$ (h) $x = -1, y = 0$ (i) $x = \frac{3}{2}, y = \frac{1}{2}$
 (j) $x = -2, y = \frac{3}{2}$ (k) $x = -\frac{1}{4}, y = \frac{6}{5}$ (l) $x = \frac{5}{4}, y = \frac{1}{4}$
 (m) $x = 9, y = 11$ (n) $x = \frac{12}{5}, y = \frac{9}{5}$
2. (a) $A = \frac{3}{5}, B = -\frac{4}{5}$ (b) $A = \frac{3}{22}, B = \frac{5}{22}$
3. (a) $I_1 = \frac{14}{13}A, I_2 = \frac{8}{13}A$
 (b) $I_1 = \frac{5}{4}A, I_2 = -\frac{1}{3}A$

Exercises 9.6

1. (a) $x = -\frac{2}{5}$ (b) $x = \frac{1}{5}$
2. (a) $x = 2, -2.7$ (b) $x = -2, 1.3$ (c) $x = 0, -0.67$
3. $x = 311$ m and $x = 8689$ m
4. (b) (i) 3, −1 (ii) 3.4, −1.4 (iii) 2.22, −0.22

Computer and calculator exercise 9.6

1. 1, −1, −2

Exercises 9.7

1. (a) $x = 1, y = 1$ (b) $x = 7, y = -2$ (c) $x = -4, y = -2$
2. No solutions because lines do not intersect
3. $x = -0.75$
4. −1.8, 0.8

Computer and calculator exercises 9.7

1. (a) $x = -0.91, y = 1.95$ (b) $I_1 = 1.73, I_2 = 0.09$
2. $-2, \frac{5}{3}$

Self-assessment questions 9.8

2. $-5 < t < 5, \quad -3 \le t \le 3$

Exercises 9.8

1. $x < -3$
2. (a) $x > 2$ (b) $x > -3$ (c) $x < -\frac{1}{4}$ (d) $x \ge \frac{6}{5}$
3. (a) $s \ge -5$ (b) $t \ge -\frac{18}{5}$
4. (a) $x \ge -2$ (b) $x \ge -\frac{2}{3}$ (c) $x < -\frac{7}{2}$ (d) $y \le -2$
 (e) $x \le -2$ (f) $t \ge 0$
5. (a) $-6 < x < 4$ (b) $-6 \le y \le 2$ (c) $-8 < y < 10$
 (d) $-3 < x < -\frac{5}{3}$

Exercises 9.9

1. (a) $x < -1$ (b) $x < 5$ (c) $t < -7$ (d) $t > 3$
2. (a) $-\frac{1}{2} < x < 2$ (b) $x > 2$ or $x < -\frac{1}{2}$ (c) $-6 < x < 4$
 (d) all x
3. (a) $x > 4$ and $x < -1$ (b) $x > -5$ (c) $-1 < x < 4$ and $x < -5$
4. (a) $-1 < x < \frac{1}{2}$ (b) $x < -2$ and $x > -1$ (c) $-3 < x < 3$
 (d) $-3 < x < 3$ (e) $x > 3$ or $x < -3$ (f) $-3 < x < 3$
 (g) $-\frac{7}{2} < x < -\frac{3}{2}$
5. $x < -2, \ -\frac{1}{2} < x < \frac{1}{2}$ and $x > 2$
6. (a) $1 < x < 2$ (b) $-2 < x < -\frac{1}{2}$ (c) $-1 < x < \frac{2}{3}$
 (d) $-1 < x < -\frac{1}{2}$ (e) $-\frac{4}{3} < x < 2$ (f) $-6 < x < -\frac{5}{2}$
 (g) $1 < x < 2, \ 3 < x < 4$

Computer and calculator exercise 9.9

1. $x < -8, \ 4 < x < 7$

Review exercises 9

1. (a) $t = \frac{11}{16}$ (b) $v = \frac{5}{11}$ (c) $s = \frac{11}{5}$ (d) $t = 20$

2. (a) $x=-\frac{7}{2}$, 5 (b) $x=\frac{7}{2}$ (c) no real roots (d) $x=\frac{1}{7}$, $-\frac{7}{2}$
   (e) -1.145, 0.145 (f) 1.230, -1.897
3. (a) $x<\frac{8}{19}$ (b) $x\leqslant 1$ (c) $-5<x<5$ (d) $x>1$ and $x<-1$
4. $x=\pm 3$, ± 2
5. (a) $x=2$, $y=8$ (b) $s=-1$, $t=1$ (c) $a=-1$, $b=-3$
6. 0.618
7. 2.76, 0.56
8. (c) (i) ± 2.24 (ii) -1.93, 2.59 (iii) 1.78, -2.58

Exercises 10.2

1. (a) 60° (b) 72° (c) 22.9° (d) 401.1° (e) 6.88°
2. (a) $3\pi/4=2.36$ radians (b) $5\pi/3=5.24$ radians
   (c) $4\pi/3=4.19$ radians (d) 1.17 radians (e) 3.74 radians
3. (a) 1.5 radians (b) 7.2 cm
4. (a) 100° (b) 30°

Exercises 10.3

1. (a) 0.6018 (b) 0.3624 (c) 1.1918 (d) 0.6816 (e) 0.2588
   (f) 0.0300 (g) 1.2294 (h) 9.5668 (i) 0.6421
2. (a) 25.22° (b) 71.28° (c) 68.20° (d) 40.64° (e) 62.01°
   (f) 26.84°
4. (a) 28.96° (b) 14.04° (c) 36.87° (d) 23.58°
5. (a) 41.11° (b) 29.05° (c) 36.87°

Exercises 10.4

3. 4th
4. 3rd

Self-assessment questions 10.5

1. 1st, 2nd
2. 3rd, 4th
3. 2nd, 3rd
4. 1st, 4th

5. 1st, 3rd
6. 2nd, 4th

Exercises 10.5

1. (a) 51.57°, 128.43° (b) 193.83°, 346.17° (c) 53.69°, 306.31°
   (d) 153.39°, 206.61° (e) 57.49°, 237.49° (f) 160.91°, 340.91°
2. (a) 0.7343, 2.4072 (b) 3.3657, 6.0591 (c) 1.2216, 5.0615
   (d) 2.6754, 3.6078 (e) 1.1903, 4.3319 (f) 2.7996, 5.9412
3. (a) 0.4819, 1.6125, 2.5763, 3.7069, 4.6707, 5.8013
   (b) 0.4392, 2.0100, 3.5808, 5.1516
   (c) 0.1309, 0.6545, 1.7017, 2.2253 (d) 0.4308, 1.9497, 2.5252
   (e) 0.4310, 2.8522 (f) 0.2237 (g) 1.0614, 1.5079
4. (a) 21.34°, 68.66°, 201.34°, 248.66° (b) 199.816°, 340.184°
   (c) 47.3765°, 72.6235°, 167.3765°, 192.6235°, 287.3765°, 312.6235°
   (d) 19.181°, 160.819°, 199.181°, 340.819°
   (e) 103.718°, 216.282°
   (f) 3.8915°, 29.4418°, 123.8915°, 149.4418°, 243.8915°, 269.4418°
   (g) 13.928°, 58.928°, 103.928°, 148.928°, 193.928°, 238.928°,
       283.928°, 328.928°
   (h) no solutions in the interval [0, 360°]
   (i) 5.095°, 65.095°, 125.095°, 185.095°, 245.095°, 305.095°

Review exercises 10

1. (a) 0.1745 (b) 3.7525 (c) 17.4533 (d) -2.2689
   (e) 8.7266
2. (a) 206.26° (b) $-180°$ (c) $-45°$ (d) 123.76°
   (e) 286.48°
3. 1.6667
4. 10 cm
5. $A=41.8103°$, $C=48.1897°$
6. (a) 0.3218, 1.2490 (b) 2.6779, 0.4636 (c) 0.2702, 1.8410
   (d) 1.0472 (e) no solutions in the given interval
   (f) 0.9273 (g) 0.1411, 4.7421

7. (a) $44.43°$, $135.57°$ (b) $224.43°$, $315.57°$ (c) $45.57°$, $314.43°$
(d) $134.43°$, $225.57°$ (e) $34.99°$, $214.99°$ (f) $145.01°$, $325.01°$
(g) $24.43°$, $115.57°$ (h) $104.43°$, $195.57°$ (i) $17.50°$, $107.50°$,
$197.50°$, $287.50°$

Self-assessment questions 11.2

1. No
2. (a) improper (b) could be either

Exercise 11.2

1. (a) $n=1$, $d=2$, proper (b) $n=3$, $d=3$, improper
(c) $n=0$, $d=1$, proper (d) $n=2$, $d=1$, improper
(e) $n=2$, $d=3$, proper (f) $n=1$, $d=0$, improper
(g) $n=1$, $d=1$, improper

Self-assessment question 11.3

1. True

Exercise 11.3

1. (a) $\dfrac{2}{x} - \dfrac{1}{x+4}$ (b) $\dfrac{3}{y+2} - \dfrac{2}{y-2}$ (c) $\dfrac{2}{x+2} + \dfrac{7}{x+1}$

(d) $\dfrac{3}{x-1} - \dfrac{2}{x+4}$ (e) $\dfrac{9}{x+2} + \dfrac{2}{x-3}$ (f) $\dfrac{3}{4x+3} - \dfrac{1}{2x+3}$

(g) $\dfrac{37}{11(3x-1)} - \dfrac{21}{11(2x+3)}$ (h) $\dfrac{1}{2(s+1)} + \dfrac{5}{2(s-3)}$

(i) $\dfrac{2}{t-1} - \dfrac{1}{t+1}$ (j) $\dfrac{1}{s+1} - \dfrac{2}{2s+1}$ (k) $\dfrac{2}{x+3} - \dfrac{9}{x+2}$

(l) $\dfrac{4}{3y+2} + \dfrac{1}{y-2}$ (m) $\dfrac{2}{x+2} + \dfrac{1}{x+1} + \dfrac{3}{x}$

Exercise 11.4

1. (a) $\dfrac{5}{(t+3)^2} + \dfrac{2}{t+3}$ (b) $\dfrac{11}{x+2} + \dfrac{5}{(x+1)^2} - \dfrac{9}{x+1}$

(c) $\dfrac{3}{x+2} - \dfrac{2}{(x+2)^2}$ (d) $\dfrac{7}{(x+5)^2} - \dfrac{1}{x+5}$

(e) $\dfrac{1}{4(x+1)} + \dfrac{3}{2(x-1)^2} + \dfrac{1}{4(x-1)}$ (f) $\dfrac{1}{s-2} - \dfrac{2}{(s-2)^2}$

(g) $\dfrac{2}{(y-3)^2} - \dfrac{3}{y-3}$ (h) $\dfrac{1}{(s-4)^2} + \dfrac{3}{s-4}$ (i) $\dfrac{6}{z+7} - \dfrac{3}{(z+7)^2}$

(j) $\dfrac{2}{(x+1)^3} - \dfrac{2}{(x+1)^2} + \dfrac{1}{x+1}$ (k) $\dfrac{1}{(s-1)^3} + \dfrac{3}{(s-1)^2}$

Exercise 11.5

1. (a) $\dfrac{3x+1}{x^2+1} + \dfrac{7}{x}$ (b) $\dfrac{3x-2}{2x^2-x+4} - \dfrac{2}{x}$ (c) $\dfrac{2}{2x^2+1} + \dfrac{3}{x^2+1}$

(d) $\dfrac{x}{x^2+3} + \dfrac{2x-1}{x^2+x+1}$ (e) $\dfrac{x-7}{3x^2+4} + \dfrac{2x+1}{2x^2+x+9}$

(f) $\dfrac{s}{s^2+3} + \dfrac{1}{s^2+1}$ (g) $\dfrac{y}{2y^2+1} + \dfrac{2}{y+3}$ (h) $\dfrac{2x+1}{x^2-x-1} + \dfrac{1}{x-1}$

(i) $\dfrac{8t-1}{5(t^2+1)} + \dfrac{2}{5(t+2)}$ (j) $\dfrac{1}{x^2+1} - \dfrac{1}{(x+2)^2} + \dfrac{1}{x+2}$

Exercise 11.6

1. (a) $1 + \dfrac{4}{y+2} - \dfrac{1}{(y+2)^2}$ (b) $\dfrac{23}{8(2z+1)^2} + \dfrac{3}{8(2z+1)} + \dfrac{z}{4} - \dfrac{1}{4}$

(c) $1 + \dfrac{2}{x+3} - \dfrac{3}{(x+3)^2}$ (d) $\dfrac{5}{(y+1)^2} + \dfrac{4}{y+1} + y - 2$

(e) $1 + \dfrac{1-2x}{x^2+1} + 2x$ (f) $1 + \dfrac{1}{4(x-1)} - \dfrac{1}{2(x^2+1)}$

(g) $3 - \dfrac{8t+9}{t^2+3t+3}$ (h) $3 + \dfrac{3t+23}{t^2+3t-12}$ (i) $2+3t+\dfrac{1}{3t+2}$

(j) $2+2x+\dfrac{1}{x+2}+\dfrac{2}{x+3}$

Review exercise 11

1. (a) $\dfrac{1}{2(s-1)} - \dfrac{4}{s-2} + \dfrac{9}{2(s-3)}$ (b) $-\dfrac{2}{x+2} - \dfrac{1}{(x+1)^2} + \dfrac{2}{x+1}$

(c) $\dfrac{1}{x-1} + x + 2$ (d) $\dfrac{1}{s-2} - \dfrac{1}{s-1}$ (e) $\dfrac{1}{(s-1)^2}$

(f) $\dfrac{1}{3(x^2+2)} - \dfrac{2}{3(x+1)}$ (g) $\dfrac{1}{(t+1)^2} + \dfrac{2}{t+1} - 1$

(h) $\dfrac{1}{2(y+1)} + \dfrac{1}{2(y-1)} + y$ (i) $\dfrac{6}{(x-4)^2} + \dfrac{1}{x-4} + 1$

(j) $\dfrac{4}{x+3} + \dfrac{2}{x}$ (k) $\dfrac{3}{2s+1} - \dfrac{1}{s+2}$ (l) $\dfrac{1}{x+1} - \dfrac{1}{x^2+1}$

(m) $\dfrac{1}{x^2+1} + x^2 - 1$ (n) $-\dfrac{2}{t+1} + t + 1$ (o) $\dfrac{2}{x-1} + 1$

Computer and calculator exercises 11

1. (a) $\dfrac{1}{x-1} + \dfrac{1}{3(x-3)} - \dfrac{4}{3x}$

(b) $\dfrac{1}{27(x+2)} + \dfrac{1}{(x-1)^4} + \dfrac{2}{3(x-1)^3} + \dfrac{1}{9(x-1)^2} - \dfrac{1}{27(x-1)}$

(c) $\dfrac{x}{(x^2+1)^2} + \dfrac{1}{(x^2+1)^2} + \dfrac{x}{x^2+1} + \dfrac{1}{x^2+1}$

(d) $\dfrac{-x}{6(x^2+x+1)} + \dfrac{1}{6(x^2+x+1)} + \dfrac{x}{6(x^2-x+1)} + \dfrac{1}{6(x^2-x+1)} +$

$\dfrac{1}{6(x-1)} - \dfrac{1}{6(x+1)}$

(e) $-\dfrac{s}{s^2+1} + \dfrac{2}{s^2+1} - \dfrac{2}{s^6} + \dfrac{1}{s^5} + \dfrac{2}{s^4} - \dfrac{1}{s^3} - \dfrac{2}{s^2} + \dfrac{1}{s}$

2. $(x+1)^3(x^2-4x+1)$,

$\dfrac{x}{9(x^2-4x+1)} - \dfrac{7}{18(x^2-4x+1)} - \dfrac{1}{3(x+1)^3} - \dfrac{1}{6(x+1)^2} - \dfrac{1}{9(x+1)}$

Exercises 12.2

1. (a) rough (b) accurate (c) rough (d) accurate
(e) rough

Exercises 12.3

1. (a) $C_1 = C_2$ and $C = \frac{1}{2}C_1$ (b) If $C_2 \ll C_1$ then $C \approx C_2$
(i) exact 0.98 μF, approx. 1 μF, good approx.
(ii) exact 0.91 μF, approx. 1 μF, good approx.
(iii) exact 1.7 μF, approx. 3 μF, poor approx.

2. If $R_1 \ll R_2$, R_3 then $R \approx R_1$. Good estimate if $R_1 \ll R_2$, R_3.
3. (a) 900 (b) 450 (c) 1800

Exercises 12.4

1. (a) × factor of 4 (b) × factor of 9 (c) increase by 6%
(d) increase by 14% (only a rough estimate)
2. (a) increase by 50% (b) decrease by 50%
3. (a) × factor of 4 (b) × factor of $\frac{1}{2}$ (c) × factor of 8

Review exercises 12

1. (a) accurate (b) rough (c) accurate (d) rough
3. (a) ×2 (b) ×$\sqrt{2}$ (c) increase by ≈5%
(d) increasing the independent variable by k% leads to a $\frac{1}{2}k\%$ increase in the dependent variable if k is not too big
4. (a) ×8 (b) ×27 (c) increase of 15%
(d) a k% increase leads to a 3k% increase if k is not too big

Self-assessment questions 13.2

2. F   3. F   4. F

Exercises 13.2

1. (a) $\log_{10} 100 = 2$ (b) $\log_4 256 = 4$ (c) $\log_{16} 256 = 2$
   (d) $\log_{10} 10000 = 4$ (e) $\log_{12} 1728 = 3$ (f) $\log_5 \frac{1}{25} = -2$
   (g) $\log_3 \frac{1}{3} = -1$ (h) $\log_{10} \frac{1}{100} = -2$ (i) $\log_{27} 3 = \frac{1}{3}$
   (j) $\log_{27} \frac{1}{3} = -\frac{1}{3}$ (k) $\log_{10} 10 = 1$ (l) $\log_{10} 1 = 0$
   (m) $\log_x 1 = 0$

2. (a) $2^5 = 32$ (b) $3^3 = 27$ (c) $2^9 = 512$ (d) $9^2 = 81$
   (e) $10^3 = 1000$ (f) $5^3 = 125$ (g) $6^2 = 36$ (h) $2^7 = 128$

3. (a) $\log_2 4 = 2, \log_2 16 = 4, \log_2 64 = 6$
   (b) $\log_2 4 + \log_2 16 = \log_2 64$

4. $\log_a 1 = 0$

Exercises 13.3

1. (a) 4.605   (b) 1   (c) 1   (d) 0.434   (e) 2.322   (f) 3.611

2. (a) 0.477   (b) -0.477 (c) 1.946   (d) -1.946

3. (a) 1.792, 1.609, 3.401, 0.182, -0.182 (b) $\ln 30 = \ln 6 + \ln 5$
   (c) $\ln 6 - \ln 5 = \ln \frac{6}{5}$ (d) $\ln 5 - \ln 6 = \ln \frac{5}{6}$

4. 795 m s^{-1}

5. (a) 0.301, 0.602, 0.903, 1.204, 1.505   (b) $\log_{10} 2^n = n \log_{10} 2$

Self-assessment questions 13.4

2. T

Exercises 13.4

1. (a) $\log_{10} 50$ (b) $\log 2$ (c) $\ln \frac{1}{2}$ (d) $\log 2.5$ (e) $\ln 500$
   (f) $\log_{10} 1.25$ (g) $\ln 10$ (h) $\log \frac{30\,34^2}{5^4}$ (i) $\log \frac{6^2 10^3}{2^6}$

2. (a) $\ln xyz$ (b) $\log_{10} x$ (c) $\log x^2 y^3 z^4$ (d) $\ln x^5$
   (e) $\ln t^{12}$ (f) $\log_{10} 12x$ (g) $\log z$ (h) $\ln \frac{x^8}{x^3}$
   (i) $\log x^5 y^5 z^2$ (j) $\ln 1 = 0$ (k) $\ln (1/z)$ (l) $\log 2t^2$
   (m) $\log_{10} (d^5/c)$ (n) $\log (a^7/b^7)$ (o) $\ln (a^{10}/b^7)$

---

3. (a) $K = \log 3$ (b) no
4. 30 dB

Self-assessment question 13.5

1. F

Exercises 13.5

1. (a) 2   (b) 0.7211   (c) 1   (d) 1.6356   (e) 4   (f) 2
   (g) 1.3333   (h) 0.6990

2. $K = 2.303$

Exercises 13.6

1. F
2. 26.021 dB
3. 26.021 dB, 12.041 dB, 38.062 dB

Exercises 13.7

1. (a) 78 125   (b) 122.89   (c) 0.354   (d) 63.096   (e) 0.005
   (f) 7.389   (g) 0.607

2. (a) e^6 (b) e^{12} (c) $e^{5.8}$ (d) $e^{3.5}$ (e) $6e^9$ (f) $10e^{-5}$
   (g) $12e^{-4}$ (h) e   (i) e^7 (j) 3e   (k) e^{-4} (l) $e^{9.8}$
   (m) $\frac{4}{3}e^{-4}$ (n) $\frac{2}{5}e$ (o) $e^5/6$ (p) 6   (q) 16e   (r) $6e^4$
   (s) 24e   (t) $\frac{5}{2}e$

3. (a) e^6 (b) $2^5 e^5$ (c) $3^4 e^{-4}$ (d) e^{20} (e) e^5 (f) e^{-5}
   (g) e^{-5} (h) $144e^2$ (i) $e^8/9$ (j) $\frac{3}{2}e^{-1}$

4. (a) e^{6t} (b) $\frac{5e^x}{6}$ (c) $12e^{5z}$ (d) e^{-x} (e) $e^t/2$ (f) 1
   (g) e^{-x} (h) $\frac{e^{2x}}{2}$ (i) e^{x-1}

Self-assessment questions 13.8

4. F

Exercises 13.8

1. 0.00176 A

Computer and calculator exercises 13.8

1. (b) −0.37
2. (b) 1.87

Exercises 13.9

1. (a) 6.397 (b) −0.223 (c) 1.398 (d) −0.602
2. $x \geqslant 13.816$
3. $x \geqslant -12.206$
4. (a) 0.4771 (b) −0.4771

Exercises 13.10

1. (a) 2.303 (b) 0.693 (c) 1.672 (d) −0.114 (e) −0.349
   (f) −0.302 (g) 1.099 (h) 0.401 (i) −0.002 (j) 4.523
   (k) 1.342 (l) 0.847 (m) −0.973
2. (a) 39.811 (b) 20.086 (c) 202.21 (d) 63.25 (e) 3.687
   (f) ±1.920 (g) 4.130 (h) 0.0498 (i) −0.01 (j) ±2.528
   (k) ±5.274

Computer and calculator exercises 13.10

1. Approx. (a) 0.20 (b) 1.86 (c) 1.24, −2.21 (d) −0.52, 2.99
2. Approx. (a) 3.24 (b) 6.31 (c) 1.69, −3.7 (d) 0.16
3. Approx. 1.1, −1.8

Review exercises 13

1. (a) $\ln xy^2$ (b) 10^{3x} (c) $\log_{10} b$ (c) $\ln y^5$ (e) $\log B^4$
   (f) $\frac{1}{3}e^t$ (g) e^n (h) $\log xy$ (i) $\ln \dfrac{a^5 c^7}{b^3}$ (j) 10^{2x-y}
   (k) $1+e^t$ (l) $\ln(A^2/B^2)$
   (m) not possible to simplify, but could be written as
   $\ln x \left(1 + \dfrac{1}{\ln 10}\right)$

2. (a) 2.485 (b) 0.768 (c) −0.334 (d) −0.916 (e) $-\frac{2}{3}$
   (f) 1.20 (g) 0.527 (h) 1.451 (i) 1.245 (j) ±2.146
   (k) ±1.357 (l) ±1.904 (m) ±2.531
3. (a) 7.099 (b) 199.5 (c) 1.977 (d) 0.533 (e) 28.048
   (f) ±7.934 (g) ±2.032 (h) 1.411 (i) 3.160 (j) 4.137
   (k) −0.412

Self-assessment questions 14.2

4. Yes
5. Yes
6. False

Exercises 14.2

1. 45°, 45°, 90°
2. (a) 32° (b) 90° (c) 116°

Exercises 14.3

1. (a) 7.141 m (b) 10.817 mm (c) 9 cm
2. BC = 3 cm, AB = 4.243 cm
3. (a) $\sqrt{2}$ cm (b) $\sqrt{2}$ cm (c) 1 cm (d) 1 cm
4. (a) 15 m (b) 4.031 m

Self-assessment questions 14.4

1. Need one more angle and length of one side *or* length of two sides
2. Need length of one side

Exercises 14.4

1. (a) B=65°, AB=2.891 cm, BC=6.841 cm
   (b) AC=7.331 cm, B=37.66°, C=52.34°
   (c) C=48°, AC=6.753 m, BC=10.092 m
   (d) BC=9.080 cm, B=36.49°, C=53.51°
   (e) B=53°, AC=12.474 cm, BC=15.619 cm

Exercises 14.5

1. All sides 2 cm, all angles 60°
2. $A = C = 65°$, AC = 17.750 mm
3. $B = C = 80°$, AB = AC = 74.864 mm
4. $B = 100°$, $C = 40°$, AC = 18.385 cm

Exercises 14.6

1. (a) $C = 63°$, AC = 4.45 cm, AB = 13.57 cm
   (b) $A = 47.219°$, $B = 52.781°$, AC = 8.571 cm
   (c) $A = 77°$, AC = 7.936 cm, BC = 12.029 cm
   (d) Two solutions: $A = 20°$, $B = 24.606°$, $C = 135.394°$,
   AB = 14.167 cm; and $A = 20°$, $B = 155.394°$, $C = 4.606°$,
   AB = 1.620 cm
   (e) Two solutions: $A = 109.073°$, $B = 25°$, $C = 45.927°$,
   BC = 22.363 cm; and $A = 20.927°$, $B = 25°$, $C = 134.073°$,
   BC = 8.451 cm   (f) $A = 152°$, BC = 16.472 cm, AB = 4.883 cm

2. Longest side must be opposite largest angle.
3. $A = 3.9°$, AB = 2938.7 m
4. Need two sides and a nonincluded angle.

Exercises 14.7

1. (a) $a = 13.60$ cm, $B = 37.53°$, $C = 75.47°$
   (b) $B = 50.73°$, $C = 90.53°$, $A = 38.74°$
   (c) $b = 21.89$ cm, $A = 28.48°$, $C = 51.52°$
   (d) $c = 3.074$ cm, $B = 37.72°$, $A = 122.28°$
   (e) $B = 74.41°$, $A = 44.47°$, $C = 61.12°$

Review exercises 14

1. AC = 23.26 cm, 25.46°
2. (a) $C = 107.02°$, $B = 32.01°$, $A = 40.97°$
   (b) $A = 73°$, $b = 9.537$ cm, $a = 12.272$ cm

(c) Two solutions: $B = 23.963°$, $C = 135.037°$, $c = 29.578$ cm; and
$B = 156.037°$, $C = 2.963°$, $c = 2.164$ cm
(d) $a = 17.785$ cm, $B = 53.73°$, $C = 91.268°$
(e) $B = 43.32°$, $A = 58.14°$, $C = 78.54°$
(f) $C = 115°$, $a = 9.54$ cm, $b = 10.95$ cm
(g) $C = 54.43°$, $B = 50.57°$, $b = 15.19$ cm
(h) $b = 21.74$ mm, $A = 82.73°$, $C = 44.27°$

3. CD = 1 cm, AD = $\sqrt{3}$ = 1.732 cm

Exercises 15.2

3. Answer = 1

Computer and calculator exercises 15.3

1. Max. = 2 at $\theta = 90°$
2. (a) Max. = 1 at $\theta = 0$, 120°, 240°, 360°
4. $\theta = 45°$, 225°

Exercises 15.4

1. (a) 0.2   (b) 0.1   (c) 0.15   (d) 0.05   (e) $\frac{1}{9}$
2. $\sin\theta \approx \theta$, $\cos\theta \approx 1$, $\tan\theta \approx \theta$

Review exercises 15

1. (a) $\tan^2 A$ (b) 1   (c) $\tan 2A$ (d) $-2\cos\theta$ (e) 0
   (f) $2\cos\theta$
4. 0.25
5. (a) 0.067   (b) 0.125   (c) $\frac{1}{9}$

Computer and calculator exercises 15

1. Approx. $\theta = 0.25$, 2.6, 5.3
2. 30°, 150°, 210°, 330°

Exercises 16.2

1. (a) $y = 2\sin\dfrac{\pi t}{5}$ (b) $y = 0.7\sin 4\pi t$

2.

	A	ω	f	Time disp.	T
(a)	3	6	$3/\pi$	0	$\pi/3$
(b)	.4	5	$5/2\pi$	0	$2\pi/5$
(c)	5	4	$2/\pi$	0	$\pi/2$
(d)	$\frac{1}{6}$	1	$1/2\pi$	0	2π
(e)	6	$\frac{1}{6}$	$1/12\pi$	0	12π
(f)	$\frac{2}{3}$	7	$7/2\pi$	0	$2\pi/7$
(g)	0.6	$\frac{2}{3}$	$1/3\pi$	0	3π
(h)	1	1	$1/2\pi$	2	2π
(i)	1	2	$1/\pi$	$-\frac{1}{2}$	π

3. (a) $3, \frac{3}{5}$ (b) $-3, -3$ (c) $3, \frac{3}{2}$ (d) $3, 6$ (e) $-3, -\frac{9}{2}$
(f) $-\frac{4}{7}, -2$ (g) $\frac{5}{2}, \frac{5}{6}$

Computer and calculator exercises 16.2

1. (a) amplitude $= 3$ (b) time displacement $= \frac{5}{4}$ s

Exercises 16.3

1. (a) $\sqrt{20}\sin(2t - 0.4636)$ (b) $\sqrt{45}\sin(4t - 5.8196)$
(c) $\sqrt{\dfrac{13}{4}}\sin(t - 5.6952)$ (d) $5\sin(\tfrac{1}{2}t - 5.3559)$

2. (a) $\sqrt{2}\sin(t + \pi/4)$ (b) $\sqrt{34}\sin(4t + 5.2528)$
(c) $\sqrt{45}\sin(2t + 2.6779)$ (d) $1.8868\sin(t + 0.5586)$

3. (a) $\sqrt{5}\cos(3t + 1.107)$ (b) $\sqrt{40}\cos(t + 5.0341)$
(c) $\sqrt{34}\cos(2t + 4.1720)$ (d) $2.0156\cos(3t + 5.2315)$

4. (a) $\sqrt{53}\cos(4t - 1.2925)$ (b) $\sqrt{2}\cos(t - 5\pi/4)$
(c) $\sqrt{13}\cos(5t - 5.6952)$ (d) $\sqrt{37}\cos(3t - 2.9764)$

Review exercises 16

1.

	A	ω	f	T	ϕ	Time disp.
(a)	3	2	$1/\pi$	π	0	0
(b)	4	1	$1/2\pi$	2π	0	0
(c)	5	3	$3/2\pi$	$2\pi/3$	-1	$-\frac{1}{3}$
(d)	6	$\frac{1}{2}$	$1/4\pi$	4π	0	0
(e)	3	$\frac{2}{3}$	$1/3\pi$	3π	1	$-\frac{3}{2}$
(f)	$\frac{2}{3}$	$\frac{2}{3}$	$1/3\pi$	3π	-1	$-\frac{3}{2}$
(g)	$\frac{3}{4}$	0.4	$0.2/\pi$	5π	0	0
(h)	1	100π	50	0.02	0	0

2. (a) $\sqrt{13}\cos(3t - 2.1588)$ (b) $1.1402\cos(5t - 0.9098)$
(c) $\sqrt{10}\cos(\pi t - 0.3218)$

3. (a) $\sqrt{5}\sin(2t + 5.8195)$ (b) $\sqrt{17}\sin(3t + 1.3258)$
(c) $1.1180\sin(\pi t + 5.1760)$

4. (a) $\sqrt{34}\sin(t - 5.7428)$ (b) $1.1180\sin(4t - 1.1071)$
(c) $\sqrt{5}\sin(3t - 4.2487)$

5. (a) $\sqrt{13}\cos(5t + 5.6952)$ (b) $\sqrt{10}\cos(t + 1.2490)$
(c) $\sqrt{17}\cos(\tfrac{1}{2}t + 4.9574)$

6. (a) $y = 6\sin 2000\pi t$ (b) $y = 0.35\sin(0.02\pi t)$
(c) $y = 0.4\sin 5\pi t$ (d) $y = 5\sin\left(\dfrac{\pi t}{6}\right)$

Exercises 17.2

1. $(0, 0)$
2. (a) y (b) x
3. (b) and (d) only
4. (c) and (e) only

Exercises 17.3

1. (a) $3, 2$ (b) $-2, -3$ (c) $\frac{1}{2}, 0$ (d) $-\frac{2}{3}, 1$ (e) $2, \frac{1}{2}$
(f) $-3, 9$ (g) $-\frac{1}{2}, \frac{19}{4}$ (h) $-1, 6$ (i) $\frac{3}{4}, -\frac{9}{4}$

2. (a) $y = 3x - 1$ (b) $y = 4x + 2$ (c) $y = 2x + 4$ (d) $y = -\frac{3}{2}x$
(e) $y = -\frac{1}{2}x + 2$

3. (a) $y=2x$ (b) $y=6-x$ (c) $y=0.5x+3.5$ (d) $y=-\frac{2}{3}x-2$
   (e) $y=-x-2$
4. (a) $x=3$ (b) $x=-4$ (c) $x=2$ (d) $x=-2$ (e) $x=12$
5. (a) yes (b) no (c) yes (d) no (e) yes
6. (4, 11)
7. $y=3$
8. (a) $y=3x+7$ (b) $y=-2x+6$ (c) $y=3x$ (d) $y=kx$

Exercise 17.4

1. (a) $\sqrt{53}$ (b) $\sqrt{50}$ (c) $\sqrt{13}$ (d) $\sqrt{26}$ (e) $\sqrt{5}$
   (f) $\sqrt{x^2+y^2}$ (g) $\sqrt{(x_2-x_1)^2+(y_2-y_1)^2}$

Exercise 17.5

1. (a) (4, 5) (b) (2.5, 5) (c) $\left(-\frac{1}{2}, \frac{1}{2}\right)$ (d) $\left(\frac{a}{2}, \frac{b}{2}\right)$
   (e) $\left(\frac{a+c}{2}, \frac{b+d}{2}\right)$

Review exercises 17

1. (a) $x=1$ (b) $y=-2$
2. (a) $y=2x$ (b) $y=-\frac{1}{2}x+\frac{7}{2}$ (c) $y=3x+6$ (d) $y=-3$
   (e) $y=3x+7$ (f) $y=x$ (g) $y=-2x-8$
3. (a) $2, -1, \frac{1}{2}$ (b) $-\frac{1}{2}, 3, 6$ (c) $\frac{1}{2}, 3, -6$ (d) $\frac{3}{4}, 4, -\frac{16}{3}$
   (e) $-\frac{2}{7}, \frac{16}{7}, 8$
4. (a) $\sqrt{5}$ (b) $\sqrt{40}$ (c) $\sqrt{45}$ (d) $\sqrt{a^2+b^2}$
   (e) $\left(\frac{3}{5}, -\frac{2}{5}\right)$
5. (a) (4.5, 8.5) (b) (-1, 2.5) (c) (-1.5, 1.5)
6.
7. (a) $f(t)=\begin{cases} -t & -1\leqslant t<0 \\ t & 0\leqslant t<1 \end{cases}$ period $=2$
   (b) $g(t)=\begin{cases} -2t & -1\leqslant t<0 \\ 2t & 0\leqslant t<1 \end{cases}$ period $=2$

(c) $h(t)=\begin{cases} -4t & -1\leqslant t<0 \\ 2t & 0\leqslant t<2 \end{cases}$ period $=3$

Exercises 18.3

1. (a) $(x-3)^2+(y+1)^2=4$ (b) $(x+1)^2+(y-0.5)^2=25$
   (c) $x^2+y^2=49$ (d) $x^2+y^2=r^2$ (e) $(x+1)^2+(y-1)^2=13$
   (f) $(x-3)^2+(y-3)^2=5$
2. (a) $x=2.732, -0.732, y=2.732, -0.732$ (b) $x=\pm10, y=\pm10$
   (c) $x=2.899, -6.899, y=-5.583, 3.583$
3. (0.894, 0.894), (-0.894, -0.894)
4. (1.044, 2.088), (-1.044, -2.088)
5. (a) $(x-3)^2+(y-2)^2=81$ (b) $(x-4)^2+(y-1)^2=5$
6. (a) (7, -4), 6 (b) (-3, -1), 2 (c) (-0.5, 0.5), 1

Computer and calculator exercises 18.3

1. Approx. (2, 2.2) and (3, 0.3)
2. Approx. $x=\pm3.830, y=\pm2.309$

Exercises 18.4

1. (a) $x=0.268, 3.732$ (b) $y=1$
2. $(3, \sqrt{7}), (3, -\sqrt{7}), (-3, \sqrt{7}), (-3, -\sqrt{7})$

Exercises 18.5

1. (a) (2.70, 4.21) (b) (0.28, -0.96) (c) (1.08, -1.68)
   (d) (-2.57, 1.55) (e) (-1.66, -3.64) (f) (2.30, 1.93)
   (g) (-1.74, 9.85) (h) (1.03, 2.82) (i) (3.46, -2)
2. (a) 4th (b) 4th (c) 2nd (d) 2nd (e) 3rd (f) 2nd
3. (a) $\sqrt{58}\angle 1.166$ (b) $\sqrt{29}\angle 1.951$ (c) $\sqrt{20}\angle(-2.678)$
   (d) $\sqrt{65}\angle(-0.519)$ (e) $3\angle 0$ (f) $3\angle\pi$ (g) $3\angle\frac{1}{2}\pi$
   (h) $3\angle(-\frac{1}{2}\pi)$

Review exercises 18

1. (a) $(2, 2)$, $\sqrt{6}$ (b) $(-2, 3)$, $\sqrt{12}$ (c) $(0, 0)$, $\sqrt{10}$
   (d) $(-3, 0)$, $\dfrac{\sqrt{12}}{\sqrt{12}}$
2. (a) $(x-2)^2 + (y-1)^2 = 16$ (b) $5.87, -1.87$ (c) $4.46, -2.46$
3. (a) $\sqrt{2}\angle\frac{1}{4}\pi$ (b) $\sqrt{2}\angle(-\frac{1}{4}\pi)$ (c) $\sqrt{8}\angle\frac{3}{4}\pi$ (d) $3\angle\frac{1}{2}\pi$
   (e) $2\angle\pi$
4. (a) $(1.621, 2.524)$ (b) $(1.732, -1)$ (c) $(0.283, 3.990)$
   (d) $(3.999, -0.105)$ (e) $(-0.25, 0.433)$
5. $(2.87, -0.87), (-0.87, 2.87)$
6. (b) $(\sqrt{2}, 1), (\sqrt{2}, -1), (-\sqrt{2}, 1), (-\sqrt{2}, -1)$ (c) $(1, 0), (-1, 0)$
7. $(\sqrt{4/3}, \sqrt{4/3}), (-\sqrt{4/3}, -\sqrt{4/3})$

Exercises 19.2

1. (a) $80\,424.8 \text{ cm}^2 = 8.04 \text{ m}^2$ (b) 78.54 m^2
2. 28.27 m^2, 113.10 m^2
3. 112.8 mm
4. $r = \dfrac{l}{\sqrt{\pi}}\,\text{m}$
5. 206.35 cm^2
6. (a) 201.06 cm^2 (b) 27.93 cm^2 (c) 173.13 cm^2 (d) 3.42 cm^2
7. (a) 112 cm^2 (b) 28 cm^2 (c) 42 cm^2 (d) 42 cm^2
8. $\dfrac{\sqrt{3}}{4}l^2$
9. (a) 101.67 cm^2 (b) 97.428 cm^2 (c) 45.14 cm^2
   (d) 25.98 cm^2

Exercises 19.3

1. 1.8963 m^2
2. 615.75 cm^2
3. 3.77 m^2
4. 5.09 m^2

5. 6.158 m^2
6. $48\,773.2 \text{ mm}^2$

Exercises 19.4

1. (a) 113.10 cm^3 (b) $20\,579.5 \text{ cm}^3$ (c) $38\,792.4 \text{ mm}^3$
   (d) 50.965 m^3
2. 0.255 m^3
3. 0.01005 m^3
4. 1017.88 cm^3
5. $1.5 \times 10^7 \text{ mm}^3$
6. 375 cm^3
7. 0.002744 m^3
8. $47\,713 \text{ mm}^3$
9. 137.258 m^3
10. (a) 0.045 m^2 (b) $0.015\,08 \text{ m}^3$

Review exercises 19

1. (a) 1.4368 cm^3 (b) 7.2382 m^3 (c) $28\,730.9 \text{ mm}^3$
2. 270.6 cm^2
3. $r = 2R$
4. 0.3474 m^3
5. 32 cm
6. 5.365 cm^2
7. $r = 0.0608 \text{ m}$
8. 20.25 cm

Exercises 20.2

1. (a) $0, 3, 6, 9, 12$ (b) $\frac{1}{2}, 1, \frac{3}{2}, 2, \frac{5}{2}$ (c) $2, 5, 8, 11, 14$
   (d) $0, 1, 4, 9, 16$ (e) $1, 4, 9, 16, 25$
   (f) $16, 64, 256, 1024, 4096$ (g) $\frac{1}{4}, 1, 4, 16, 64$
   (h) $3, 6, 9, 12, 15$ (i) $-1, 4, 13, 26, 43$
2. $20, 7, 0, -1, 4, 15, 32$
3. Dependent variable, g; independent variable, k

4. Figure 20.2.4

5. Figure 20.2.5

6. $f[k] = \dfrac{1}{k}$ $k = 1, 2, 3, 4, 5, \ldots$

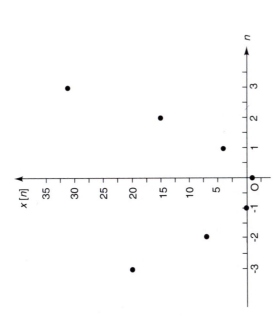

Figure 20.2.4

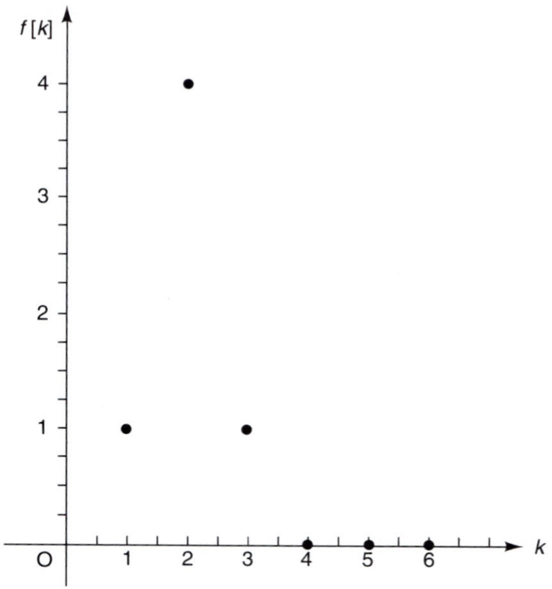

Figure 20.2.5

Exercises 20.3

1. 0, 1, 8, 27, 64

2. $T = 0.01$; 3, -3, 3, -3, 3

3. 1, 2.7183, 7.3891, 20.0855, 54.5982

4. (a) $\frac{1}{4}\pi$ (b) 0, 1, 0, -1, 0, 1   (c) $g[k] = \sin\frac{1}{2}\pi k$ $k = 0, 1, 2, \ldots$

5. (a) 0   (b) 0.0995   (c) $\frac{1}{10}k\cos\frac{1}{10}k$

Exercises 20.4

1. 1, 5, 9, 13, 17

2. 2, 5, 8, 11, 14, 17

3. 7, 4, 1, -2, -5

4. First term $= 19$; common difference $= -3$

5. (a) -25, -213 (b) -60, -436

6. All the terms are identical.

7. $\frac{71}{19}$

8. (a) $\frac{5}{6}$ (b) 0   (c) $\frac{40}{3}$

Exercises 20.5

1. (a) 64 (b) 4096
2. $2, \frac{2}{3}, \frac{2}{9}, \frac{2}{27}, \frac{2}{81}, \frac{2}{243}$
3. $-\frac{1}{2}$
4. When $0<r<1$, the terms are all of the same sign and get smaller. When $-1<r<0$, the terms alternate in sign and get smaller in magnitude.
5. All the terms are the same.
6. $\frac{49}{6}$
7. Either $r=1$ or $r=-1$.

Exercises 20.6

1. (a) diverges (b) 0 (c) diverges (d) 0 (e) 2
   (f) diverges (g) diverges (h) 0 (i) 0
2. Q1(g) is such a sequence.

Exercises 20.7

1. (a) 34 (b) 130 (c) 34
2. (a) 26 (b) 74 (c) 100
3. (a) 210 (b) 2870

Exercises 20.8

1. 255
2. -222
3. 35
5. -50
6. 145

Exercises 20.9

1. 189
2. 2730

3. 5
4. 1.5

Exercises 20.10

1. (a) not possible (b) $\frac{8}{3}$ (c) $-\frac{2}{3}$ (d) 17 (e) 20
   (f) not possible (g) not possible (h) $\frac{21}{4}$
2. (a) $\frac{4}{3}$ (b) 3 (c) $\frac{3}{2}$ (d) 32
3. 85

Exercises 20.11

1. $a^7 + 7a^6b + 21a^5b^2 + 35a^4b^3 + 35a^3b^4 + 21a^2b^5 + 7ab^6 + b^7$
2. (a) $a^4 + 4a^3b + 6a^2b^2 + 4ab^3 + b^4$
   (b) $1 + 12y + 54y^2 + 108y^3 + 81y^4$
3. $a^3 + 3a^2b + 3ab^2 + b^3$; $27x^3 + 108x^2y + 144xy^2 + 64y^3$
4. $1 + 5y + 10y^2 + 10y^3 + 5y^4 + y^5$;
   $1 + 20x + 160x^2 + 640x^3 + 1280x^4 + 1024x^5$
5. $16x^4 + 32x^3y + 24x^2y^2 + 8xy^3 + y^4$
6. $1 - 2x + 3x^2 - 4x^3 + 5x^4$, valid for $-1<x<1$
7. $1 - \frac{3}{2}x + \frac{3}{2}x^2 - \frac{5}{4}x^3$, valid for $-2 < x < 2$

Review exercises 20

1. (a) 1, 3, 9, 27, 81 (b) 0, 1, 8, 27, 64 (c) 1, -2, 4, -8, 16
   (d) 5, 5, 5, 5, 5
2. (a) arithmetic (b) geometric (c) arithmetic (d) neither
   (e) geometric (f) geometric (h) arithmetic and geometric
3. (b), (e) and (g) are convergent
4. $1 - 9x + 54x^2 - 270x^3$; valid for $-\frac{1}{3}<x<\frac{1}{3}$
5. (a) 33 (b) 1193
6. (a) 175 (b) 2875
7. 8.4566×10^{-3}
8. (a) 339.99 (b) 3.8257×10^9 (c) 2.5497×10^9
9. (a) -2 (b) -36
10. (a) 0.5 (b) 1, 2.25, 5, 9.25, 15
11. (a) 11440 (b) 171

Self-assessment questions 21.2

2. Negative change

3. Positive change

4. No. It means the values at $t=0$ and $t=1$ are the same.

5. No. It could decrease for some of the time.

Exercises 21.2

1. $-16\,°\mathrm{C}$

2. $-21\,\mathrm{V}$

3. $16\,\mathrm{m\,s^{-1}}$

4. 633

5. 1

6. (a) 14.5   (b) 17.5   (c) -1 (d) -7

7. $\frac{37}{40}$

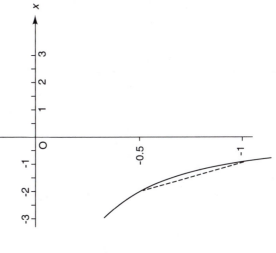

Figure 21.3.4

Self-assessment questions 21.3

1. Yes

2. No

3. No

4. For at least part of the interval

Exercises 21.3

1. $y=12x-10$, rate of change $=12$

2. 0.5, 0, 1, 0, -2

3. (a) 36   (b) 60

4. $-\frac{1}{2}$, $-\frac{1}{4}$. Figure 21.3.4

5. 26 m.p.h.

6. 992

7. 25

Exercises 21.4

1. (a) Figure 21.4.1 (b) (i) 10 (ii) 8.50 (c) 7
2. (a) Figure 21.4.2 (b) $-\frac{1}{6}$, $-\frac{2}{15}$ (c) $-\frac{1}{9}$

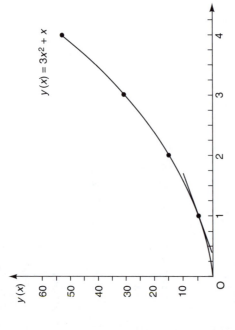

$y(x) = 3x^2 + x$

Figure 21.4.1

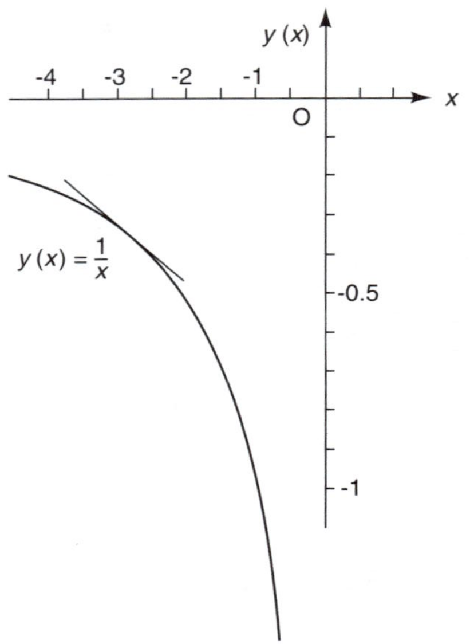

Exercises 21.5

1. (a) 6x (b) 18
2. 28
3. $6x^2$
4. -3
5. 0

Review exercises 21

1. $0.667\,\mathrm{m\,s^{-2}}$
2. 21
3. (a) 6x (b) $15x^2$ (c) $2x+1$

Figure 21.4.2

$y(x) = \frac{1}{x}$

4. 1.1752
5. (a) 6x (b) 6 (c) $y = 6x+1$
6. (a) -1 (b) 1 (c) 3

Exercises 22.2

1. (a) $16x^{15}$ (b) $-2x^{-3}$ (c) $\frac{1}{3}x^{-\frac{2}{3}}$ (d) $-2x^{-3}$ (e) $\frac{1}{3}x^{-\frac{2}{3}}$
   (f) $\frac{7}{2}x^{\frac{5}{2}}$
2. 0

Exercises 22.3

1. (a) $8x^7$ (b) $-8x^{-9}$ (c) $-8x^{-9}$ (d) $-\frac{1}{4}x^{-\frac{5}{4}}$ (e) $9x^8$
   (f) $-9x^{-10}$ (g) $1.5x^{0.5}$

Exercises 22.4

1. (a) $14t^{13}$ (b) $-z^{-2}$ (c) $-\frac{1}{2}s^{-2}$ (d) $\frac{1}{2}t^{-\frac{1}{2}}$
2. (a) $\frac{1}{3}t^{-\frac{2}{3}}$ (b) $-3s^{-4}$
3. $17t^{16}$
4. $-2s^{-3}$
5. (a) $1.7t^{0.7}$ (b) $-4n^{-5}$ (c) $4y^3$

Exercises 22.5

1. (a) $-3\sin 3x$ (b) $3x^2$ (c) $3\sec^2 3x$ (d) $\frac{1}{x}$ (e) e^x
   (f) $3e^{3x}$ (g) $-2e^{-2x}$ (h) $-4e^{-4x}$
2. 1210.3
3. 0.5403
4. 7
5. $\ldots, \dfrac{-5\pi}{2}, \dfrac{-3\pi}{2}, \dfrac{-\pi}{2}, \dfrac{\pi}{2}, \dfrac{3\pi}{2}, \ldots$

Exercises 22.6

1. (a) $12x^2+4x$ (b) -2 (c) $\frac{9}{2}x^{-\frac{1}{2}}-12x^3$
2. (a) $6t-12t^3+7$ (b) $\cos t+e^t$ (c) $4\cos t$
3. (a) $e^x-\sin x$ (b) $18t^2-6t$ (c) $\cos t+8t$ (d) 0
   (e) $3\cos x+\sin x$ (f) $\dfrac{e^x-2\cos x}{3}$
4. (a) 1.7183 (b) -0.8415 (c) 15.7758 (d) -0.5 (e) 1
   (f) 4
5. (a) $\dfrac{2}{x}$ (b) $\dfrac{1}{2x}$
6. (a) $\dfrac{3}{x}$ (b) $\dfrac{2}{x}$ (c) $\dfrac{3}{x}$
7. (a) 0.5466 (b) 22.1672 (c) -0.4060 (d) -0.0451
   (e) -7.4363×10^{-3}

Self-assessment questions 22.7

3. (a) and (c) only

Exercises 22.7

1. (a) $x^3\sec^2 x+3x^2\tan x$ (b) $(x^2+x)(-2\sin 2x)+(2x+1)\cos 2x$
   (c) $-2e^{-x}\sin 2x-e^{-x}\cos 2x$ (d) $-\dfrac{\sin x}{x}-\dfrac{\cos x}{x^2}$
   (e) $3x^{\frac{1}{2}}\cos 3x+\dfrac{x^{-\frac{1}{2}}\sin 3x}{2}$ (f) $-3t\sin 3t+\cos 3t$
   (g) $\dfrac{2+\ln x}{2\sqrt{x}}$
2. (a) $(3x^3+2x^2)(-\sin x)+(9x^2+4x)\cos x$
   (b) $(8\cos 2x)e^{2x}-(8\sin 2x)e^{2x}$
   (c) $-x^{-2}\sin x-2x^{-3}\cos x$
   (d) $x^{\frac{3}{2}}3\sec^2 3x+\frac{3}{2}x^{\frac{1}{2}}\tan 3x$
3. 1
4. 4.0774
5. -0.3012

Exercises 22.8

1. (a) $\dfrac{2x}{(x^2+1)^2}$ (b) $\dfrac{2x\cos 2x-\sin 2x}{3x^2}$ (c) $\dfrac{-4t^2+8t-1}{e^t}$
   (d) $\dfrac{-2(2t+1)\sin 2t-2\cos 2t}{(2t+1)^2}$ (e) $\dfrac{e^x+1-xe^x}{(e^x+1)^2}$
   (f) $\dfrac{1-\ln x}{x^2}$ (g) $\dfrac{\ln x-1}{(\ln x)^2}$
   (h) $\dfrac{2e^{-x}+2+2xe^{-x}}{(e^{-x}+1)^2}$ $\dfrac{xe^{-x}+e^{-x}+1}{2x^2}$ (i) $\dfrac{-8}{(4+t)^2}$
   (j) $\dfrac{3z^2+6z-1}{(z+1)^2}$

2.
(a) $\dfrac{x \sec^2 x - \tan x}{2x^2}$ (b) $\dfrac{6x \sin x - 3x^2 \cos x}{\sin^2 x}$

(c) $\dfrac{-x^2 - x + 1}{e^x}$

3.
(a) $-\sqrt{x}\sin x + \dfrac{\cos x}{2\sqrt{x}}$ (b) $-(\cos x + \sin x)e^{-x}$

(c) $-(\sin x + \cos x)e^{-x}$ (d) $1 + \ln t$ (e) $\dfrac{(x+1)\cos x - \sin x}{(x+1)^2}$

4.
(a) $\dfrac{-12}{(3x-2)^2}$ (b) $\dfrac{2(x^2-2x-2)}{3(x-1)^2}$

5.
(a) 0 (b) 0 (c) 1

7.
(a) $\dfrac{x\cos x + \sin x - x\sin x}{e^x}$ (b) $\dfrac{e^x(x\cos x + x\sin x - \sin x)}{x^2}$

(c) $\dfrac{e^x(x\sin x + \sin x - x\cos x)}{\sin^2 x}$ (d) $\dfrac{\sin x - x\cos x - x\sin x}{e^x \sin^2 x}$

(e) $\dfrac{e^x(x\sin x - x\cos x - \sin x)}{x^2 \sin^2 x}$ (f) $\dfrac{x\cos x - x\sin x - \sin x}{e^x x^2}$

Self-assessment questions 22.9

1. (a) product (b) chain (c) quotient (d) chain
(e) quotient

Exercises 22.9

1. (a) $-8x\sin 4x^2$ (b) $-2\sin(2x+1)$
(c) $(2x+3)\cos(x^2+3x+4)$
(d) $-\dfrac{1}{x^2}\sec^2\left(\dfrac{1}{x}\right)$ (e) $2xe^{x^2}$ (f) $(2x+3)e^{x^2+3x+2}$

2. (a) $30(6t+1)^4$ (b) $-8(4t+3)^{-3}$ (c) $e^t\sec^2 e^t$
(d) $\dfrac{(t+1)^{-\frac{1}{2}}}{2}$ (e) $(6t+2)\cos(3t^2+2t)$

3. (a) $12t$ (b) 3 (c) $36t$

4.
(a) $-\dfrac{2}{27}$ (b) 10.2851 (c) 3.8407 (d) 218.3926
(e) 593.6526 (f) 0.8944 (g) 1.6641

5.
(a) $\dfrac{2x+1}{x^2+x+3}$ (b) $\dfrac{7t^6+4}{t^7+4t}$ (c) $\cot t$

Exercises 22.10

1. (a) (b) Figure 22.10.1(a, b)
2. (a) min. at (0, 0) (b) max. at (0, 0)
3. (a) max. at (1, 6), min. at (3, 2) (b) max. at (0, 0), min. at (4, −32)
(c) point of inflection at (0, 0)
4. (a) $k=2$ (b) min. at (0.6934, 4.3267)
5. min. at (1, −10), max. at (−4, 115)
6. max. at (1, 11), min. at (4, −16)

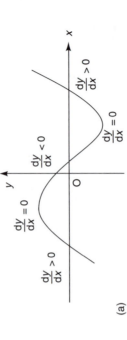

(a)

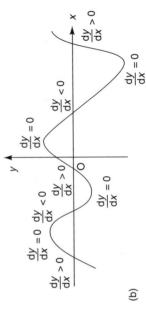

(b)

Figure 22.10.1

Exercises 22.11

1. (a) $6x + 6$ (b) 0 (c) 0 (d) $30x^4 - 42x$ (e) $108x^2$
(f) 14

2. (a) $\dfrac{2}{x^3}$ (b) $\dfrac{6}{x^4}$ (c) $\dfrac{6}{x^3}$ (d) $-\sin x$ (e) $-\sin x - \cos x$
(f) $90x^4 + \dfrac{2}{x^3}$ (g) $2\sec^2 x \tan x$ (h) $8\sec^2 2x \tan 2x$
(i) e^x (j) $4e^{2x}$ (k) $k^2 e^{kx}$ (l) $-\dfrac{1}{x^2}$

3. (a) $e^t(t+2)$ (b) $2\cos x - x\sin x$ (c) $\dfrac{2}{(t+1)^3}$ (d) $-\dfrac{(t+1)^{-\frac{3}{2}}}{4}$
(e) $-k^2 \sin kx$ (f) $\dfrac{e^x(x^2 - 2x + 2)}{x^3}$

4. (a) max. at $(0, 0)$, min. at $(\sqrt{\frac{1}{8}}, -\frac{3}{16})$, min. at $(-\sqrt{\frac{1}{8}}, -\frac{3}{16})$
(b) point of inflection at $(0, 0)$

5. $y' = -xe^{-x} + e^{-x}, \; y'' = -xe^{-x} - 2e^{-x}$

Computer and calculator exercises 22.11

1. Figure 22.11.1 Minimum at $(-3, -1.34)$
2. Figure 22.11.2 Maximum at $(1.65, 0.18)$

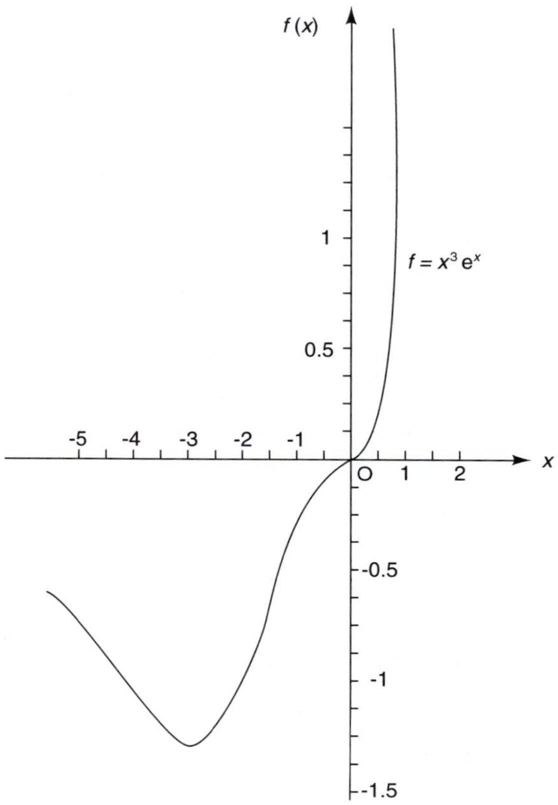

Review exercises 22

1. -27

2. (a) $11x^{10}$ (b) 0 (c) $-\dfrac{20}{x^6}$ (d) $5x^{-6}$ (e) $-\dfrac{3}{2}t^{-\frac{1}{2}}$
(f) $\dfrac{3}{2}\sqrt{7}x^{\frac{1}{2}}$ (g) $35x^{2.5}$

3. (a) $-5e^{-5x}$ (b) $6\cos 2x$ (c) $-42x^{-8}$ (d) $\dfrac{1}{2x}$
(e) $-1.5x^{-2.5}$ (f) 0

4. $4(1 + 2x)$

5. (a) $2t - 6t^2$ (b) $20e^{4t}$ (c) $-4\sin 4z + 4\cos 4z$
(d) $3\sec^2 3t$

7. $2, 4/3$

8. $-\dfrac{K}{V^2}$

Figure 22.11.1

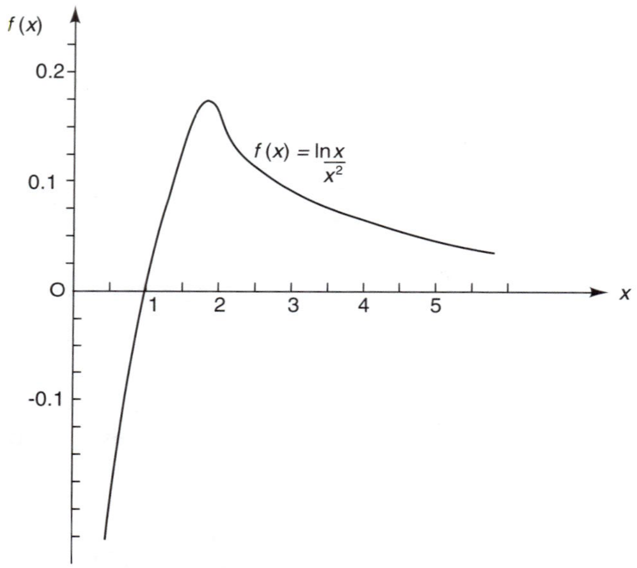

19. (a) point of inflection at (2, 0), min. at $(\frac{1}{2}, -\frac{27}{16})$
(b) min. at $(-18, -\frac{1}{36})$
20. (a) 3 (b) 7 (c) $y = 7x - 11$
21. (a) $2e^{2t}t^3(2+t)$ (b) $2e^{2t}(t^4 + 2t^3 + 1)$ (c) $\cos^2 x - \sin^2 x$
(d) $5\cos t(1+\sin t)^4$ (e) $\dfrac{r^2 + 2r - 1}{(r+1)^2}$ (f) $\dfrac{6c}{3c^2+2}$

Exercises 23.2

1. (a) $\dfrac{x^8}{8} + c$ (b) $\dfrac{x^{10}}{10} + c$ (c) $\dfrac{2x^{\frac{3}{2}}}{3} + c$ (d) $\dfrac{5x^{1.2}}{6} + c$
(e) $-\dfrac{x^{-2}}{2} + c$ (f) $-\dfrac{x^{-3}}{3} + c$ (g) $-\dfrac{2x^{-\frac{3}{2}}}{3} + c$ (h) $4x + c$
(i) $4t + c$ (j) $5z + c$

2. (a) $\dfrac{2x^{\frac{3}{2}}}{3} + c$ (b) $-\dfrac{1}{x} + c$ (c) $2\sqrt{x} + c$

3. (a) $\dfrac{e^{4x}}{4} + c$ (b) $\dfrac{e^{5x}}{5} + c$ (c) $-\dfrac{e^{-6x}}{6} + c$ (d) $-\dfrac{e^{-7x}}{7} + c$
(e) $e^x + c$ (f) $-e^{-x} + c$

4. (a) $\dfrac{\sin 3x}{3} + c$ (b) $\dfrac{\cos(-2x)}{2} + c$ (c) $2e^{0.5x} + c$

5. (a) $\dfrac{t^3}{3} + c$ (b) $\frac{1}{3}\ln|\sec 3t| + c$ (c) $\ln|t| + c$ (d) $-\dfrac{\cos 7z}{7} + c$
(e) $2\sin\left(\dfrac{t}{2}\right) + c$ (f) $-\dfrac{2}{3}\cos\dfrac{3\theta}{2} + c$ (g) $-\sin(-x) + c$

6. (a) $\dfrac{x^5}{5} + c$ (b) $-\dfrac{x^{-3}}{3} + c$ (c) $\sin x + c$ (d) $\dfrac{3x^{\frac{4}{3}}}{4} + c$
(e) $5x + c$ (f) $\dfrac{x}{2} + c$ (g) $\pi x + c$ (h) $\dfrac{2x^{\frac{5}{2}}}{5} + c$

Exercises 23.3

1. (a) $\dfrac{3x^4}{4} + c$ (b) $\dfrac{4x^3}{3} + c$ (c) $\dfrac{10x^{\frac{3}{2}}}{3} + c$ (d) $-\dfrac{7}{x} + c$

Figure 22.11.2

9. $\dfrac{\pi}{\sqrt{gl}}$

10. $k(\alpha + 2\beta i)$

11. (a) 12 (b) -0.5 (c) -384 (d) 0.5

12. 1

13. $14 - 4(b-a)$

14. $\dfrac{8}{5}e^{-t/50}$

15. (a) $V_0 \sin\theta - gt$ (b) $-g$

16. $\dfrac{Ve^{-t/RC}}{R}$

17. Max. at (1.176, 13)

18. Min. at (0, 0), max. at $(-2, 0.5413)$

2.
(a) $\dfrac{2x^5}{5} - \dfrac{7x^3}{3} + \dfrac{3x^2}{2} - 2x + c$ (b) $\dfrac{x^3}{3} + 3x^2 + 9x + c$
(c) $3x^3 - 6x^2 + 4x + c$

3.
(a) $\dfrac{2x^3}{3} + \dfrac{3x^2}{2} + c$ (b) $\dfrac{3x^8}{8} - \dfrac{2x^3}{3} + c$ (c) $\sin x - \cos x + c$
(d) $\dfrac{7\sin 3x}{3} + c$ (e) $-4\cos 2x + c$ (f) $-8e^{-2x} + c$
(g) $\dfrac{\sin 3x}{3} + \dfrac{\sin 2x}{2} + c$ (h) $\dfrac{5e^{3x}}{3} + c$

4. $\omega = \omega_0 + \alpha t,\ \theta = \omega_0 t + \dfrac{\alpha t^2}{2} + c$

5.
(a) $3\ln|x| - x^2 + c$ (b) $-\dfrac{x^2}{12} + c$ (c) $3\ln|x| - 2x + c$
(d) $9x - 6x^2 + \dfrac{4x^3}{3} + c$

6.
(a) $-\dfrac{3\cos 4x}{4} + c$ (b) $-\dfrac{4\cos 3x}{3} + c$ (c) $-\dfrac{\cos 4x}{12} + c$
(d) $-\dfrac{\cos 5x}{20} + c$

Review exercises 23

1.
(a) $\dfrac{x^9}{9} + c$ (b) $\dfrac{8x^{\frac{9}{8}}}{9} + c$ (c) $-\dfrac{x^{-7}}{7} + c$ (d) $\dfrac{4x^{1.25}}{5} + c$
(e) $-\dfrac{\cos 3x}{3} + c$ (f) $-3\cos\left(\dfrac{x}{3}\right) + c$ (g) $4\sin 0.25x + c$

2.
(a) $\dfrac{x^5}{5} - \dfrac{x^{-3}}{3} + c$ (b) $\dfrac{x^3}{3} - \dfrac{1}{x} + c$ (c) $\dfrac{x^2}{2} + \ln|x| + c$
(d) $\dfrac{4x^{\frac{5}{4}}}{5} + \ln|x| + c$

3.
(a) $\dfrac{x^{10}}{10} + c$ (b) $\dfrac{t^4}{4} + \dfrac{t^3}{3} + c$ (c) $\dfrac{x^3}{3} + \dfrac{\sin 3x}{3} + c$
(d) $\dfrac{x^8}{8} - 7e^x + c$

4.
(a) $\dfrac{e^{3x}}{3} - \dfrac{e^{-3x}}{3} + c$ (b) $7x + c$ (c) $-7x + c$ (d) $ex + c$

5.
(a) $\dfrac{t^2}{6} + c$ (b) $\dfrac{t^2}{6} + c$ (c) $\dfrac{t}{3} + c$ (d) $\dfrac{t}{3\pi} + c$ (e) $\dfrac{\pi t}{3} + c$

6. $y = \dfrac{x^2}{2} + 3x + 5$

7. $y = e^t - 1$

8.
(a) $-\dfrac{x^{-2}}{2} + c$ (b) $-\dfrac{7}{x} + c$ (c) $-\dfrac{4e^{-3x}}{3} + c$
(d) $11\ln|x| + c$

9.
(a) $\dfrac{3t^2}{2} + 7t + c$ (b) $\dfrac{4t^3}{3} + \dfrac{3t^2}{2} - 2t + c$ (c) $2\ln|t| + c$
(d) $-\dfrac{\cos 3t}{3} + \dfrac{\sin 3t}{3} + c$

Exercises 24.2

1. (a) 6.25 (b) 5.0625
2. 6.6563. This calculation underestimates the area.
3. (a) 0.8333 (b) 0.95

Exercises 24.3

1. 46
2. $\dfrac{52}{3}$
3. $\dfrac{148}{3}$

Exercises 24.5

1. (a) $\dfrac{1}{3}$ (b) -2 (c) $\dfrac{4}{3}$ (d) $\dfrac{1}{6}$
2. (a) $\dfrac{5}{6}$

3. $\frac{3x^4}{4} - \frac{7x^3}{3} + 1$

4. (a) 576.4 (b) 2 (c) 0.6931 (d) 3.4641 (e) 0

7. (a) 4.6708 (b) 2.3504 (c) 27.2314

8. (a) 0 (b) $\frac{2}{3}$ (c) 0 (d) $-\frac{1}{5}$ (e) 1

Exercises 24.6

1. $\frac{1}{6}$
2. 23.4133
3. 203.33
4. (a) $\frac{16}{3}$ (b) 9 (c) 4

Review exercises 24

1. 2.5333
2. $\frac{1}{4}\left(1 + \frac{2}{n} + \frac{1}{n^2}\right)$, 0.25
3. (a) 44 (b) $\frac{14}{3}$ (c) 301.605 (d) $\frac{26}{3}$ (e) $\frac{1}{2}$ (f) 0.3098
   (g) 19.0855 (h) 2.3504 (i) 1.4725
4. (a) $\frac{4}{3}$ (b) 2 (c) $\frac{4}{3}$ (d) $\frac{8}{3}$
5. 0, 4
6. (a) 0.9194 (b) 0.7081 (c) -0.0931 (d) 2.7279
   (e) 3.1945 (f) 3.4366 (g) $\frac{10}{3}$ (h) $\frac{19}{3}$ (i) $\frac{3}{4}$ (j) 1.3863
7. (a) 1 (b) 1.2720

Exercises 25.2

1. (a) $4e^x(x-1)+c$ (b) $\frac{1}{3}e^{3x}(x-\frac{1}{3})+c$ (c) $\frac{5}{4}e^{2x}(2x-1)+c$
   (d) $\frac{\sin 3x}{9} - \frac{x\cos 3x}{3} + c$ (e) $2x^2\ln 5x - x^2 + c$
   (f) $\ln|\cos x| + x\tan x + c$
2. $x\ln x - x + c$
3. (a) 1 (b) 0.3863 (c) 0 (d) -39.4784
4. 40.1710

Exercises 25.3

1. (a) $\frac{(7x+4)^3}{21} + c$ (b) $\frac{\sin(2x+1)}{2} + c$ (c) $\frac{(7x^2+3x-2)^2}{2} + c$
   (d) $\frac{e^{3x-9}}{3} + c$

2. (a) $\frac{2(x+1)^{\frac{3}{2}}}{3} + c$ (b) $\frac{(2x+1)^{\frac{3}{2}}}{3} + c$ (c) $\frac{2(ax+b)^{\frac{3}{2}}}{3a} + c$
   (d) $\ln|x+1| + c$ (e) $\frac{-1}{x+1} + c$ (f) $\frac{(x+1)^{-n+1}}{-n+1} + c$
   (g) $\frac{1}{a}\ln|ax+b| + c$ (h) $\frac{(ax+b)^{-n+1}}{a(-n+1)} + c$

3. (a) $\frac{1}{2}\ln|2x+5| + c$ (b) $\ln|2x-1| + c$ (c) $\ln|2x+7| + c$
   (d) $\ln|x-1| + c$ (e) $\frac{1}{2}\ln|2x+1| + c$ (f) $\ln|\ln|x|| + c$

4. $\ln|\sin x| + c$
5. (a) 182 (b) 0 (c) 0.1567 (d) 0.2974

6. (a) $-\frac{1}{a}\cos(ax+b)+c$ (b) $-\frac{1}{a}\sin(ax+b)+c$ (c) $\frac{e^{ax+b}}{a} + c$

7. $\frac{7}{2}e^{x^2} + c$
8. $\ln|10 + e^{3x}| + c$

Exercises 25.4

1. (a) $\frac{4}{3}\ln|x-2| - \frac{4}{3}\ln|x+1| + c$ (b) $\frac{8}{3}\ln|x+2| + \frac{7}{3}\ln|x-1| + c$
   (c) $9\ln|x+3| - 7\ln|x+2| + c$

2. $\frac{1}{(x+1)^2} - \frac{2}{x+1} + 1, \; -2\ln|x+1| - \frac{1}{x+1} + x + c$

3. $2\ln|x+3| - \frac{1}{x+3} + c$

4. (a) 0.5232 (b) 0.6322 (c) -1.2274

Exercises 25.5

1. (a) $\dfrac{x}{2} + \dfrac{\sin 2x}{4} + c$ (b) $\dfrac{\pi}{4}$

2. (a) $\dfrac{3\pi}{2}$ (b) 0.4281

3. $x + c$

4. $-\dfrac{\cos 5x}{10} - \dfrac{\cos x}{2} + c$

5. (a) $-\dfrac{\cos 4x}{8} + c$ (b) $-\dfrac{\cos^2 2x}{4} + c$

6. $\tan x + c$

Exercises 25.6

1. (a) 15   (b) 0.6366   (c) 40.3333
2. (a) 3   (b) 9.3333
3. (a) 5   (b) 2
4. 495.583

Exercises 25.7

1. 1.7889
3. 1.4142
4. 0.7071
5. 1.7873
6. (a) 3.5355   (b) 3.4641

Computer and calculator exercises 25.8

1. $e^x(x^3 - 3x^2 + 6x - 6)$
2. $\frac{1}{128}e^{4x}(32x^4 - 32x^3 + 24x^2 - 12x + 3)$
3. 2.8608
4. 0.6421

Review exercises 25

1. (a) 1.5945   (b) 1.0650   (c) 2.7854   (d) 2.0942
2. (a) 0   (b) −0.1111   (c) 1.1188×10^5 (d) 2.9438
   (e) 6.9986
3. (a) 2952.33   (b) 0   (c) 0.1908   (d) 0.5728
4. (a) $\frac{1}{3}$ (b) 0.4413
5. 7.0946

Self-assessment questions 26.2

1. The number of rows comes first

Exercises 26.2

1. (a) 1×4 (b) 2×3 (c) 5×1 (d) 4×3 (e) 3×2
   (f) 3×4
2. 15, mn
3.
$$\begin{pmatrix} 1 & 0 & 0 & 0 \\ 0 & 1 & 0 & 0 \\ 0 & 0 & 1 & 0 \\ 0 & 0 & 0 & 1 \end{pmatrix}$$

4. (a) 0   (b) 6   (c) 4   (d) 3   (e) 0.5   (f) −1

Exercises 26.3

1. (a) $\begin{pmatrix} 10 & 8 & 13 \\ 21 & -4 & 10 \end{pmatrix}$ (b) $\begin{pmatrix} 8 & 4 & -3 \\ 15 & 0 & 0 \end{pmatrix}$ (c) $\begin{pmatrix} 4 & 10 & -1 \\ 3 & 5 & 7 \end{pmatrix}$

(d) $\begin{pmatrix} -2 & -6 & 17 \\ 3 & -9 & 3 \end{pmatrix}$ (e) $\begin{pmatrix} 12 & 14 & -4 \\ 18 & 5 & 7 \end{pmatrix}$

(f) $\begin{pmatrix} -6 & 2 & -14 \\ -18 & 9 & -3 \end{pmatrix}$ (g) $\begin{pmatrix} 13 & 16 & 4 \\ 21 & 3 & 12 \end{pmatrix}$

(h) $\begin{pmatrix} 7 & 0 & 22 \\ 21 & -11 & 8 \end{pmatrix}$ (i) $\begin{pmatrix} 5 & -4 & 6 \\ 15 & -7 & -2 \end{pmatrix}$

2.

(a) $\begin{pmatrix} 16 & 19 \\ -11 & -2 \end{pmatrix}$ (b) $\begin{pmatrix} 16 & 19 \\ -11 & -2 \end{pmatrix}$ (c) $\begin{pmatrix} -2 & -3 \\ 7 & 16 \end{pmatrix}$

(d) $\begin{pmatrix} 2 & 3 \\ -7 & -16 \end{pmatrix}$

Exercises 26.4

1.

(a) $\begin{pmatrix} 5\alpha & 5\beta \\ 5\gamma & 5\delta \end{pmatrix}$ (b) $\begin{pmatrix} 28 & -21 \\ 35 & 35 \end{pmatrix}$ (c) $\begin{pmatrix} 28-5\alpha & -21-5\beta \\ 35-5\gamma & 35-5\delta \end{pmatrix}$

(d) $\begin{pmatrix} 5\alpha+28 & 5\beta-21 \\ 5\gamma+35 & 5\delta+35 \end{pmatrix}$

3.

(a) $\begin{pmatrix} 7 & 0 \\ 0 & 7 \end{pmatrix}$ (b) $\begin{pmatrix} -3 & 0 \\ 0 & -3 \end{pmatrix}$ (c) $\begin{pmatrix} 0.5 & 0 \\ 0 & 0.5 \end{pmatrix}$ (d) $\begin{pmatrix} \lambda & 0 \\ 0 & \lambda \end{pmatrix}$

3.

(a) $\begin{pmatrix} 5 & 10 & 40 \\ 15 & -10 & 25 \end{pmatrix}$ (b) $\begin{pmatrix} 12 & 32 & -36 \\ 0 & 28 & 8 \end{pmatrix}$

(c) $\begin{pmatrix} -7 & -22 & 76 \\ 15 & -38 & 17 \end{pmatrix}$ (d) $\begin{pmatrix} 4.5 & 3 & 2.5 \\ 9 & -1 & 2.5 \end{pmatrix}$

4.

(a) $\begin{pmatrix} 4 & 4 \\ 6 & -1 \end{pmatrix}$ (b) not possible   (c) $\begin{pmatrix} 7 & -3 \\ 8 & 2 \end{pmatrix}$

(d) $\begin{pmatrix} -16 & -10 & 4 \\ 3 & -11 & -26 \end{pmatrix}$ (e) not possible   (f) not possible

Computer and calculator questions 26.4

1.

(a) $\begin{pmatrix} 36 & 24 & 20 \\ 72 & -8 & 20 \end{pmatrix}$ (b) $\begin{pmatrix} 6 & -38 & 78 \\ 54 & -55 & 1 \end{pmatrix}$

(c) $\begin{pmatrix} -108 & -36 & 141 \\ -189 & -24 & 60 \end{pmatrix}$

Self-assessment questions 26.5

4. No.

5. A and B are square matrices of the same size.

6. A is square.

Exercises 26.5

1. $AB = \begin{pmatrix} -15 & 7 \\ -15 & 21 \end{pmatrix}$, $BA = \begin{pmatrix} 24 & 37 \\ -6 & -18 \end{pmatrix}$, hence $AB \neq BA$.

2. $\begin{pmatrix} 14 & -18 \\ 9 & 23 \end{pmatrix}$

3. $AB = \begin{pmatrix} 1 & 0 \\ 0 & 1 \end{pmatrix}$, $BA = \begin{pmatrix} 1 & 0 \\ 0 & 1 \end{pmatrix}$, A is the inverse of B, and B is the inverse of A.

4. $MN = \begin{pmatrix} 62 \\ 33 \end{pmatrix}$

It is not possible to calculate NM.

5. $PQ = \begin{pmatrix} 3\alpha+7\gamma \\ 2\alpha+\gamma \\ -2\alpha+7\beta \end{pmatrix}$

It is not possible to calculate QP.

7. $\begin{pmatrix} 3x \\ 4y \\ 5z \end{pmatrix}$

Computer and calculator exercises 26.5

1. $AB = \begin{pmatrix} 13 & 14 & -29 \\ 4 & 0 & -17 \\ 9 & 5 & -39 \end{pmatrix}$, $ABC = \begin{pmatrix} 116 & 170 & -149 \\ 19 & 34 & -62 \\ 52 & 118 & -141 \end{pmatrix}$

2. $AB = BA = \begin{pmatrix} 1 & 0 & 0 \\ 0 & 1 & 0 \\ 0 & 0 & 1 \end{pmatrix}$

A is the inverse of B, and B is the inverse of A.

Exercises 26.6

1. (a) 6 (b) 5 (c) 0 (d) 9 (e) -3
2. (a) 16 (b) -25 (c) 2.25 (d) 222 (e) -28
3. 7
4. $\begin{pmatrix} 1 & 1 \\ 1 & 1 \end{pmatrix}$ for example
5. The determinant is zero.
6. 1, 6

Computer and calculator exercises 26.6

2. (a) $|A|=42$, $|B|=68$ (b) $AB=\begin{pmatrix} 7 & -2 & 10 \\ 12 & 52 & -10 \\ 2 & 6 & 7 \end{pmatrix}$

   (c) $|AB|=2856$
3. -652
4. 60

Exercises 26.7

2. (a) $\frac{1}{26}\begin{pmatrix} -2 & 7 \\ 4 & -1 \end{pmatrix}$ (b) $\frac{1}{2}\begin{pmatrix} 3 & -1 \\ 4 & -2 \end{pmatrix}$ (c) $\frac{1}{94}\begin{pmatrix} -1 & 7 \\ 14 & -4 \end{pmatrix}$

   (d) singular (e) $\frac{1}{50}\begin{pmatrix} 3 & -7 \\ 2 & 12 \end{pmatrix}$ (f) $\frac{1}{42}\begin{pmatrix} 0 & 14 \\ 3 & 0 \end{pmatrix}$
3. $\begin{pmatrix} 1 & 0 \\ 0 & 1 \end{pmatrix}$, $\begin{pmatrix} 0 & 1 \\ 1 & 0 \end{pmatrix}$
4. $\dfrac{1}{\alpha\delta - \beta\gamma}\begin{pmatrix} \delta & -\beta \\ -\gamma & \alpha \end{pmatrix}$

   The inverse does not exist if $\alpha\delta - \beta\gamma = 0$.

Computer and calculator exercises 26.7

1. (a) $\frac{1}{26}\begin{pmatrix} -2 & 7 \\ 4 & -1 \end{pmatrix}$ (b) $\frac{1}{2}\begin{pmatrix} 3 & -1 \\ 4 & -2 \end{pmatrix}$ (c) $\frac{1}{94}\begin{pmatrix} -1 & 7 \\ 14 & -4 \end{pmatrix}$

(d) no inverse (e) $\frac{1}{50}\begin{pmatrix} 3 & -7 \\ 2 & 12 \end{pmatrix}$ (f) $\frac{1}{42}\begin{pmatrix} 0 & 14 \\ 3 & 0 \end{pmatrix}$

2. (a) $\begin{pmatrix} 1 & 0 & 0 \\ 0 & \frac{1}{2} & 0 \\ 0 & 0 & \frac{1}{3} \end{pmatrix}$ (b) $\begin{pmatrix} 3 & 0 & 0 \\ -2 & 1 & 0 \\ -6 & -3 & 3 \end{pmatrix}$ (c) $\begin{pmatrix} 0 & 0 & 1 \\ 0 & 1 & 5 \\ 1 & -4 & -25 \end{pmatrix}$

   (d) $\begin{pmatrix} 0 & 0 & 1 \\ 0 & 1 & -17 \\ 1 & -8 & 129 \end{pmatrix}$ (e) $\begin{pmatrix} 17 & 3 & 7 \\ 12 & 2 & 5 \\ 1 & 0 & 0 \end{pmatrix}$

Exercises 26.8

1. (a) $x=2$, $y=1$ (b) $x=1$, $y=-1$ (c) $x=3$, $y=4$
   (d) $x=1$, $y=0$ (e) $x=8$, $y=4$ (f) $x=9$, $y=-3$
   (g) $x=-4$, $y=-2$
2. (a) $I_1 = \frac{118}{263}$, $I_2 = \frac{4}{263}$ (b) $I_1 = \frac{27}{151}$, $I_2 = -\frac{26}{151}$

Review exercises 26

1. $PQ=\begin{pmatrix} 0 & 5 & 10 & 15 \\ -10 & -5 & 0 & 5 \\ -16 & -7 & 2 & 11 \end{pmatrix}$, QP does not exist

3. (a) $\frac{1}{7}\begin{pmatrix} 3 & -4 \\ 1 & 1 \end{pmatrix}$ (b) $\frac{1}{10}\begin{pmatrix} -4 & 2 \\ 3 & 1 \end{pmatrix}$ (c) does not exist
   (d) does not exist

4. (a) $\begin{pmatrix} 4 & 2 & 10 \\ 6 & -2 & 5 \end{pmatrix}$ (b) does not exist (c) $\begin{pmatrix} 16 \\ 9 \end{pmatrix}$
   (d) $\begin{pmatrix} -2 \\ 2 \end{pmatrix}$ (e) does not exist (f) does not exist

5. (a) $\begin{pmatrix} 9 & -18 \\ 6 & 18 \end{pmatrix}$ (b) $\begin{pmatrix} -5 & 10 \\ 20 & 0 \end{pmatrix}$ (c) $\begin{pmatrix} 4 & -8 \\ -2 & 6 \end{pmatrix}$
   (d) $\begin{pmatrix} 13 & -26 \\ -10 & 18 \end{pmatrix}$ (e) $\begin{pmatrix} -23 & 46 \\ -6 & -42 \end{pmatrix}$

6. BA, AC and CB can be formed.
7. (a) $x=-1$, $y=3$ (b) $x=0$, $y=4$ (c) $x=-2$, $y=-1$

Exercises 27.3

1. Figure 27.3.1

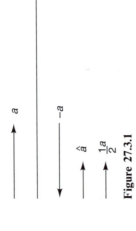

Figure 27.3.1

2. Figure 27.3.2

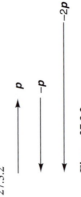

Figure 27.3.2

3. (a) $2a$ (b) $-a$ (b) $2c$, $a+b$, $b-c$
4. (a) Figure 27.3.4   (b) $\overrightarrow{BC} = 2\overrightarrow{PQ}$
5. (a) Figure 27.3.5

Exercises 27.4

1. $9i + 2j$, 9.2195
2. $-16i + 12j$
3. (a) $12i - j$ (b) $12i - j$ (c) $2i + 5j$ (d) $-2i - 5j$
       $a + b = b + a$, $a - b = -(b - a)$
4. (a) $-72i + 6j$ (b) $59i + 68j$ (c) $-11i - 72j$
5. (a) 9   (b) 9.0554   (c) 3.6056   (d) 0   (e) 5   (f) 0.7071
6. (a) 7.2801   (b) 2.2361   (c) 2   (d) 5.8310   (e) 1
       (f) 7.0178

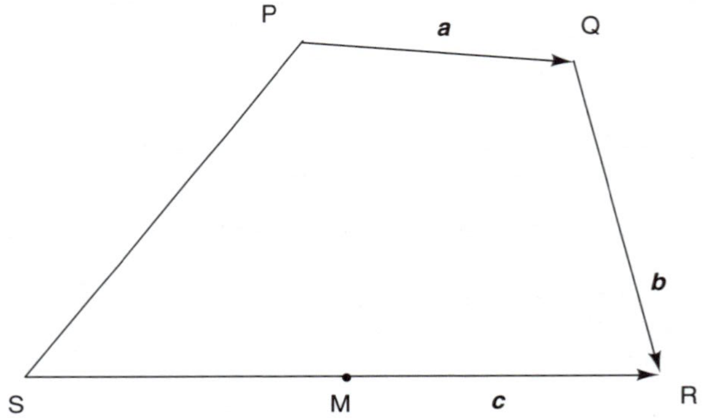

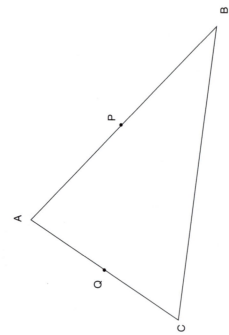

Figure 27.3.4

Figure 27.3.5

Exercises 27.5

1. (a) 47 (b) 135 (c) 97 (d) 34 (e) 202
2. Scalar product is 0, and so $\cos\theta = 0$ and hence $\theta = 90°$.
3. 0
4. (a) 14 (b) 14 (c) 10 (d) 52
5. (a) $i - 5j$, $4i - 7j$ (b) 39 (c) 18.43°

Self-assessment question 27.6

1. The identity matrix

Exercises 27.6

1. (a) $(1, -1), (3, -1), (3, -4)$ (b) $(-1, 1), (-3, 1), (-3, 4)$
2. $\begin{pmatrix} 213 \\ 291 \end{pmatrix}$
3. $(1, -4), (8, -2)$

Review exercises 27

1. Figure R27.1

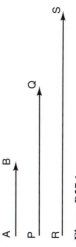

Figure R27.1

2. $-5i - 12j$
3. (a) $-19i - 19j$ (b) $5i + 11j$
4. (a) $i + 7j$ (b) $\dfrac{1}{\sqrt{50}}(i + 7j)$
5. 7, 52.13°

6. $\begin{pmatrix} 6 \\ 9 \end{pmatrix}$ and $\begin{pmatrix} 18 \\ 27 \end{pmatrix}$, $\begin{pmatrix} 4 \\ -2 \end{pmatrix}$ and $\begin{pmatrix} -4 \\ 2 \end{pmatrix}$
7. (a) 8.602 (b) 9.055 (c) $0.116(5i + 7j)$ (d) $0.110(i - 9j)$
   (e) -58 (f) 16.492
8. 168.69°
9. $\begin{pmatrix} 2 \\ 0 \end{pmatrix}$
10. (a) $\sqrt{5}$ (b) $\frac{9}{2}i + 3j$ (c) 5.4083 (d) 2 (e) 9
    (f) $28i - 20j$ (g) 1.25

Exercises 28.2

1. (a) -3 (b) -4 (c) -2 (d) $-4j$ (e) 1 (f) -3
   (g) 2
2. (a) 9 (b) $9j$ (c) $12j$ (d) 0.5 (e) $0.5j$ (f) $\frac{1}{4}j$

Self-assessment questions 28.3

3. When $b^2 - 4ac < 0$.

Exercises 28.3

1. (a) real part 3, imaginary part 7 (b) $5, -2$ (c) 7, 0
   (d) 0, 5 (e) α, β (f) $2x, 3y$ (g) -1, 0 (h) $\cos\theta$, $\sin\theta$
2. (a) $-4, 1$ (b) $2.6375, -1.1375$ (c) $4j, -4j$ (d) $-2j, 2j$
   (e) $2, -2$ (f) $1 + 2j, 1 - 2j$ (g) $7 + 2j, 7 - 2j$ (h) $\frac{1}{2} + \frac{1}{2}j, \frac{1}{2} - \frac{1}{2}j$
   (i) $3 + 9j, 3 - 9j$

Exercises 28.4

1. (a) $6 + 10j$ (b) $2 + 4j$ (c) $-2 - 4j$ (d) $-13 + 26j$
   (e) $\frac{29}{13} + \frac{2}{13}j$ (f) $\frac{29}{65} - \frac{2}{65}j$ (g) 65 (h) $-33 + 56j$
2. (a) $8 + 9j$ (b) $72j$ (c) $-\frac{8}{9}j$ (d) $-\frac{1}{9}j$
3. (a) $15 - 9j$ (b) $3 + 5j$ (c) $5 + 3j$ (d) $-3 + 5j$ (e) $16 - 30j$
4. 5

(e) $\dfrac{x_1x_2 + y_1y_2 + (x_2y_1 - x_1y_2)j}{x_2^2 + y_2^2}$ (f) $\dfrac{x_1x_2 + y_1y_2 + (x_1y_2 - x_2y_1)j}{x_1^2 + y_1^2}$

7.

8. $2 + 3j$
(a) $x^2 - 49 = 0$ (b) $x^2 + 49 = 0$

9. $x^2 - 6x + 13 = 0$

Exercise 28.5

1. Figure 28.5.1

Exercises 28.6

1. $|z_1| = \sqrt{5}$, $\arg(z_1) = 63.43°$, $|z_2| = \sqrt{5}$, $\arg(z_2) = -116.57°$
Figure 28.6.1

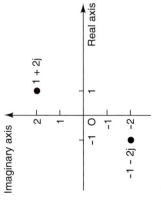

Figure 28.6.1

2. (a) modulus $= 3$, $\arg = 90°$ (b) $4, 0°$ (c) $1, -90°$ (d) $1, 180°$
(e) $0.5, 90°$ (f) $1, 180°$ (g) $1, -90°$ (h) $1, -90°$
(i) $1, 180°$

3. Real part $= r\cos\theta$, imaginary part $= r\sin\theta$. Figure 28.6.3

4. Figure 28.6.4
(a) $3j$ (b) $4.3301 + 2.5j$ (c) -2 (d) $-3.5355 - 3.5355j$

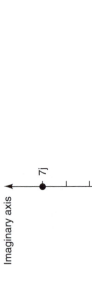

Figure 28.5.1

5. (a) $\frac{1}{3}j$ (b) $\dfrac{2+j}{5}$

6. (a) $x_1 + x_2 + (y_1 + y_2)j$ (b) $x_1 - x_2 + (y_1 - y_2)j$
(c) $x_2 - x_1 + (y_2 - y_1)j$ (d) $x_1x_2 - y_1y_2 + (x_1y_2 + x_2y_1)j$

Review exercises 28

1. (a) real part = 19, imaginary part = −3   (b) −1, 2   (c) 6y, 2x
   (d) cos 3θ, sin 3θ   (e) cos ωt, sin ωt   (f) cos ωt, −sin ωt

2. (a) 13 + 5j   (b) 3 + 5j   (c) −1206 − 242j

3. (a) 15 + 16j   (b) $\dfrac{-21 + 22j}{37}$ (c) −2 + 4j   (d) 455 − 130j

   (e) $\dfrac{1 - j}{2}$

4. x = 2 + 4j, x = 2 − 4j. Figure R28.4.
   4.4721 ∠ 1.1071, 4.4721 ∠ (−1.1071)

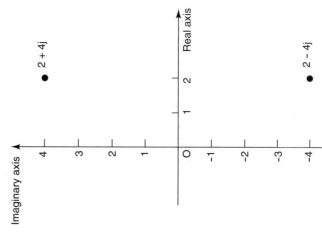

Figure R28.4

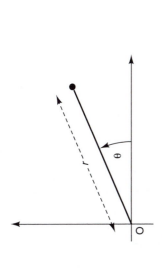

Figure 28.6.3

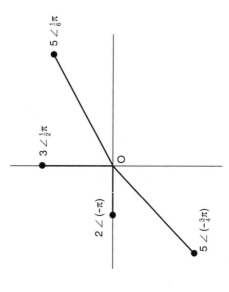

Figure 28.6.4

Computer and calculator exercises 28.6

1. (a) −4   (b) 6706 − 5758j   (c) 0.2233 − 0.2770j
2. (a) −0.5 − 2.5981j, −0.5 + 2.5981j   (b) −1, −j, j

751

5. (a) $\sqrt{18}\angle\frac{1}{4}\pi$ (b) $\sqrt{32}\angle(-\frac{1}{4}\pi)$ (c) $1\angle(-\frac{1}{2}\pi)$ (d) $9\angle 0$
    (e) $\sqrt{2}\angle(-\frac{3}{4}\pi)$ (f) $1\angle\pi$ (g) $\sqrt{41}\angle 0.8961$ (h) $4\angle(\pi)$
    (i) $1\angle 0$

6. Figure R28.6
    (a) $2.8284+2.8284j$ (b) $6j$ (c) $1.7321-j$

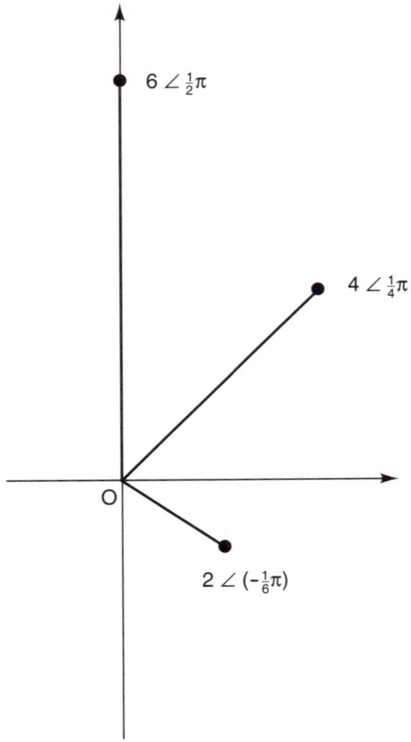

Figure R28.6

$6\angle\frac{1}{2}\pi$

$4\angle\frac{1}{4}\pi$

$2\angle(-\frac{1}{6}\pi)$

O

7. (a) $1\angle\frac{1}{3}\pi$ (b) $3\angle\frac{1}{3}\pi$

Exercise 29.2

1. (a) continuous   (b) continuous   (c) discrete
    (d) continuous   (e) discrete

Exercise 29.3

1. (a) 500   (b) 3200, 3700, 4200, 4700, 5200

Exercises 29.4

1. (a) (b) Figure 29.4.1(a, b)

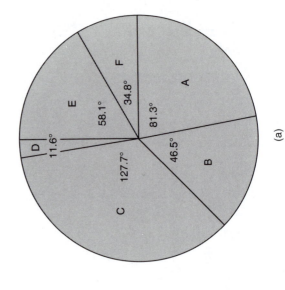

(a)

Figure 29.4.1(a)

2. Figure 29.4.2 (a, b)

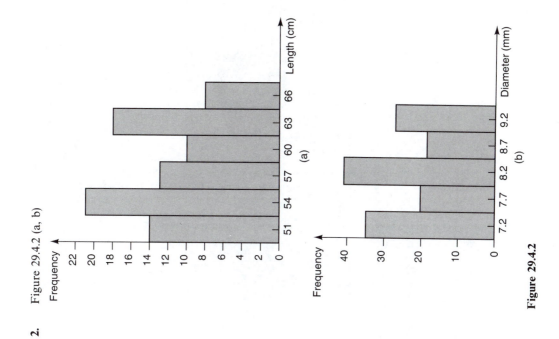

(a)

(b)

Figure 29.4.2

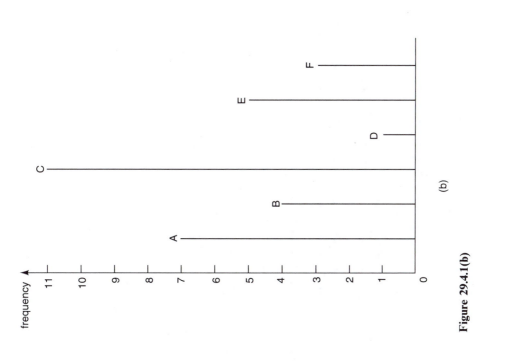

(b)

Figure 29.4.1(b)

Exercises 29.5

1. -1.4
2. 223.35
3. (a) mean = 3.5, median = 3.5, no mode
   (b) mean = 2.5, median = 2.5, no mode
   (c) mean = 5.43, median = 6, mode = 6
   (d) mean = 0.29, median = 1, mode = -2 and 1

Exercises 29.6

1. (a) 1.7078   (b) 3.1314   (c) 39.5656   (d) 6.5680
2. (a) 2   (b) $2k$ (c) 0.8165   (d) 0.8165k
   (e) If all data values are multiplied by k, the mean and standard deviation are also multiplied by k.

Self-assessment questions 29.7

3. (b) and (c) cannot be probabilities.

Exercises 29.7

1. 0.35
2. 0.25
3. 0.342
4. 0.925
5. (a) $\frac{1}{52}$ (b) $\frac{1}{13}$ (c) 0
6. 0.2559
7. 0.9776

Exercises 29.8

1. (a) $\frac{1}{2}$ (b) $\frac{2}{3}$ (c) $\frac{1}{6}$ (d) 0   (e) 1
2. (a) 1   (b) $\frac{2}{3}$ (c) $\frac{5}{6}$ (d) 0   (e) $\frac{1}{3}$ (f) $\frac{1}{3}$ (g) $\frac{5}{6}$ (h) $\frac{1}{6}$
   (i) $\frac{5}{6}$ (j) $\frac{1}{3}$ (k) 1   (l) $\frac{1}{6}$

Exercises 29.9

1. $\frac{1}{12}$
2. 0.011
3. (a) states independent events
4. (a) 0.94   (b) 0.8306

Exercises 29.10

1. Figure 29.10.1

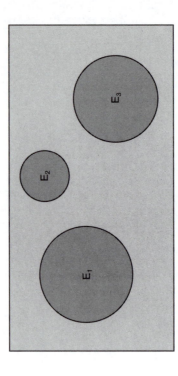

Figure 29.10.1

2. (a) The component is not within the manufacturer's tolerance.
   (b) 0.97
3. (a) A component is either acceptable or unacceptable.
   (b) A component is either first-grade or unacceptable.
   (c) A component is either first-grade or acceptable.
   (d) 0.05   (e) 0.95   (f) 0.9

Exercises 29.11

1. (a) E_1 and E_3, E_1 and E_4, E_2 and E_3
   (b) (i) 1   (ii) $\frac{2}{3}$ (iii) $\frac{1}{2}$ (iv) $\frac{2}{3}$ (v) $\frac{2}{3}$ (vi) $\frac{1}{2}$

2. (a) 0.679   (b) 0.988   (c) 0.012   (d) 0.321

Review exercises 29

1. (a) $\frac{1}{18}$ (b) $\frac{10}{179}$
2. (a) continuous   (b) continuous   (c) discrete
   (d) continuous
3. (a) (b) Figure R29.3(a, b)

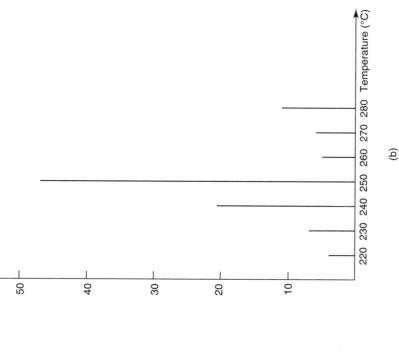

Frequency

Figure R29.3(b)

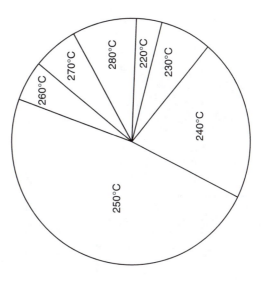

(a)

Figure R29.3(a)

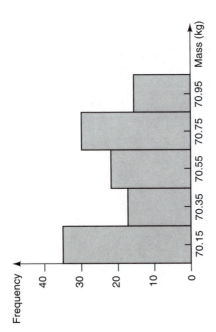

Figure R29.4

4. Figure R29.4

5. (a) mean −3, median −2, mode 0
   (b) mean 0.304, median 0.31, mode 0.36

6. (a) 3.9279   (b) 0.051 61

7. (a) The machine will not break down in the next three months.
   (b) 0.8

8. (a) 0.010 85   (b) 1.1772 × 10⁻⁴   (c) 0.9784   (d) 0.021 46

9. (a) 0.25   (b) 0.6   (c) 0.422   (d) 0.372   (e) 0.75
   (f) 0.343   (g) 0.715

10. (a) 0.9412
    (b) At least one of the lines will develop a fault within the next three months.

Index